COSPAR COLLOQUIA SERIES

VOLUME 1

PHYSICS OF THE OUTER HELIOSPHERE

Sponsors

THE COMMITTEE ON SPACE RESEARCH (COSPAR)
THE INTERNATIONAL ASTRONOMICAL UNION (IAU)
and
THE POLISH ACADEMY OF SCIENCES*

Program Committee

D.E. Page (ESA/JPL) (Chairman)
S. Grzedzielski, Poland (Executive Chairman)
W.I. Axford (F.R.G.)
V.B. Baranov (U.S.S.R.)
J.L. Bertaux (France)
S. Bowyer (U.S.A.)
H.J. Fahr (F.R.G.)
L.A. Fisk (U.S.A.)
P.C. Frisch (U.S.A.)
K.I. Ciringauz (U.S.S.R.)
A. Hewish (U.K.)
J. Kóta (Hungary)
R.R. Meier (U.S.A.)
M. Nicolet (Belgium)
A. Nishida (Japan)
E.J. Smith (U.S.A.)
E.C. Stone (U.S.A.)
K.-P. Wenzel (ESA/ESTEC)

Local Organizing Committee

S. Grzedzielski*
R. Ratkiewicz*
D. Ruciński*
J. Ziemkiewicz*

Editorial Assistance

A. Czechowski*

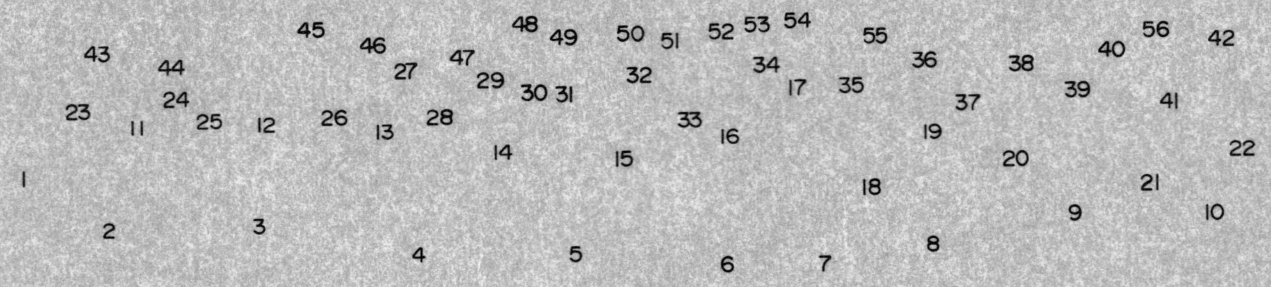

KEY TO PHOTOGRAPH OF PARTICIPANTS

1. V.B. Baranov; 2. K. Scherer; 3. W.I. Axford; 4. D.E. Page; 5. S. Grzedzielski; 6. S.A. Grib; 7. M.S. Ruderman; 8. J. Ziemkiewicz; 9. M. Verigin; 10. G. Cini Castagnoli; 11. H.J. Fahr; 12. E.C. Roelof; 13. R.B. McKibben; 14. K. Sauer; 15. W.S. Kurth; 16. H. Lehmann; 17. S.V. Chalov; 18. Badruddin; 19. A. Czechowski; 20. K.-P. Wenzel; 21. M. Serio; 22. B. Popielawska; 23. K. Baumgärtel; 24. N. Kömle; 25. J.F. McKenzie; 26. R.A. Burger; 27. E. Möbius; 28. E.J. Smith; 29. I.S. Veselovsky; 30. F. Mariani; 31. O.V. Likin; 32. A.G. Krainiuk; 33. A. Somogyi; 34. G.N. Zastenker; 35. A.J. Lazarus; 36. M. Gruntman; 37. P. Kiraly; 38. Z. Kobylinski; 39. H. Pérez-de-Tejada; 40. M. Ostrowski; 41. M. Banaszkiewicz; 42. J. Juchniewicz; 43. W.-H. Ip; 44. A. Barnes; 45. M.S. Potgieter; 46. P. Dyal; 47. Y.C. Whang; 48. M. Lee; 49. J.R. Jokipii; 50. D.L. Judge; 51. M.V. Alaniya; 52. —; 53. J. Kóta; 54. M. Bzowski; 55. S. Bowyer; 56. M.K. Bird.

PHYSICS OF THE OUTER HELIOSPHERE

Proceedings of the 1st COSPAR Colloquium held in Warsaw, Poland, 19–22 September 1989

Edited by

S. GRZEDZIELSKI

Space Research Centre, Polish Academy of Sciences, Ordona 21, 01–237 Warsaw, Poland

and

D.E. PAGE

European Space Agency and Jet Propulsion Laboratory, 4800 Oak Grove Drive, MS 169/506, Pasadena, CA 94035, U.S.A.

Published for
THE COMMITTEE ON SPACE RESEARCH
by
PERGAMON PRESS

Oxford · New York · Beijing · Frankfurt · São Paulo · Sydney · Tokyo · Toronto

U.K.	Pergamon Press plc, Headington Hill Hall, Oxford OX3 0BW, England
U.S.A.	Pergamon Press, Inc., Maxwell House, Fairview Park, Elmsford, New York 10523, U.S.A.
PEOPLE'S REPUBLIC OF CHINA	Pergamon Press, Room 4037, Qianmen Hotel, Beijing, People's Republic of China
FEDERAL REPUBLIC OF GERMANY	Pergamon Press GmbH, Hammerweg 6, D-6242 Kronberg, Federal Republic of Germany
BRAZIL	Pergamon Editora Ltda, Rua Eça de Queiros, 346, CEP 04011, Paraiso, São Paulo, Brazil
AUSTRALIA	Pergamon Press Australia Pty Ltd., P.O. Box 544, Potts Point, N.S.W. 2011, Australia
JAPAN	Pergamon Press, 8th Floor, Matsuoka Central Building, 1-7-1 Nishishinjuku, Shinjuku-ku, Tokyo 160, Japan
CANADA	Pergamon Press Canada Ltd., Suite No. 271, 253 College Street, Toronto, Ontario, Canada M5T 1R5

First edition 1990

Library of Congress Cataloging in Publication Data

COSPAR Colloquium (1st: 1989: Warsaw, Poland)
Physics of the outer heliosphere: proceedings of the 1st COSPAR Colloquium held in Warsaw, Poland, 19–22 September, 1989/edited by S. Grzedzielski and D.E. Page.

Sponsors, the Committee on Space Research (COSPAR), The International Astronomical Union (IAU), and the Polish Academy of Sciences.

1. Heliosphere—Congresses 2. Astrophysics—Congresses.
3. Solar Wind—Congresses. 4. Interstellar matter—Congresses. 5. Galactic cosmic rays—Congresses.
I. Grzedzielski, S. II. Page, D. Edgar. III. COSPAR.
IV. International Astronomical Union. V. Polska Akademia Nauk. VI. Title.
QC881.2.H43C67 1989 523.7—dc20 90–7920

ISBN 0–08–040780–3

In order to make this volume available as economically and as rapidly as possible the author's typescript has been reproduced in its original form. This method unfortunately has its typographical limitations but it is hoped that they in no way distract the reader.

Printed in Great Britain by BPCC Wheatons Ltd, Exeter

CONTENTS

THE INTERNATIONAL HELIOSPHERIC STUDY AND THE FIRST COSPAR COLLOQUIUM — PREFACE

The International Heliospheric Study (IHS) was approved by COSPAR as long ago as 1982. At the time it was expected that the study would centre on the ESA/NASA Ulysses spacecraft but unfortunately the start of that mission has been delayed until October 1990. It was decided that the study should in any case begin in 1986 in order to exploit the existence of the Pioneer 10 and 11 and the Voyager 1 and 2 spacecraft in the outer heliosphere, and results from other spacecraft will of course be included as they become available.

The scientific topics of greatest interest to the IHS are the three dimensional structure of the solar wind and the interplanetary magnetic field carried by it, the propagation of energetic solar particles throughout the heliosphere and the modulation of galactic cosmic rays by the heliosphere. The Pioneer and Voyager spacecraft at great distances from the Sun present the possibility of detecting a boundary between our solar system and the local interstellar medium.

The IHS task group set up by COSPAR aims to strengthen links between those scientists interested in the heliosphere so that the best possible use can be made of the missions , facilities and data that become available. The intention is not to create any sort of bureaucratic coordinating machinery but only to facilitate the flow of information. The members of the Task Group are W. I. Axford (Germany), R. V. Bhonsle (India), K. I. Gringauz (U.S.S.R.), S. Grzedzielski (Poland), M. Oda (Japan), D. E. Page (ESA/JPL) as Chairman, E. J. Smith (U.S.A.), E. C. Stone (U.S.A.) and K. P. Wenzel (ESA).

This task group was given the honour of organising the very first COSPAR Colloquium and called this "Physics of the Outer Heliosphere". They were happy to accept the invitation of Professor S. Grzedzielski, Director of the Space Research Centre, Polish Academy of Sciences, to hold the meeting in Warsaw and would like to thank the local organising committee (S. Grzedzielski, R. Ratkiewicz, D. Rucinski and J. Ziemkiewicz) for an excellent job carried out in less than ideal circumstances. The Colloquium programme committee was S. Grzedzielski (Poland, Executive Chairman), W. I. Axford, (F.R.G), V. A. Baranov (U.S.S.R), J. L. Bertaux (France), R. V. Bhonsle (India), S. Bowyer (U.S.A.), H. J. Fahr (F.R.G.), L. A. Fisk (U.S.A.), P. C. Frisch (U.S.A.), K. I. Gringauz (U.S.S.R.), A. Hewish (U.K.), J. Kota (Hungary), R. R. Meier (U.S.A.), M. Nicolet (Belgium), A. Nishida (Japan), D. E. Page (ESA, Chairman), E. J. Smith (U.S.A.), E. C. Stone (U.S.A.), K. P. Wenzel (Netherlands).

Thanks are due to those referees who helped collect and edit the papers arriving from the five Colloquium sessions. Particular mention should be made of Andrzej Czechowski (Space Research Centre, Warsaw) who assisted in a major way in bringing together the material for the Colloquium Proceedings which are to be published by Pergamon Press in September 1990.

A highlight of the Colloquium was the report on the encounter of the Voyager 2 spacecraft with Neptune which had occurred just a few weeks earlier. We are grateful to N.F. Ness who stepped in at short notice when it was found that the Voyager project scientist was unable to be present. The report is not available for the Colloquium Proceedings.

D. Edgar Page
Chairman COSPAR IHS Task Group

PRESIDENTIAL ADDRESS

This is the first volume in the series "COSPAR Colloquia" which, it is anticipated, will become a permanent and vigorous component of COSPAR's programme of publications on all aspects of scientific space research. It should be a useful adjunct to "Advances in Space Research" which records the proceedings of the biennial COSPAR Assemblies. The Colloquia are intended to be specialized meetings to be held as a need is perceived and generally in countries where an Assembly may not be expected to take place during the following few years. It is hoped that on the one hand, the influence of COSPAR, in a positive sense, can thereby be more widely felt than it might be otherwise, and, on the other hand, that communities with specialized interests can also be better recognized and served in COSPAR.

The first Colloquium held in Poland has been a success in every respect. The topic chosen was a timely one, in view of the continuation of the Voyager and Pioneer missions beyond their last planetary encounters and eventually perhaps into the undisturbed interstellar medium, and also because of the planned launch of the Ulysses mission in 1990. The topic has also long been of special interest to space scientists in Poland who have made many important contributions to its study, some of which are described in this volume.

This Colloquium will serve as a model for all subsequent Colloquia both in terms of the quality of the scientific presentations and the efficient organization. I wish to record, on behalf of COSPAR, our gratitude to Professor Grzedzielski and his colleagues and also to Dr. Page, representing the International Heliospheric Study, for the great efforts they have made in ensuring its success and in providing COSPAR with an excellent start in a new venture.

W.I. Axford
President of COSPAR

Opening Session

INTRODUCTORY LECTURE — THE HELIOSPHERE

W.I. Axford

Max-Planck-Institut für Aeronomie, D-3411 Katlenburg-Lindau, F.R.G.

ABSTRACT

A brief review is given of the historical ideas dealing with the development of our present concept of the heliosphere, together with a summary of the present status, particularly with regard to the distance to the solar wind shock termination.

INTRODUCTION

The "heliosphere" is defined to-day as the region of space around the Sun where the solar wind plasma and magnetic field have a controlling effect on the state and behaviour of the medium. The heliosphere must therefore contain a shock which marks the termination of the solar wind as a supersonic flow and there should be an outer boundary, termed the "heliopause", separating plasma of interstellar and solar origins. In fact the definition may be over simple in view of the probable importance of reconnection between the solar and interstellar magnetic fields and the transparency of the boundary to neutral particles coming from either side. Nevertheless the definition is easily understood and should be retained as long as it remains useful. The likely configuration of the heliosphere is sketched in Figure 1. By analogy with magnetospheric terminology the region between the termination shock and the heliopause is termed the "heliosheath".

This is not intended as an exhaustive review of the literature of heliospheric physics but rather as an introduction to some of the more interesting early history of the subject, which is now being gradually forgotten, together with a brief update on some recent developments which seem to be of particular significance. The reader is referred to the remaining papers in this volume for more detailed accounts and also to the reviews by Holzer, Suess, Baranov /1/ and, from a more historical perspective, Simpson /2/.

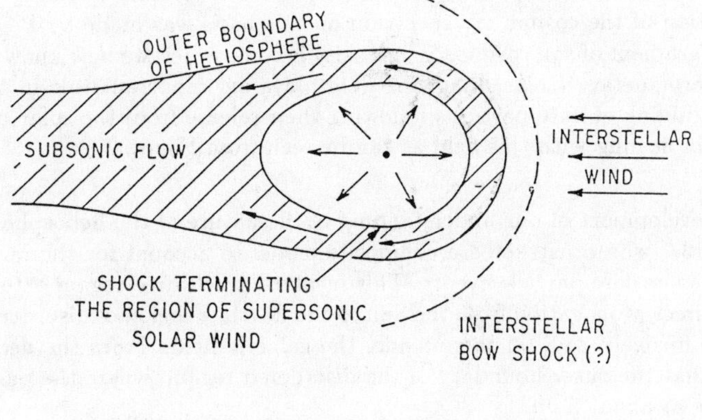

Fig. 1 The configuration of the heliosphere shown schematically.

THE FIRST PHASE OF STUDIES OF THE HELIOSPHERE

The first suggestion that the heliosphere (then not defined as such) might have important observable effects was made by Alfvén /3/ who, following Teller, proposed that cosmic rays have a local (solar) origin and are accelerated somehow by the irregular solar "corpuscular streams" believed to cause geomagnetic storms. A first version of the required acceleration mechanism /4/ involved radial diffusion in the solar dipole magnetic field in a manner similar to that believed to occur in a magnetosphere; this would of course lead to the particles having their highest energies close to the Sun and therefore a negative radial gradient which is not observed. A second version involved pitch angle scattering and "magnetic pumping" and was successful in producing a power law spectrum.

In fact, it is obvious from a closer consideration of the nature of cosmic rays that they cannot be of local origin, although there are components of the energetic particle population observed near the Earth that have been accelerated in the solar system (e.g. solar flare particles, the "anomalous" component, particles accelerated by interplanetary shocks and in planetary magnetospheres). Alfvén however does not give up so easily as is evident from an amusing exchange with Ginzburg /5/:

Ginzburg: "The arguments against a local/solar origin of cosmic rays are numerous .... I have worked in this field for some years and I can say in the course of time the argument for a galactic origin slowly improves. So I hope during my lifetime I shall see the full victory of this theory."
Alfvén: "I hope you will live very long!"

The first serious discussion of the nature of the heliosphere as such was made by Davis /6,7/ whose ideas are effectively summarized in Figure 2. Davis pointed out that the solar corpuscular radiation described by Biermann /8/ and now termed the solar wind, would produce a cavity in the galactic magnetic field permitting energetic solar flare particles to reach the Earth. (A magnetic field of galactic origin between the Earth and the Sun would shield the former from solar particles having energies less than 45 GeV if the field strength were 1 nT.) In order to estimate the size of the field-free cavity Davis simply equated the dynamic pressure of the solar wind at R au to the interstellar magnetic pressure:

$$n_e m V^2/R^2 = B_g^2/8\pi \qquad (1)$$

and took $B_g = 1\,nT$ and $V = 1000\ km/sec$. The values of n_e, the electron density at 1 au, derived by Biermann on the basis of an incorrect idea concerning the interaction between solar wind and cometary plasma (viz. $n_e = 10^3 - 10^5 cm^{-3}$) gave uncomfortably large values for R but Davis nevertheless took $R = 200\ au$ as a working value. This later proved acceptable on considerations of the scattering of sunlight by interplanetary electrons by Kiepenheuer /9/ and of the amplitude of geomagnetic storm sudden commencements (i.e. $n_e = 1 - 10\ cm^{-3}$). Even so Davis was fairly cautious in his second paper /7/ and suggested that the boundary might be beyond the orbit of Jupiter or even of Uranus. Our present estimates, which involve modifications of (1), suggest that R is of the order of 100 au with an uncertainty of perhaps 30% and that this refers only to the minimum distance to the solar wind termination shock rather than to the heliopause.

Davis was most interested in the effects of the solar wind cavity on galactic cosmic rays. He suggested that cosmic rays with energies less than about $10^{12}eV$ could be trapped in the cavity for long periods and accordingly that changes in the size of the cavity could produce the observed solar cycle modulation effect and that the electron component would be removed by its Compton scattering of sunlight /10/. An attempt to explain the time variation of the cosmic ray spectrum on this basis was made by Beiser /11/; such a theory implies that the radial gradient of the cosmic ray intensity is zero, which we now know is not the case. In fact, it is clear that the interplanetary (solar) magnetic field plays an important role in the modulation process /12/ as well as in the guiding of flare particles following their release from the solar corona; Davis had some conceptual difficulties in dealing with this field as can be seen from Figure 2.

The next step in the development of our understanding of the nature of the heliosphere was made by Meyer, Parker and Simpson /13/, who constructed a composite model to account for the initial anisotropy and the temporal behaviour of solar flare particle events. This involved a field-free region extending beyond the orbit of the Earth allowing direct propagation from the Sun to the Earth, a region of disordered fields through which the particles are lost diffusively and, on the outside, the galactic field. From the decay phase of the events it could be estimated that the outer boundary of the disordered region, where the particles were assumed to escape freely, should be at about 5 au.

Following his account of the true nature of the solar wind, Parker /14/ was able to modify this model in a significant way by including a smooth spiral interplanetary magnetic field in the inner region, thus providing a simple explanation of the east-west asymmetry in the behaviour of flare particle events. He also suggested that the disordered field could be considered as being the consequence of hydromagnetic (firehose, mirror, cyclotron) instabilities which might naturally arise in a collisionless solar wind having a spiral magnetic field. In fact, with the introduction of the concepts of coronal transport /15/ and anisotropic diffusion in the largely ordered interplanetary magnetic field /16/ the need for the main feature of the Meyer et al. model is obviated and indeed, subsequent observations have shown that there is no isotropically diffusing region beyond the orbit of the earth, at least out to 40 *au*.

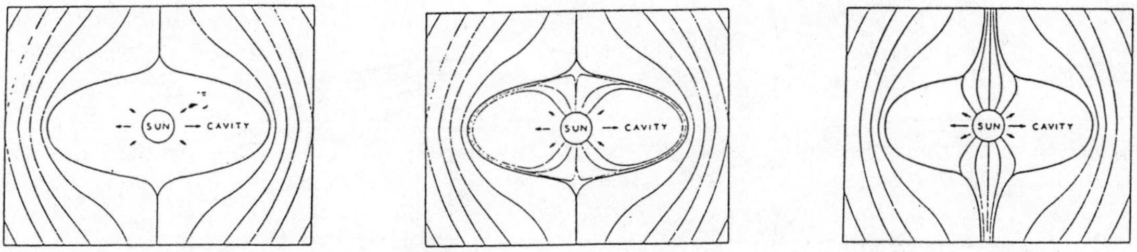

(a) (b) (c)

Fig. 2 Various versions of the heliosphere according to Davis /6,7/. It is argued that the region between the Earth and the Sun is free of magnetic field, although a general weak solar field could be accommodated as shown in (b) and (c).

With the development of the concept of the solar wind as a supersonic flow resulting from the continual heating of the solar corona, it was natural for a quite different approach to be made to the problem of cosmic ray modulation /17/. It became possible to account for the solar cycle and irregular short term variations (Forbush decreases) on the basis of diffusion in the expanding solar wind with occasional modifications induced by expanding shock associated with flares, for example /12,17,18/, and the diurnal variations on the basis of anisotropic diffusion in this spiral interplanetary magnetic field pattern which must corotate with the Sun /19/.

It is quite obvious to those with a background in gas dynamics that the solar wind, which was shown by Parker to have a certain analogy to flow in a Laval nozzle, must terminate in a shock front at a point where its dynamic pressure has dropped to that of the external medium /20/. This condition is, with minor modifications /21/ that given by equation (1), which should not therefore be regarded as giving the distance to the heliopause as tacitly assumed by Davis. With more up-to-date values for the various parameters involved it is difficult to place the termination of the supersonic solar wind closer than about 100 *au* if $B_g = 0.3\ nT$ /22/. Nonetheless arguments have been advanced for putting it much closer: for example, Brandt and Michie /23/, using an incorrect interpretation of the nature of comet tails argued that the supersonic wind would change to a solar breeze at about 2 *au* in rough agreement with the inner boundary of the Meyer et al. diffusing region, while Lanzerotti and Schulz /24/, Jovian radio emissions, argued that 5 *au* might be appropriate. There has never been any basis for assuming that the pressure in the interstellar medium could be high enough to produce such small values of R (equivalent to $B_g = 300\ nT$ and $50\ nT$, respectively) and we know now from observations from the Pioneer and Voyager spacecraft that R exceeds 40 *au*.

If the Sun had always been at rest with respect to the interstellar medium, the position of the termination shock would always have been stationary but the heliopause would have grown in radius as $t^{-1/3}$. At the present time the radius of the spherical region occupied by solar plasma would be a few parsecs if there had been no great variations in the solar wind in the past. If all stars were equally privileged it would be expected that the whole of interstellar space would be filled by hot bubbles of stellar wind plasma with little room left for the rest of the interstellar medium. In fact this result does not depend on the stars being stationary since the plasma tails of moving stars would occupy roughly the same volume. Cooling as a consequence of interaction with neutral component of the interstellar medium changes this as far as our heliosphere is concerned, although in the case of more powerful stellar winds the hot plasma may be much more resistant and therefore make an important contribution to the very hot ($10^6 K$) component of the interstellar medium.

With the assumption that the region of supersonic flow in the heliosphere is small, Parker /17/ has examined the possible shape of the heliopause in the case where the Sun moves subsonically with respect to the external medium and the contribution of the magnetic field to the pressure of this medium is small (Figure 3a). In the case of supersonic motion one would expect a similar shape but there would be a bow shock in the interstellar medium upstream. The presence of a weak magnetic field would lead to the tail of the heliosphere being noncircular in cross section as a consequence of the curvature stresses exerted by the field. At the other extreme, when the motion of the Sun can be neglected but the interstellar magnetic field is strong, the subsonic solar wind plasma divides and flows in two circular channels parallel to the field (see Figure 3b). In the intermediate case these two channels would be tilted backwards from the direction of motion of the Sun perpendicular to the field and the cross sections would again become noncircular.

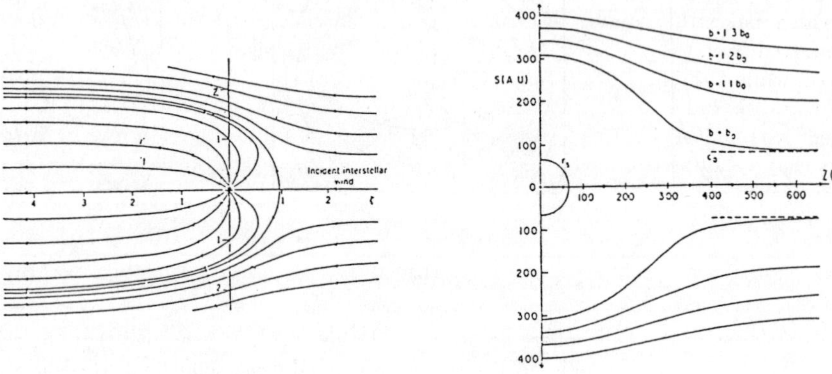

(a) (b)

Fig. 3 Two versions of the heliosphere and heliopause according to Parker /18/: (a) a source in a flowing, incompressible external medium, and (b) a source in a stationary medium with a strong magnetic field.

THE SECOND PHASE: THE NEUTRAL INTERSTELLAR GAS

A fresh element was introduced into the study of the nature of the heliosphere with the suggestion that interstellar hydrogen atoms could play an important role in dissipating the solar wind energy, thereby avoiding the possibility that stellar wind bubbles completely dominate the interstellar medium /25/. The proper motion of the Sun is such that if it provides a reasonable estimate of the relative speed of such atoms with respect to the Sun they are able to penetrate to within a distance much smaller than R without being photoionized. (Gravity would also help but, as pointed out by Brandt /26/ it is effectively cancelled by the pressure of solar Lyman alpha radiation.) The main consequence of the presence of these neutral H atoms, as far as the solar wind is concerned, is that the plasma in the subsonic region beyond the termination shock is rapidly cooled to temperatures of the order of $10^4 K$, the magnetic field becomes subject to reconection and reconfiguration and a shower of fast H atoms is emitted more-or-less isotropically.

This last idea was taken up by Paterson et al. /27/ who argued that the observations of broadened Lyman alpha emission in the interplanetary medium could be explained as scattering of solar Lyman alpha by such fast atoms as penetrate reasonably close to the Sun. On this basis they estimated that $R = 20\ au$, taking the interstellar neutral H density to be $0.02\ cm^{-3}$. This result was criticized by Hundhausen /28/ on the grounds that it was assumed that the fast H atoms are produced just beyond the shock and, assuming that the flow there is incompressible, derived a corrected value for R of about 5 au. In fact, it has been pointed out by Tinsley /29/ that the interpretation in terms of scattering from fast secondary atoms may be incorrect as the filter used in the observations was in reality quite narrow (equivalent to about 25 km/sec).

A qualitative change in our perception of the behaviour of interstellar atoms in the solar neighbourhood was induced by Blum and Fahr /30/ in a series of papers in which they developed the ideas which are now favoured in treating this problem. In particular they considered the behaviour of a cold beam of neutral H or He atoms subject to loss by photoionization and interaction with the solar wind (charge exchange or impact ionization). (The procedure is very similar to that used in a now abandoned theory of cometary origin proposed by Lyttleton /31/.) It is found that on the upwind side the H atoms penetrate to within about 4 au from the Sun before losses become serious while He atoms penetrate to closer than 0.5 au. On the downwind side there is a focussing so that, for a cold beam, the density becomes (logarithmically) infinite on the axis of symmetry. Similar results were obtained, without at first knowing of the work of Blum and Fahr, by Holzer, Johnson and Gregory /32/ who took into account the effects of radiation pressure due to solar Lyman alpha, calculated the resulting He^+ ion production in the solar wind and also the pattern of resonantly scattered solar HI(1216), HeI(584) and HeII(304) radiation. It was also shown by Forman /unpublished/ that the time variations of the solar Lyman alpha flux associated with active regions and modulated by solar rotation, may disturb the trajectories of H atoms to such an extent that a measurable random component is induced which could be detectable by measuring the line shapes of the scattered radiation.

The anisotropic background of Lyman alpha around the Sun was discovered by Blamont and Bertaux and Thomas and Krassa in 1971 /33/ and interpreted according to the above scheme. Many subsequent observations of both the neutral H and He scattered radiation have since lead to the conclusions concerning the state of the local (neutral) interstellar medium given in Table 1. The close coincidence between the temperatures and velocities of the two components is significant and gives us some confidence in the correctness of the interpretation, which makes the assumption that the solar helium line is red shifted by a small amount (as are most UV lines emitted from the transition region) /34/. At a temperature of about 8000 K helium is almost completely neutral while hydrogen may be partly ionized. Accordingly, assuming that there is 8% of helium in the interstellar gas we find that N_e, the density of electrons and protons, may be $30(0-80) \times 10^{-3} cm^{-3}$, which is consistent with pulsar dispersion measurements (made over a much greater range) and with thermal equilibrium or perhaps photoionization by interstellar EUV radiation. For the electron density to be significantly larger than this range, there would have to be a major departure from equilibrium which could perhaps be the case if there had been a supernova in the neighbourhood in the last 10^6 years.

TABLE 1

$N(H)$	$=$	$65 \pm 10 \times 10^{-3}\ cm^{-3}$	$N(He)$	$=$	$8 \pm 4 \times 10^{-3}\ cm^{-3}$
$T(H)$	$=$	$8000 \pm 1000\ K$	$T(He)$	$=$	$7000 \pm 2000\ K$
$V(H)$	$=$	$20 \pm 1\ km/sec$	$V(He)$	$=$	$21.5 \pm 2.5\ km/sec$

It should be noted that the parameters derived for neutral He from the measurement of scattered He(584) radiation have recently been confirmed from interpretations of measurements of singly-ionized helium ions in the solar wind (reported elsewhere in these Proceedings). The neutral hydrogen measurements may be affected by multiple scattering which would cause the density to be overestimated by 5 - 35% /35/ and hence allow for a slightly higher electron density than given by the range suggested.

With the above parameters characterizing the state of the very local interstellar medium (VLISM) and assuming that the solar wind termination shock occurs at a distance $R < 100\ au$ (which is the interesting range), we can draw several important conclusions:

1. The supersonic solar wind flow is not significantly affected by the drag associated with pick-up of ions originating in the neutral interstellar gas /36/ and, in particular, R is not reduced by more than a few percent. (Nevertheless there may be observable effects in the form of an additional component of turbulence associated with the unstable velocity distributions resulting from the pick-up process /37/.)

2. "Shielding" of the inner heliosphere from the full neutral H atom flux as a result of coupling (via charge exchange) with the plasma which is diverted around the heliopause is also probably negligible since the mean free path for the H on H^+ interaction exceeds 600 au if $N_e < N(H)$ /38/.

3. The inverse interaction (H^+ on H) is however very important in the "heliosheath" between the heliopause and the termination shock, since the lifetime against charge exchange of the hot thermalized solar wind protons is only about $10^8 sec$. This corresponds to only $6-60\ au$ if the flow speed is $10-100\ km/sec$, which is much smaller than the dimensions of the heliosphere. Accordingly the heliosheath should be prone to collapse, at least as far as the proton component is concerned. Helium ions and electrons would remain largely unaffected and would then provide most of the plasma pressure.

4. Taking the interstellar magnetic field strength to be 0.35 nT it can be shown that the transverse (Alfvén) waves in the VLISM which have wavelengths comparable to, or less than, the typical scale of the heliosphere (say 300 au) have group speeds greater than 20 km/sec (i.e. $V(H)$) if $N_e < N(H)$ /39/. This implies that the motion of the VLISM relative to the heliosphere is "subsonic" with respect to fast mode waves unless N_e is larger than expected. The waves concerned have sufficiently high frequencies that the neutral component of the VLISM does not participate in the motion and therefore does not affect the wave speed.

If the distance to the shock termination is very much larger than 100 au and the degree of ionization of the VLISM is greater than 50%, some of these conclusions would of course have to be changed.

THE THIRD PHASE: BACK TO COSMIC RAYS

A very significant discovery was made in 1973, namely the "anomalous" component of cosmic rays /40/. These particles are prominent in the energy range $1 - 10^2 MeV/nucleon$ and appear to be restricted to elements with high first ionization potentials. O and He are the most abundant but C, N, Ne, Ar and H are also present, suggesting that those species that can be photoionized by Lyman alpha are perhaps excluded. Fisk et al. /41/ suggested that the anomalous component is in fact the result of acceleration of ions of interstellar origin within the heliosphere and that they penetrate initially as neutrals, becoming ionized by interaction with the solar wind plasma and solar ultraviolet radiation. Various possible acceleration schemes have been considered but the most promising appears to be shock acceleration at the solar wind termination shock /42/.

The anomalous component has been intensively investigated in the meantime and there is now strong evidence to suggest that the particles are to a large extent singly ionized and have correspondingly large rigidities /43/. Furthermore, the radial gradients are quite large ($5 - 10\%/au$) and since the combined pressure is about $0.07\ eV/cm^3$ at 21 au it could be expected that this would be sufficient to dominate the solar wind at $70 - 100\ au$. This is the conclusion one would draw from modern cosmic ray modulation theory /12/ and if borne out by observations from the Voyager spacecraft during the next 10 - 20 years we can expect that large scale heliospheric physics will become a matter for direct investigation as well as speculation.

Observations of the radial gradient of galactic cosmic rays lead to a similar conclusion concerning the value of R /44/. The gradients in the region $1 - 40\ au$ from the Sun are $1.25 - 2.0\%/au$ for particles with a mean energy of a few GeV/nucleon /45/. According to a simple model of the behaviour of cosmic rays in the galaxy /46/ the total modulation at this energy is about a factor 2, which when combined with these gradients, suggests that R should be in the range $50 - 100\ au$ if they are maintained to the termination shock.

The possible effects of galactic cosmic rays on the solar wnd have been considered by numerous authors in the past /47/ and it is agreed that they are not likely to be important if R does not greatly exceed 100 au. However the anomalous component may be capable of changing the characteristics of the solar wind so much that the termination shock is not only weakened as a result of the effective speed of sound increasing but it may be completely spread out and its thickness determined by the diffusion of the anomalous cosmic rays rather than the usual microscopic plasma properties. In this case the transition to subsonic flow in the heliosheath involves a "cosmic ray" shock /48/ which is of great interest in connection with theories of cosmic ray acceleration but not normally realisable in the interplanetary medium. Such shocks are subject to severe internal instabilities which may play a role in the diffusion of the particles in addition to the turbulence induced in the solar wind by the ion pick-up process /49/.

CONCLUSIONS

On the basis of the deductions made from the observed radial gradients of the anomalous component of the cosmic radiation and also because we are optimistic it is interesting to re-examine the implications of a solar wind shock termination occurring at $R = 100\ au$. It is important to note that the anomalous component itself cannot cause the transition to occur at a particular distance from the Sun because it is, in a sense, part of the solar wind and does not exert a pressure from outside. The anomalous component is simply an indicator of change in the solar wind and suggests that the transition lies in the range $R = 70 - 100\ au$ without implying a reason other than that the total external pressure must be sufficient to make the shock occur in this region. (There is some loss of solar wind ram pressure as a consequence of charge exchange with interstellar H atoms and the modulation of galactic cosmic rays, but this is minor if R lies in the range suggested.)

Contributions to the external pessure exerted on the heliopause by the interstellar medium arise from the magnetic field (perhaps amplified by being wrapped around the surface), the flowing interstellar plasma and possibly, low energy cosmic rays which are unobservable within the heliosphere. It is most unlikely that a sufficient flux of low energy galactic cosmic rays exists /22/ but reasonable constraints can be placed on the pressure of the interstellar plasma since it must have the same temperature and flow speed as the neutral H and He and its density is likely to be less than $0.1\ cm^{-3}$, as argued above. The strength of the interstellar magnetic field in the VLISM is unknown, however measurements of the dispersion and Faraday rotation of pulsar signals indicate that a value of the order of $0.2 - 0.3\ nT$ might be appropriate for the region within about 100 *parsecs* from the Sun /50/.

A suitably enlarged version of an old diagram is given in Figure 4 /21/. This shows the expected value of R for various combinations of the plasma density and magnetic field strength in the VLISM. For $0 < N_e < 0.1\ cm^{-3}$ the magnetic field strength must lie in the range $0.23 - 0.4\ nT$ if $R = 100\ au$ and in the range $0.45 - 0.6\ nT$ if $R = 70\ au$. These are quite acceptable values for the field strength bearing in mind that substantial fluctuations from its large scale mean value might be expected. It is quite difficult however to achieve the value $R = 46\ au$ suggested by Kurth et al. /51/ on the basis of low frequency radio emissions; this would require a magnetic field strength of $\sim\ 1\ nT$. Nevertheless we can therefore be rather optimistic as far as the continuation of the Voyager mission is concerned and can look forward to determining the position of the termination shock during the next 20 years as well as making in situ observations of the structure of a cosmic ray shock for the first time.

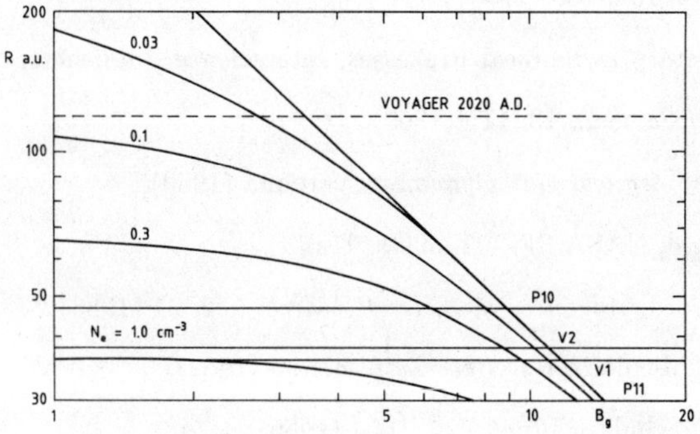

Fig. 4 The minimum distance to the solar wind shock terminator as a function of interstellar electron density N_e and magnetic field strength B_g. The positions of the spacecraft Pioneers 10, 11 and Voyagers 1, 2 in 1990 as well as the limit of the Voyager mission (2020 AD) are shown.

REFERENCES

1. T.E. Holzer, *Rev. Geophys. Space Sci.*, 15, 467 (1977), and *Ann. Rev. Astron. Astrophys.*, 27, 199 (1989);
S.T. Suess, *Rev. Geophys.*, in press (1990);
V.B. Baranov, *Space Sci. Rev.*, 52, 89 (1990).

2. J.A. Simpson, *Advances in Space Res.*, 9/4, 5 (1989) and *Proc. Int. Cosmic Ray Conf.*, Moscow, 9, 84 (1987).

3. H. Alfvén, Cosmic Electrodynamics, Clarendon Press, Oxford (1950);
H. Alfvén and E. Teller, *Phys. Rev.*, 75, 892 (1949);
R.D. Richtmeyer and E. Teller, *Phys. Rev.*, 75, 1729 (1949);
H. Alfvén, *Phys. Rev.*, 75, 11 (1949) and 77, 3 (1950).

4. H. Alfvén, *Tellus*, 6, 232 (1954) and 11, 106 (1959);
H. Alfvén and E. Äström, *Nature*, 181, 330 (1958;
C.-G. Fälthammar, *Proc. Enrico Fermi School No.19*, 99, Academic Press, N.Y., (1963).

5. V.L. Ginzburg, in "Early History of Cosmic Ray Studies", eds. Y. Sekido and H. Elliot, Reidel, Dordrecht (1985).

6. L.E. Davis, Jr., *Phys. Rev.*, 100, 1440 (1955).

7. L.E. Davis, Jr., *J. Phys. Soc. Japan*, 17, A-II, 543 (1962).

8. L. Biermann, *Z. Astrophysik*, 29, 274 (1951) and *Mem. Roy. Soc. Liege*, 4th Series, 13, 291 (1953).

9. K.O. Kiepenheuer, "The Sun", ed. G.P. Kuiper, Univ. Chicago Press, Chicago, 437 (1953).

10. E. Feenberg and H. Primakoff, *Phys. Rev.*, 73, 449 (1948).

11. A. Beiser, *J. Geophys. Res.*, 63, 1 (1958).

12. J.R. Jokipii, in "The Sun and Heliosphere in Three Dimensions", ed. R. Marsden, Reidel, Dordrecht, 375 (1986);
J.R. Jokipii and J. Kóta, *Proc. Int. Conf. Cosmic Rays*, Adelaide, 6, 198 (1990).

13. P. Meyer, E.N. Parker and J.A. Simpson, *Phys. Rev.*, 104, 768 (1962).

14. E.N. Parker, *Astrophys. J.*, 128, 664 (1958).

15. G.C. Reid, *J. Geophys. Res.*, 69, 2659 (1964).

16. W.I. Axford, *Planet. Space Sci.*, 13, 1301 (1965).

17. P. Morrison, *Phys. Rev.*, 101, 1397 (1956).

18. E.N. Parker, Interplanetary Dynamical Processes, Interscience Publishers, New York (1963).

19. W.I. Axford, *Planet. Space Sci.*, 13, 115 (1965).

20. T. Clauser, *4th Symp. Cosmical Gas Dynamics*, Varenna (1960).

21. W.I. Axford, Solar Wind, NASA SP-308, 609 (1972).

22. W.-H. Ip and W.I. Axford, *Advances in Space Research*, 6/2, 27 (1986).

23. J.C. Brandt and R.W. Michie, *Phys. Rev. Lett.*, 8, 195 (1962).

24. L.J. Lanzerotti and M. Schulz, *Nature*, 222, 1054 (1969).

25. W.I. Axford, A.J. Dessler and B. Gottlieb, *Astrophys. J.*, 137, 1268 (1963).

26. J.C. Brandt, *Icarus*, 3, 253 (1964).

27. T.N.L. Patterson, F.S. Johnson and W.B. Hanson, *Planet. Space Sci.*, 11, 767 (1967).

28. A.J. Hundhausen, *Planet. Space Sci.*, 16, 783 (1968).

29. B.A. Tinsley, *Rev. Geophys. Space Phys.*, 9, 89 (1971).

30. P.W. Blum and H.J. Fahr, *Nature*, 223, 936 (1969); *Astron. Astrophys.*, 4, 280 (1970); *Astrophys. Lett.*, 5, 127 (1970).

31. R.A. Lyttleton, "The Comets and Their Origin", University Press, Cambridge (1953).

32. C.T. Gregory, *Planet. Space Sci.*, 20, 841 (1972);
H.E. Johnson, *Planet. Space Sci.*, 20, 829 and 1784 (1972);
T.E. Holzer, Ph.D.Thesis, Univ.California, San Diego (1970).

33. J.L. Bertaux and J.E. Blamont, *Astron. Astrophys.*, 11, 200 (1971);
G.E. Thomas and R.F. Krassa, *Astron. Astrophys.*, 11, 218 (1971).

34. E. Chassefière, J.L. Bertaux, R. Lallement, B.R. Sandell and L. Broadfoot, *Astron. Astrophys.*, 199, 304 (1988);
E. Chassefière, F. Dalaudier and J.L. Bertaux, *Astron. Astrophys.*, 201, 113 (1988);
R. Ratkiewicz, D. Rucinski and W.-H. Ip, *Astron. Astrophys.*, 230, 227 (1990).

35. H.-U. Keller, K. Richter and G.E. Thomas, *Astron. Astrophys.*, 102, 415 (1981).

36. T.E. Holzer, *J. Geophys. Res.*, 77, 5407 (1972).

37. C.S. Wu and R.C. Davidson, *J. Geophys. Res.*, 77, 5399 (1972);
M.A. Lee and W.-H. Ip, *J. Geophys. Res.*, 92, 11041 (1987);
H.J. Fahr and J. Ziemkiewicz, *Astron. Astrophys.*, 202, 295 (1988).

38. H.J. Fahr and H.W. Ripken, *Astron. Astrophys.* 139, 551 (1984);
M.K. Wallis, *Astron. Astrophys.*, 130, 200 (1984);
S. Bleszinsky, *Astron. Astrophys.*, 180, 201 (1987).

39. W.I. Axford, M.K. Dougherty and J.F. McKenzie, *Proc. Int. Cosmic Ray Conf.*, Adelaide, 3, 311 (1990).

40. M. Garcia-Munoz, G.M. Mason and J.A. Simpson, *Astrophys. J. Lett.*, 182, L81 (1973);
D. Hovestadt, O. Vollmer, G. Gloeckler and C.Y. Fan, *Phys. Rev. Lett.*, 31, 650 (1973);
F.B. McDonald, B.J. Teegarden, J.H. Trainor and W.R. Webber, *Astrophys. J. Lett.*, 249, L71 (1974).

41. L.A. Fisk, B. Kozlovsky and R. Ramaty, *Astrophys. J. Lett.*, 190, L35 (1974).

42. M.E. Pesses, J.R. Jokipii and D. Eichler, *Astrophys. J. Lett.*, 13, L628 (1981).

43. F.B. McDonald, N. Lal, T.T. van Rosevinge and W.R. Webber, *Proc. Int. Cosmic Ray Conf.*, Moscow, 9, 171 (1987);
S. Biswas, N. Durgaprasad, A. Dutta, J.N. Goswani, D. Lal, R.K. Singh, M.N. Vakia and J.S. Yadav, *Proc. Int. Cosmic Ray Conf.*, Moscow, 3, 451 (1987).

44. B.A. Randall and J.A. Van Allen, *Geophys. Res. Lett.*, 13, 628 (1986);
W.R. Webber and J.A. Lockwood, *Astrophys. J.*, 317, 534 (1987).

45. W. Fillius, *Advances in Space Res.*, 9/4, 209 (1986).

46. W.-H. Ip and W.I. Axford, *Astron. Astrophys.*, 149, 7 (1985).

47. W.I. Axford and R.C. Newman, *Proc. Int. Cosmic Ray Conf.*, Phys. Soc. London, 173 (1965);
S.F. Sousk and A.M. Lenchek, *Astrophys. J.*, 158, 781 (1969);
I.V. Dorman and L.I. Dorman, *Geomag. Aeron.(USSR)*, 8, 652 (1968);
C.M. Ko and G.M. Webb, *Astrophys. J.*, 323, 657 (1987), and 325, 296 (1988);
C.M. Ko, R.J. Jokipii and G.M. Webb, *Astrophys. J.*, 326, 761 (1988).

48. W.I. Axford, E. Leer and G.A. Skadron, *Proc. Int. Cosmic Ray Conf.*, Plovdiv, 11, 132 (1977);
L.O'C. Drury and H.J. Völk, *Astrophys. J.*, 248, 344 (1981);
W.I. Axford, E. Leer and J.F. McKenzie, *Astron. Astrophys.*, 111, 317 (1982).

49. G.P. Zank, J.F. McKenzie and W.I. Axford, *Astron. Astrophys.*, in press (1990).

50. A.A. Ruzmaikin, A.M. Shukurov and D.D. Sokoloff, "Magnetic Fields of Galaxies", Astrophysics and Space Science Library, Kluwer, Dordrecht (1988).

51. W.S. Kurth, D.A. Gurnett, F.L. Scarf and R.L. Poynter, *Nature*, 312, 27 (1984).

Session I

Spectroscopic Data on the Local Interstellar Medium and the Related XUV Radiation Background

(Curator — P. C. Frisch)

CHARACTERISTICS OF THE LOCAL INTERSTELLAR MEDIUM

Priscilla C. Frisch

Department of Astronomy and Astrophysics, University of Chicago, 5640 S. Ellis Avenue, Chicago, IL 60637, U.S.A.

ABSTRACT

Observations of stars at distances up to 50 parsecs from the Sun yield a picture of the environment for the solar system, whereby the Sun is located inside a warm, low density, cloud (log T$\sim$ 4 °K, log $n_{total}$$\sim$–1 cm^{-3}). Different methods of determining the ionization are consistent for $n_e/n_{total} \sim$ 5–10%, providing that the ambient gas is in ionization equilibrium. However, since the evaporation time for the interstellar cloud surrounding the solar system is $\lesssim$ 1 million years, which is about equal to the cooling time and the recombination time between an electron and proton, the assumption of equilibrium may be invalid. At the position of the Sun, column densities of log N(H°) $\gtrsim$ 18 cm^{-2} are seen in all observed directions. The solar system appears to be positioned near the boundary of a cloud system flowing past the Sun so that nearby stars located in the hemisphere between galactic longitudes of 270° to 0°, and 0° to 90°, typically show higher column densities than nearby stars in the opposite hemisphere. An interstellar magnetic field is found within a few parsecs of the Sun in the general direction of the galactic center, pointing towards the direction of l$\sim$90°. Given our current understanding of interstellar magnetic fields, a realistic guess for the strength of this field is 1.6 μG.

INTRODUCTION

The study of the outer heliosphere of the solar system presents a unique opportunity to study the classic problems of encounters between stars and interstellar clouds from a perspective not usually available to astronomers, i. e., from the inside out. So far, such interactions have only been observed at large distances, where high ambient interstellar densities (n>>100 cm^{-3}) and UV fluxes are required to make bow shocks observable, and where 100 au structures are only barely resolvable under excellent observing conditions. Thus, aside from our anthropomorphic biases, the *in situ* studies of the outer heliosphere now possible with the Pioneer and Voyager spacecraft offer the best opportunity we have to collect detailed data on a star-cloud interaction.

The location of the helioshock and heliopause, and the properties of the boundary region between the solar wind and the interstellar medium, depend critically on the properties of the interstellar gas closest to the Sun. Thus, this review discusses the spatial density, temperature, ionization of, and magnetic field strength in, nearby interstellar gas, as well as cloud velocity and equilibrium. These parameters, together, yield the pressure in the cloud surrounding the Sun. Since this was a cross-disciplinary talk, and addressed to an audience with limited background in astronomy, I define some basic astronomical terms in an Appendix.

BACKGROUND

The distance scale of the nearest stars is vastly different from the distance scale of the solar system, with the nearest star $\sim$200,000 times more distant than the planet Pluto. For reference, the nearest stars, α CenA and α CenB, are at distances of 1.3 parsecs ("pc"). The outer boundary of the hypothetical Oort cloud is $\lesssim$ 50,000 au, and galactic tides equal the gravitational attraction of the Sun at distances of $\sim$150,000 au in the plane of the galaxy /43/. Using observations of relatively distant stars to define the properties of the ambient interstellar gas requires a little chutzpah, since this implicitly assumes that the properties of interstellar matter do not vary across scale sizes which differ by three orders of magnitude.

The nearest interstellar material has been alternately referred to as the "local fluff", "very local interstellar gas", "very very local interstellar cloud", "nearby interstellar medium", the "local interstellar cloud", and acronyms for all of these terms. The confusion of names reflects somewhat the confusion in determining the physical extent of the cloud surrounding the solar system. Here, the term "local fluff" (LF) is used to refer to the nearest 3 pc of interstellar material, while recognizing that the local fluff is part of a more extended cloud system. A 3 pc radius is chosen since the minimum H° column density seen in any direction is logN(H°)$\sim$18 cm^{-2}, which corresponds to a distance of 3 pc for an average spatial density of 0.1 cm^{-3}. This is a geocentered view of reality, since it is shown later (fig. 3) that the Sun is not centered inside of a small isolated interstellar cloud.

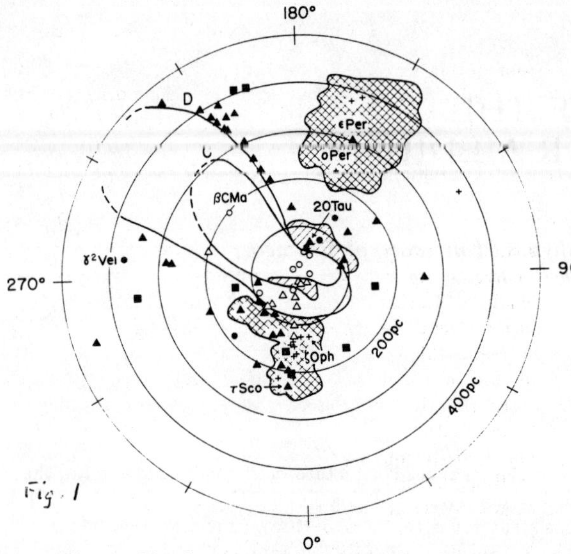

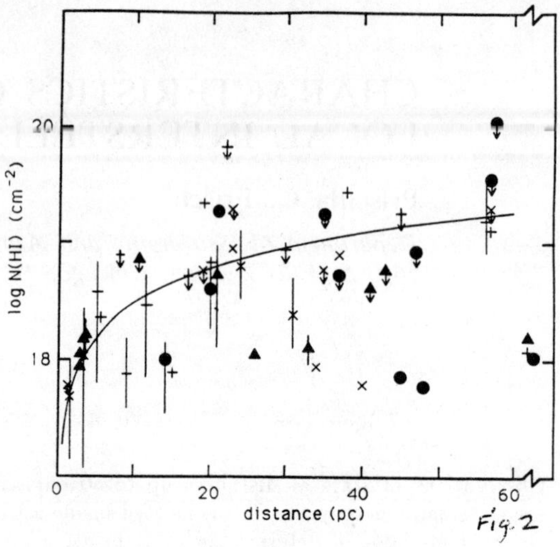

Fig. 1: Nearby neutral hydrogen distribution as viewed looking on the galactic plane. Contours C and D correspond to N(H)=0.5 and 5×10^{19} cm^{-2}, respectively. (See /17/ for more information.)

Fig. 2: Log N(H°) (cm^{-2}) plotted against the distance of the star for stars at distances $d < 60$ pc. Stars with distances greater then 60 pc have been added to the plot if log N(H°) < 18.5 cm^{-2}. Stars are coded so that stars in the galactic center hemisphere of the sky are plotted as crosses and "x's", and stars in the anti-center hemisphere of the sky are plotted as circles and triangles. The hypothetical column density that would result from a uniform cloud of infinite extent and 0.1 cm^{-3} average spatial density is plotted as a line. (This figure is taken from /15/, where references used in compiling these data are listed.)

To identify the properties of the interstellar cloud surrounding the solar system, it is helpful to understand our galactic environment. It is now well known that the Sun is moving through a region of space with a low ambient density of interstellar matter (ISM) /17/. Density contrasts between the LF cloud and, for instance, the molecular clouds in the Orion star forming region, span five orders of magnitude. Observations summarized in fig. 1 show that one must go to greater than 80-100 pc in any direction before H° accumulates a column density greater than 5×10^{20} cm^{-2}. This large "hole" in the distribution of the ISM persists to distances of $\geq$500 pc between galactic longitudes of 210° and 250°, and there is a "window" near l=245°±5, b=−3°±3, with N(H°)$<10^{21}$ cm^{-2}, where it is optically thin out to the edge of our galaxy in H° 21-cm radiation /41/.

The ultimate goal of a review such as this is to define the properties of the interstellar matter within 10^3 au, since that is the information required to estimate the confinement of the heliosphere /3,25/. However, astronomical observations have not yet successfully observed this region. Thus, this review focuses mainly on the region within the nearest 50 pc, where interstellar matter is unevenly distributed around the Sun. Other reviews discussing nearby interstellar matter are found in refs. /9,15,17,18,32,45/.

DENSITY AND TEMPERATURE

Spectral observations of nearby stars give information on interstellar gas properties averaged over the nearest 1−10 pc. The composition, average spatial density, kinetic temperature, and velocities of interstellar clouds are obtained from data on absorption lines formed by H° and trace elements (e. g. C°, N°, Na°, Mg^+, Fe^+, Ar°, Zn^+, etc.) in interstellar clouds. Hydrogen column densities derived from these data are shown in fig. 2. Data points for the half of our view towards the galactic center are plotted as crosses and "x's", and those in the opposite half (the anti-center) are plotted as circles and triangles. With the exception of two stars, λ And and 2 And, H° column densities for stars within 35 pc of the Sun in the anti-center hemisphere do not exceed $\sim3\times10^{18}$ cm^{-2}. In contrast, in the galactic center hemisphere, neutral hydrogen column densities of stars with $d<35$ pc exceed 3×10^{19} cm^{-2}. The solid line plotted in fig. 2 shows the predicted column density for a local fluff average spatial density of 0.1 cm^{-3}, and this approximation fits the data fairly well. Fig. 3 shows the shape, as defined by observations of nearby stars, that the interstellar cloud surrounding the solar system *would have if* it has an average neutral hydrogen density of $n_{HI}\sim0.1$ cm^{-3} (i. e. ignoring the *ad hoc* boundary of the LF). The asymmetrical distribution of interstellar matter at the position of the Sun is immediately apparent from figs. 2 and 3. Fig. 3b suggests that the LF is a part of a "curtain" of material which is mainly on the galactic center side of the Sun. This fits in with the suggestion that the LF is part of the expanding Loop I supernova shell (see below). Currently observations do not distinguish whether this feature is homogeneous, with $n_{total}=0.1$ cm^{-3} in all locations, or whether it is a cloud complex with average density 0.1 cm^{-3}, and higher spatial density in some portions. Since the Sun is moving at 20 km s^{-1} (~20 pc per million years), it has entered the LF relatively recently.

Three primary kinds of observations have been used in constructing figs. 1—3: spectral observations of H° and trace elements seen in absorption in nearby hot stars (where stellar and interstellar lines should be distinguishable from each other), observations of H° Lyα absorption superimposed on Lα emission features found in cooler stars (A, F, G, K types), and observations of EUV (λ<912 Å) fluxes from nearby white dwarfs which have been modeled with theoretical atmospheres convolved with interstellar absorption in order to derive interstellar column densities.

Each of these three categories of stars introduces problems that compromise the accuracy of interstellar parameters derived from stellar spectral data. For instance, correctly modeling underlying stellar Lα emission features is problematical in deriving interstellar column densities in stars with active chromospheres. Also, A stars are cool enough (T_{eff}~10^4 °K) that interstellar trace elements (e. g. Ca^+, Na°, Mg^+, etc.) absorption features may also be formed in the stellar atmosphere. To avoid this second problem, rapidly rotating target stars are selected so that stellar features are broadened and can, in principal, be distinguished from interstellar features which have FWHM<10 km s^{-1} in nearby gas. In practice, non-linear effects in the atmosphere, including possible chromospheres, mass-loss, starspots and shell features, can introduce microstructure into the stellar absorption lines that can be confused with interstellar features (e. g. /13,35,44/). Last, the presence of heavy elements observed in white dwarf atmospheres, as well as the atmosphere structures, are poorly understood, so that theoretical models of white dwarf EUV fluxes are uncertain /29/.

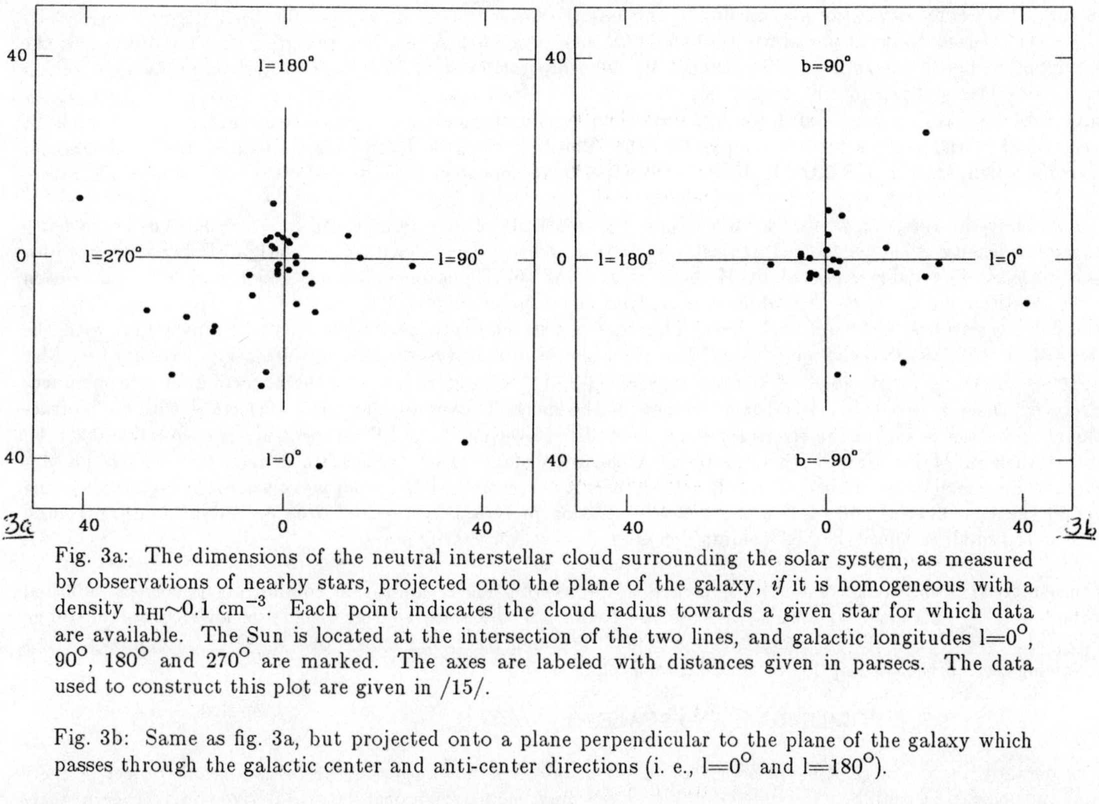

Fig. 3a: The dimensions of the neutral interstellar cloud surrounding the solar system, as measured by observations of nearby stars, projected onto the plane of the galaxy, *if* it is homogeneous with a density n_{HI}~0.1 cm^{-3}. Each point indicates the cloud radius towards a given star for which data are available. The Sun is located at the intersection of the two lines, and galactic longitudes l=0°, 90°, 180° and 270° are marked. The axes are labeled with distances given in parsecs. The data used to construct this plot are given in /15/.

Fig. 3b: Same as fig. 3a, but projected onto a plane perpendicular to the plane of the galaxy which passes through the galactic center and anti-center directions (i. e., l=0° and l=180°).

Absorption lines also provide the temperature of nearby interstellar gas, the velocity of the cloud, and a measure of turbulence through the Doppler shift and broadening of the absorption profiles. The velocity of nearby interstellar matter is discussed below. The best temperature for the LF derived from spectral observations of stars comes from the *Copernicus* satellite observations of gas in Canis Majoris (in the anti-center hemisphere). In this region, the absorption lines from trace elements with different atomic weights are shown to be thermally broadened in a gas with temperature 11,750±750 °K /20/. This temperature for the LF is consistent with temperatures derived from other spectroscopic data of nearby stars (see fig. 4). Lallement et al. /27/ also find a thermal width of 12,000 °K for Ca^+ absorption lines in nearby stars. This temperature of the LF is 50% greater then the 8,000 °K temperature found from H° Lα backscattered observations /28/. If real, this difference may be due to inhomogeneities in the cloud; if not real it may reflect difficulties in modeling the interaction of H° and the solar system or interpreting the UV absorption data. Turbulence can not explain the temperature discrepancy since towards Canis Majoris the line widths are proportional to $A^{-1/2}$ (A=atomic number) as expected for purely thermal broadening.

IONIZATION

Since the errors on observations of the backscattered H° $L\alpha$ and He° λ584 emission permit the possibility that the LF is either neutral or ionized, several attempts have been made to derive n_e/n_{total} from astronomical data. These estimates make the *ad hoc* assumption that the LF is a stable object in ionization equilibrium. The physics of the LF is poorly understood, and this assumption may be wrong. Thus, the results of this section, where values for n_e/n_{total} which explicitly assume a stable cloud in ionization equilibrium are derived, must be accepted only with qualifications. In a following section, limits on the assumption of equilibrium are discussed. The net result will be that although the two ionization methods discussed here are consistent with a fractional ionization of ~5–10%, it is unlikely that the fundamental assumptions of cloud equilibrium and stability are satisfied.

Interstellar H° is ionized in low density clouds through either photoionization by photons with $\lambda<912$ Å, or collisional ionization with electrons. Thus, in neutral clouds, most material is shielded from radiation with $\lambda<912$ Å, the Lyman limit. For example, a cloud with N(H°)$\sim10^{18}$ cm^{-3} has an optical thickness at 910 Å of $\tau_{910}\sim6.3$, and at 400 Å of $\tau_{400}\sim1$. Nevertheless, observations have shown that some EUV photons, mainly from nearby white dwarfs, do penetrate the local fluff to the position of the solar system (see refs. in /15,32/), and are thus available to ionize H°. Reynolds makes an *in situ* estimate of n_e/n_{total} based on hydrogen ionization and recombination at the position of the Sun. He derives local ionization rates from observed and estimated UV and soft X-ray photon fluxes, and infers a lower lmit of n_e/n_{total}=0.09 for n_{total}=0.1 cm^{-3} /34/. Frisch et al. used observations of Mg° (which is ionized by $\lambda>912$ Å ionizing radiation and not subject to the same optical depth effects in the LF), assumed a LF temperature of 11,750 °K, and derived an average ionization using interstellar Mg° absorption lines seen in the nearest stars. The values found are $n_e/n_{total}<0.05$ for n_{total}=0.1 cm^{-3} and $n_e/n_{total}<0.1$ for n_{total}=0.06 cm^{-3}. Furthermore, any region with $n_e/n_{total}>0.3$ must be very small (with a radius less then 1 parsec) /16/. The Mg° electron density estimates increase by a factor of four for a temperature of 8,000 °K due to the temperature dependence of the dielectronic recombination rate.

To evaluate the range of overlap between these two methods of determining the LF electron density, in fig. 5 the total density is shown plotted against the electron density derived with each method. Using the errors for the ~825 Å flux value reported by Holberg *et al.* /24/ (67% photon statistics errors, and 20% calibration uncertainties), the hydrogen recombination analysis at temperature of 8,000 °K gives the error range shown as the envelope around the bold dash line. This analysis is relatively insensitive to gas temperature, with the derived electron density changing by ~10% when the temperatures changes by 50%. The Frisch *et al.* Mg° analysis shown is for a cloud of temperature of 11,750 °K, and radiation field uncertainties are included. From fig. 5, it is seen that the error envelopes on the methods overlap, and $n_{total}\sim0.1$ cm^{-3} with a 6% fractional ionization is within the errors for both methods. However, if the LF temperature is correctly 8,000 °K, rather then the value derived from the Canis Majoris UV data, then the electron density derived by the Mg° method increases by a factor of four. In either case, a very small (d<<1 pc) weakly ionized region may surround the solar system and explain the higher ionization preferred by the hydrogen recombination arguments, but be too small to affect the Mg° column densities observed in nearby stars.

Consistent with the errors of both arguments, if the assumption of ionization equilibrium is correct, nominal values of $n_{total}\sim0.1$ cm^{-3} and $n_e/n_{total}\sim0.06$ are found. If the total spatial density is lowered to the range $n_{total}\sim0.06$ cm^{-3}, or if the LF temperature is 8,000 °K, then fractional ionizations of ≥10% are allowed. In a following section the assumption of equilibrium is challenged.

CLOUD EQUILIBRIUM AND SCALES

The detailed physics of the LF are poorly understood, and it is not known if the LF is a stable object in thermal and ionization equilibrium. Indeed, although we may use astronomical data to derive the average properties of relatively long sightlines, small scale inhomogeneities may persist undetected. The discussion here assumes a LF density of 0.1 cm^{-3}, temperature 11,750 °K, fractional ionization of 6%–10%, and magnetic field strength of 1.6 μG.

Most astronomical observations average over long sightlines, so that structure in interstellar clouds of scales less than 10^{-2} pc are difficult to detect. However, several independent sets of data allow the possibility that small scale inhomogeneities exist in low density neutral and ionized interstellar gas. Spangler and Gwinn argue /39/, based on radio studies of the refraction of radio sources by interstellar plasmas, that the turbulent density spectrum of the interstellar matter has an inner scale of about 100 km, comparable to plasma microscales. The plasma microscale (Lamor radius) of the LF is ~1000 km, suggesting that density inhomogeneities of this scale may be present in the LF. Independent evidence for au scale structures in neutral interstellar clouds has also been found from radio interferometric observations of H° 21-cm emission using 200-600 km baselines. These data show evidence for small scale variations in neutral interstellar clouds, with dimensions 25 au and n$\sim10^4$–10^5 cm^{-3}, and column densities N(H°)$\sim10^{18.5}$cm^{-2} /11/.

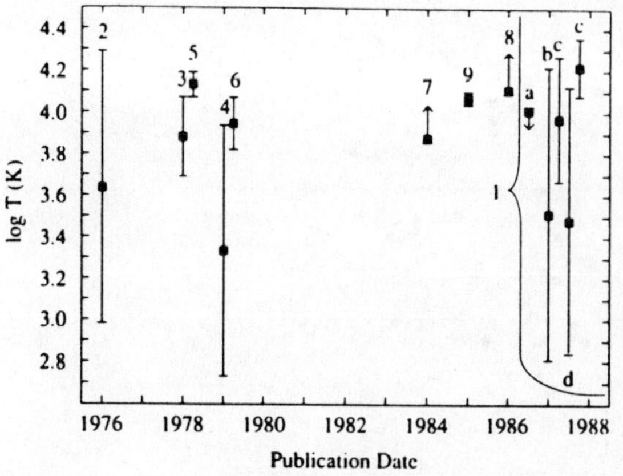

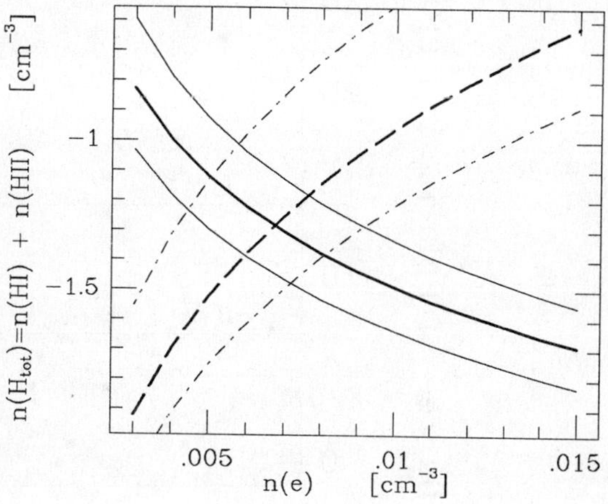

Fig. 4: Temperatures for the LF that have been derived for spectral data on nearby stars. Point number 9 is the Gry et al. "best" temperature T=11,750±750 °K referred to in the text. (This plot is taken from /16/, where references for the data are given.)

Fig. 5: Comparison of the Reynolds /34/ and Frisch et al. /16/ determinations of the ionization level in the LF (see text).

All sky surveys of soft X-ray emission ($\sim$ 0.15 keV) are generally interpreted to indicate that the LF is embedded in a hot low density plasma (T$\sim 10^6$ °K, n$\sim 10^{-3}$–10^{-4} cm^{-3}), although details of these models are not agreed on (e. g. /37/). The characteristics of this plasma determine the properties of the interface between the LF and the hot material. In addition, the interface will be ionized since the mean free path of a $\lambda\sim$ 912 Å photon in a 0.1 cm^{-3} neutral cloud is 0.5 pc, giving an interface layer at the boundary of thickness $\gtrsim$0.5 pc where the cloud is ionized. The properties of the LF interface are discussed in /36/.

In the LF, the time scales for the cloud to evaporate into the surrounding hot plasma, for it to cool, for protons and electrons to recombine, and the probable age of the cloud are all approximately 1 million years. Thus, detailed cloud models which assume equilibrium conditions are unlikely to be correct. The evaporation time is derived by modeling the LF as a spherical cloud with radius 3 pc and average spatial density of 0.1 cm^{-3} (implying a mass of 0.3 solar masses) and assuming the McKee and Ostriker three phase ISM model /31/ for cloud evaporation. This gives an evaporation time scale of 0.2 to 1.3 million years, and the evaporation time increases linearly with the cloud mass. Cooling is mainly by electron excitation of ions and atoms. At 10^4 °K a gas cools faster then it recombines, and for 5% fractional ionization, the LF will cool in 1 million years; increasing the fractional ionization increases the cooling rate. The radiative recombination rate for hydrogen in the LF is $\gtrsim 10^6$ years. During this time the proton travels an average of $\gtrsim$10 pc in the absence of a magnetic field; however, the proton gyroradius in a 1.6 μG magnetic field is $\sim$700 km, confining the warm proton to the cloud unless it is at the cloud surface. Finally, the hot plasma surrounding the LF indicates that the solar system is immersed in a supernova remnant, which is between 200,000 and 10^6 years old. With these timescales, the conditions for equilibrium are not satisfied.

MAGNETIC FIELD

There are presently no data that yield a direct estimate of the magnetic field strength in the LF, B_{LF}. The polarization of optical starlight shows that a nearby field is present, but yields no quantitative estimate for the field strength. In other regions, interstellar magnetic field strengths have been derived from Zeeman splitting in radio H° 21-cm lines /21/, and from comparison of pulsar dispersion and rotation measures /33/, but low total column densities prevent the use of Zeeman splitting to measure the magnetic field in the LF, while pulsar data average over relatively long sightlines. We can use these data on the large scale interstellar magnetic fields to guess a reasonable value for B_{LF}.

The thermal pressure in the LF is P/k=1100 cm^{-3}. The surrounding X-ray emitting plasma is modeled to have a thermal pressure of P/k$\sim$9000 cm^3 °K /37/. If the local fluff is in pressure equilibrium with this gas, and a magnetic field supplys the additional pressure, field strengths of $\sim$5 μG are required. Based on the arguments below, however, I think that the LF field is more likely to have a strength of $\sim$1.6 μG.

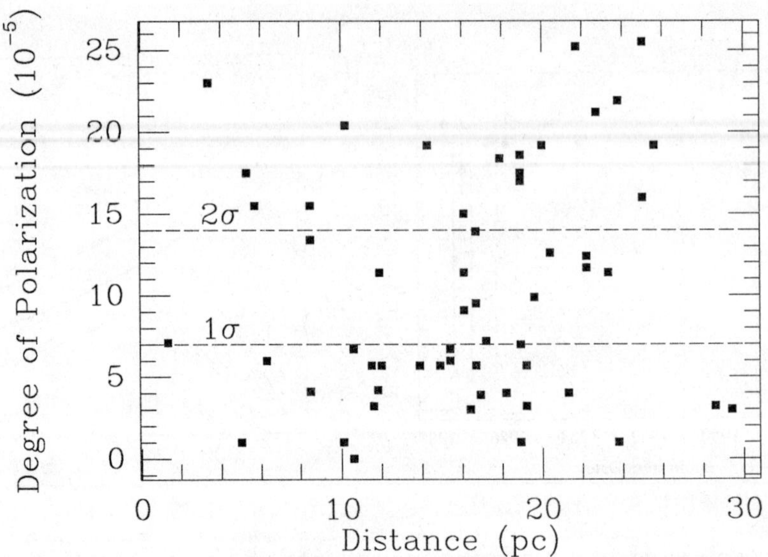

Fig. 6: A plot of the polarization found towards nearby stars by Tinbergen /42/. The 1σ and 2σ errors are shown. The conditions which cause the polarization (i. e. interstellar grains aligned by interstellar magnetic fields) are found within a few parsecs of the Sun, indicating that the LF contains a magnetic field.

On the d<500 pc scale, an overall picture of the galactic field in our region of the galaxy has been derived from pulsar rotation measures by Rand and Kulkarni /33/. Analyzing rotation measures of over 100 pulsars within 3 kiloparsecs of the Sun, they found an ordered component of the galactic magnetic field with a strength $B_O=1.6\pm0.2$ μG, pointing towards the galactic longitude of $l=96°\pm4°$ (and with a field reversal at 600 ± 80 pc towards the inner Galaxy). Superimposed on this ordered field, they found a strong magnetic field component associated with the North Polar Spur (a nearby SNR with diameter on the sky of over 100°) and a random component with an average strength of $B_R\sim 5$ μG, formed in cells with average sizes of $\sim$ 55 pc. The random component apparently is associated with segments of distant superbubble shells, while the North Polar Spur field is associated with the nearby Loop I superbubble shell. Zeeman splitting of H° in superbubble shells show field strengths of $\sim$5–6 μG /21/, supporting the view that the Rand and Kulkarni random magnetic field segment is probably formed in superbubble shell segments. In view of the fact the LF has a density a factor of 100 lower then typical superbubble shells, we choose the ordered component of the Rand and Kulkarni magnetic field as an estimate for the strength of the magnetic in the LF, $B_{LF}\sim1.6$ μG.

This estimate is supported by the polarization of light seen in nearby stars, where the polarization is caused by interstellar dust found within 30 pc of the Sun. From the weak polarization ($\gtrsim$0.05%) of the light of nearby stars by interstellar dust grains, we know an interstellar magnetic field is found in the 5-30 pc interval towards the galactic center. Tinbergen /42/ observed $\sim$ 180 stars within 35 pc of the Sun and found that $\sim$30 stars in the region $l=340^{\circ}\pm70^{\circ}$, $b=0^{\circ}\pm40^{\circ}$ systematically had weak ($>2\sigma$) polarizations due to foreground interstellar material; the polarizations are on the order of 0.02%, where $2\sigma\sim0.007$ %. Nearby stars in other directions do not show this polarization. These polarization vectors are aligned parallel to the direction of the magnetic field (see Appendix). In the Tinbergen data, the region with systematically aligned angles of polarization is centered around $l\sim0^{\circ}$, and the polarization vectors are roughly parallel to the galactic plane. The polarization direction becomes random, or polarization disappears, for regions $l>30^{\circ}$ or $l<340^{\circ}$. The mechanism which polarizes the starlight (see Appendix) requires the magnetic field to be viewed in a perpendicular direction in order to produce observable polarization, especially when the polarization is weak. Thus, it appears that the magnetic field sampled by these optical data points towards the direction of $l\sim90^{\circ}$ and $b\sim0^{\circ}$, and is consistent with the ordered magnetic field direction found by Rand and Kulkarni. Although the properties of the populations constituting interstellar grain are not definitely determined, it has been argued that 1000 Å diameter silicate particles with superparamagnetic inclusions can be aligned by magnetic fields on the order of 1 μG /30/.

The nearest polarized star observed by Tinbergen is 5 pc distant. In principal, if the interstellar cloud giving rise to the polarization is located at a given distance, then all stars beyond that distance would be polarized, while closer stars would be unpolarized. Looking for significant increases in the polarizations of stars towards Tinbergen's patch, the polarizations of stars in this direction are plotted against the distance of the star in fig. 6 (where stars identified by Tinbergen as having abnormal polarizations have been omitted from the plot, even if the polarizations may possibly be interstellar). No apparent correlation between polarization and the distance of the star is evident. A second plot (fig. 7) histograms this data by assuming that the absence of polarization within the patch correlates to the absence of interstellar dust, rather than to the absence of a magnetic field, and bins only stars showing significant polarization (i. e. $>2\sigma$) into 10 parsec bins, and plots the average polarization for the stars in each bin. This histogram shows no statistically significant increase in the polarization with distance. From this we conclude that the interstellar magnetic field which causes the starlight polarization found by Tinbergen must extend to within 5 pc of the Sun in the galactic center region.

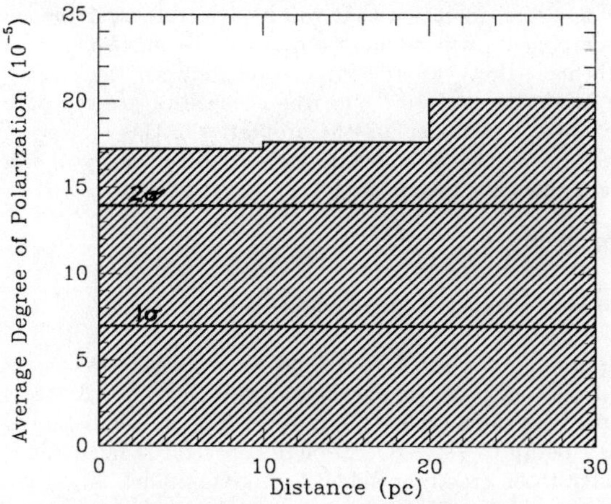

Fig. 7: A histogram of the average polarizations of only those stars in fig. 6 which have statistically significant (i. e. $>2\sigma$) polarizations. The average polarizations are calculated for stars binned into 10 pc intervals. The 1σ and 2σ error levels are shown as horizontal lines.

A third indirect estimate for the ambient magnetic field comes from information on the magnetic field in the filaments of neutral hydrogen which form the shell of the Loop I superbubble. The Loop I superbubble is centered at l$\sim$329^o, b$\sim$+18^o, with a radius of 58^o and centered at a distance of 130 pc /4,5/. Spoelstra /40/ finds that the Loop I rim is 70 pc from the Sun, giving a distance of 36 pc for the nearside of Loop I if it is spherical. Alternatively, if Loop I expanded asymmetrically into the low density region surrounding the solar system, the Sun may be embedded either in the shell of the supernova or the supernova itself. Heiles also /21/ concluded that the nearside of Loop I was seen in expansion at V_{LSR}=−30 km s^{-1}, near (l,b)$\propto$(300°,-10°), although he inferred higher column densities then found for the LF. He observed the H^o Zeeman splitting within the Loop I shell, and found the parallel component of the magnetic field is $\leq$6 μG. (Looking only at the negative velocity Zeeman components, /15/ found a similar magnetic field strength.) If a factor of four compression in density occurs between the gas ahead of the expanding shell (which is expanding towards the Sun from the direction of the center of Loop I), and the post-shell gas, and if the magnetic flux is compressed by this same factor, then the magnetic field ahead of the front side of the expanding shell of Loop I, towards λ Sco, would have a value $\sim$1.5 μG, consistent with the strength of the ordered component of the local galactic magnetic field.

NEARBY ISM VELOCITY FIELD

Over the last 15 years, there have been many discussions of the velocity field of the interstellar matter seen in front of nearby stars, and several attempts to relate it to the LF velocity traced out by the Lα backscattered radiation /1,6,7,10,14,17,26/, with mixed results. Although d$<$90 pc neutral gas and the LF show flow vectors from the same general region of the sky, the actual flow vector derived from least squares fits to the data depends on which star sample is used to derived the vector /8/. Two other papers in this issue by Lallement et al. /27/ and Bzowski /7/ also discuss the velocity vector of d$<$100 pc interstellar gas.

Most attempts have tried to fit the velocity component projected in the line of sight (the "radial component") to a cloud system moving with a uniform velocity past the Sun. Overall, both optical and UV absorption line data can be modeled roughly by a linear flow of gas past the solar system, extending to include relatively cool interstellar clouds seen in all directions out to distances of $\sim$80 pc. From the optical data, the upwind direction in the local standard of rest, with an assumed single linear flow is l$\sim$354^o, b$\sim$+3°, V_{LSR}=−12 km s^{-1}. It has been suggested that this flow is driven by the expanding Loop I supernova remnant centered in the Sco-Oph association (e. g. /14,17/). If so, particularly in view of the curtain structure seen in fig. 3b, then the LF may be part of the remnant shell, and would then explain the enhanced abundances of trace elements in the gas phase seen in this nearby gas /14/.

The model with the linear flow vector has short comings. A linear flow vector does not give a perfect fit to all of the data. This property has led to models that invoke many individual clouds to explain the velocity field of nearby gas /26/. When UV data is included a somewhat different picture emerges since high velocity highly ionized gas is seen /46/, and strong lines such as the Mg^+ h and k lines are displaced from weaker lines showing that single or multiple clouds with significant velocity dispersion are present /19/. If the LF is part of an expanding shell then the velocity vectors of nearby interstellar cloud components would not be adequately fit by a single linear flow vector.

CONCLUSION

Most astronomical observations sample the macroscopic properties of interstellar matter over scale sizes of 1—50 pc. With the improved UV spectral resolution that will be available after the launch of the *Hubble Space Telescope*, weak LF absorption components will be more accurately distinguished from stronger and more distant features blended in the sightline. However, the interaction between the solar system and the ambient cloud is determined by material in the nearest 10^{-3} pc, where densities are comparatively low. Traditional astronomical observing techniques do not easily isolate the properties of this material. Thus, ultimately, a reliable determination of the properties of nearby interstellar gas must be made by *in situ* sampling of the nearby interstellar cloud as spacecraft leave the solar system.

APPENDIX: BASIC INFORMATION

Here are a few basic definitions to help non-astronomers read this paper.

Distance Scales: One *parsec* is approximately equal to 3 light years or 200,000 astronomical units. Distances smaller than 20 parsecs are based on trigonometric parallaxes, while larger distances are usually based on the spectral class of the star, which predicts the absolute luminosity. For stellar distances d>20 parsecs, distance errors based on spectral types may be up to 15%. Occasionally spectral peculiarities in a star or circumstellar matter cause its absolute luminosity to be grossly misidentified, giving much larger errors. The nearest star is ~200,000 times more distant than the planet Pluto at 39 au. For reference, the nearest stars, α CenA and α CenB, are at distances of 1.3 parsecs. The outer boundary of the hypothetical Oort cloud is $\lesssim$ 50,000 au, and galactic tides equal the gravitational attraction of the Sun at distances of $\lesssim$ 150,000 au in the plane of the Galaxy (values are slightly different out of the plane).

Column Densities: A *column density* of species X represents the number of atoms or ions of that species found in a column with a 1 cm^{-2} face and a length determined by the distance of the observed target. This quantity is denoted "N(X)", and has units cm^{-2}. A *spatial density* is the number of atoms or ions per cubic centimeter, is denoted "n_x", and has units cm^{-3}.

Coordinate Systems: The *galactic coordinate system* is used in this review. The galactic center fixes the longitude meridian of $l=0^{\circ}$, and our neighborhood of the galaxy corotates towards $l=90^{\circ}$. The galactic latitudes of $b=\pm 90^{\circ}$ correspond to a perpendicular through the plane of the galaxy at the position of the Sun, with $b+90^{\circ}$, the "north galactic pole", or $\alpha = 12^h.8, = +27^{\circ}$. The "north galactic pole" is so named because it is visible from the northern terrestrial hemisphere, and not because of the handedness of the ambient galactic magnetic field.

Velocities: The *local standard of rest* (LSR) velocity coordinate system is at rest with respect to the average motion of stars near the Sun orbiting around the center of the galaxy. Solar motion is such that the Sun is moving through the local standard of rest with a velocity of ~20 km s^{-1} towards galactic coordinates $l=57^{\circ}$, $b=+22^{\circ}$ /2/. This motion corresponds to 20 parsecs per million years, or 4 au per year. Since velocities for interstellar clouds which have been measured optically are often given in the rest frame of the Sun (referred to as "heliocentric coordinates"), while cloud velocities determined from radio data (e. g. H° and CO lines) are given in LSR coordinates, velocity transformations often have to be made when comparing optical and radio data.

Interstellar Gas: The standard three-phase model for interstellar clouds by McKee and Ostriker /31/ presents clouds as cool neutral concentrations ($n_{total}\sim 40$ cm^{-3}, $T\sim 80$ °K, $n_e/n_{total}\sim 10^{-3}$) embedded in an hot ionized medium ($n_{total}\sim 0.004$ cm^{-3}, $T\sim 5\times 10^5$ °K, $n_e/n_{total}\sim 1$) which has been heated by mechanical energy released by supernova explosions. Between these phases, are warm neutral and ionized layers ($T\sim 8{,}000$ °K, $n_{total}\sim 0.3$ cm^{-3}, $n_e/n_{total}\sim 0.1-0.7$). About 3% of space is filled with the cold cloud core phase, 23% with the interface layers, and the remaining with hot tenuous gas, although these filling factors have not been confirmed. Typical equilibrium pressures given for diffuse interstellar clouds are 2,000-3,000 cm^{-3} °K (e. g. /38/). *Pressures* in interstellar clouds are usually quoted as P/k=nT, with units of cm^{-3} °K (n=spatial density, cm^{-3}; T=gas temperature, °K). This picture does not include magnetic fields. However, near the Sun most H° is found in cool filaments which are generally aligned with magnetic fields /8/. These filaments usually have column densities of $>10^{20}$ cm^{-2}, and probably result from shock front activity. In the filaments, magnetic pressure (since B~5 μG typically) dominates gas pressure /21/. A combination of supernova activity and stellar wind activity associated with associations of massive young stars often inflates interstellar material within the association into what is known as a "superbubble", or large H° shell observed at its tangential points on the sky. Typical densities in the neutral hydrogen portion of superbubble shells are 5 cm^{-3}, and many such shells are observed to have parallel magnetic fields with typical field strengths of ~5 μG aligned parallel to the axis of the filaments. An interstellar cloud is said to be in *ionization equilibrium* if the number of ionizations equals the number of recombinations per unit time and space.

Interstellar Dust: Mixed with interstellar gas, roughly with a 100/1 gas to dust mass ratio, are a population of *interstellar dust grains* which have been modeled as various $\lesssim$0.01-0.3 μ diameter mixes of graphite, amorphous carbon, silicate and large carbon molecules (e. g. /30/ and references therein). These dust grains redden the light of background stars; typically gas and dust are mixed in a cloud such that the cloud becomes optically thick in the visual due to dust, $\tau_V \sim 1$, when H° column densities build up to log N(H°)$\sim$21.2 cm^{-2}. Included in this population are dielectric grains which polarize the light of background stars. These grains are elongated and aligned so that the long axis twirls in a plane *perpendicular* to the direction of the magnetic field, causing the light of background stars to be polarized *parallel* to the magnetic field /30,12/.

Stars: Stars have *spectral types* named O, B, A, F, G, K, M with mass and temperature decreasing towards the end of the sequence (e. g. see /2/).

Magnetic Fields: The interstellar *magnetic field* strength is usually derived from one of two ways: (a) from Zeeman splitting of H° 21-cm emission features, which gives the component of the magnetic field which is is perpendicular to the line of sight (the circular Zeeman components have not been detected); (b) by comparing pulsar dispersion and rotation measures: The "dispersion measure" of a pulsar refers to the frequency dependent broadening of the pulsar pulse caused by refraction in interstellar plasmas and is proportional to the electron density. The "rotation measure" gives the electron density weighted parallel component of the magnetic field, and is from the Faraday rotation of the polarization angle of the pulsar wave packet. Thus, the ratio of the rotation measure to the dispersion measure gives the parallel component of the galactic magnetic field, as weighted by the electron density, from which the total field is inferred. For reviews of the interstellar magnetic field see /22,23,38/.

REFERENCES

1. Adams, T. F. and Frisch, P. C. *Ap. J.*, **212**, 300 (1976).
2. Allen, C. W. *Astrophysical Quantities, Third Edition* (London: William Clowes and Sons, Limited) (1973).
3. Axford, W. and Ip, W. *Adv. Sp. Res.*, **6**, 27 (1986).
4. Berkhuijsen, E. M. *Astr. Ap. Supl.*, **5**, 205 (1972).
5. Berkhuijsen, E. M. *Astr. Ap.*, **14**, 359 (1971).
6. Bohm-Vitense, E. *Ap. J.*, **244**, 504 (1981).
7. Brzowski, M. this issue (1990)
8. Cleary, M. N., Heiles, C. and Haslam, C. G. T. *Astr. Ap. Supl.*, **36**, 95 (1979).
9. Cox, D.P., and Reynolds, R. J. *A. R. A. A.*, **25**, 303 (1987).
10. Crutcher, R. M. *Ap. J.*, **254**, 82 (1982).
11. Diamond, P. J., Goss, W. M., Romney, J. D., Bpoth, R. S., Kalberla, P. M., and Mebold, U. *Ap. J.*, **347**, 301 (1989).
12. Duley, W. W., *Ap. J. Letters*, **219**, L129 (1978).
13. Freire, R., Czarny, J., Felenbok, P., and Praderie, F. *Astr. Ap.*, **61**, 785 (1977)
14. Frisch, P. C. *Nature*, **293**, 377 (1981).
15. Frisch, P. C. and York, D. G. in *Extreme Ultraviolet Astronomy*, ed. R. F. Malina and S. Bowyer, Pergamon Press (1989).
16. Frisch, P. C., Welty, D. G., York, D. G., and Fowler, J. R. *Ap. J.*, July 10 edition (1990).
17. Frisch, P. C., York, D. G., in *The Galaxy and the Solar System*, eds. Smoluchowski, Bahcall, Matthews, p. 83, University of Arizona Press (1986).
18. Frisch. P. C., *Adv. Sp. Res.*, **6**, 345, (1986).
19. Frisch, P. C., Welty, and D. G., York, D. G. in preparation (1990).
20. Gry, C., York, D. G., and Vidal-Madjar, A. *Ap. J.*, **296**, 593 (1985).
21. Heiles, C., *Ap. J.*, **336**, 808 (1989).
22. Heiles, C., *Interstellar Processes*, ed. D. J. Hollenbach, H. A. Thronson, Jr. p. 171 (D. Reidel Publ. - Holland) (1987)
23. Heiles, C. *A. R. A. A.*, **14**, 1 (1976).
24. Holberg, J. B., Sandel, B. R., Forrester, W. T., Broadfoot, A. L., Shipman, H. L., Barry, d. C. *Ap. J. Letters*, **242**, 119 (1980).
25. Holzer, T. E., *A. R. A. A.*, **27**, 199 (1989).
26. Lallement, R., Vidal-Madjar, A., Ferlet, R. *Astr. Ap.*, **168**, 225 (1986).
27. Lallement, R., Ferlet, R., Vidal-Madjar, A., and Gry, C. this issue (1990).
28. Lallement, R. this issue (1990).
29. Liebert, J. *A. R. A. A.*, **18**, 363 (1980).
30. Mathis, J. S., *Ap. J.*, **308**, 281 (1986).
31. McKee, C. F., and Ostriker, M. P., *Ap. J.*, **218**, 148 (1977).

32. Paresce, F. *Astr. J.*, **89**, 1022 (1984).
33. Rand, R. J. and Kulkarni, J. R. *Ap. J.*, **343**, 760 (1989).
34. Reynolds, R. J. this issue (1990)
35. Simon, T. and Landsman, W., *Proceedings of the fifth Cambridge Workshop on Cool Stars, Stellar Systems and the Sun*, eds. J. L. Linsky, R. E. Stencel. **265** (1987).
36. Slavin, J. D. *Ap. J.*, **346**, 718 (1989).
37. Snowden, S. L., Cox, D. P., McCammon, D., and Sanders, W. T. *Ap. J.*, **354** (1990).
38. Spitzer, L. *Physical Processes in the Interstellar Medium* (John Wiley and Sons, New York) (1978).
39. Spangler, S. R., Gwinn, C. R. *Ap. J. Letters*, in press (1990).
40. Spoelstra, T. A. *Astr. Ap.*, **21**, 61 (1972).
41. Stacy, J. G. and Jackson, P. D. 1982, *Astr. Ap. Supl. Ser.*,**50**, 377.
42. Tinbergen, J., *Astr. Ap.*, **105**, 53 (1982).
43. Torbett, M. V., *The Galaxy and the Solar System*, eds. Smoluchowski, Bahcall, Matthews, p. 147 (University of Arizona Press) (1986).
44. Vogt, S. S., Penrod, D. and Hatzes, A. P. *Ap. J.*, **321**, 496 (1987).
45. York, D. G., and Frisch, P. C. in *Local Interstellar Medium: NASA CP-2345*, ed. Y. Kondo, F. C. Bruhweiler, B. D. Savage, P. 51 (1984).
46. York, D. G. *Ap. J.*, **264**, 172 (1983).

A NEW OBSERVATIONAL APPROACH TO INVESTIGATE THE HELIOSPHERIC INTERSTELLAR WIND INTERFACE: THE STUDY OF EXTREME AND FAR ULTRAVIOLET RESONANTLY SCATTERED SOLAR RADIATION FROM NEON, OXYGEN, CARBON AND NITROGEN

Stuart Bowyer* and Hans J. Fahr**

**Center for EUV Astrophysics, University of California, Berkeley, CA 94720, U.S.A.*

***Institut für Astrophysik, Universität Bonn, D-5300 Bonn 1, F.R.G.*

ABSTRACT

One of the outstanding uncertainties in our understanding of the heliosphere concerns the character of the interaction between the outflowing solar wind and the interstellar medium. A new possibility for obtaining information on this topic is suggested. The cosmically abundant elements neon, oxygen, carbon and nitrogen will be affected differently at their interface passage depending upon the character of this region. Consequently, the distribution of these atoms and their ions will vary within the inner heliosphere. The study of resonantly scattered solar radiation from these species will then provide information on the nature of the interface. A preliminary evaluation of this approach has been carried out, and the results are encouraging. The relevant lines to be studied are in the extreme and far ultraviolet. The existing data in these bands are reviewed; unfortunately, past instrumentation has had insufficient resolution and sensitivity to provide useful information. The capabilities of future approved missions with capabilities in this area are evaluated.

INTRODUCTION

The character of the interaction between the shocked solar wind and the inflowing local interstellar medium (LISM) is a long-standing unresolved question. Different configurations for plasma interfaces resulting from different assumptions concerning the state of the LISM have been discussed in the literature, i.e., subsonic LISM flow /1/, supersonic LISM flow /2, 3/, magnetically confined solar wind outflow /4/, or submagnetosonic LISM flow /5–9/. The principal outcome in all of these interfaces is a general deflection of the LISM plasma at its passage over the heliospheric obstacle. Within the context of the most promising of these models, crucial parameters, such as the thickness of the disturbed plasma upwind of the heliosphere and the proton density within this region, are not well known.

Additional uncertainty is introduced because of our lack of knowledge of key parameters of the inflowing LISM. In particular, the degree of ionization of the LISM is poorly known, as is the relative degree of ionization of hydrogen and helium. In principle, one could derive the LISM ratio of neutral hydrogen to helium from measurements of resonantly scattered solar radiation in the inner heliosphere. In fact, interactions at the interface are quite likely to alter this parameter with respect to its unperturbed value.

We have considered a new observational test to determine the character of the heliosphere/LISM interface. We propose to trace the flow of cosmically abundant, low-atomic-weight neutrals through various models of this interface. We will then compute the distribution of these species within the inner heliosphere. A comparison of the resonantly scattered solar radiation pattern from these species can then be made with observational data, thereby testing the validity of the interface models.

In the following we outline our initial efforts in this area, evaluate the relevant existing data, and discuss potential future sources of data.

OVERVIEW OF THE APPROACH

Besides H and He, we consider only the more abundant species in the LISM, i.e., neon, oxygen, carbon, and nitrogen. Neutrals penetrating through the interface region can interact, in principle, by charge exchange with solar wind ions and by electron impact ionization. Since the temperature of the interface plasma is relatively low ($\sim 10^5$K) /2, 10/, the corresponding electron ionization rates are low and can be neglected. Neutrals will be affected only if they can effectively exchange charge with the solar wind plasma. An examination of the relevant charge exchange cross sections /11/ shows that He, Ne, and C atoms can penetrate the interface nearly unaffected, whereas H, O, and N atoms may be substantially depleted, depending on the average column density of protons in the interface plasma.

To evaluate this depletion, we compute the effective transmission of these species following the method of Ripken and Fahr /10/, Fahr and Ripken /12/, and Fahr /6, 7/, who use an appropriately constructed Boltzman-Vlasou equation to account for the fact that the interface plasma not only serves as a sink of inflowing atoms but can also serve as a source. This is because interface protons with appropriate velocity vectors pointing towards the inner heliosphere at the moment of neutralization via charge exchange act as a source of neutral hydrogen. Fahr and Ripken have shown that the effective transmission of LISM hydrogen atoms through the interface is $Tr_H \cong 0.6$ for a LISM proton density of 10^{-2} cm^{-3}. For oxygen and nitrogen atoms, the transmissivities will be lower since the interface plasma is a substantial sink for these species but is only a poor source. A reasonable estimate of the effective transmission of these atoms can be derived with the assumption that the interface plasma is a negligible source for oxygen and nitrogen atoms. In view of the fact that the charge exchange cross sections of O-H$^+$ collisions are nearly identical with those of H-H$^+$ collisions, one obtains /12/

$$Tr_O \cong \exp(-\tau_{ex,O}) \cong \exp(-\tau_{ex,H}) \cong 0.2 \ .$$

Considering the effective charge-exchange rates of oxygen and nitrogen atoms with protons given by Siscoe and Mukherjee /11/, we find that the associated velocity-averaged charge exchange cross sections are related by

$$< \sigma_{O,H^+} > \cong 1.83 < \sigma_{N,H^+} >$$

and consequently obtain

$$Tr_N \cong \exp\left(-\frac{1}{1.83}\tau_{ex,O}\right) = (Tr_O)^{0.55} = 0.41 \ .$$

We assume local thermodynamic equilibrium for conditions in the local cloud of the ISM. While this is clearly not appropriate for the ISM in general and may not be appropriate for the local cloud, it allows us to derive nonarbitrary number densities for each of the interstellar neutral species through the use of the Saha-Eggert equations. We estimate the fractional ionization, $\Sigma = (1 - \xi)$, where ξ is defined by $n_1 = (1-\xi)(n_1+n_{1+})$.

The ionization state of the LISM is poorly determined. Estimates of this number for the *overall* ISM have been made based on measurements of the diffuse Hα astronomical background /13/ and on radio dispersion measurements from pulsars in globular clusters /14/. An estimate for the LISM electron density has been obtained from the lack of neutral magnesium absorption in nearby stars /15/. This later result, though based on an ingenious idea, is rather dependent upon several underlying assumptions. A direct measurement of neutral helium and neutral hydrogen in the LISM has been obtained by a Berkeley sounding rocket experiment which measured the extreme ultraviolet absorption of these species in the spectrum of a nearby star /16/. Following the discussion in that paper, we use an electron number density of the LISM of 10^{-2} cm^{-3} in this work. We assume a temperature for the LISM of 10^4 K obtained from helium solar resonant scattering studies /17/ and list our results in Table 1.

After passing through the heliospheric interface region, neutral atoms will undergo ionization in the inner heliosphere by charge exchange with solar wind protons, electron impact ionization with solar wind electrons, and photoionization by solar extreme ultraviolet photons. Neutrals will be gravitationally focused by the Sun in the downwind region, and ions will be convected outwards with the solar wind plasma as pickup ions /18–20/. The summed effect of all these processes for neutral species is reasonably well understood, and in the case of hydrogen and helium, theoretical predictions have undergone direct comparison with observational data.

Our analysis indicates that neon atoms will be the dominant neutral species (after helium) in the inner heliosphere since (1) their cosmic abundance is relatively high, (2) their fractional ionization in the LISM is about

1.0, (3) their interface transmissivity is about 1.0, and (4) their extinction rate in the heliosphere is fairly low. Oxygen and nitrogen atoms will be reasonably well represented. Neon, oxygen, and nitrogen will form pronounced cones of condensed neutral matter in the downwind region. Neutral carbon atoms will not be present in the inner heliosphere because of prior ionization in the nearby ISM and in the outer heliosphere. However, singly ionized carbon may be a useful discriminator of interface conditions.

TABLE 1 Fractional Ionization and Transmissivity of Inflowing ISM Neutral Atoms at the Heliospheric Interface

Element	Fractional Ionization	Transmissivity
H	0.54	0.6
He	1.0	1.0
Ne	1.0	1.0
O	0.68	0.2
N	0.75	0.41
C	0.01	1.0

The distribution of these neutrals and ions in the inner heliosphere can be determined by their resonant scattering of solar radiation. The modeling of this effect is well established /6, 7, 17, 21, 22/. The validity of this modeling approach has been verified by direct comparison with observations for the cases of hydrogen and helium with data obtained with a number of satellite missions (see /23, 24/ and references therein).

In Table 2 we list primary resonance lines which can, in principle, trace the distribution of neutrals and ions in the inner heliosphere. We include hydrogen and helium for comparison purposes. The cosmic abundances are from Cameron /25/. We also provide an estimate of the solar flux at these lines (solar spectrum information taken from Hinteregger et al. /26, 27/ and Torr and Torr /28/) as this is one of the prime factors affecting the expected intensity of the scattered line.

TABLE 2 Candidate Lines for Solar Resonant Scattering Studies

Element	Cosmic Abundance	Primary Resonant Line (Å)	Solar Flux at Line (10^9 ph cm^{-2} s^{-1})*
H	1.0	1216	251
He	0.07	584	1.3
He II	—	304	6.9
Ne	1.08 10^{-4}	736	—
Ne II	—	461	—
O	6.76 10^{-4}	1304	3.46
O II	—	834	—
N	1.18 10^{-4}	1200	—
N II	—	1085	—
C	3.71 10^{-4}	1657	8.5
C II	—	1335	4.3

As can be seen from the data in Table 2, the relevant lines occur in the extreme and far ultraviolet bands of the spectrum. In the following section we review our current knowledge of the diffuse background in these bands.

THE CURRENT STATUS OF MEASUREMENTS OF THE FAR AND EXTREME ULTRAVIOLET BACKGROUND

The Far Ultraviolet Background

From the beginnings of space research, attempts were made to measure the diffuse far ultraviolet (FUV, λλ 1000–3000 Å) background. The first twenty years of measurements, carried out by a number of groups in

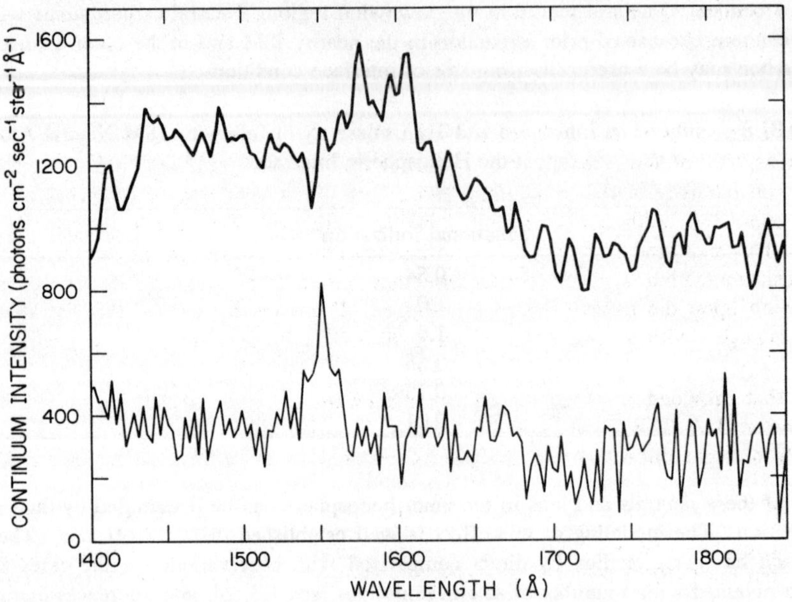

Fig. 1. The cosmic FUV flux from 1350 to 1800 Å. The lower curve shows the spectrum in a direction of low neutral hydrogen column. The line at 1550 Å is the C IV line from hot interstellar gas obtained with an instrument with 15 Å resolution /32/. The upper curve shows the spectrum in a direction of higher neutral hydrogen column. The majority of this radiation is starlight scattered by interstellar dust, but other processes also contribute to this flux /33/.

Spectroscopy of the diffuse EUV background below 500 Å is in its infancy. Grazing incidence optics must be used in order to obtain any reasonable throughput, and grazing incidence spectrometers for diffuse radiation are technically difficult. Labov and Bowyer /39/ designed and built a high throughput, modest resolution (10 to 30 Å) spectrometer that was sensitive from 80 to 650 Å. This instrument was flown on a sounding rocket, and interplanetary He I 584 Å emission and geocoronal He II 304 Å emission were clearly detected. Several other features consistent with line blends from a hot (10^5–10^6 K) ISM were also observed. A line was detected at ~630 Å at a strength greater than the upper limit obtained by Holberg /38/. This discrepancy may be an artifact, but it could be a real effect due to the different view directions of the two measurements.

The existing data on the diffuse EUV background are shown in Figure 2.

PROSPECTS FOR FUTURE MISSIONS

Several missions with EUV capabilities are scheduled for deployment in space in the next four years, including the British Wide Field Camera on the *ROSAT* satellite /40/; the Hopkins Ultraviolet Telescope, which will be flown on the Space Shuttle as part of the *ASTRO* mission /41/; the Delta-launched *Extreme Ultraviolet Explorer* mission /42/; the Berkeley EUV/FUV spectrometer /43/ on the German/USA *ORFEUS* mission to be flown on the Space Shuttle /44/; and the Los Alamos *ALEXIS* project /45/, which will be lofted into orbit by an airplane-launched rocket.

The *ALEXIS* and Wide Field Camera experiments employ photometric instruments with moderate- to broad-bandpass capabilities in the EUV. Each of these instruments will be capable of establishing the total flux in its relative bandpasses.

Since the sources of the EUV background are at this point conjectural, the possibility of obtaining spectroscopic data on this radiation is especially interesting. Several instruments to be flown in the early 1990s will have spectroscopic capabilities in the EUV. The spectrometer on the Hopkins Ultraviolet Telescope is designed to work primarily above 912 Å, but it does have EUV capabilities by use of the grating in second order. However, the instrument has been carefully designed to optimize work on *point sources,* and it is not useful for diffuse measurements. The spectrometers on the *Extreme Ultraviolet Explorer* were also designed at least five countries, indicated that the flux was uniform across the celestial sphere and hence was cosmological in origin. Estimates of the intensity of this flux, however, varied by *three orders of magnitude* with no clustering around a mean. In retrospect, there are a number of reasons for these discrepant results; the reader is referred to Davidsen, Bowyer, and Lampton /29/ and Paresce and Jakobsen /30/ for reviews of this earlier work.

A turning point in the study of this background occurred in 1980 when data obtained with a FUV channel of a telescope flown as part of the *Apollo-Soyuz* mission were analyzed and published. These data exhibited a correlation between intensity of the background and galactic neutral hydrogen column as derived from 21 cm radio measurements /31/. Although these initial results were criticized on a variety of grounds, they were quickly confirmed by independent rocket and satellite measurements. The intensities obtained in these experiments were consistent within a factor of two, and they varied from roughly 200 to 2000 photons cm^{-2} s^{-1} sr^{-1} $Å^{-1}$, depending upon the galactic hydrogen column. These results showed that the vast majority of the FUV flux was connected with processes in our Milky Way Galaxy and was not, in fact, extragalactic in origin.

An experiment that provided a variety of new data relevant to this field was a Space Shuttle–borne FUV diffuse spectrometer /32/. This instrument was sensitive from ~1350 to 1850 Å with ~15 Å resolution. Data were obtained from eight different directions in the celestial sphere with a wide range of neutral hydrogen column densities.

Several important results came from the study of the data obtained with this instrument. Emission lines from a collisionally excited hot (~10^5 K) gas in the interstellar medium were discovered /32/. Data obtained in directions with relatively high column densities of neutral hydrogen showed that a component of cosmic dust was present; dust has a relatively low albedo ($\omega \lesssim 0.2$) and is scattered almost isotropically, indicating that the dust is relatively small and roughly spherical. There are strong indications that at least traces of this dust are present everywhere, including regions in directions of very low hydrogen column density /33/.

The FUV background from 1350 to 1800 Å is shown in Figure 1. The lower curve shows the spectrum in a direction of low neutral hydrogen column. A variety of processes contributes to this flux; the spectral lines noted are emitted by the hot ISM. The upper curve shows the spectrum in a direction of higher neutral hydrogen column. The majority of this radiation is starlight scattered by dust in the ISM, but other processes also contribute to this flux.

The Extreme Ultraviolet Background

Early measurements of the extreme ultraviolet (EUV, λλ 100–1000 Å) background were made with broad-band filters that covered the wavelengths of the resonantly scattered solar lines of He I (584 Å) and He II (304 Å). These lines dominated the data and made it extremely difficult to estimate the intensity of any other flux that might be present. The first estimates that were useful in this regard were obtained with the Berkeley EUV telescope flown on the *Apollo-Soyuz* mission. These results were also obtained with rather broad-band filters. However, the geocoronal ionized helium responsible for scattering the solar He II 304 Å line is confined to a few Earth radii by the Earth's magnetic field. In the anti-solar direction, the Earth's shadow leaves the helium in darkness, and since the line is optically thin, its intensity is greatly reduced in this direction. Stern and Bowyer /34/ analyzed the extensive *Apollo-Soyuz* observations in the anti-solar direction to obtain significant upper limits to this flux. Paresce and Stern /35/ argued that the residual signal in the shorter *Apollo-Soyuz* bandpass was the result of off-band sensitivity of the instrument to He I 584 Å radiation, and through a differencing process they obtained a still lower upper limit for this flux.

Several broad-band upper limit measurements have been made at longer wavelengths. Bixler, Bowyer, and Grewing /36/ obtained an upper limit to the diffuse flux from 1040 to 1080 Å with an instrument with a narrow (~ ¼ deg^2) field of view that was used to ensure no stellar signals were contributing to the flux. A similar limit was obtained in the 912 to 1050 Å band with a much wider field of view detector /37/.

Two relatively coarse spectrographic measurements have been carried out in the EUV. Holberg /38/ obtained data at wavelengths longward of 600 Å by analyzing 1,508,198 s of observations with the *Voyager 2* ultraviolet spectrometer. These high galactic latitude observations were obtained with 40 Å resolution during the interplanetary cruise phase. The EUV part of this observation was dominated by resonantly scattered solar He I 584 Å flux radiation. Holberg removed this bright line from his spectra along with several other sources of background to derive upper limits to the diffuse EUV background.

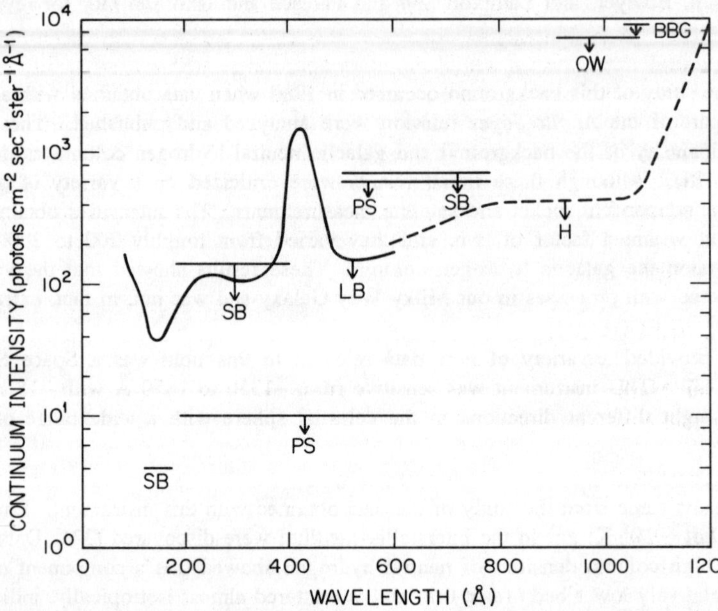

Fig. 2. The cosmic EUV flux from 100 to 1200 Å. The measurements shown by solid horizontal lines were made with broad-band detectors. Most are upper limits to the diffuse flux (SB = Stern and Bowyer /34/; PS = Paresce and Stern /35/; OW = Opal and Weller /37/; BBG = Bixler, Bowyer, and Grewing /36/). The dashed curve is an upper limit to the diffuse flux obtained with a low resolution (~ 40 Å) spectrometer (H = Holberg /38/). The solid curved line was also obtained with a low resolution (~ 20 Å) spectrometer; some lines and line complexes appear to have been detected in this measurement (LB = Labov and Bowyer /39/).

for the study of point sources, and although they possess some capability for studying diffuse sources /46/, this capability is rather limited.

***ORFEUS*, the Orbiting Retrievable Far and Extreme Ultraviolet Spectrometer, utilizes a 1 m normal incidence EUV/FUV telescope to feed two spectrometers /44/. The Berkeley spectrograph for this mission /43/ will obtain high resolution (λ/Δλ > 7000) spectra of point sources between 390 and 1200 Å. However, this spectrograph has intriguing capabilities for EUV spectroscopy of diffuse flux. The Berkeley instrument incorporates a set of four novel, spherically figured, variable line-space gratings used in a geometry that is superficially similar to that of the classic Rowland mount. Unlike the conventional Rowland spectrograph, however, the groove spacing of each grating varies as a fourth order function of position on the optic. Furthermore, the distances from the telescope image to the grating center, and from the grating center to the detector surface, do not correspond to those of a Rowland spectrometer. The astigmatism and aberrations of this spectrograph are substantially lower than what would be achievable with a conventional Rowland spectrograph with optimized toroidal gratings.**

The *ORFEUS* telescope has three apertures that can be used at the observer's discretion. The use of a larger aperture will increase the sensitivity to diffuse flux at the expense of spectral resolution. In Table 3 we show the sensitivity and resolution of the Berkeley spectrometer to diffuse flux.

TABLE 3 *ORFEUS* Sensitivity and Resolution for Diffuse Flux

Aperture	λ/Δλ	Counts produced by a 1 Rayleigh line in 20 minutes
10′ ϕ	40	~5000
20″ ϕ	1200	~5
10″ ϕ	2400	~2

SUMMARY

A new method to investigate the heliospheric-interstellar wind interface is suggested. This interface will affect inflowing LISM carbon, nitrogen, oxygen, and neon atoms differently, depending upon the character of the interaction at the interface. A preliminary evaluation of this effect has been carried out, and the results indicate that the distribution of these species within the inner heliosphere will vary with different model interface regions. The study of resonantly scattered solar radiation from these species will then provide a means to discriminate among these models. The lines relevant for these studies fall in the extreme and far ultraviolet. Current knowledge of the diffuse background in these wave bands is limited. Instrumentation which can provide data relevant to these studies and which will be available in the early 1990s is discussed; the *ORFEUS* instrumentation shows substantial promise for use in this work. A more detailed evaluation of the effects to be expected from various proposed models of the interface region and of their observational consequences will be the subject of further studies.

ACKNOWLEDGMENTS

We would like to thank Mark Hurwitz and Dr. Supriya Chakrabarti for their assistance. This work has been funded by NASA contract NAS5-30180.

REFERENCES

1. E. N. Parker, in: *Interplanetary Dynamical Processes,* Sec. IX, Interscience Publishers, New York, 1963, p. 113.

2. V. B. Baranov, M. B. Lebedev, and M. S. Ruderman, *Astrophys. Space Sci.*, 41, 491 (1976).

3. V. B. Baranov, M. B. Lebedev, and M. S. Ruderman, *Astrophys. Space Sci.*, 66, 441 (1979).

4. W. Neutsch, and H. J. Fahr, *Mon. Not. Roy. Astron. Soc.*, 202, 753 (1983).

5. H. J. Fahr, W. Neutsch, S. Grzedzielski, W. Macek, and R. Ratkiewicz, *Space Sci. Rev.*, 43, 329 (1986).

6. H. J. Fahr, *Adv. Space Res.*, 6(2), 13 (1986).

7. H. J. Fahr, in: *The Sun and the Heliosphere in Three Dimensions,* D. Reidel Publ. Company, Dordrecht, 1986, p. 421.

8. H. J. Fahr, S. Grzedzielski, and R. Ratkiewicz, *Annales Geophysicae,* 6(4), 337 (1988).

9. M. A. Lee, in: *Solar Wind VI, Boulder 1987,* NCAR-TN-306, vol. 2, 1988, p. 635.

10. H. W. Ripken, and H. J. Fahr, *Astron. Astrophys.*, 122, 181 (1983).

11. G. L. Siscoe, and N. R. Mukherjee, *J. Geophys. Res.*, 77, 6042 (1972).

12. H. J. Fahr, and H. W. Ripken, *Astron. Astrophys.*, 139, 551 (1984).

13. R. J. Reynolds, in: *Proc. IAU Symposium 139, Galactic and Extragalactic Background Radiation,* ed. S. Bowyer and Ch. Leinert, Kluwer Academic Publishers, Dordrecht, in press (1990).

14. R. J. Reynolds, *Ap. J. (Letters),* 339, L29 (1989).

15. P. C. Frisch, D. G. York, D. E. Walty, and J. R. Fowler, *Ap. J.*, submitted (1989).

16. J. Green, P. Jelinsky, and S. Bowyer, *Ap. J.*, in press (1990).

17. J. L. Bertaux, in: *IAU Colloquium No. 81,* NASA CP 2345, 1984, p. 3. Also see references therein.

18. E. Möbius, D. Hovestadt, B. Klecker, M. Scholer, G. Gloeckler, and F. M. Ipavich, *Nature*, 318, 6045 (1985).

19. E. Möbius, B. Klecker, D. Hovestadt, and M. Scholer, *Astrophys Space Sci.*, 144, 487 (1988).

20. H. J. Fahr, and D. Rucinski, *Planet. Space Sci.*, 37, 555 (1989).

21. S. Bowyer and F. Paresce, *Astron. Astrophys.*, 27, 399 (1973).

22. H. J. Fahr, *Space Sci. Rev.*, 15, 483 (1974).

23. J. L. Bertaux and R. Lallement, *Astron. Astrophys.*, 140, 230 (1984).

24. E. Chassefière, J. L. Bertaux, R. Lallement, and V. B. Kurt, *Astron. Astrophys.*, 160, 229 (1986).

25. A. G. W. Cameron, *Space Sci. Rev.*, 14, 392 (1973).

26. H. Hinteregger, *EUV Solar Irradiance from AE-E: Data File SC/210BS* (1981).

27. H. Hinteregger, *Solar UV Irradiance at Wavelengths below 185 Nanometers Observed for Sunspot Cycle 21: EGS, Uppsala* (1981).

28. D. G. Torr and M. R. Torr, *J. Atmos. Terr. Phys.*, 41, 797 (1979).

29. A. Davidsen, S. Bowyer, and M. Lampton, *Nature*, 247, 513 (1974).

30. F. Paresce, and P. Jakobsen, *Nature*, 288, 119 (1980).

31. F. Paresce, C. McKee, and S. Bowyer, *Ap. J.*, 240, 387 (1980).

32. C. Martin, and S. Bowyer, *Ap.* J., in press (1990).

33. M. Hurwitz, S. Bowyer, and C. Martin, private communication (1990).

34. R. Stern, and S. Bowyer, *Ap. J.*, 230, 755 (1979).

35. F. Paresce, and R. Stern, *Ap. J.*, 247, 89 (1981).

36. J. Bixler, S. Bowyer, and M. Grewing, *Astron. Astrophys.*, 141, 422 (1984).

37. C. B. Opal and C. S. Weller, *Ap. J.*, 282, 445 (1984).

38. J. B. Holberg, *Ap. J.*, 311, 969 (1986).

39. S. Labov, and S. Bowyer, *Ap. J.*, submitted (1990).

40. J. P. Pye, M. G. Watson, K. A. Pounds, and A. Wells, in: *Extreme Ultraviolet Astronomy*, ed. R. F. Malina and S. Bowyer, Pergamon Press, New York, in press (1990).

41. A. Davidsen, R. Kimble, S. Durrance, and C. W. Bowers, in: *Extreme Ultraviolet Astronomy*, ed. R. F. Malina and S. Bowyer, Pergamon Press, New York, in press (1990).

42. R. F. Malina, and S. Bowyer, in: *Extreme Ultraviolet Astronomy*, ed. R. F. Malina and S. Bowyer, Pergamon Press, New York, in press (1990).

43. M. Hurwitz, and S. Bowyer, in: *Extreme Ultraviolet Astronomy*, ed. R. F. Malina and S. Bowyer, Pergamon Press, New York, in press (1990).

44. M. Grewing, G. Krämer, I. Appenzeller, N. Eberhard, W. Gringel, A. Kälble, N. Kappelmann, J. Krautter, H. Mandel, R. Oestreicher, J. Petrick, J. Riegger, in: *Extreme Ultraviolet Astronomy*, ed. R. F. Malina and S. Bowyer, Pergamon Press, New York, in press (1990).

45. W. C. Priedhorsky, J. J. Bloch, F. Córdova, B. W. Smith, M. D. Ulibarri, O. H. W. Siegmund, H. L. Marshall, J. V. Vallerga, and P. W. Vedder, in: *Extreme Ultraviolet Astronomy*, ed. R. F. Malina and S. Bowyer, Pergamon Press, New York, in press (1990).

46. J. V. Vallerga, P. Jelinsky, P. W. Vedder, and S. Bowyer, in: *Extreme Ultraviolet Astronomy*, ed. R. F. Malina and S. Bowyer, Pergamon Press, New York, in press (1990).

VELOCITY STRUCTURE OF THE LOCAL INTERSTELLAR MEDIUM

R. Lallement,* R. Ferlet,** A. Vidal-Madjar** and C. Gry***

*Service d'Aéronomie du CNRS, Verrières-le-Buisson, France
**Institut d'Astrophysique de Paris (CNRS), France
***Laboratoire d'Astronomie Spatiale de CNRS, Marseille, France

ABSTRACT

In the last five years, a series of high quality ground-based observations of CaII and NaI interstellar absorption lines towards nearby stars were recorded at the ESO-La Silla 1.4m and the CFH-Mauna-Kea 3.6m telescopes. Observed absorption lines are used to disentangle the different motions and characteristics of the local interstellar matter , as close as possible to the sun. The method is presentedbriefly. There is a complex velocity structure around the sun with at least 4 different motions (clouds?) within 20 pc. One of the aims of this work is the search for the local cloud surrounding the sun. A possible candidate has been found, moving at 29 km/s with respect to the sun in the same direction as the solar system interstellar flow.

INTRODUCTION

The sun is moving through a tenuous interstellar cloud, termed the local interstellar cloud (hereafter LIC). This is demonstrated by observations of a flow of neutral hydrogen and helium atoms of interstellar origin entering the solar system and detected through resonance scattering of solar photons. Modeling of the Lyman-a resonance glow is used to determine the flow characteristics: relative velocity of 20 km/s, temperature of 8000 K, and neutral hydrogen density of 0.07 atoms per cubic centimeter. However, these parameters are derived from observations of a volume whose size is of the order of 50 AU around the sun, the size of the emissive region. That is to say, they refer to the gas flow just before its close interaction with the sun and they are not necessarily identical to the initial parameters of the LIC. Since the interstellar medium is partially ionized, the flow of interstellar plasma interacts with the solar wind flow, and one expects perturbations of the neutral atoms via charge-exchange reactions while traveling through the interface region between the two plasmas . The distance to the heliopause, or solar wind boundary is still unknown. It is at least greater than 40-50 AU, since Voyager and Pioneer spacecraft at this distance are still in the supersonic solar wind. The structure of the interface and the physical processes at work are still under discussion (Axford and Ip, 1986 /1/, Lee and Axford, 1988 /2/, Holzer, 1989 /3/). The interface properties and the expected perturbations of neutrals have been computed by Baranov (1986 /4/,1989 /5/) , Fahr and Ripken (1984) /6/, Fahr et al (1986) /7/. They show that the nature of the heliospheric boundary and its distance to the sun are closely dependent on the local interstellar parameters, mainly the electron density, the magnetic field strength, the neutral density, and the bulk velocity of the LIC. Knowledge of any of these parameters will introduce large constraints on the heliospheric interface structure. In addition the comparison, if possible, between the external parameters (the LIC parameters) and the observed values for the neutrals inside the heliosphere will also bring stringent constraints on the type of the interface. In what follows we will call $\mathbf{V}_w$ (20 km/s, λ_{ecl}=71°, β_{ecl}=-7.5°) the velocity vector characterizing the interstellar flow in the inner solar system as deduced from H Ly-α solar backscattered radiation (Bertaux et al, 1985 /8/).

One can then ask whether the characteristics of the LIC can be determined with the same method which is usually used to derive physical properties of the interstellar clouds, that is to say by using absorption lines from the clouds towards some appropriate targets, mainly hot stars (see fig 1). This method introduces the problem of the size of the LIC. Is the sun in a small isolated patch or in a large cloud? More generally what is the degree of homogeneity of the interstellar medium over distances of the order of the distance to the target stars (the nearest star is at 1.3 pc with 1parsec =200,000 AU)? The physical and kinematical structure of the diffuse interstellar clouds and their typical sizes are not well known. The general picture arising from observations and theoretical calculations (MacKee and Ostriker,1977 /9/) consists of cold dense cores (80 K, 40 at.cm^{-3}), surrounded by a warm (8000 K), partially ionized envelope (see fig 2) imbedded in a hot, ionized, tenuous and soft X-rays emitting medium. The typical size of such a cloud is of the order of 4 pc but smaller clouds without dense cores could also exist. In fact detailed studies and comparisons with these type of models are possible only in the immediate vicinity of the sun. When using very distant targets only average properties can be reached due to overlapping of the different absorptions features. It appears that all the ingredients for a MacKee and Ostriker structure are present in the sun vicinity, a cavity of hot rarefied plasma of about 80 pc (Arnaud et al, 1984 /10/), and low density (<<1 at.cm-3) cold and warm regions very close to the sun, the so-called Local Fluff (Cox and Snowden, 1986 /11/, Frisch and York, 1983 /12/, Frisch, 1986 /13/). This unique opportunity to extend the description of the interfaces between the different phases is one of the

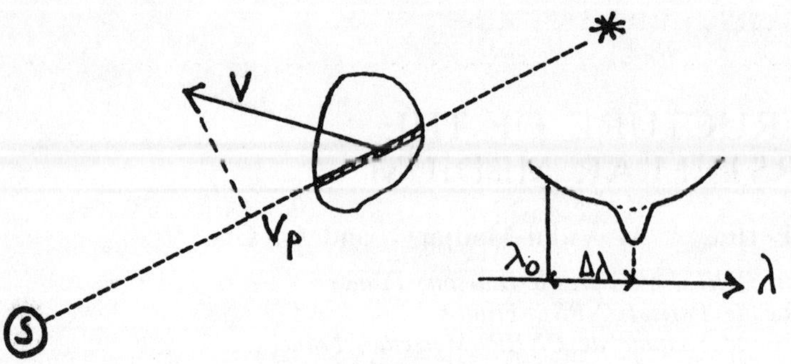

Fig. 1: An interstellar cloud moving at velocity **V** with respect to the observer produces an absorption line in the spectrum of a star located behind or inside the cloud: this line is shifted by $\Delta\lambda=\lambda-\lambda_0=\lambda_0.V_p/c$ (Vp projection of V onto the direction of the star).The depth of the line is depending on the column-density of the absorbing element i.e. the number of atoms in a column of section 1 cm^2 (represented by the hatched area). The line broadening is the sum of thermal and turbulent motions.

main reasons for which a study of the local interstellar medium (within 50 pc) was undertaken a few years ago, using ground-based observations at high-resolution and high signal-to-noise ratio of the hot nearby stars (Vidal-Madjar et al,1978 /14/, Ferlet et al,1986 /15/). High sensitivity is required since one tries to detect very small amounts of material in front of the nearest stars, while the resolving power of 10^5 permits separation of relative motions of 1 km/s. The error in the absolute velocity scale is smaller than 1 km/s. Doppler shifts of the observed lines can be used to disentangle the different motions of the absorbing material (Crutcher, 1982 /16/, Lallement et al ,1986 /17/, hereafter LVF, Bzowski, 1988 /18/ for UV lines) and to locate the corresponding matter. This kinematic and topographical study, which is valid only very close to the sun, is a first and in our opinion mandatory step to the physical study of the nearby interstellar clouds. Identification of the LIC , if it is not too small as discussed in the next part, is one of the pieces of the puzzle.

GENERAL CHARACTERISTICS OF THE GROUND-BASED OBSERVATIONS

Visible lines employed are the strong CaII (Ca^+) line at 393.3 nm and the NaI (Na^o) doublet at 588.9 -589.5 nm. Provided the corresponding stellar feature is broad enough (hot stars, fast rotators), the superimposed interstellar absorption line is easily distinguished. Examples of the recorded spectra are shown in Fig3. With the help of a profile-fitting program (Vidal-Madjar et al, 1977 /19/), each absorption is characterized by the velocity shift , the column-density of the absorbing element (see Fig1) and an equivalent temperature, "sum" of the thermal and turbulent motions. In the Alpha Aquilae (Altair) spectrum three different motions are detected with very low column-densities , 1.5 mA total equivalent width requiring a signal/noise ratio>1000, and separated by a few km/s. This spectrum is representative of the very high quality of some of the spectra. Note that IUE or Copernicus spectral resolution (15 and 30 km/s) would not have allowed separation of the three motions. In what follows we will call cloud or cloudlet any of the detected velocity components. Turbulent motions with a spatial scale smaller than the lines-of-sight distances are precluded by the observations since the observed lines show a small non-thermal broadening, see end of part 3 . On the other hand, if there are turbulent motions at a spatial scale of the order of the clouds size, shock splitting or distorsion, that is to say if the cloud can be schematically divided into a few parts with different motions, then the different parts are designated as clouds.

Fig. 2: From McKee and Ostriker (1977) /9/ : A schematic view of a typical diffuse interstellar cloud imbedded in the hot and rarefied medium of a supernova remnant. Some of the clouds are too small to have a cold core. The expected typical size is of the order of a parsec and then comparable to the mean distance between the stars in the sun environment.

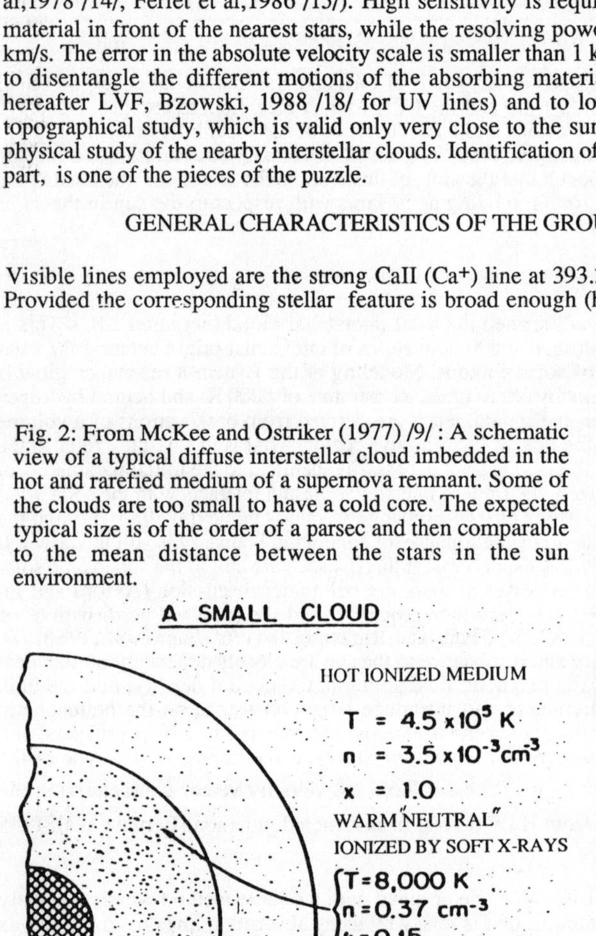

Using a single spectrum, it is not possible to know if the observed absorption is due to the local cloud or to an external cloud, surrounding the target star or in front of it(see Fig4). Also it is not possible to derive a volumic density of the observed cloud, since the path-length through the cloud is unknown. One can make at least some estimates about the detectability of the local cloud by using the following arguments: 1) The neutral He density inside the solar system being ≈ 0.01 at.cm^{-3} the total H density in the LIC around the sun should be of the order of 0.1 at.cm^{-3} , for a cosmological ratio H/He ratio of 10 and assuming that helium is not ionized , nor filtrated at the heliopause.

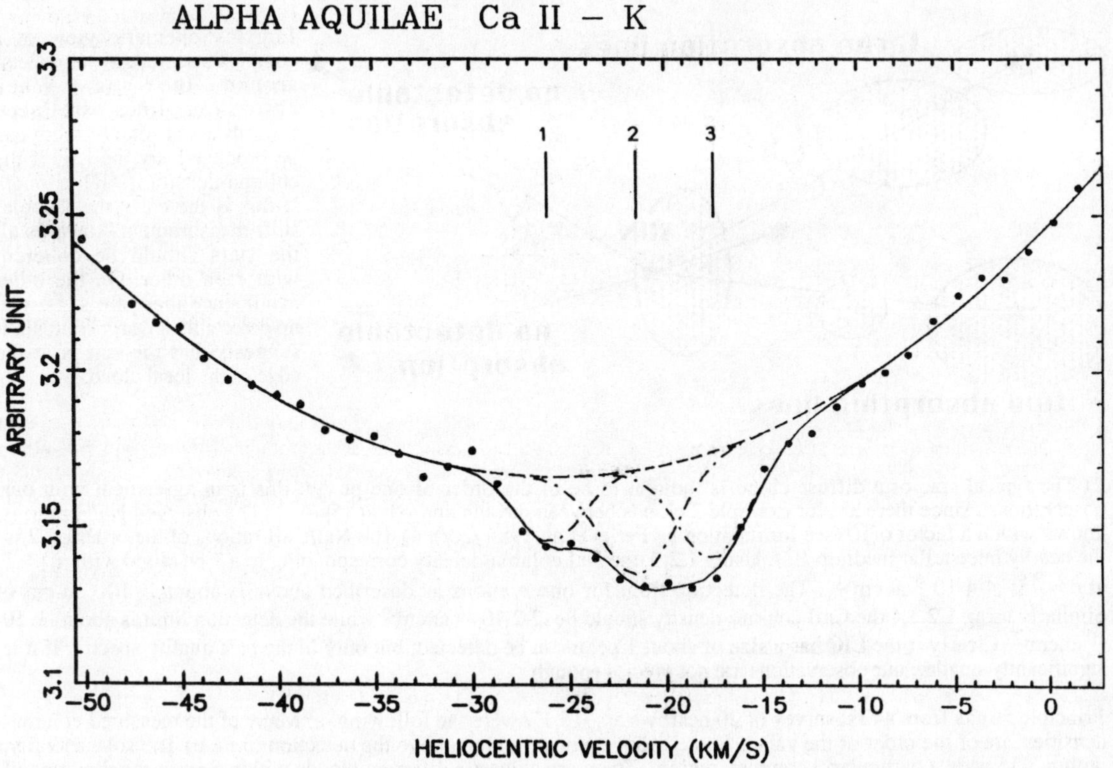

Fig. 3: Examples of CaII-K and NaI-D2 recorded spectra (dots) towards Altair (5pc) and HR4023 (30 pc) respectively. Best fit models are shown (solid line) with three velocity components (3 "clouds") superimposed on the star line (dashed line) .

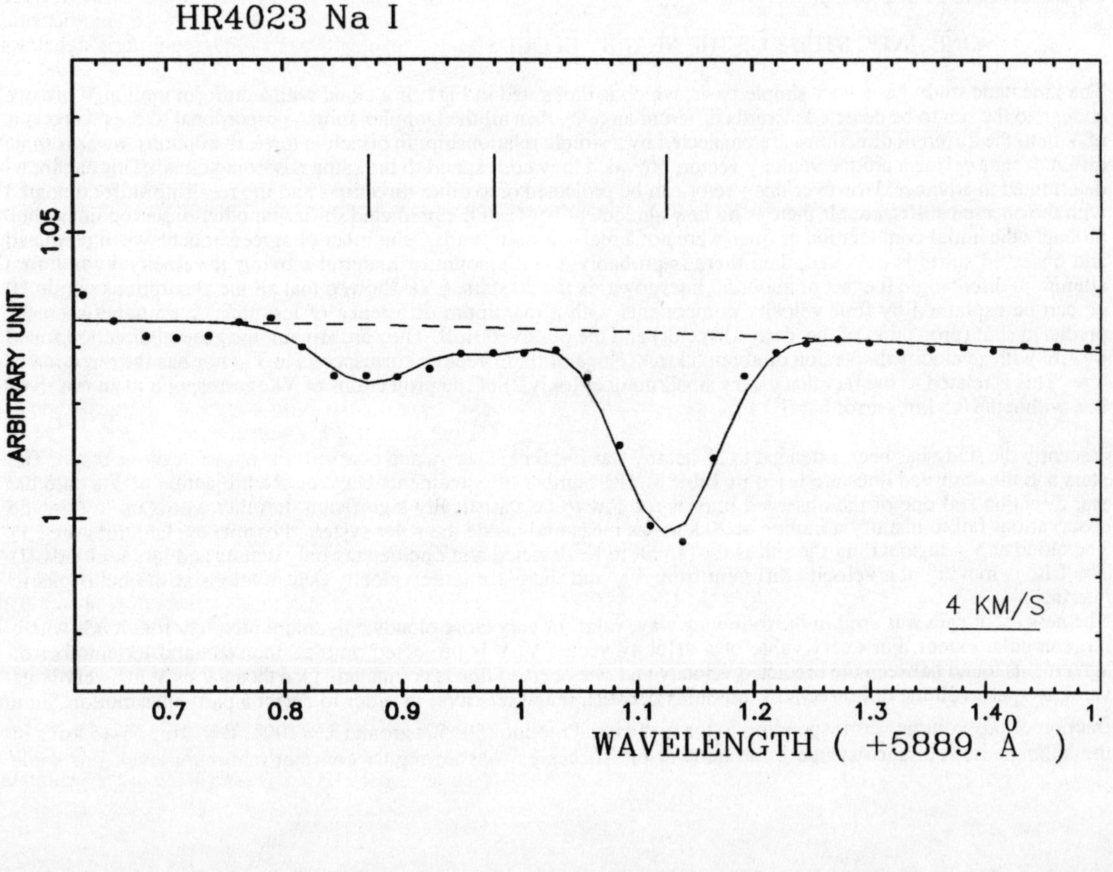

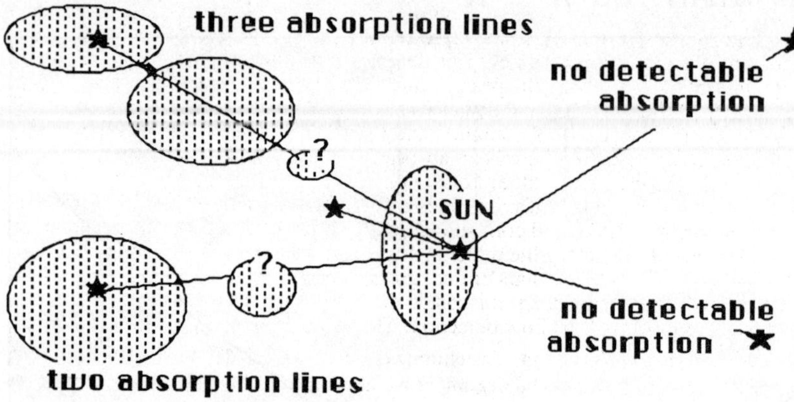

Fig. 4: Schematic view of a complex structure of "cloudlets" around the sun. When absorption lines are seen towards a star, one of them can be produced by the LIC if the column-density is high enough. If this is the case, the Doppler shift measurements towards all the stars should be coherent with each other. On the other hand, since there are stars with no detectable absorption at all, it suggests that the sun is at the edge of the local cloud.

2) The typical size of a diffuse cloud is thought to be of the order of one pc /9/: this is in agreement with our observations , since there are for example 3 clouds between the sun and Altair (5pc). 3) The observed Na/H ratio is known within a factor of10 (see formulation by Ferlet et al, 1985 /20/). 4) The NaI/CaII ratio is of the order of .2 in the nearby interstellar medium /15/. Using 1,2,3, the NaI column/density corresponding to a 1 pc cloud with n_H =.1 at.cm^{-3} is .4-4 10 ^{9} at.cm^{-2} . The detection limit for observations as described above is about 1. 10 ^{9} at.cm^{-2}. Similarly using 1,2,3,4 the CaII column-density should be .2-2 10 ^{10} at.cm^{-2} while the detection limit is about .3 10 ^{10} at.cm^{-2}. Clearly if the LIC has a size of about 1 pc, it can be detected, but only in the best quality spectra. If it is significantly smaller, our observations are not precise enough .

Principle results from a first survey of 20 nearby stars /15,17/ were the following. a) Many of the measured column-densities are of the order of the values quoted above, not very high above the detection limit. b) The solar vicinity, within 5-15 pc is a particularly complex region . There are at least 3 different clouds within 5 pc in the direction of Altair and many nearby stars show multiple absorptions. Almost all stars at less than 20 pc show 2 or three "clouds". c) It is confirmed that the "dense" region, now composed of a few different "cloudlets", is located primarily in one hemisphere, centered in the Scorpius-Ophiucus direction, while nearby stars in the opposite hemisphere show no or only very weak absorptions. The Sun is at the edge of the "dense" region. d) There is no significant increase of the column-density between ≈20 pc and ≈50 pc, which suggests that the "dense" region is closer to the sun than 20 pc. e) The mean equivalent temperature is of the order of 10^4 K (FWHM for the CaII line of 4km/s), while some cold material (T < 3000K) is also present. Absorptions at those wavelengths by the hot rarefied medium are too low and too broadened to be detected.

KINEMATIC STUDY OF THE NEARBY CLOUDS

The kinematic study has a very simple basis, which is illustrated in Fig1. If a cloud with a uniform motion **V** is close enough to the sun to be detected towards different targets , then all the Doppler-shifts , proportional to the projections of **V** onto the different directions, are connected by a simple relationship. In principle three absorptions towards three different stars define a unique velocity vector, provided they correspond to the same absorbing cloud. This cannot be determined in advance. However this vector can be projected onto other directions and the resulting shift compared with the oberved shifts, etc.. If there is no new agreement between the predicted shifts and other observed lines, then probably the initial combination of lines were not a real solution. If a large number of agreements between predicted and observed shifts is obtained, then there is probably a real amount of material moving at velocity **V**. In a first attempt to disentangle the set of about 50 lines towards the 20 stars, LVF showed that all the absorptions within 20 pc can be explained by four velocity components with a maximum difference of less than .5 km/s between the predicted shift (projection of the derived vector) and the observed shift. They are around the general direction found by /16/ with a velocity dispersion of about 5 km/s. None of these vectors corresponds to $\mathbf{V}_w$, nor has the direction of $\mathbf{V}_w$. This is related to the fact that a very small number (only 2) of the projections of $\mathbf{V}_w$ correspond to an observed line within the 0.5 km/s error bar (LVF).

Recently the study has been extended to 38 nearby star (5-50 pc) spectra and confirms the above negative result. The stars and the observed lines are listed in Table 1. The number of agreements between the projection of $\mathbf{V}_w$ onto the star direction and one of the observed lines is too low to be statistically significant. In other words up to now the observations fail to identify a motion at 20 km/s as measured inside the solar system. Possible explanations are : 1) The cloud at $\mathbf{V}_w$ surrounding the sun is too small to be detected and one detects only denser and larger clouds. 2) The LIC is moving at a velocity different from $\mathbf{V}_w$ and there are large velocity perturbations at the heliospheric interface.

The new set of data was used in the following way, valid for very close clouds only, more precisely for clouds with a large angular extent. For every value of a velocity vector **V**, **V** is projected onto all the stars and the number of agreements found between the predicted velocity and one oberved line is computed . One then varies V in a very large velocity space volume (all directions, modulus between 0 and 60 km/s) in order to detect a particular motion. In a fraction of this volume, corresponding to the mean local motion (50*50° around λ = 100° , β = -10°, 15-45 km/s in the heliocentric reference system), the number of "successes" has a roughly constant minimum level. For some particular values of **V** there are some "peaks" above the "noise" . The "noise" is due to the fact that owing to the large number of lines, there are always some coincidences between the projections of **V** and the observed lines, which can be an indication of a cloud with a small angular extent, provided the corresponding stars are located in the same part of the sky, but can also be non significant fortuitous coincidences. This deserves a detailed discussion of the entire velocity volume, but that will not be presented here. For each "peak" value of **V**, the corresponding selected stars have been located on a map, and an analysis of their location and distances has been completed. In only one case are the stars located in a definite part of the sky (a "connexe" solid angle) , and there is no "hole" inside this region, in the sense that for all the stars inside the solid angle either there is an absorption at the predicted velocity (20 stars show an absorption at less than 1 km/s from the predicted value) , or the observed spectrum does not preclude an absorption comparable to the observed ones (blending of different lines or too noisy spectrum). Fig 5 shows the location of the stars and the contour of the hypothetical cloud corresponding to this peak value of **V**. The contour is surrounded by some close stars for which there is no detectable material at all. This fact, the extent of the cloud (≈ one hemisphere), and the proximity of some of the selected stars are in agreement either with the cloud being very close to the sun , or with the geometry of Fig4. In this case the sun is inside the cloud but at the edge of it, with path-lengths through it in the opposite hemisphere too small to allow detection. The velocity vector $\mathbf{V}_l$ of this emerging feature is characterized by a modulus of 29.4 ± .5 km/s, and the direction (λecl=71.5°, βecl=-8°) is identical to the solar system flow direction within 3°. This gives an additional argument in favor of a very close feature. If this is really the LIC, then one can consider a deceleration of the neutral atoms crossing the heliospheric interface from 29 to 20 km/s as an explanation of the discrepancy between $\mathbf{V}_w$ and $\mathbf{V}_l$. As a matter of fact, there are two arguments which tend to preclude turbulence as an explanation for a region at 9km/s velocity difference from the bulk motion: 1) All the CaII observed lines are compatible with a very low non-thermal broadening. The average FWHM ≈ 5 km/s corresponds to an apparent temperature of 12000 K, which implies a 1-2 km/s non-thermal broadening if the real temperature is 8000K. 2) The turbulent motion would have the same direction as the bulk motion : this would be a peculiar coincidence. However the existence of a small peculiar patch of gas different from the homogeneous bulk is not precluded.

Table1: Characteristics of the target stars.

STAR	HR	lII	bII	d	TYPE
		° '	° '	pc	
αAql	7557	4743	-08 54	5	A7V
αPsa	8728	20 30	-64 54	7	A3V
βLeo	4534	250 39	70 48	12	A3V
αOph	6556	35 54	22 35	15	A5III
ιCen	5028	309 25	25 48	16	A2V
τ3Eri	919	213 32	-60 17	17	A4IV
αGru	8425	350 00	-52 28	18	B7IV
1Cen	5168	315 51	28 27	19	F3IV
δVel	3485	273 05	-07 22	20	AIV
βCrt	4343	274 46	34 32	20	A2III
αHyi	591	289 28	-53 46	21	FOV
ζAql	7235	046 51	03 15	22	FOV
βPer	936	148 59	-14 54	22	B8V
αLeo	3982	226 26	48 56	22	B7V
δHer	6410	046 50	31 25	23	A3IV
51 Oph	6519	002 31	05 20	25	B95V
γTri	664	142 39	-32 50	26	A1V
αPeg	8781	088 53	-40 23	26	B9V
δCas	403	127 11	-02 21	27	A5V
τPeg	8880	097 30	-34 36	27	A5V
109Vir	5511	355 17	52 41	27	AOV
κLep	1705	213 53	-27 33	29	B9V
HR4023	4023	274 20	11 53	30	A2V
αAnd	15	111 44	-32 50	31	B8IV
ρCen	4638	296 47	10 02	31	B3V
β1Tuc	126	306 47	-54 01	33	B9V
ωOph	6153	356 18	17 50	36	A7
δCen	4621	296 00	1134	38	B2IV
αEri	472	290 50	-58 47	40	B3V
ηAur	1641	165 21	00 16	45	B3V
βCar	3685	285 59	-14 24	46	A2IV
αVir	5056	316 07	50 50	50	B1II-IV
βCMa	2294	226 03	-14 16	52	B1II-III
HR5174	5174	315 17	25 17	53	AOV
αCru	4730	300 08	-00 22	100	B1IV
λSco	6527	351 44	-02 12	100	B2IV
βCen	5267	311 46	01 15	100	B1II
βCru	4853	302 27	03 11	100	BO.5III

PERSPECTIVES AND DISCUSSION

The possible detection of a very close cloud , (surrounding the sun ?) at an heliocentric velocity of 29 km/s in the direction of the solar system flow is a promising step in the search for a link betwen the solar system interstellar gas and the local interstellar medium. This approach and these results deserve further work. Steps include improvment of many of the spectra (some are too noisy), elimination of possible sporadic stellar features by doing new observations, use of new targets both inside the contour in order to confirm the existence of an homogeneous velocity, and near the contour. Observations near the contour are important, because if the cloud is seen in more than 2*pi steradians it increases the probability of the sun being immersed in it . Couples of stars in two opposite directions would also be crucial tests. Identification of the local cloud cannot be completely disconnected from the search for the whole structure of the nearby cloud complex. When one increases the number of targets, and goes further with the whole structure, fortuitous coincidences are recognized as such and one avoids ambiguities.

There is a great hope for improvements in the knowledge of the local structure from the Hubble Space Telescope (some programs are devoted to the local gas): many UV lines have the advantage of being stronger than the visible ones and smaller amounts of material can be detected in front of the nearest stars (see the recent review by Bzowski/18/). One expects to derive electron densities and temperatures by using lines ratios. The resolution and absolute accuracy are expected to be nearly identical to the ground measurements, which has never been possible in space (Resolution of 15-30 km/s, absolute velocity scale within ±5 km/s or more for IUE or Copernicus) . As an example , H and D absorptions at Ly-a by the gas towards Alpha Centauri (1.3 pc) (Landsman et al, 1986 /21/) do not allow to discriminate between projections of the above $\mathbf{V}_l$ vector or of the

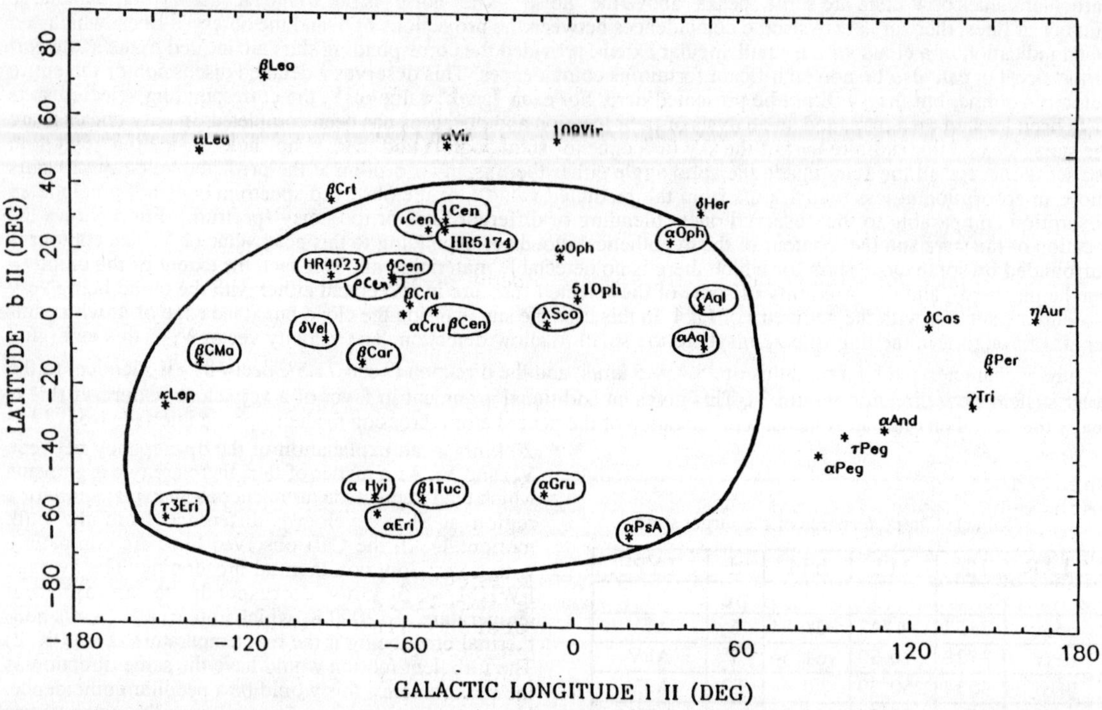

Fig. 5: Sky map of the observed target stars: those showing an absorption line compatible with a cloud moving at velocity V_l (see text) are encircled. A contour of the detectable part of the cloud has been drawn. Non encircled stars inside the contour have ambiguous spectra, for which compatible absorptions are not precluded. If real, this cloud is very close to the sun, and possibly surrounding it. V_l has about the same direction than the solar system interstellar flow and a modulus of 29 km/s.

vector found by Bzowski /18/ at 23 km/s from UV measurements only. Hopefully the HST should give an answer. However precise velocity structures derived from the ground-based optical observations are of great help to understand the UV lines. Since many of the UV lines are saturated, their interpretation is much easier if the line-of-sight velocity structure is known in advance.

References

1. Axford W.I, Ip W.H., ***Adv. Space Res., 6***(2), 27 (1986)
2. Lee M. A., Axford W.I., ***Astron. Astrophys.*, 194**, 297 (1988)
3. Holzer T.E., ***Ann. Rev. Astron. Astophys.*, 27**, 199 (1989)
4. Baranov V. B., ***Adv. Space Res., 6***, 5 (1986)
5. Baranov V. B., this issue (1989)
6. Fahr H.J., Ripken H.W., ***Astron. Astrophys.* 139**, 551 (1984)
7. Fahr H.J., Neutsh W., Grzedzielski S., Macek W., Ratkiewicz-Landowska R., ***Space Science Reviews*, 43**, 329 (1986)
8. Bertaux J. L., Lallement R., Kurt V.G., Mironova E.N., ***Astron. Astrophys.* ,150**, 82 (1985)
9. Mckee C.F, Ostriker J.P, ***Astrophys. J.*, 218**, 148 (1977)
10. Arnaud M., Rothenflug, Rocchia R., in The Local Interstellar Medium, IAU Colloquium 81, NASA Conf. Pub. 23455 (1984)
11. Cox D. P., Snowden S.L., ***Adv. Space Res.*, 6.2**, 97 (1986)
12. Frisch P.C., York D.G., ***Astrophys. J. Letters*, 271**, 159 (1983)
13. Frisch P.C., ***Adv. Space Res.*, 6.1**, 345 (1986)
14. Vidal-Madjar A., Laurent C., Bruston P., Audouze J., ***Astrophys. J.*, 223**, 589 (1978)
15. Ferlet R., Lallement R., Vidal-Madjar A., ***Astron. Astrophys.* ,163**, 204 (1986)
16. Crutcher R.N. Ap. J., **254**, 82 (1982)
17. Lallement R., Vidal-Madjar A., Ferlet R., ***Astron. Astrophys.* 168**, 225 (1986)
18. Bzowski M., ***Acta Astron.*, 38**, 443 (1988)
19. Vidal-Madjar A., Laurent C., Bonnet R.M., York D.G., ***Astrophys. J.*, 211**, 91 (1977)
20. Ferlet R., Vidal-Madjar A., Gry C., ***Astrophys. J.*, 298**, 838 (1985)
21. Landsman W.B., Murthy J., Henry R.C., Moos H.W., Linsky J.L., Russel J.L., ***Astrophys. J.*, 303**, 791 (1986)

A STUDY OF THE LOCAL INTERSTELLAR MATTER FLOW IN THE (IMMEDIATE) SOLAR NEIGHBOURHOOD

Maciej Bzowski

Space Research Centre of the Polish Academy of Sciences, 01–237 Warszawa, ul. Ordona 21, Poland

ABSTRACT

Results of various nearby interstellar matter flow studies are compared with each other and with the density structure in the solar neighbourhood. Into account are taken results of backscattered radiation analysis in Lyman-α and in He I 584 Å, and models based on fits to radial velocities of interstellar (ionised and neutral) gas observed in the spectra of nearby (~100 pc) and very nearby (~10 pc) stars. The flow in the immediate solar neighbourhood seems to agree with the results of the backscattered solar radiation analysis, while the flow of more distant gas, although far from being homogeneous, is in agreement with the Crutcher /1/ model. Thus the hypothesis of the Sun embedded in a local cloud seems confirmed. The local cloud's velocity with respect to an ambient interstellar medium which moved with the Crutcher vector would be $v \simeq 10$ km/s, $l \simeq 70^{\circ}$ and $b \simeq -10^{\circ}$. Need for extreme care when interpreting the kinematics of possible nearby interstellar clouds is stressed.

INTRODUCTION

In recent years a number of studies have been conducted on the interstellar matter density distribution and flow in the solar neighbourhood (for review see /2,3,4,5/). Various experimental techniques have been used, and a picture of the local interstellar medium has been constructed. An analysis of kinematic components of interstellar (I/S) matter absorption lines observed in the spectra of stars within ~100 pc from the Sun /1,4/ yields the velocity of an average flow of I/S matter in the solar neighbourhood. This flow is by no means homogeneous, as implied by the detection of numerous kinematic components of I/S features /6,7/. On the other hand, results of backscattered solar UV radiation analysis /8,9,10/ show in the most immediate solar neighbourhood a flow of different speed and direction. There has been a longstanding gap in understanding the LISM beyond the heliopause but within 15 parsecs of the Sun. This is because most observation techniques do not sample this region: it is too close for polarimetric or radioastronomic studies of gas distribution, and the column density of material is so low that data from spectral analysis of interstellar lines are difficult to obtain and therefore far from being exhaustive. Lines of sight towards most nearby stars permit only spaceborne UV observations. The aim of this work is to fill at least partly the gap by extending an earlier study /11/ and presenting some results of I/S matter flow in the immediate solar neighbourhood.

DATA AND CALCULATIONS

Data selection and calculations have been described in the earlier work /11/. Briefly speaking, the Crutcher's method has been applied to various sets of kinematic components of interstellar matter observed in spectra of nearby and some distant stars. The data are in Lyman-α and Mg II 2800 Å as well as in optical domain (for references see /11/).

RESULTS AND DISCUSSION

Results of fits to line-of-sight velocities from various subsets of stars are presented in Fig.1. They have been published earlier in /11/ in the heliocentric frame. The results show that while fits to several nearest stars cluster in the vicinity of the values derived from the analysis of backscattered radiation (*Bertaux Wind - BeW*), fits to more distant stars reproduce the Crutcher Wind vector (*CrW*). Due to limitation in the accuracy of the data, however, this difference in velocity vectors might be not real; for a more thorough discussion see /11/. Table 1 presents the LSR upwind velocities and directions of some relevant flow vectors.

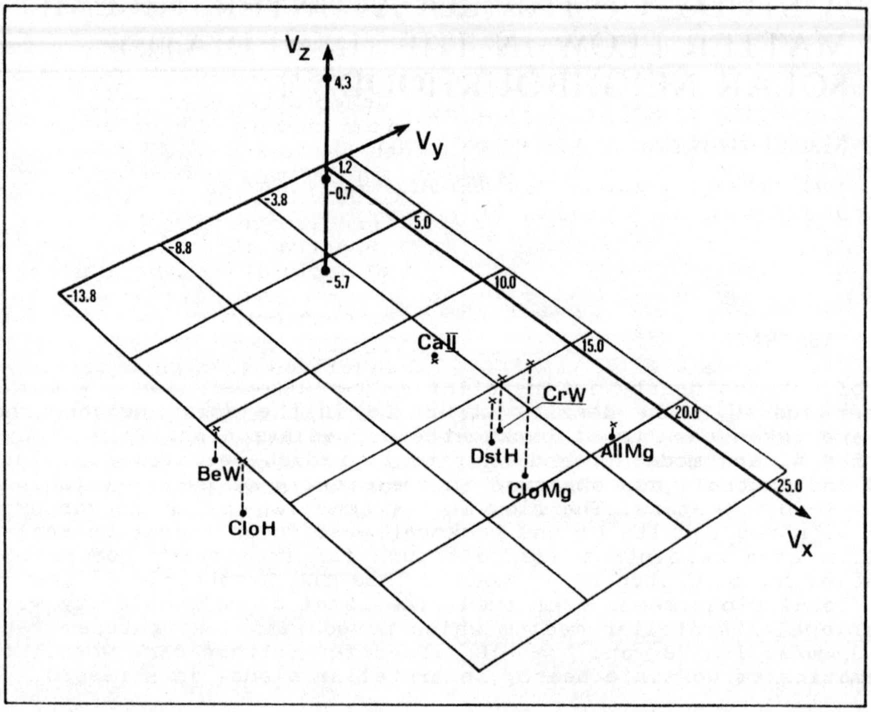

Fig.1. Comparison of various upwind velocity vectors in 3-D LSR velocity space. For symbols refer to Table 1. Scale is in km/s, v_x- axis points towards galactic centre and v_z-axis towards northern galactic pole. Thick dots mark ends of vectors, "x"'s their projections on the v_x-v_y plane. To get the right impression connect the origin of the coordinate system with the dots of ends of vectors. All vectors point below galactic plane.

TABLE 1 LSR Velocities And Directions of I/S Matter Flows*

Denotation	Symbol	v[km/s]	l[deg]	b[deg]
Bertaux wind (hydrogen)		16.6	301.2	-2.7
Bertaux wind (helium)	BeW	18.3	310.3	-3.5
Crutcher wind	CrW	15.5	345.5	-9.7
Crutcher wind from /4/		11	356	-2
Nearby stars with H in LOS**	Clo H	18.9	310.7	-7.9
Distant stars with H in LOS	Dst H	16.4	342.2	-7.1
Nearby stars with Mg in LOS	Clo Mg	16.0	351.0	-21.0
Distant stars with Mg in LOS		22.9	351.5	11.7
All stars with Mg in LOS	All Mg	19.3	353.2	-2.2
Lallement *et al.* /7/ stars	Ca II	13.4	337.7	0.1
(BeW_{He} - CrW)***		~10	~70	~-10
Solar motion apex velocity****		20.0	56.5	21.9

*From /11/
**LOS - line of sight.
***Relative velocity of local cloud and LISM.
****From /12/, α = 271.5°, δ = 30°.

Stars in the set denoted as *Clo H* are marked with "x"'s in Fig.2, while stars in the set denoted as *Clo Mg* are marked with thick dots. A look at Fig.2 shows that the stars which gave *Clo H* are in a different region than the stars which gave *Clo Mg*. This picture is in favour of a hypothesis presented in /5/ according to which the Sun is embedded in a very small cloud 3 to 5 pc in radius, extending in all directions. This cloud would have the LSR velocity consistent with *BeW*, and its velocity relative to the surrounding LISM with a motion defined by the Crutcher flow would be $v \simeq 10$ km/s, $l \simeq 70^\circ$, $b \simeq -10^\circ$. Whether or not this cloud extends towards galactic anticentre as far away as β Gem and α Aur remains unclear due to the low weights of kinematic data for these stars.

This small, local cloud, although denser than the surrounding LISM material, would be too rarefied to produce easily detectable kinematic features in the nearest stars when observed in the optical domain. This is implied by the fact that it has not been detected in both Ca II and Na lines either in front of the stars closer than 5 pc /7,13/ or in front of other stars within 5 to ~15 pc. Interstellar features discovered optically in the sightlines to some of these stars do not coincide with the predictions for *BeW*. Thus the cloud should not be easily detectable in the spectra of other distant stars and possible kinematic components observed at velocities predicted by the *BeW* flow found in the spectra of some of them may be coincidences, unless the local cloud extends much farther in some directions.

On the other hand, the UV spectra of sightlines towards stars more distant than ~2 to 4 pc should show multiplicity. Unfortunately, the spectral resolution and sensitivity of the COPERNICUS and IUE satellites have been too low to reveal them: separation of components from *BeW* and *CrW* should be smaller than 6.5 km/s. The line of sight toward α Aql observed in Ca II K /6/ shows a weak feature at the velocity predicted for *CrW* and two stronger features, of which one is in marginal agreement with the predictions for *BeW*. There is, however, a possibility that some of the observed components are stellar /5/.

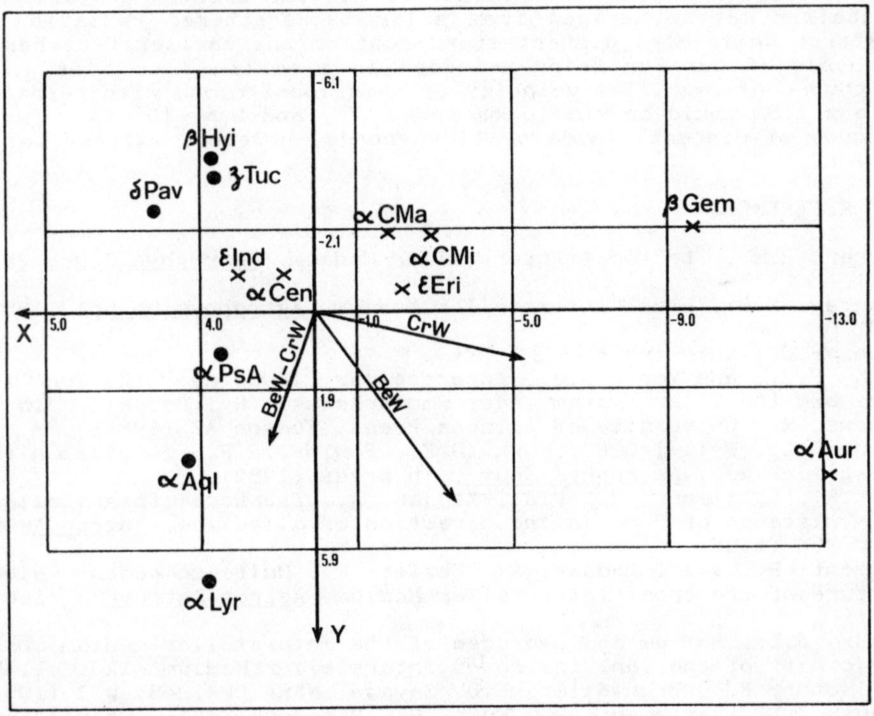

Fig.2 Projections at the galactic plane of the nearest stars with interstellar matter features in the spectra. Crosses mark stars which give a flow denoted as *Clo H* (similar to *BeW*) and thick dots stars which give a flow denoted as *Clo Mg* (similar to *CrW*). Axes are in parsecs, Sun is in the origin, *x*-axis points towards galactic centre, *y*-axis towards $l=90^\circ$. Projections of downwind *BeW*, *CrW* and their differences (i.e. their relative velocity) are also plotted. For a 3-D picture refer to /11/.

In commenting on the reliability of the method used in this work one has to stress that the results can only be trusted when many stars covering an extensive fragment of the sky have been used. Even in that case, information on the average flow properties only can be extracted. If one wants to discuss detection of individual clouds or other I/S matter structures, it is possible to talk only about indications favouring one or another hypothesis after introduction of additional premises.

For example, in the most interesting paper /7/ the authors claim the detection of several interstellar clouds which they link with weak kinematic components observed in stellar spectra. It is possible, however, to select a subset of kinematic components from "certain detections" of different postulated clouds and to fit a new "cloud" without affecting significantly the results and the quality of the original fits. Conclusions on "detection" or "non-detection" in that work are very sensitive to the adopted models: only small modification of *VLISM* flow parameters (e.g. v = 23 km/s instead of 20 km/s, which seems feasible from /10/) significantly changes the picture of "*VLISM* confirmations" (see also the remark above on feasibility of *VLISM* detection in optical data). *CrW*, *BeW* and the postulated flows do not differ dramatically, so differences in their projections on the lines of sight are also not big, making thus the task of linking the observed features with the postulated multiple clouds or flows even more questionable.

If the cloud containing the Sun is typical, chances of detection of other such clouds by kinematic studies are small. A spherical cloud of ~6 pc in diameter with the centre only 15 pc from the Sun would have an angular diameter of ~23°. Thus it would be in front of few observable stars, preventing any reliable fit to the observed kinematic components. A search for clouds or flows of I/S matter based only on the analysis of kinematic components does not seem possible to the author. A "kinematic check" of large clouds derived from other premises (e.g. /14/) might be, however, worth trying.

CONCLUSIONS

This analysis of I/S Doppler components of stellar spectra confirms the flow of interstellar matter derived from solar backscattered radiation for the nearest stars, while more distant stars confirm the earlier Crutcher result. The hypothesis of our Sun being embedded in a small (3 to 5 pc in radius) cloud is thus confirmed. The velocity of this local cloud with respect to the more distant LISM would be $v \simeq 10$ km/s, $l \simeq 70°$ and $b \simeq -10°$. Any further kinematic study of distant clouds or flows needs, however, extreme care in interpretation.

REFERENCES

1. Crutcher, R.M., *The Local Interstellar Medium*, Astrophys.Jour. 254, p 82 (1982)
2. Bochkarev, N.G., *Local Interstellar Medium*, Astrophys.Sp.Sci. 138, p 229 (1987)
3. Frisch, P.C., this issue
4. Frisch, P.C., and York, D.G., *Interstellar clouds near the Sun*, from: *The Galaxy and the Solar System*, eds. Smoluchowski, R., Bahcall, J.N., Matthews, M., University of Arizona Press, Tucson AZ (1986)
5. Frisch, P.C., Welty, D.E., York, D.E., Fowler, J.R., *Ionization in nearby interstellar gas*, Astrophys.Jour., in press (1989)
6. Ferlet,R., Lallement, R., Vidal-Madjar, A., *The Local Interstellar Medium over a distance of 5 pc in the direction of α Aquilae*, Astron.Astrophys. 163, p 204 (1986)
7. Lallement, R., Vidal-Madjar, A., Ferlet, R., *Multi-component velocity structure of the Local Interstellar Medium*, Astron.Astrophys. 168, p 225 (1986)
8. Bertaux, J.L., *Helium and hydrogen of the interstellar medium observed in the vicinity of the sun*, in: *Local Interstellar Medium*, IAU Coll No 81, ed. Y.Kondo, F.C. Bruhweiler, B.D. Savage, NASA CP-2345, p 3 (1984)
9. Bertaux, J.L., Lallement, R., Kurt, R., Mironova, E.N., *Characteristics of the local interstellar hydrogen determined from PROGNOZ 5 and 6 interplanetary Lyman-α line profile measurements with hydrogen absorption cell*, Astron.Astrophys. 150, p 1 (1985)
10. Chassefière, E., Bertaux, J.L., Lallement, R., Sandel, B.R., Broadfoot, L., *A new estimate of interstellar helium parameters from measurements of Voyager 1/2 ultraviolet spectrometers*, Astron.Astrophys. 199, p 304 (1988)
11. Bzowski, M., *Local interstellar wind velocity from Doppler shifts of interstellar matter lines*, Acta Astron. 38, p 443 (1988)
12. Robinson, J.H., *Astronomy Data Book*, Halsted Press, New York (1982)
13. Lallement,R., Ferlet, R., Vidal-Madjar,A., Gry, C., *Velocity structure of the Local Interstellar Medium*, this issue (1990)
14. Frisch, P.C., York, D.G., *The distribution of nearby HI and HII gas*, in: *Extreme ultraviolet astronomy*, eds. R.F. Malina and S. Bowyear, Pergamon Press (1989)

Session II

Solar UV Backscatter on Neutral Galactic Gases

(Curator — P. C. Frisch)

SCATTERING OF SOLAR UV ON LOCAL NEUTRAL GASES

R. Lallement

Service d'Aéronomie du CNRS, F-91371 Verrières-le-Buisson, France

ABSTRACT

The solar system is imbedded in a glow of the resonance lines of H and He atoms, through scattering of solar photons by flowing interstellar neutral gas. The flow of neutral galactic gas is modified by solar gravitation, ionization, radiation pressure, and possibly by interaction with ionized species around the heliopause. The study of intensity distribution allows the derivation of estimates of densities,velocity and temperature of neutral H and He in the VLISM (Very Local Interstellar Medium). High resolution spectroscopic observations allow the derivation of the flow velocity of hydrogen, which is found to be 20 ± 1 km/s, whereas the shape of the helium cone would favor a value of 27 km/s. A Doppler shift of the solar helium line /1/ can reconcile both results. A review of the different ways of deriving parameters from conventional models and of their shortcomings is presented, including a detailed presentation of the Chassefiere et al /1/ solution. Particular attention is paid to the remaining unsolved questions, mainly unexplained features of the Ly-α glow which deserve further attention. They result from the absence of two types of model refinements: - a more precise description of the physics of the interaction of neutrals with the sun, including radiative transfer of scattered solar photons, solar line shape effects, collisions etc.. - a detailed modeling of the density and velocity distribution of the neutrals having entered the heliosphere. One of the main problems is to identify the two types of effects.
Some recent observations (Pioneer-Venus, Voyager) are discussed, as well as their possible implications.

GENERAL CHARACTERISTICS OF THE OBSERVATIONS

Up to now, the study of the resonance glow of the interstellar H and He approaching the sun has been the major source of information on the interaction between the sun and the local interstellar cloud it is moving through.

A) Observations from the inner solar system

An important series of intensity maps of H and He glows has been obtained (for detailed information on the basic methods and results the reader is invited to refer to reviews by Bertaux, 1984 /2/, Ajello et al, 1987 /3/). They allow us to derive the main characteristics of the interaction: the flow direction, the bulk velocity and the temperature of the gas before close interaction with the sun. The emission pattern is a function of these three parameters, and of solar parameters. For H, the main parameters are the lifetime against ionization τ (polar and equatorial values), and the ratio of radiation pressure to gravitation (μ). For He, the main parameters are the lifetime against ionization τ, and the solar line-width $\Delta\lambda$ (plus the line-shift $\delta\lambda$, see part 2). However these results deserve some remarks and criticisms:

- The sun-neutral interstellar gas interaction occurs at two levels: around the heliospheric boundary at an unknown distance (50-500 AU?) as a result of coupling with the interstellar and solar plasmas (region A of fig1), and later on, closer to the sun as a result of intense ionization and dynamical processes (region B). Observations made from the inner solar system are concerned with a glow from a "small "region (90% of the glow comes from a distance smaller or much smaller than 50 AU) around the sun and thus mainly within region B. "Classical" models assume that there is an initial simple neutral distribution emerging from A (the "intermediate" state) and reaching B, providing a set of corresponding "intermediate" parameters for this distribution as well as solar interaction parameters. These "classical" models do not provide, of course, the characteristics of the "initial" unperturbed interstellar gas before reaching A. We do not know yet to what extent 1) the "intermediate" parameters differ from the "initial" ones 2) the assumption of a homogeneous flow from A affects the determination of the "intermediate" parameters. A full modeling of the evolution of neutrals (A and B) requires one to solve self-consistently for the plasma-plasma interaction, as well as the neutral-plasma coupling, in addition to the interaction with the sun. It is a huge work (Fahr and Ripken, 1984 /4/, Bleszinski, 1987 /5/, Suess et al, 1987 /6/, Gangopadhyay and Judge, 1989 /7/, Baranov, 1986 /8/ and 1989 /9/) which has not yet reached the point where a comparison with glow observations is possible.
- The whole emissivity pattern of both H and He (an emissive region on the upwind side for H, and on the downwind side for He, as illustrated in Fig 2 and 3) can be reproduced by several combinations of the different parameters quoted above (V,T,μ,τ for H and V,T,$\Delta\lambda$,$\delta\lambda$,τ for He). That is the reason why there is a crucial need for not only photometric observations, but other diagnostics such as those provided by spectrometry or by in-situ detection of neutrals or pick-up ions. This is well known for H and illustrated in Figs 4 and 5, but this is also true for He since Chassefière et al /1/ found a way to reconcile the H and He observations by invoking a shift $\delta\lambda$ of the solar helium line (see part 2), while simultaneously showing that there is a whole class of solutions to reproduce the He glow. Certainly He spectrometry or direct detection of particles (e.g. the neutral helium experiment on board

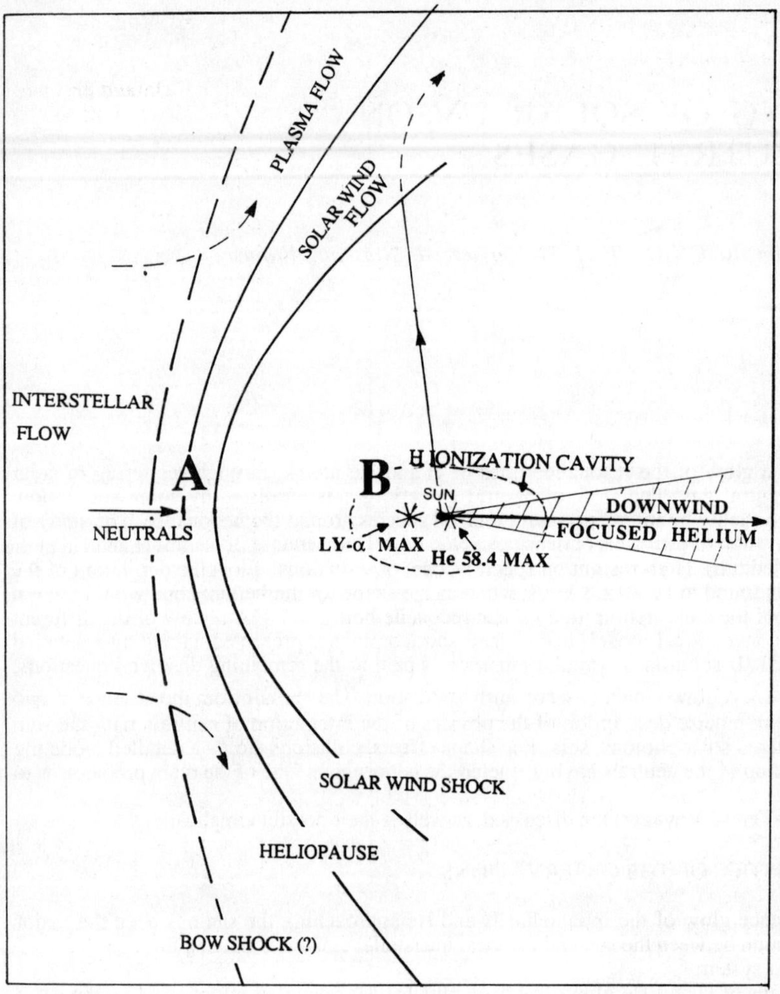

Fig 1: Schematic view of interstellar neutral helium and hydrogen within the heliosphere. The interaction with the sun occurs at two levels, at the heliospheric interface by coupling with the plasma (A) and closer to the sun by ionization and gravitation mainly (B). H ionization is so strong that atoms are prevented from approaching the sun, while helium is focused in the downwind direction with a density peak very close to the sun (at ≈.1 AU). H and He glows are characterized by most emissive regions upwind and downwind, respectively. Conventional models of the glows assume a homogeneous density and velocity distribution far from the sun issued from region A.

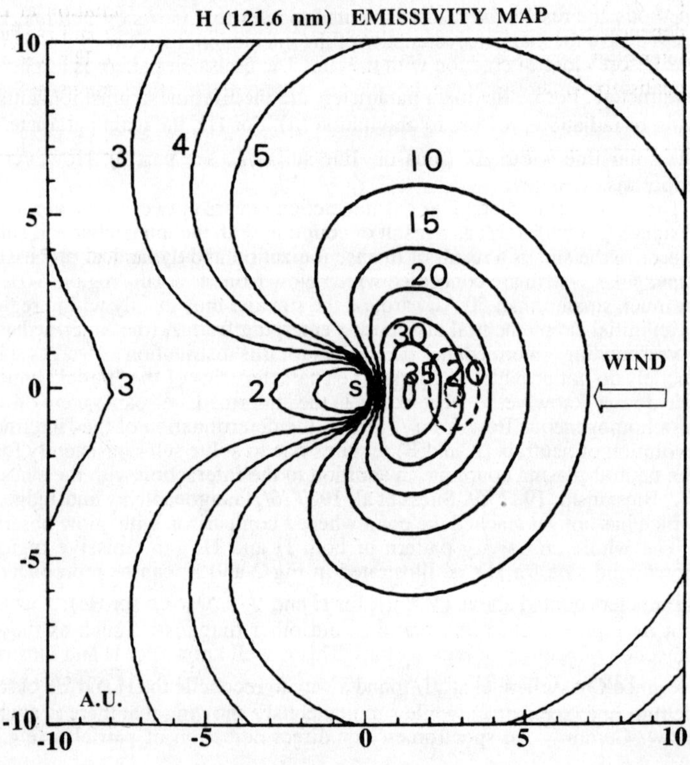

Fig 2: Model of Ly-α emissivity map (arbitrary units): the Maximum Emission Region (MER) is located at 1.5 -3 AU on the upwind axis. When the ionization rate is increased by about a factor of 1.5 (or the lifetime τ decreased) the MER moves away from the sun (dashed lines) and the maximum emissivity decreases.

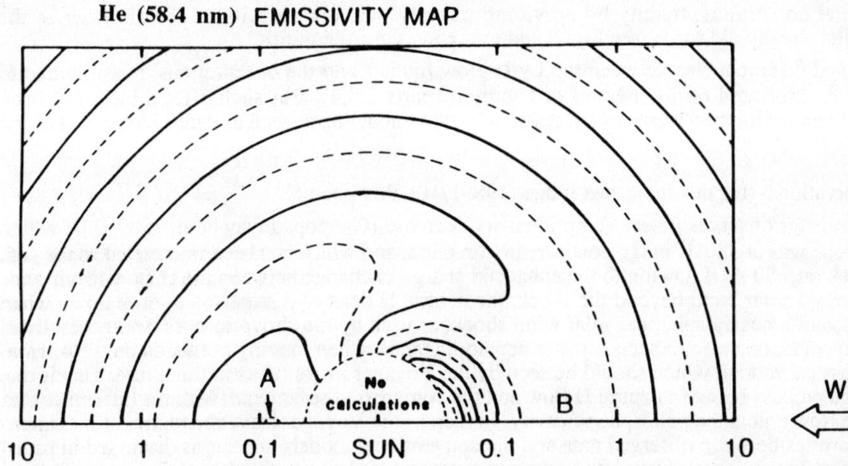

Fig 3: He 58.4 nm iso-emissivity contours (from Dalaudier et al /16/) in a half plane (arbitrary units) . A log scale on the distance to the sun is used and there is a factor of ten between two adjacent solid lines or dashed lines. The absolute maximum is at point A.

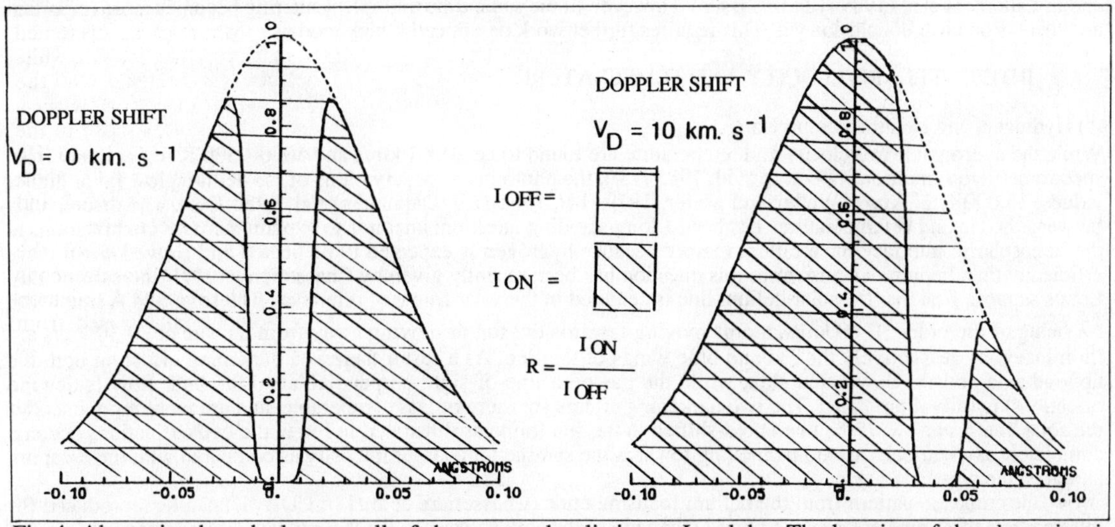

Fig 4: Absorption by a hydrogen cell of the scattered radiation at Ly-alpha. The location of the absorption is depending on the relative motion V_D between the observer and the flowing gas. I_{ON} and I_{OFF} are the mesured intensities, when the cell is activated or inactive, respectively. The reduction factor is the ratio between the remaining unabsorbed intensity when the cell is activated and the total intensity $R = I_{ON} / I_{OFF}$.

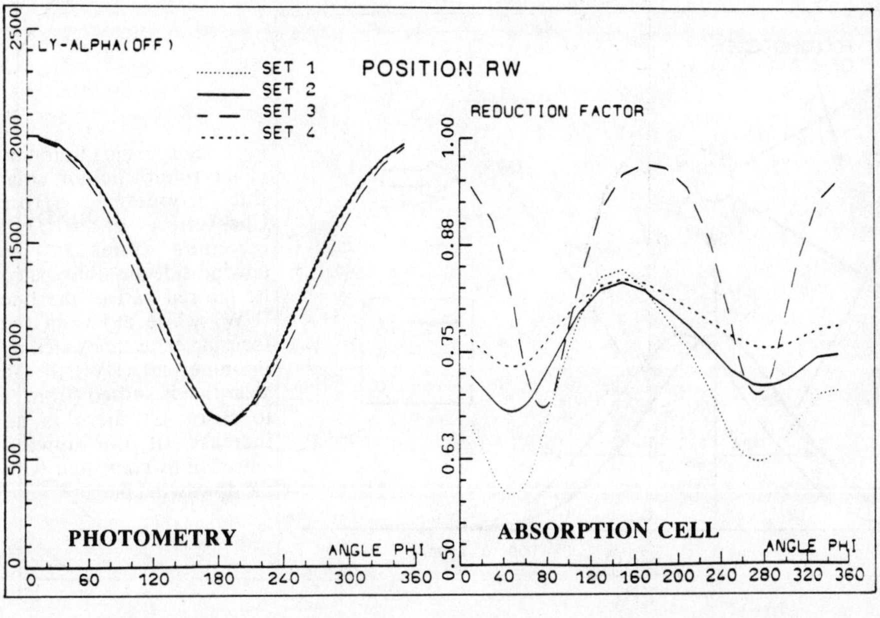

Fig 5: Illustration of the discriminant role of the H cell (from Lallement et al /20/). For four different sets of parameters the glow pattern is nearly the same, while there is a large differentiation of the reduction factor curves. Phi is the line-of-sight rotation angle in a scan plane perpendicular both to the ecliptic and the sun-earth line, when this line is perpendicular to the flow.

Set 1: T= 5000K, μ= .73, V= 20 km/s

Set 2: T= 10000K, μ= .85 , V= 20 km/s

Set 3: T= 10000K, μ= 1.17, V= 10 km/s

Set 4: T= 15000K, μ= 1.0 , V= 20 km/s

Ulysses) would bring tighter constraints, mainly by providing directly the bulk velocity of the He flow as the Prognoz hydrogen cell results already did for H (see Fig 2) and removing the ambiguities.

- There are some unexplained discrepancies between the Ly-α glow models and the observations, possibly due to perturbations from region A imprinted on the neutral distribution (parts 2,3,4). But such effects have not been carefully examined yet, and require improvements of the resonance glow modeling as well as careful treatment of the multiple scattering effects.

B) Outer solar system observations (for this topic, see Judge (1989) /10/, this issue).

The Pioneer-10 Ly-α photometer observes a very sharp intensity decrease (Gangopadhyay et al, 1989 /11/) which started in 1986 when the probe was at 38 AU in the downstream direction, and which has been interpreted as the sign of a nearby solar wind shock (at ≈50 AU). Owing to the enhanced charge-exchange between the H flow leaving the heliosphere and the compressed solar wind beyond the shock, the neutral H density is expected to drop down when moving outwards. In the case of a nearly-spherical solar wind shock, as used by the above authors, and generally in all the heliospheric shock models, one also expects a faster density increase when moving outwards on the upwind side as compared with a classical model, which should be seen by the Voyager instruments in the future. This is due to the same charge-exchange process between neutral H flow and the compressed solar wind, with the difference that on the upwind side now the flow enters the heliospheric cavity. In fact, as the Voyager leave the region of maximum H and He glow, some departures between observed data and the conventional models appear, as discussed in part 4 and in Lallement et al (1989 /12/, this issue). However, at the same time multiple scattering becomes more important and there is no clear conclusion yet. This requires further work on optically thick models.

INTERSTELLAR VELOCITY AND TEMPERATURE

A) Hydrogen and Helium comparison

While the hydrogen flow velocity and temperature are found to be 20 ± 1 km/s and 8000 ± 500 K respectively from spectrometric observations (Bertaux et al, 1985 /13/), the photometric observations of the helium glow favor higher values: 15,000K, 27 km/s (Weller and Meier, 1979 /14/, 1981 /15/, Dalaudier et al, 1984 /16/). The discrepancy between the He and H temperatures has been a long-standing question: attempts to explain them by modifications at the heliospheric interface have failed, mainly because hydrogen is expected to be heated and slowed down more efficiently than helium. An answer to this question has been recently given by Chassefière et al /1/, hereafter CDB. Let us suppose first that the solar helium line is centered in the solar frame of reference. The solar 584 Å line width $\Delta\lambda$ being of the order 30-60 km/s, atoms moving towards the sun or moving away from the sun at 20-30 km/s are illuminated by respectively the red and blue wings of the line. As a result there is a darkening effect on both the upwind side and the downwind side, while the gas on a line-of-sight perpendicular to the bulk flow (sidewind directions) is fully illuminated. The two darkening effects are more or less comparable, and are more pronounced if the solar line is narrow. If the line is red-shifted in the sun frame as illustrated in Fig 6, the two extinctions become completely asymetrical: fast atoms moving towards the sun scatter a larger intensity as compared with the escaping atoms on the downwind side.

Now, the emission pattern from the helium focusing cone (Chassefiere et al /17/, CDB /1/) can be reproduced for different sets of parameters, provided that 1) the ratio of the bulk velocity to the thermal velocity is maintained around a fixed value, because this ratio governs the filling of the cone, and 2) a relationship between the solar line-width and

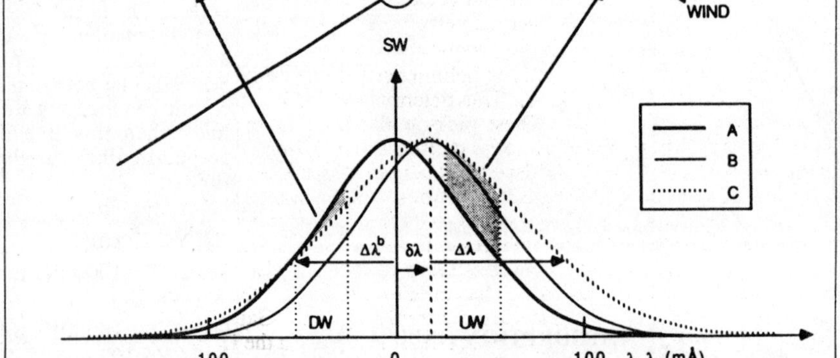

Fig 6 : Schematic illustration of the role of helium solar line asymetries (from Chassefière et al /1/): incoming atoms in the upwind side are illuminated by the red part of the line (UW) while atoms in the focusing cone are excited by the blue part (DW). If the solar line is shifted (from A to B or C) there is an increase of the upwind emission as compared with the downwind one.

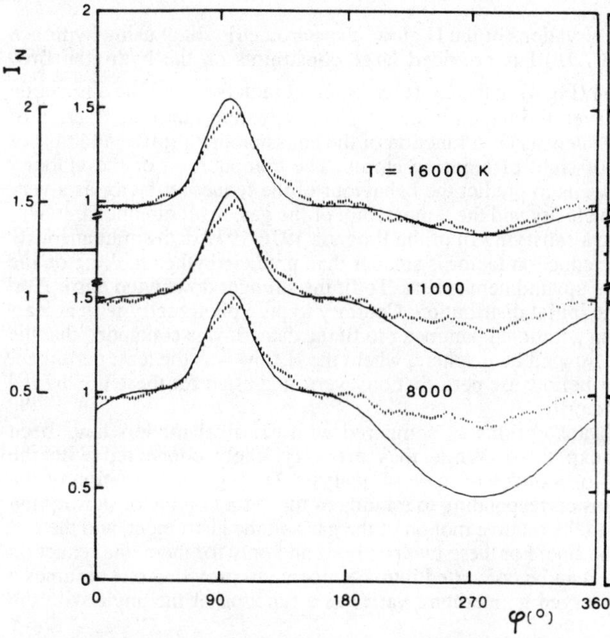

Fig 7: Observed and modeled He58.4 nm intensities (from Chassefière et al /1/). ϕ is the rotation angle in a scan plane perpendicular to the ecliptic and to the sun-earth line (earth longitude 8 °). It is possible to fit the helium cone with a temperature of 8000K, but in this case the upwind intensity (around 270°) is larger than the model predicts.

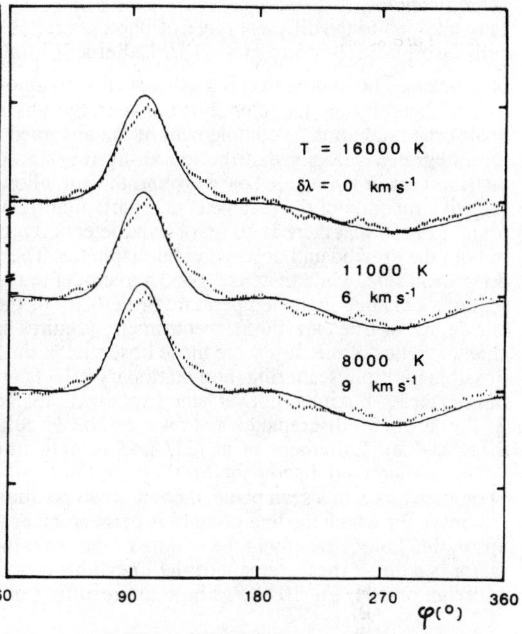

Fig 8 : Same as Fig 7 when introducing a redshift of the solar line (from Chassefière et al, /1/). In this case, the whole intensity pattern can be fitted with the parameters which are found for hydrogen (8000K, 21km/s).

the temperature is also maintained. The second requirement follows from the following reasons. The ratio between the sidewind density (close to the density at "infinity") and the downwind density (inside the dense cone) is a decreasing function of the temperature T only (Feldman et al /18/). Consequently, when the line-of-sight moves off from the sidewind direction to the cone axis, the corresponding intensity increase results from combination of both the density increase (and then a decreasing function of T) and the previously described darkening effect (which is a decreasing function of $\Delta\lambda$). What happens is that if T is as small as 8000K, the density enhancement in the cone is so large that to fit the observations of the cone requires a strong darkening effect or a small $\Delta\lambda$. However in this case the model predicts a very weak illumination on the upwind side too, which is not observed. This is illustrated in Fig7 which shows three different models for three velocity-temperature pairs. In each case, the solar line has been choosen to give the best data-model fit in the downwind region. Clearly a large temperature is required to fit the cone and the upwind side simultaneously. If a small shift of the solar line is introduced (favoured by theory and observations as discussed by CDB), then the upwind darkening is reduced (while the downwind darkening is increased) resulting in a larger upwind modeled intensity. In this case the temperature can be decreased. CDB have shown that different combinations of shift, line-width, velocity and temperature can provide an equivalent good fit. A temperature of 8,000 K and a velocity of 21 km/s are among the solutions, that is to say the hydrogen and helium glow can be fitted with the same parameters (Fig 8).

There are at least three arguments for equality of temperature and velocity of H and He. 1) The shape of the helium cone is sensitive to the ratio of the bulk velocity over the thermal velocity. By coïncidence this is approximately the same for the two sets of parameters: 20 km/s and 8,000 K for the measured hydrogen parameters, and 27 km/s and 15,000 K for the previously inferred helium parameters. 2) The helium temperature derived from He^+ pick-up ions distribution is 6,000K - 13,000K (Möbius /19/, this issue). This determination is compatible with the hydrogen temperature of 8,000 K, but incompatible with 15,000K. These pick-up ions result from charge exchange between neutral interstellar He and solar protons. They are found to be denser in the helium cone and their distribution depends on the initial He distribution.This provides an independant way to measure the helium parameters. 3) A small shift of the helium line is enough to reconcile the H and He observations, as mentioned above. This is of course a very important result. As a matter of fact it implies a very weak differentiation between hydrogen and helium at the heliospheric interface (assuming the hydrogen and helium flows have the same "initial" state before entering the heliosphere which can hardly be questioned) as well as large constraints on the nature of the interface. However, in view of the important role of the solar helium line shape which has been revealed by CDB, it should be noticed that, while the linewidth and the shift are so crucial, perhaps departures from a symetric Gaussian (as assumed by the model) could also play a large role and should be taken into account in future work on the He glow. Maybe the solar helium line structure will become a by-product of the interstellar gas study if good constraints on the helium flow come from future measurements, the interplanetary helium serving as a gigantic spectrometer ?

B) Unexplained characteristics of the H Ly-α line profiles possibly affecting the velocity and temperature determinations.

1) In addition to the different types of photometric observations of the H glow, a spectrometric study using hydrogen cell absorption (Bertaux et al /13/, Lallement et al /20/) has provided large constraints on the hydrogen flow parameters. The hydrogen cell produces an absorption (Fig 4) in the Ly-α background radiation, and the absorption is modulated by the Doppler shift between the observer and the emitting gas. The reduction factor or fraction of unabsorbed radiation (= complement of the absorbed intensity) is a function of the emission line profile, and thus of the integrated velocity distribution along this line-of-sight of emitting atoms. The computation of the velocity distributions in the whole space around the sun allows us to predict the behaviour of the reduction factor as a very sensitive function of the parameters, in particular the velocity and the temperature of the gas, in all directions. It was found (Fig 9) that there is no set of parameters giving a satisfying fit of the Prognoz 1976-1977 data simultaneously in both the upwind and downwind hemispheres. The reduction factor is smaller than predicted when looking on the downwind side, while there is a good agreement in the upwind hemisphere. To fit the stronger downwind absorption requires a change in the temperature or velocity of the initial distribution. Contrary to pure photometry with at least one degree of freedom (Fig5), spectrometry requires an additional parameter to fit the data. It was concluded that the characteristics of the flow were those observed in the upwind hemisphere, where the H flow was the least perturbed. Possible multiple scattering, non-stationary effects or heliopause perturbations were suggested for these downwind discrepancies, but it has not yet been explained satisfactorily.

2) Some other discrepancies shown by the H cell absorptions as compared with classical models have been discussed by Lallement et al /21/ and remain unexplained. While they are very likely connected with the upwind/downwind discrepancies, they are the result of a different type of analysis. During a full rotation of the photometer axis in a scan plane, there are two positions corresponding to minima of the reduction factor (absorption maxima), for which the line-of-sight is perpendicular to the relative motion of the gas and the instrument, and the cell absorption hole is exactly in the middle of the emission line. For these two positions and only for them, the reduction factor is a direct measurement of the linewidth which can be converted into a temperature provided one assumes a gaussian profile. Fig 10 shows how the resulting observed temperature varies as a function of the angle with the

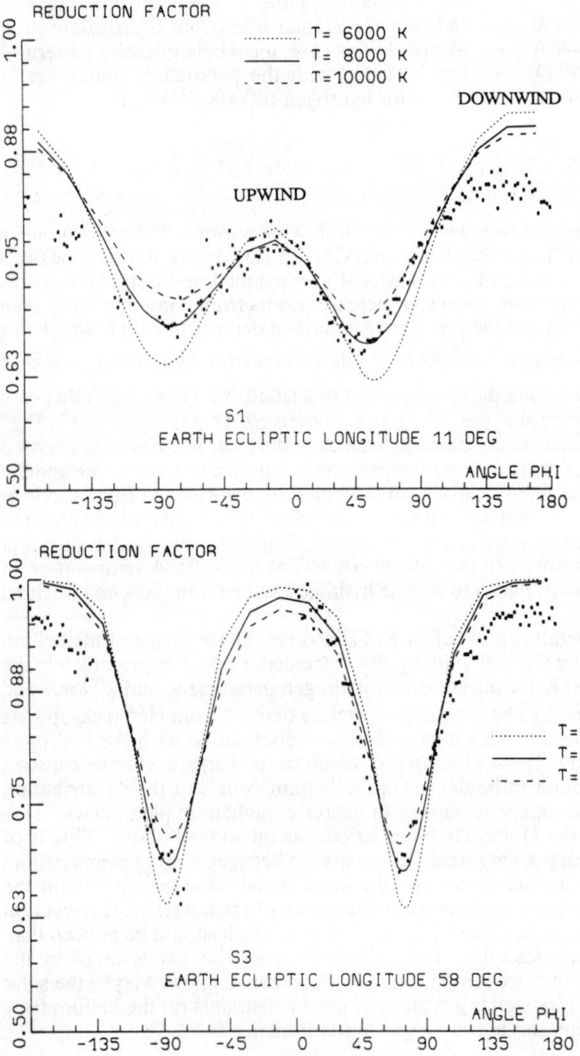

Fig 9: Hydrogen cell data and models (from Bertaux et al /13/) for two different scan planes perpendicular to the ecliptic. The Prognoz data could not be fitted with one set of parameters in both the upwind and downwind side. When the model is adjusted in the upwind hemisphere (-90,90°), the observed absorption is larger than predicted by the model in the downwind region (-180,-90°; 90°,180°).

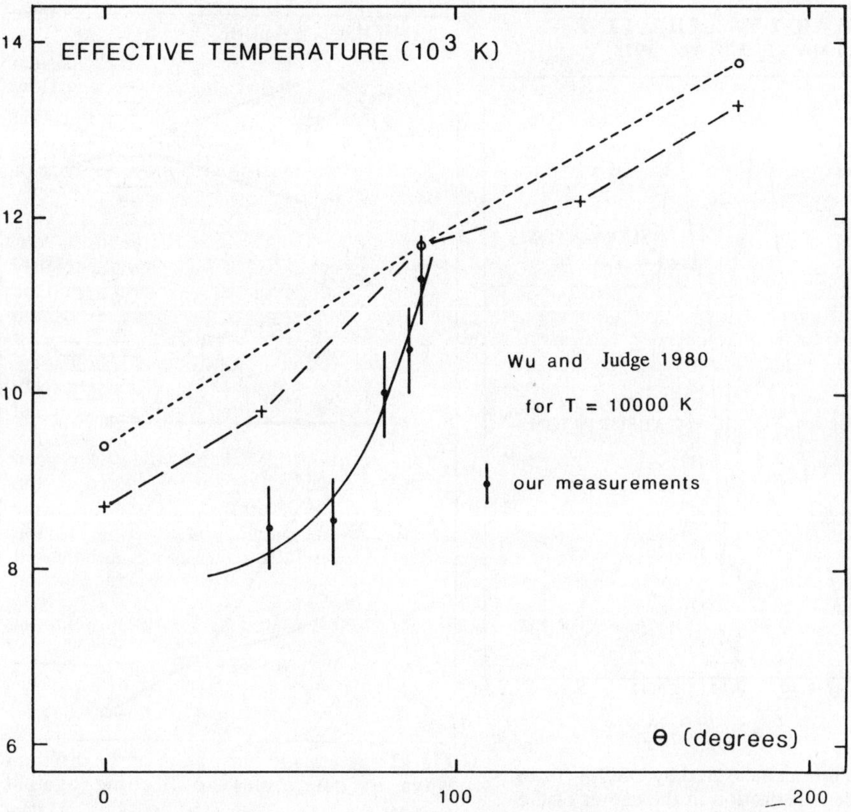

Fig 10: Observed and modeled effective temperatures derived from the cell absorption (from Lallement et al /21/).The maximum absorption by the cell occurs when there is no Doppler shift (absorption in the center of the emitted line) and provides a direct measurement of the line-width (or effective temperature, assuming a gaussian line). θ is the angle with the upwind axis. The observed temperature gradient is larger than predicted by models from Wu and Judge / 33/ around 90°.

upwind axis θ (in an interval of about 45 °). Both the model and the data show an increase of the linewidth when approaching the downwind direction as due to dynamical effects and differential extinction of atoms (faster atoms are less ionized than slower atoms). However the data show a much stronger gradient around 90 degrees than predicted by the model. One of the suggested possible explanations is a linewidth increase by multiple-scattering, but it seems difficult to imagine how such a sharp gradient can be the consequence of a "smooth" effect such as the evolution of the radiative transfer inside the cavity. It remains that most likely we do not understand some physical processes becoming important in the downwind cavity.

C) Direct observations of the emission line

Up to now two direct measurements of the Ly-α background spectrum have been performed: the first observation was done with Copernicus by Adams and Frisch /22/ and the observed velocity was found to be 22 ± 3 km/s. The second measurement (Clarke et al /23/ led to a larger value : 25.5 ± 5 km/s and was performed with IUE. The error bars reflect the resolution of respectively 15 and 40 km/s of the spectrometers, while the IUE instrument has a higher sensitivity. The common interval is 20.5-25 km/s and is (marginally) compatible with the results of the H-cell analysis (20±1 km/s). Measurements with a higher resolution are required to remove the remaining uncertainties, as could perhaps be obtained with the Hubble Space Telescope.

SOLAR PARAMETERS DETERMINATION

1) General characteristics

We are concerned here with the solar ionization processes and the solar lines for both H and He, and with the radiation pressure for H (negligible for He). In principle all these solar parameters are measured independently of interplanetary glow observations. However owing to their temporal variations and the significant uncertainties of the measurements, they have to be considered as free parameters of the models. They all have an influence on the upwind to downwind intensity ratio.

The ionization processes govern the extent and the shape of the hydrogen cavity, while their effects are smaller for helium. On the contrary, as already discussed in part 2, the helium line shape (Fig 6) is of primary importance as far as the helium glow is concerned, while the hydrogen line shape has a smaller influence (fig 11). Neutral H is illuminated by the central part of the self reversal of the Ly-alpha line and there could be some small effects as 1) a weakened intensity from the sidewind regions (zero Doppler shift), or, 2) a small increase of the downwind intensity if the central core is asymetrical as is illustrated in Fig 11.

While being negligible for helium, the repulsive force due to radiation pressure has a large effect on the hydrogen distribution since it prevents the filling of the cavity. As we already said, the role of the ionization, of the radiation pressure and of the velocity and temperature are difficult to disentangle with pure photometry of the hydrogen glow since they all influence the upwind/downwind density contrast. The radiation pressure, however, can be derived from spectrometry.

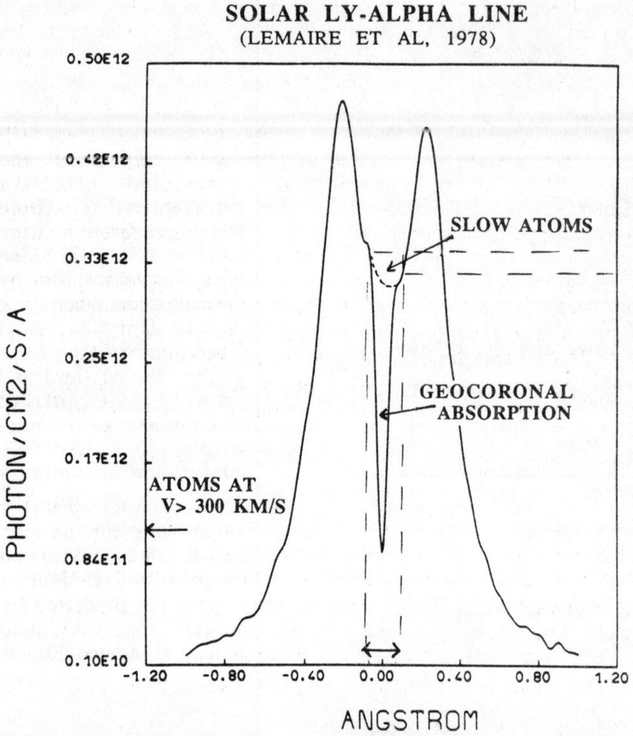

Fig 11 : The solar Ly-alpha line as measured by OSO-8 (from Lemaire et al /34 /).The deep absorption in the center of the line is of terrestrial origin. Inflowing interstellar neutrals are illuminated by the central core. An asymetry of this core could slightly modify the upwind and downwind intensity ratios. Fast atoms at 400 km/s resulting from charge-exchange between neutral H and solar protons are very weakly illuminated by the far blue wing of the line.

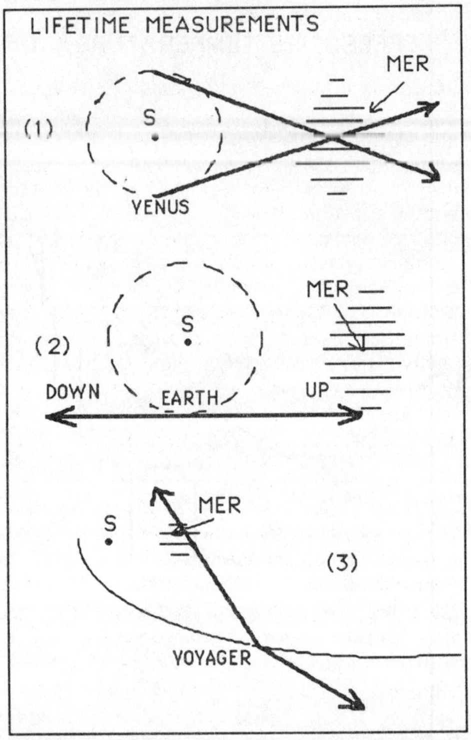

Fig 12 : Schematic illustration of the different ways to determine the lifetime against ionization τ of the interstellar H atoms. 1) The parallax effect, since the location of the MER depends on τ 2)The upwind/downwind intensity ratio, which depends primarily on τ 3) The ratio between the maximum emission region intensity (depending on τ) and the antisolar directions (independant on τ for an observer far from the sun).

2) Uncertainties on the determination of the hydrogen ionization rate

In fig 2 is represented a typical emissivity map of H Lyman-α glow. It is characterized by an emissive region at 2-3 AU on the upwind side, the maximum emission region or MER, where H atoms having escaped ionization can approach closest to the sun. As is illustrated in fig 2, when the ionization becomes stronger, the MER moves away from the sun as a result of a stronger density gradient and simultaneously its emissivity is weakened (due to the r^{-2} illumination and the smaller density everywhere around the sun) as compared with the emission from the outer regions where the density remains nearly constant. Schematically there are three different ways to measure the ionization rate (or the lifetime) as illustrated in Fig12. a) The distance to the sun of the MER or the parallax effect (Thomas /24/, Bertaux and Blamont /25/) since this region is viewed in different directions as a function of the position of the observation in the solar system. This is probably the most reliable method, since shortcomings of the modeling as ignorance of multiple scattering effects have no influence on it. The Voyager trajectories are particularly suitable for such a measurement when the spacecraft are at a few AU at right angle with the wind direction and then have the best direct view on this region. b) The upwind/downwind intensity ratio is the most sensitive function of the ionization rate. However this ratio is also a function of V, T and μ, and can be affected by non-stationary or multiple-scattering effects. c) The ratio between the emission from the MER and the emission from external regions, where the density is nearly identical to the density "at infinity" is also very sensitive to the ionization rate (Fig 2). When looking from the inner solar system it is not possible to estimate this ratio, since the largest parts of the line-of-sights are in the depleted region. However this ratio becomes a better diagnostic when moving away from the sun. The gas the Voyager spacecraft travel through at more than 15 AU on the upwind side has a density which differs from the unperturbed "intermediate" density by at most 10-15 %. The Ly-α emission from the antisolar direction is then almost independent of all the solar and interstellar parameters other than the interstellar density and the solar flux. The ratio between this emission and the MER emission is a direct measurement of the depletion all around the sun and depends almost entirely on the lifetime (and slightly on μ).

What we would like to point out is that these three methods seem to give different results on the ionization rate, and in a systematic way. The Prognoz measurements do not contain the MER region but provided a good upwind/downwind ratio: the derived lifetime (equatorial value) is of the order of 2.10^6 s. (at 1AU), which was considered a large value when compared with expectations from the in-situ solar wind measurements (Lallement et al /26/). Pioneer-Venus observations used by Lallement and Stewart /27/ to study solar wind anisotropies are characterized by a good coverage of the MER region, and of the upwind and downwind directions . The derived lifetime is 1.5 10^6 s. In the study of the whole set of Pioneer-Venus ecliptic measurements, Ajello et al /3/ show a series of comparisons between the data and models with different lifetimes. Examples are shown in fig 13. It can be noticed that for a small lifetime of 1.2 10^6s, the MER (UPWIND) is well fitted, or the parallax effect well represented. However, the downwind intensity is underestimated by the model and the downwind region shows discrepancies from the model. For a large lifetime, the downwind cavity is well fitted, but the predicted parallax is not the good one. It seems that there is a contradiction between the location of the MER and the upwind/downwind contrast. Besides, these authors already multiply the downwind intensity by a factor of about 1.3 according to multiple scattering simulations of Keller et al /28/. If one admits that the parallax effect is the most valid way to derive the lifetime, then it would mean that some effect (multiple scattering or other) increases the downwind emission by more than 30%. Now the first analysis of Voyager maps taken at large distances from the sun are also in favor of a small lifetime of the order of 1.- 1.2 10^6 s (Lallement et al /12/) while showing that at the same time the downwind cavity is more emissive than expected from the model. From these three sets of data, the general conclusion is that the lifetime is probably as small as 1.2 10^6 s , and that we should be careful in using the downwind/upwind ratio of the intensity maps for its measurement, in the absence of a reliable way to estimate the multiple scattering contribution and of a full understanding of the maps.

3) Solar wind anisotropies

If the ionization by solar wind and solar UV had a spherical symmetry, He and H distributions should be axisymmetric around the sun-wind axis. While departures from the axisymmetry have not been detected in the helium distribution, the hydrogen glow shows large enhancements at medium and high latitudes (Broadfoot and Kumar /29/). The main ionization process for H is charge-exchange with the solar protons: $H_{slow} + H^+_{fast} \rightarrow H_{fast} + H^+_{slow}$. The newly created neutrals contribute for a negligible part to the glow for the two following reasons: 1) for an equivalent flux their density would be about 20 times smaller than the density of the slow neutrals (inverse of the velocity ratio 400 km/s / 20 km/s), and 2) they are illuminated by the far blue wing of the solar Ly-α line (fig 11). Then the charge-exchange process corresponds to a disappearance of the neutrals. The charge exchange rate ν is the product of the proton flux and the charge-exchange cross-section: $\nu\ (sec^{-1}) = \sigma(V_p) \times N_p \times V_p$. Observations can

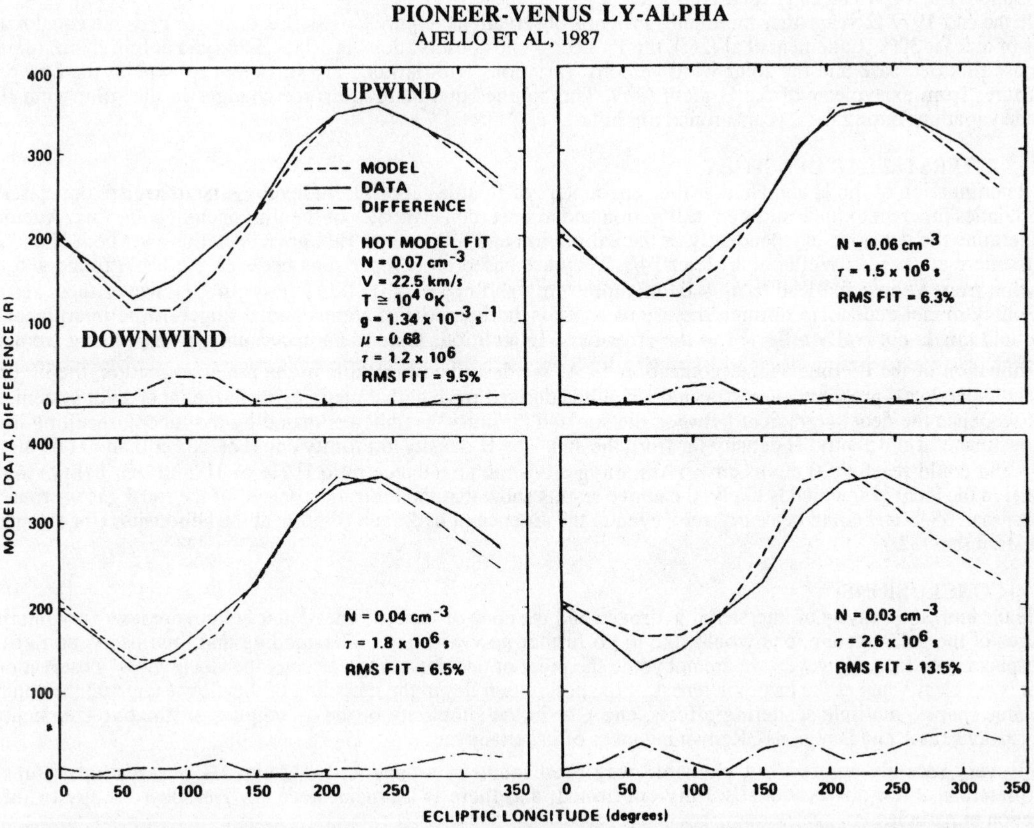

Fig 13 : Pioneer-Venus data and models (from Ajello et al /3/). For a small lifetime, the MER is well fitted and the parallax effect well represented, while the downwind cavity not.
For a large lifetime, the downwind side is well fitted (and the rms smaller), while the observed parallax shift is different from what is predicted.

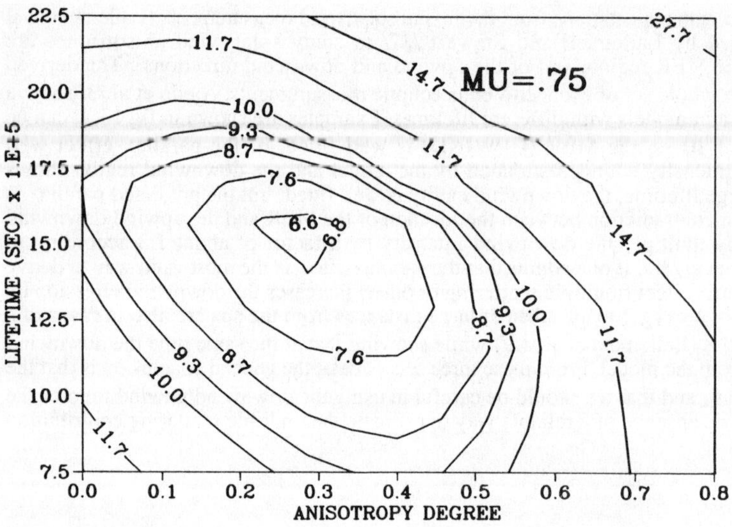

Fig 14: Pioneer-Venus out-of-ecliptic observations (from Lallement and Stewart /27/) , Curves are iso-RMS contours when varying two model parameters, the lifetime in the solar equatorial plane and the ionization anisotropy degree (or decrease of the ionization rate between the solar equator and the poles). Whatever is the value of μ (here .75), the best fit corresponds to 30% decrease of ionization with the heliographic latitude.

be very well reproduced by assuming that ν is a decreasing function of the heliographic latitude . As an example, 1985-6 Pioneer-Venus data are best fitted with a model assuming 30% decrease of the ionization rate as shown in Fig. 14 /27/. A fraction of this decrease is due to the cross-section decrease with the relative velocity ($\approx$ the proton velocity), since it is well established now that the high latitude solar wind is faster than the equatorial one around solar minimum /30/. A second part of this effect is attributed to solar wind mass flux decrease with the heliographic latitude. One expects a solar cycle variation of these anisotropies, the sun being more isotropic around solar maximum. An attempt to detect such an evolution shows that the anisotropy increases during and after solar minimum (Lallement /31/) . There is a time-delay of 1-2 years (depending on the location: the reaction is faster upwind where the ionization is strong and slower downwind) between the solar wind changes and the H density distribution reaction which can explain this tendency.

While the end-1977 (2 years after minimum) Prognoz observations require a mass flux decrease between equator and poles of at least 30% (Lallement et al /26/), the Pioneer-Venus observations in 1985-1986 (just at minimum) require no mass flux decrease and the solar wind velocity variations with latitude are sufficient to explain the observed departures from axisymetry of the H glow /27 /. This implies that there are some changes in the solar wind flux latitude variations during the 2 years around minimum.

INTERSTELLAR DENSITIES

The determination of the H and He densities encounters difficulties at different levels. First, there are the classical uncertainties inherent to the instrument calibration and to uncertain estimates of the illuminating solar flux. Attempts to determine the densities independently of the calibration and of the solar flux have been done for both H and He (Chassefiere et al /32/ , Weller and Meier/14/). They are based on comparisons between the background and the emission from Venus, and lead to $n_H \approx 0.065$ atoms cm^{-3} and $n_{He} \approx .010 \pm .003$ atoms cm^{-3}. However these results are doubly model-dependent through the use of an exospheric model of Venus and a model of the interplanetary flow, and this is not really reflected in the error bars. In addition there is an uncertainty on n_H arising from the determination of the lifetime against ionization τ. As we described in details in the previous section, the lifetime against ionization is probably overestimated when it is derived from global mapping from the inner solar system. As a consequence the density gradient between the sun and "infinity" is underestimated by the model, resulting in an underestimate of the neutral H density far from the sun. The H density at infinity could be larger than 0.065 atoms cm^{-3}, and could reach 0.10 atoms.cm $^{-3}$. Assuming a cosmic abundance ratio H/He of 10 and that helium is not ionized in the local gas, which is likely, the above results show that the ionization degree of the local gas is probably smaller than 35 % and could be nearly zero, even in the absence of hydrogen filtering at the heliopause (for extensive discussion see /32/).

CONCLUSIONS

While the ionization cavity of interstellar hydrogen and the cone of focused interstellar helium are now very familiar features of the solar system, one would like to go further now with the understanding and the observation of the heliospheric interface. However, we are not yet to the point of understanding perfectly the "local glow" observations, and it is not clear whether the incompleteness of the description lies in the modeling of the glow (uncertainties on the solar line shapes, multiple scattering effects, etc..), or in the simplicity of the assumption of the two-step neutral perturbations (in A and B regions). Remaining areas of uncertainty are:

- Only very recently has found a plausible way been found to reconcile the H and He observations. But this interpretation is not observationally fully confirmed, and there is a crucial need for He spectroscopy or direct detection of particles.
- There are still large uncertainties on the H and He densities.
- The knowledge of the H and He differentiation processes arising from coupling with the plasma at the heliospheric interface would be an important contribution to the determination of the interface properties. This knowledge requires a better accuracy on the H and He flow parameters (including densities), and then new observations and analysis.

- There are still unexplained properties of the observed H glow. They could be a sign of departures from a simple Boltzman distribution of the inflowing gas, and contain information on the interstellar plasma.
- Improvement of the modeling may reveal possible "signs" of the presence of a heliospheric shock on the upwind side. If there is a compressed and heated solar wind behind a "nearby" heliospheric shock, as predicted by some models, and possibly observed by Pioneer 10, then one of the first signs on the upwind side (before entering the subsonic wind) could be the detection of a larger Ly-α intensity as compared with conventional models by the Voyager UVS spectrometer.

References

1. Chassefière E., Dalaudier F., Bertaux J.L. :1988, Astron. Astrophys. **174**, 239
2. Bertaux J.L., 1984: IAU Colloquium 81 NASA Publ. CP2345, 3
3. Ajello J. M., Stewart A. I., Thomas G. E., Graps A.: 1987, Astrophys. J., **317**, 964
4. Fahr H.J., Ripken H.W.: 1984, Astron. Astrophys. **139**, 551.
5. Bleszinski S., Astron. Astrophys. , 1987, **180** , 201
6. Suess S.T., Hathaway D.H., Dessler A.J.: 1987, Geophys. Res. Let, **14-9**, 977.
7. Gangopadhyay P., Judge, D.L.: 1989, Astrophys. J., **336**, 999.
8. Baranov,V.B.: 1986, Adv. Space Res., **6** ,5
9. Baranov,V.B.: 1989, this issue.
10. Judge D.L.: 1989, this issue.
11. Gangopadhyay P., Ogawa H.S., Judge, D.L.: 1989, Astrophys. J., **336**, 1012
12. Lallement R., Bertaux J.L., Chassefiere E., Sandel B.R.: 1989, this issue
13. Bertaux J.L., Lallement R., Kurt V.G., Mironova E.N.:1985, Astron. Astrophys. **150**,1.
14. Weller C. S., Meier R. R., 1979, Astrophys. J., **227** , 816
15. Weller C. S., Meier R. R., 1981 Astrophys. J., **246**, 386
16. Dalaudier F., Bertaux J. L., Kurt V. G., Mironova E. N.: 1984, Astron. Astrophys. **134**, 184
17. Chassefière E., Bertaux J.L., Lallement R., Sandel B.R., Broadfoot L.: 1988a, Astron. Astrophys. **199,** 304.
18. Feldman W.C., Lange J.J., Scherb F.: 1972, Solar wind, NASA SP_308, 684
19. Moebius E., 1988, "Cosmic gas dynamics" int. workshop, Moscow, 10-14 Oct 1988.
20. Lallement R., Bertaux J.L., Dalaudier F. , 1985, Astron. Astrophys., **150**, 21
21. Lallement R., Bertaux J.L., Kurt V.G., Mironova E.N.: 1984, Astron. Astrophys., **140**, 243
22. Adams T.F., Frisch P.C.: 1977, Astrophys. J., **212**, 300
23. Clarke J.T., Bowyer S., Fahr H.J., Lay G.: 1985, Astron. Astrophys. **139**, 389
24. Thomas G. E., 1972, Solar wind SP-308, 668 (Sonnett et al, Editors)
25. Bertaux J.L., Blamont J.E., 1971,Astron. Astroph. **11**, 200
26. Lallement R., Bertaux J.L., Kurt V.G., 1985, J. Geophys. Res., **90**, A2, 1413
27. Lallement R., Stewart A.I., 1990, Astron. Astrophys., in press .
28. Keller H.U., Richter K., Thomas G.E.: 1981, Astron. Astrophys. **102**, 415
29. Broadfoot A. L., Kumar S, 1978 Astrophys. J., **222** , 1054
30. Kojima M., Kakinuma T.: 1987, Sol. Terr. Environ. Res. Jpn. (Japan), **11**, 18
31. Lallement R., 1987, in Solar wind 6 , High Altitude Observatory, Boulder , Colorado (edt) , NCAR/TN 306 +Proc, 651.
32. Chassefière E., Bertaux J.L., Lallement R., Kurt V.G.: 1986 , Astron. Astrophys. **160**, 229.
33. Wu F.M., Judge D.L.: 1980, Astrophys. J., **239**, 389
34. Lemaire P., Charra J., Jouchoux A., Vidal-Madjar A., Artzner G., Vial J.C., Bonnet R., Skumanich A., 1978, Astrophys. J., **223**, L55.

MODEL PREDICTIONS AND REMOTE OBSERVATIONS OF THE HYDROGEN DENSITY PROFILE IN THE DISTANT HELIOSPHERE

D.L. Judge,* P. Gangopadhyay* and S. Grzedzielski**

*Department of Physics and Space Sciences Center, University of Southern California, Los Angeles, CA 80089–1341, U.S.A.
**Space Research Center, Polish Academy of Sciences, 01–237 Warszawa, ul. Ordona 21, Poland

ABSTRACT

Observations of the hydrogen Ly-α glow by the deep space ultraviolet photometers and spectrometers on board Pioneer 10 and Voyager 2 have been utilized to infer the presence of a nearby solar wind shock. It is a general feature of the available solar shock models that the inflowing interstellar hydrogen will be depleted by charge exchange scattering as it traverses the heliosphere at its transport rate of about four AU per year. This depletion occurs primarily beyond the solar shock in the hot subsonic plasma. The presence of a solar shock will lead to look direction dependent incorrect estimates of the true local interstellar density. Prior estimates of the interstellar hydrogen density must accordingly be reevaluated. The observational and theoretical basis for our current understanding of the heliospheric neutral density profile is presented.

INTRODUCTION

Since 1972, four deep space probes, Pioneers 10/11 and Voyagers 1/2, have been launched to study the outer planets as well as the medium in which they reside. There are several experiments on board these spacecraft to measure plasma parameters, magnetic field strengths, cosmic ray intensities and neutral particle parameters. The neutral particle measurements are carried out indirectly through ultraviolet photometers (Pioneer 10/11) and spectrometers (Voyager 1/2). These ultraviolet detectors measure the flux of solar Ly-α 1216 Å and He 584 Å photons which are backscattered from the neutral atoms in the heliosphere. The neutral density and temperature are inferred from these intensity measurements.

The most interesting aspect of the neutral measurements is that they can be used to infer the shape, size and location of the heliospheric boundary through their interaction with the ambient proton plasma. A number of plasma models for the heliosphere, with and without a solar shock, have been proposed since Davis /1/ first postulated the the existence of a heliosphere. Several heliospheric models which include solar/or interplanetary shocks have also been proposed /2,3,4,5/. It is in principle possible to distinguish between the shocked and non-shocked models by analyzing the ultraviolet data.

MODEL PREDICTIONS

The spatial dependence of the backscattered interplanetary Ly-α glow depends on the neutral hydrogen distribution within the heliosphere and is critically dependent on the characteristics of the heliospheric interface /6,7,8,9/. In the presence of a shock all of the above authors find a large depletion of the upstream interstellar hydrogen flow as it enters the inner solar system. As a specific example, the neutral density at various angles from the upstream direction, based on a plasma model developed by one of the present authors (S. Grzedzielski), is shown in Fig.1. In this new model a viscid interstellar plasma collides with a supersonic solar wind. The kinematic viscosity of the interstellar plasma is taken high enough to constrain the size of the heliosphere. It is postulated that the high value results from turbulent momentum exchange between adjacent streamlines of the interstellar flow. This turbulence is excited by low energy cosmic rays. The details of this new model will be published in the future. At present this model is available only in the upstream direction.

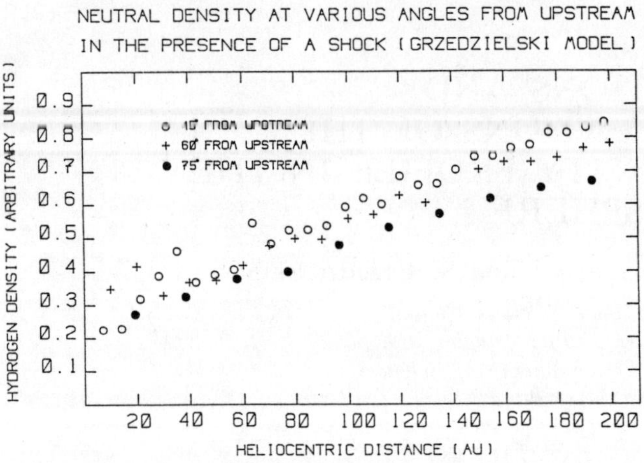

Figure 1. A Monte - Carlo calculation of the neutral hydrogen density using the Grzedzielski model at various direction, from the upstream direction.

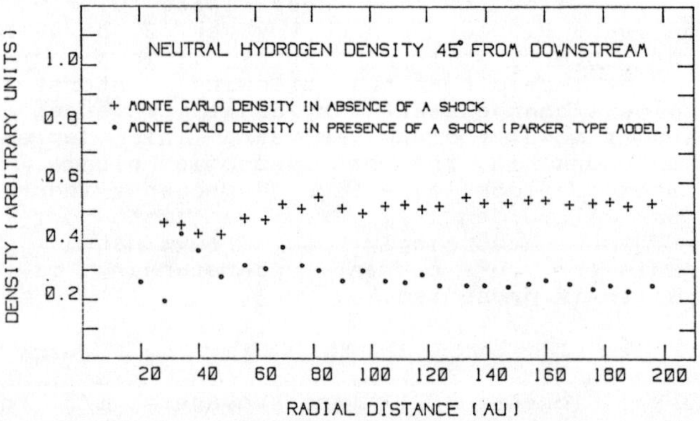

Figure 2. A Monte Carlo calculation of the neutral hydrogen density 45° from downstream in the absence (+) and in the presence of a shock (Parker type model) (o).

As shown by /9/, the hydrogen density also declines in the downstream direction (Fig.2) beyond the solar wind shock boundary. The physical mechanism for the decrease is qualitatively the same as for the upstream region. It is, unfortunately, difficult to accurately model the downstream plasma density profile. This leads to significant uncertainties in the quantitative predictions of the neutral density profile, although the results are quite useful for interpreting the downstream data.

Since the backscattered solar Ly-α spatial dependence reflects the density distribution of neutral hydrogen the radial dependence of the glow is significantly different with and without a shock. In the upstream direction the Ly-α glow should decrease slower than 1/r for a constant solar Ly-α flux, where r is the heliocentric distance. The opposite effect should be seen in the downstream direction. These effects should be especially clear if the shock is nearby (~ 50 - 60 AU). This suggests that the Ly-α intensity provides important diagnostic data on the physics of the heliosphere. In the presence of a shock, it should also be noted that any estimate of the local interstellar density obtained in the inner solar system from Ly-α glow data would be incorrect, if it did not take the effect of a shock into account.

REMOTE OBSERVATIONS

The Ly-α glow obtained from Pioneer 10 since launch is plotted against radial distance in Fig.3. A 1/r curve is superposed for comparison. The Pioneer 10 intensity multiplied by the heliocentric distance is plotted against the radial distance from the Sun in Fig. 4. If the solar Lyman-α flux and the hydrogen density at the low interstellar velocity were constant the backscattered flux multiplied by r vs. radial distance would be a straight horizontal line, if the medium were optically thin, since the backscattered flux itself should drop as 1/r for such a case. The solid and dashed lines are scaled solar fluxes which are superposed on the Pioneer 10 data points, shown as dots. The solid curve is the solar flux as deduced from CaII observations and extends from 1972 to 1982. The dashed curve is the solar flux as deduced from the He I 10830 Å absorption line width observations in the same period. The solid curve shown after 1982 is a direct measurement of solar Ly-α flux from the SME satellite. There are three distinct regions of interest; less than 10 AU, 10-37 AU, and 37 AU - 48 AU. In the region close to the Sun slow neutral atoms are converted to fast atoms by charge exchange with solar protons. These fast atoms, with speeds of the order of 400 km/sec, are Doppler shifted to the wings of the solar Ly-α line. This gives rise to the "cavity" in the backscattered Ly-α glow data in the < 10 AU region where the solar wind proton flux produces significant charge exchange reactions. Beyond 10 AU the proton flux is relatively low and the density of the low velocity component of the neutral atom flux is expected to approach its interstellar value as one moves further away from the Sun. Thus, beyond 10 AU the Ly-α data track the solar flux reasonably well. The tracking is exceptionally good since 1982 when the direct measurements of solar Ly-α flux are available. The really striking feature is in the post 37 AU region when the Pioneer 10 data begin to depart from the behaviour predicted by a non-shocked model. The qualitative nature of this decline agrees with the results of /9/. A nearby shock at ~50 to~80 AU will cause the neutral atoms to be blown away from the downstream axis. However, the magnitude of the drop suggests a moving shock /10/, in addition to perhaps other time dependent effects not yet fully explored.

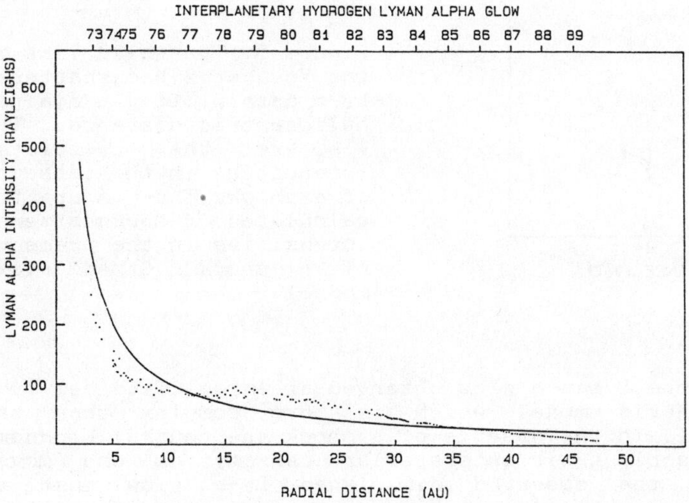

Figure 3. The Pioneer 10 backscattered Lyman-α intensity averaged over a day, plotted against time and radial distance from the Sun. The 1/r curve is normalized at 35 AU.

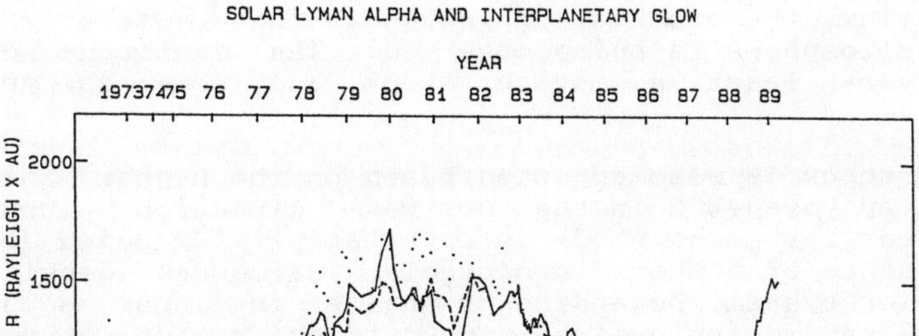

Figure 4. Comparison of the long-term solar Ly-α flux variation with the Pioneer 10 interplanetary glow observations over solar cycle 21. The observed daily averaged glow intensity is multiplied by the radial distance to remove the 1/r dependence of the backscattered radiation and is plotted against time and the radial distance (dots). The solid upper curve is the monthly averaged solar Ly-α flux as deduced from CA II K observations and extends from 1972 to 1982. The dashed curve is the monthly averaged solar flux as deduced from the He I 10830 Å absorption-line observations in the same period. The solid curve shown after 1982 is a direct measurement of solar Ly-α flux obtained with the SME satellite.

There is also some evidence from Voyager 2 data of the existence of a nearby shock. Voyager 2 data are particularly suitable for testing the upstream

Grzedzielski model since Voyager 2 is exiting the solar system in upstream direction. In Fig.5 the Voyager 2 data are plotted together with predicted intensities in the absence and presence of a shock. The shock structure used for our calculation was in accordance with the new Grzedzielski model. It is clear from Fig.5 that since about 24 AU from the Sun, Voyager 2 data are beginning to fall slower than the expected 1/r since the previously inferred interstellar densities would have been too low. The low values result from using data obtained well within the heliosphere where boundary effects are not directly observable. This rise above the expected neutral density is thus in qualitative agreement with the model.

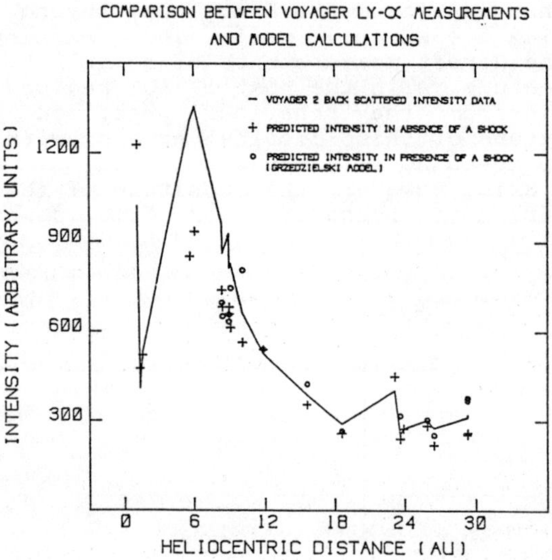

Figure 5. The solid line is the Voyager 2 backscattered Ly-α data plotted against heliocentric distance. The +'s are the calculated intensities in the absence of a shock. The o's are the calculated backscattered intensities in the presence of a shock (Grzedzielski model).

CONCLUSIONS

The rather dramatic drop in the Lyman-α glow observed at large distances from the Sun requires a heliospheric model which is more complex than the conventional nonshocked model. In the absence of a shock the neutral H atoms are expected to recover to nearly their interstellar density as one moves outwards beyond 20 AU. Since the observed diminished Ly-α glow must be accounted for by a density decrease of available scatterers, after correcting for variations in the solar flux, the observations suggest that interesting new physics in the distant heliosphere is being observed. The magnitude of the decline suggests at the very least a moving shock and some as yet unexplored effect(s).

The presence of a solar wind shock is also expected based on the higher than predicted Ly-α glow observed by Voyager 2 in the upstream direction. These data lend confidence to the earlier solar shock interpretation /10/ based on the Pioneer 10 data. The presence of a shock profoundly influences density measurements in the inner solar system. Depending on whether the inner solar system measurements were obtained in the upstream, downstream or side stream directions one would clearly obtain erroneous values for the interstellar hydrogen density unless the solar wind shock effects are considered.

REFERENCES

1. L. Davis, Phys. Rev., 100, 1440 (1955).
2. E.N. Parker, Interplanetary Dynamical Processes, Interscience, New York, 1963.
3. V.B. Baranov, K.V. Krasnobaev, and M.S. Ruderman, Ap. Space Sci., 41, 491 (1976).
4. V.B. Baranov, M.G. Lebedev, and M.S. Ruderman, Ap. Space Sci., 66, 441 (1979).
5. W. Neutsch, H.J. Fahr, M.N.R.A.S., 202, 735 (1983).
6. H.W. Ripken, H.J. Fahr, Astr. Ap., 122, 181 (1983).
7. M.K. Wallis, Astr. Ap., 130, 200 (1984).
8. S. Bleszynski, Astr. Ap., 180, 201 (1987).
9. P. Gangopadhyay, D.L. Judge, Ap. J., 336, 999 (1989).
10. P. Gangopadhyay, H.S. Ogawa and D.L. Judge, Ap. J., 336, 1012 (1989).

COMPARISON OF Ly-α AND Ly-β INTERPLANETARY GLOWS OBSERVED BY THE VOYAGER ULTRA-VIOLET SPECTROMETER

E. Chassefière,* J.C. Vial,** R. Lallement,* J.L. Bertaux* and B.R. Sandel***

**Service d'Aéronomie du CNRS, BP 3, 91371 Verrières-le-Buisson Cedex, France*

***Institut d'Astrophysique Spatiale, BP 10, 91371 Verrières-le-Buisson Cedex, France*

****Lunar and Planetary Laboratory, University of Arizona, Tucson, AZ 85721, U.S.A.*

ABSTRACT

During the first year of the cruise of Voyager 1 and 2 spacecrafts, when they were still near the Sun, many measurements of the interplanetary hydrogen glow, due to resonant scattering of solar photons, show a clear Ly-β signature at 102.6 nm. The comparison between Ly-α and Ly-β glows exhibits significant discrepancies which can not be explained by radiation transfer in the Ly-α interplanetary line. They might be due, at least partially, to the temporal variability of the solar flux, larger at Ly-α center than at Ly-β center. A study of OSO-8 high resolution Ly-α and Ly-β solar spectra shows indeed that the temporal/spatial variability of the line center solar Ly-α intensity is about three times the one of the line center solar Ly-β intensity.

INTRODUCTION

Neutral atoms from interstellar origin, mainly hydrogen and helium, enter the solar system at a relative velocity of 20 kilometers per second and temperature of 8000 K /1/. They have a density of about 0.07 atoms per cubic centimeter inside the heliosphere /2/ and can be detected near the Sun due to resonant scattering of photons emitted in H (121.6 nm) and He (58.4 nm) Ly-α solar lines. But H atoms may also be detected from the much weaker Ly-β line. Interplanetary glows at Ly-α and Ly-β have been observed by the ultra-violet spectrometer (U.V.S.) on board Voyager 1 (V1) and Voyager 2 (V2) spacecrafts from various places along their trajectories and in various directions. The U.V.S. instrument has been described in detail by Broadfoot et al /3/. Wavelengths in a range from 50 to 170 nm are recorded simultaneously in 128 contiguous intervals using a grating spectrograph (Fig. 1). A mechanical collimator defines a field of view of 0.9° by 0.1°, which leads to a spectral resolution of 3.3 nm. Because of the weakness of the Ly-β interplanetary signature, only spectra obtained at less than 4 astronomical units (AU) from the Sun, i.e. during the first year of the cruise (Sep. 1977 to Sep. 1978 for V1, Oct. 1977 to May 1978 for V2) are used. These data have been described elsewhere /4/.

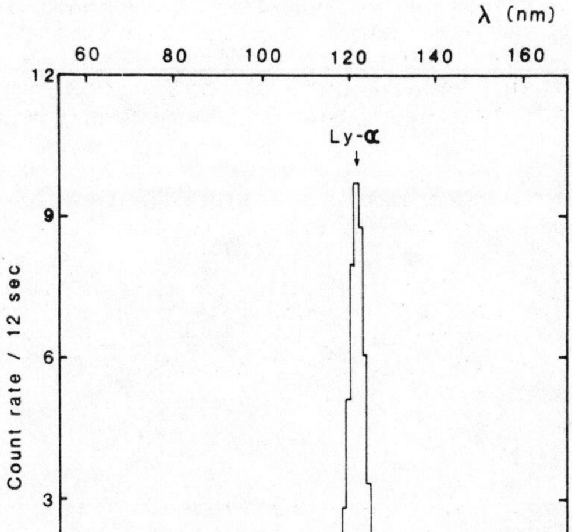

Fig. 1. Typical spectrum of the interplanetary glow obtained by the U.V.S. instrument. The x-axis is graduated both according to the channel number (lower scale) and to the wavelength (upper scale). The small bump between the He line on the left and the strong Ly-α line on the right is the Ly-β signature of interplanetary neutral hydrogen. The background continuum is due to γ-rays from the radioisotope thermoelectric generators, cosmic rays and γ-rays from the Sun. Total integration time for this spectrum is 16 h 15 mn.

Due to charge exchange between H atoms from interstellar origin passing near the Sun and protons of the solar wind, some neutral atoms are lost because extremely fast neutralized solar protons, moving radially away from the Sun at a velocity of 400 km/s, are not illuminated by the solar Ly-α and Ly-β lines which do not extend more than 0.8 Å from line center /5/, implying an upper limit on the velocity of illuminated atoms of 200 km/s. The theoretical mean life time T_D of a H atom at 1 A.U. from the Sun with respect to charge exchange is about 1.2 10^6 s but comparison between models and existing interplanetary Ly-α data favour an higher life time of 2 10^6 s. The ionization by charge exchange of H atoms passing near the Sun results in a deep depletion of neutral hydrogen downstream, called the ionization cavity of hydrogen. The attractive gravitational force F_g exerted by the Sun is partially compensated by the repulsive radiation pressure force F_p , the ratio $\mu=F_p/F_g$ varying from 0.6 to 1.1 during solar cycle. When μ=1, the two opposite forces are equal and atoms follow straight lines. A definite anisotropy of the solar wind, i.e. of the ionizing protons flux, has been discovered and estimated from the excess of interplanetary Ly-α glow in regions located far from the ecliptic plane /6/. Because there are less protons ejected at high solar latitude, neutral H atoms passing over the solar north pole or below the solar south pole undergo less charge exchange than those approaching the Sun nearer the ecliptic plane. The degree of anisotropy A is defined in such a way that the life time of H atoms at the heliographic latitude θ is obtained by multiplying the life time at equator (in the ecliptic plane) by 1-Acosθ.

The simultaneous observation of Ly-α and Ly-β interplanetary lines is of a great interest. Up to now, there are some unresolved discrepancies between computed and observed values of the Ly-α glow. For example, effects of radiation transfer in the Ly-α interplanetary emission line could slightly modify the emission pattern of H atoms as viewed from the Earth neighborhood. Because the Ly-β interplanetary line is optically thin, the comparison may bring some new information. As we will see now, there are some interesting features in the temporal evolution of the interplanetary Ly-α to Ly-β ratio as computed from Voyager observations which can not be attributed to radiation transfer. They could be related, at least partially, to solar effects which are analyzed hereafter.

REDUCTION OF DATA

Calculations of Ly-α and Ly-β intensities from U.V.S. spectra have been made in a very simple way by adding counts obtained in channels covering each line and substracting dark counts measured in a part of the spectrum where there is no signal. For Ly-α, the signal is summed over channels 70 to 78 (V1) or 72 to 80 (V2), the dark signal being taken from an average over channels 21 to 29 (V1) or 23 to 31 (V2). For Ly-β, because of the weakness of the signal and the risk in some cases that it is perturbed by small stellar or instrumental irregularities, we prefer to take the line in a very narrow interval (52-55 for V1, 55-58 for V2), the dark signal being taken from an average over the four nearest channels (50-51-56-57 for V1 and 53-54-59-60 for V2). In such a way, fluctuations over a spectral scale larger than about 6 channels are eliminated. The wings of the line are lost but we are only interested in relative values of Ly-β with respect to Ly-α intensity.

In Fig. 2, we compare Ly-α emissions observed by V1 to those modelled using the computer code constructed by Bertaux et al /1/ and Lallement et al /6/. Solar parameters have been adjusted in order to provide the best fit between model and data: T_D = 1.5 10^6 s, μ = 0.7, A = 0.5. The ratio between observed and computed values (called hereafter OCVR) is plotted versus the radial distance of the spacecraft from the Sun expressed in units of one million kilometers. There is a mean deviation of about ± 10% of the OCVR and a relative excess in observed emission of about 10 to 40% during the first three weeks of the mission, from day 257 to day 279 of the year 1977 (i.e. Sep. 14 to Oct. 6, the date of the launch being Sep. 1). This excess is of the same magnitude order as the one found previously by Donahue et al /7/ during the first period of V2 U.V.S. observations (days 246 to 258 of the year 1977, i.e. Sep. 3-15). Although Hall and Shemanski /8/ showed that this last result is based upon an erroneous transcription of tabulated data, Fig. 2 of their paper indicates that an excess of 20% around Sep. 10 can not be excluded. In the case of V1, many sets of parameters have been tried and there is no way to reconcile modelled intensities and data in the first period. At the present time, we do not have first V2 observations to compare with our model. We present now observations of the interplanetary Ly-β line.

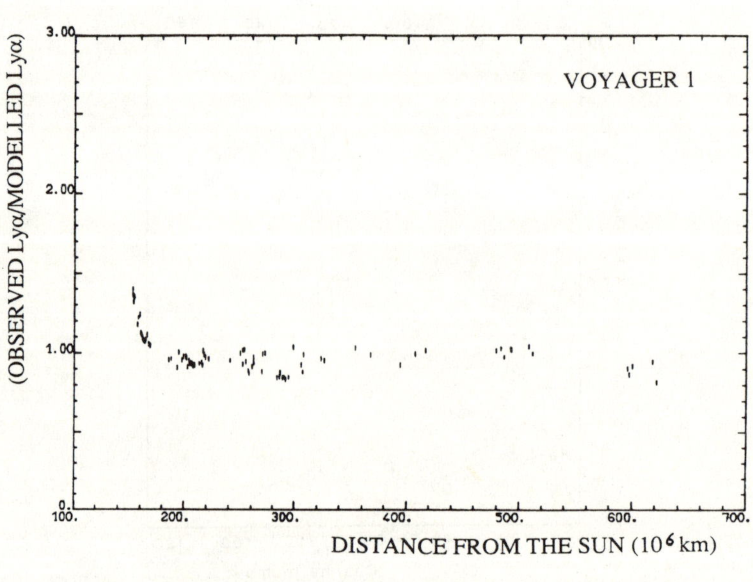

Fig. 2. Ratio between the Ly-α intensity observed by Voyager 1 U.V.S. and the modelled intensity with parameters (T_D - mean lifetime of a H atom at 1 AU from the Sun in the ecliptic plane- = 1.5 10^6 s, μ -ratio of the repulsive radiation pressure force to the attractive gravitational force- = 0.7, A -degree of latitudinal anisotropy of the solar wind as defined in the text- = 0.5) giving the best fit of computed to observed intensities as a function of the distance from the Sun to the spacecraft.

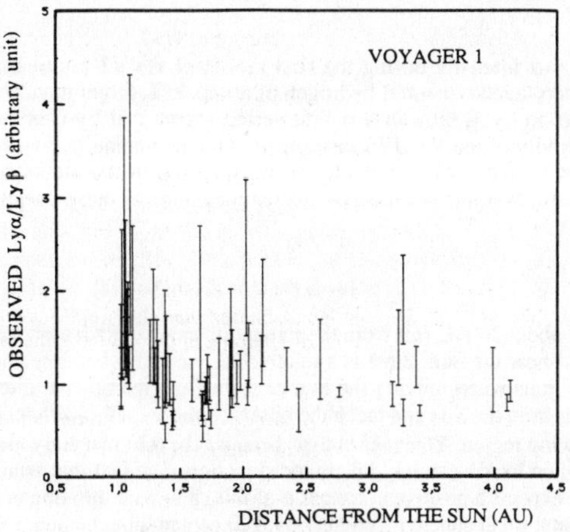

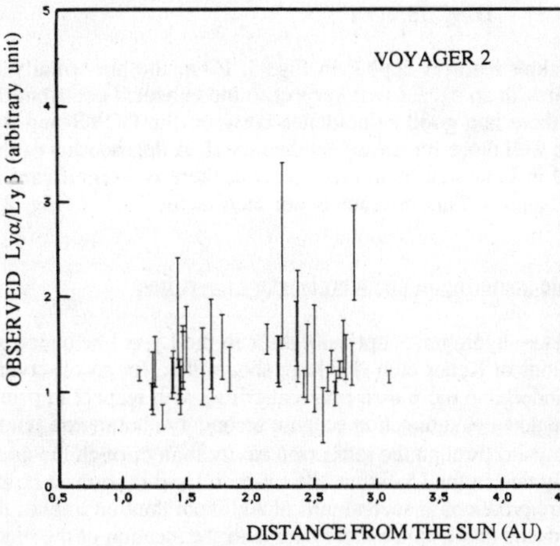

Fig. 3. Ratio between Ly-α and Ly-β intensities as a function of the distance from the Sun to the spacecraft (Voyager 1). Error bars are due to photon noise (primarily Ly-β).

Fig. 4. Same as Fig.3 for Voyager 2.

In Fig. 3 and 4, we show the Ly-α to Ly-β ratio versus the radial distance of the spacecraft from the Sun expressed in units of 1 AU for V1 and V2. Error bars represent photon noise at 1 σ, which results essentially from the denominator (Ly-β). Because of the large noise, points have been regrouped in order to reduce noise and make the temporal evolution of the ratio more readable. Inside each group, Ly-α and Ly-β intensities have been averaged in the classical way by weighting individual most likely values m_i by the inverse of the variance $1/\sigma_i^2$ (σ_i is the individual mean standard deviation due to photon noise) and calculating the variance σ^2 on the averaged signal as the inverse of $\Sigma\,(1/\sigma_i^2)$. During the first year of the cruise, the solar activity is still near minimum and variations of the ratio between total fluxes in solar Ly-α and Ly-β lines are rather weak /9/. Assuming that fluxes in the parts of these two lines illuminating interplanetary neutral H atoms (photons emitted at less than ±0.1Å from line center) are proportional to fluxes integrated over the whole lines, the U.V.S. Ly-α to Ly-β ratio can be divided by the ratio of solar fluxes in these two lines. To take into account temporal variation of solar fluxes at low solar activity has a very small effect, but nevertheless this correction is made, using Hinteregger values /9/. In Fig. 5 we show simultaneously the smoothed variation of the OCVR (Fig. 2) and values of V1 and V2 Ly-α to Ly-β ratios after having regrouped points over well chosen time intervals. The horizontal error bar indicates the length of the time interval over which averages are made, the vertical one representing photon noise. All ratios are given in an arbitrary unit. The angle between the Sun-spacecraft axis and the projection in the ecliptic plane of the downwind axis (opposite to the arrival direction of the incoming interstellar flow and nearly parallel to the ecliptic plane) is indicated on the upper x-axis. When this angle is zero, the spacecraft passes very near the downwind axis, which is also the symmetry axis of the ionization cavity of hydrogen.

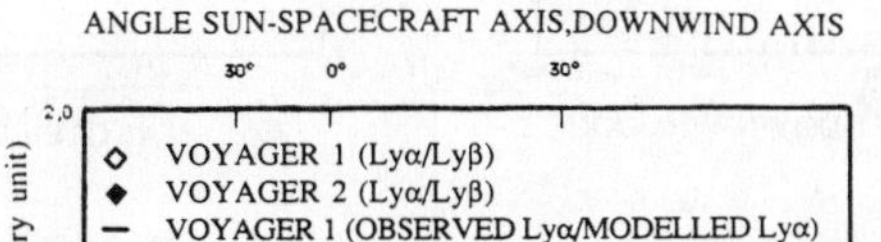

Fig. 5. Fig. 2 to 4 have been summarized in a single one. The continuous line is the temporally smoothed ratio of observed to computed intensities (Fig.1). White diamonds show the Ly-α to Ly-β ratio for Voyager 1 (Fig.3) after having regrouped points in well chosen time intervals (horizontal error bars) to reduce the photon noise (vertical error bars). Black diamonds show the same ratio for Voyager 2. The position of the spacecraft with respect to the downwind axis (symmetry axis of the ionization cavity which is nearly parallel to the ecliptic plane, at only 6° from it) is indicated on the upper x-axis by the angle between the Sun-spacecraft axis and the projection of the downwind axis on the ecliptic plane. There is a bump of the Ly-α to Ly-β ratio inside a cone of half angle 30° centered on the downwind axis.

DISCUSSION

Two puzzling features appear in Fig. 5. First, the abnormally large Ly-α intensity during the first month of the V1 cruise is quite significant with an excess with respect to the expected contribution of interplanetary neutral hydrogen of about 40%, larger than 3σ. The fact that there is a good coincidence between the OCVR and the Ly−α to Ly−β ratio in this first period shows that Ly-β intensities reproduce well those forecasted by the model. A degradation of the sensitivity of the V1 UVS instrument at the beginning of the flight is envisaged in Lallement et al /10/. Second, there is a significant increase by about 25% of the Ly-α to Ly-β ratio inside the ionization cavity (0° angle). This increase is not seen in the OCVR (Fig. 2). We now examine several possible explanations of these unexpected features:

1- Multiple scattering in the interplanetary Ly-α line:

Interplanetary hydrogen is optically thick in the Ly-α line over a scale of about 20 AU (far from the ionization cavity). Radiation transfer computations of Keller et al /11/ have shown that, for an observer located near the Sun, there is a relative increase of secondary photons (having undergone more than one scattering) with respect to primary photons when moving the line of sight from the upwind direction toward the downwind direction because secondary photons are produced far from the Sun and reach the observer more easily (without being scattered again) through the ionization cavity than through the denser upwind region. This means that, because the interplanetary medium is optically thin in the Ly-β line, the Ly-α to Ly-β ratio must be larger when looking in the downwind direction. The fact that secondary photons are produced at several tens of AU from the Sun implies that one expects a positive correlation of this ratio with the -line of sight, downwind axis- angle much more than with the location of the spacecraft in a small sphere (of radius ≈ 4 AU) centered on the Sun.

We show in Fig. 6 and 7 the Ly-α to Ly-β ratio as a function of the angle between the line of sight and the downwind axis. For each spacecraft, points indicated by a white square correspond to those belonging to the group of highest ratio in Fig. 5, points indicated by a black square to those belonging to the group of ratios just below the highest one. The V1 measurements in excess of 40% at the cruise beginning (white squares in Fig. 6) are taken at an angle of about 45° from the downwind axis (in fact about -40° of south ecliptic latitude). Those made when the spacecraft passes near the downwind axis (black squares in Fig. 6) are taken at different angles, showing rather an increase of the ratio from 0° to 90° than the decrease expected from radiation transfer effects. It comes clearly from a visual inspection of Fig. 6 and 7 that radiation transfer in the Ly-α interplanetary line does not explain the variability of the Ly-α to Ly-β ratio seen in Fig. 5, time (or variables related to time like spacecraft position) being a much more pertinent variable than the direction of observation.

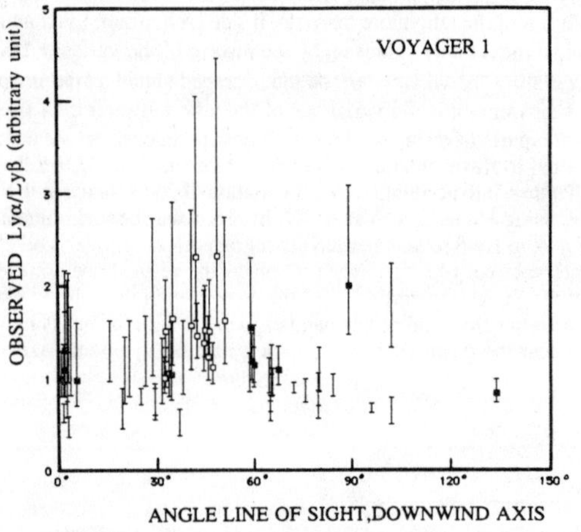

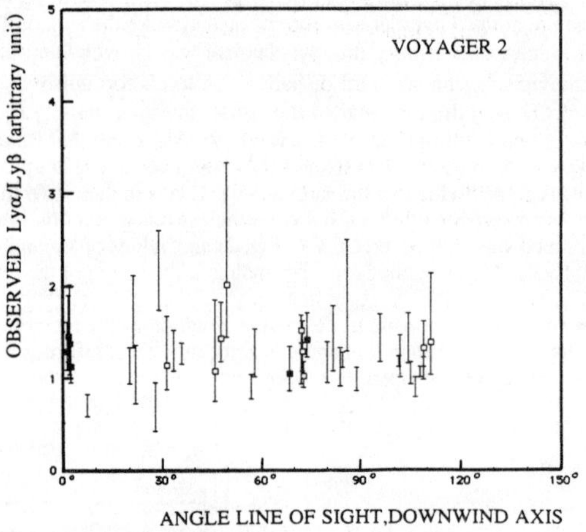

Fig. 6. Voyager 1 Ly-α to Ly-β ratio as a function of the angular distance between the line of sight and the downwind axis. White squares indicate points belonging to the group of highest ratio in Fig. 5 (at ≈1.1 AU), whereas black squares indicate those belonging to the group whose ratio is just below the highest one (at ≈2 AU).

Fig. 7. Same as Fig. 6 for Voyager 2. White squares correspond to the group at 2.3 AU in Fig. 5, black squares to the one at 1.7 AU.

2- Higher solar wind anisotropy in the south hemisphere:

The fact that first observations of V1 are performed at -40° south latitude might suggest a larger solar wind anisotropy in the south hemisphere. This explanation seems to be ruled out by the good agreement between computed and Ly-β intensities during this first period.

3- Existence of an other source of neutral hydrogen like cometesimals:

An excess of about 20-30% from 1977 Sep. 3 to Oct. 6 is compatible at 1 σ with both V1 (Fig. 5) and V2 (Fig. 2 from Hall and Shemanski /8/). So, the idea of an additional contribution of hydrogen originating from water outgassed from small comets near the Sun (Donahue et al /9/) might be considered again. Velocities of cometary hydrogen atoms do not exceed 50 km s^{-1} , so atoms are still illuminated by the Ly-β solar line. The rather good similarity of Ly-α and Ly-β profiles (Lemaire et al /5/), wavelengths being converted to Doppler velocities, should imply the same kind of excess at Ly-β during the first month, which is not the case. From the present analysis of Ly-β data, the assumption of a cometary source of neutral hydrogen can be ruled out.

4- Variability of the solar Ly-α flux in line center:

Measurements of the total and central solar Ly-α fluxes have been presented by Vidal-Madjar /12, 13/ from observations made on board the OSO-5 spacecraft from 1969 to 1975. Fig. 1 from the paper by Vidal-Madjar and Phissamay /13/, studying more specifically the solar Ly-α flux near solar minimum (1974-1975), exhibits a much stronger variability of the flux in line center (a factor of 3) than in the whole line (a factor of 1.5). To illustrate this fact, we have extracted from table 1 of the same paper a series of 16 measurements of both total and central fluxes taken from day 123 to day 223 of year 1975 and plotted in Fig. 8 the ratio of the central flux to the total flux (in an arbitrary unit) as a function of the date. We observe a relative decrease of the central flux of the same magnitude order as the decrease of the V1 Ly-α OCVR (Fig. 5). So, if the excess seen by V1 during the first month of the cruise is due to an excess of central with respect to total solar Ly-α flux, the total flux showing no strong variation during the same period, we have to understand why the central solar Ly-β flux does not present the same sharp relative variation. In order to correlate variations of central and total solar Ly-α and Ly-β fluxes, we have used spectra of Ly-α and Ly-β solar lines obtained from observations made during years 1975-1976 on board the OSO-8 satellite.

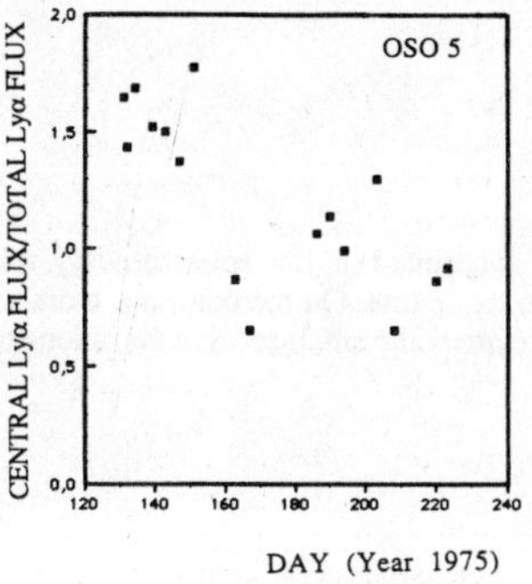

Fig. 8. Temporal fluctuations of central to total Ly-α flux ratio as seen by OSO-5 during the three last months of the mission.

The instrument is a high-resolution spectrometer coupled with a telescope with a 16 cm aperture and having, in flight, a resolution better than 3 ". The typical entrance slit has an angular field of 1" x 20", and spectral resolutions are either 20 mÅ (Ly-α) / 60 mÅ (Ly-β) or 200 mÅ (Ly-α) / 1 Å (Ly-β). Instrumentation has been described in a more detailed way by Artzner et al /14/ and Bonnet et al /15/. About 70 spectra of both lines taken in the high spectral resolution mode between 1975 Jun. 21 and 1976 May are analyzed here. An example of spectra is shown in Fig. 9. The dip seen at the center of the Ly-α line is the geocoronal signature which can be removed by fitting the spectral region from -0.15 Å to 0.15 Å (except the center where geocoronal extinction appears) by a parabola. The lower point of the parabola is taken as being the central specific intensity: In the case where the pressure radiation force exerted by Ly-α solar photons on H interplanetary atoms exactly compensates the solar gravitational force (μ=1), the velocity of H atoms with respect to the Sun is between -20 km s^{-1} (upwind) and 20 km s^{-1} (downwind), somewhat scattered due to temperature (8000 K), and H atoms are illuminated by photons emitted less than ≈80 mÅ from line center in Ly-α and ≈60 mÅ in Ly-β, or about one fifth of the line width.

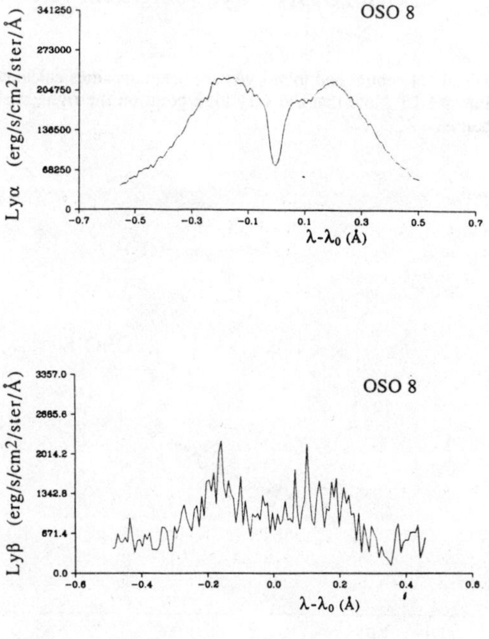

Fig. 9. Typical spectra of specific intensities in the solar Ly-α (upper part) and Ly-β (lower part) lines as seen by the high resolution spectrometer on board OSO-8 looking at small regions of the Sun (1"x20") at low solar activity (1976-1977). The dip at the center of the Ly-α line is due to geocoronal extinction. There is much more noise in the Ly-β line due to the quite smaller intensity of this line.

Because of the diversity of operation modes, limiting the wavelength range where the line is measured, we take as total intensity the one integrated from -0.35 Å to 0.35 Å for Ly-α and from -0.25 Å to 0.25 Å for Ly-β. This approximation is rather well justified for Ly-β, but leads to lose about 15% of the line for Ly-α. Because we are not interested by absolute values, but only by inter-correlations between intensities, this simple means to estimate total intensities in an standard way, without introducing any corrective factors depending on the acquisition mode, seems to us the best one. Fig. 10, 11 and 12 present correlations between total (F_α) and central (f_α) Ly-α intensities, total (F_β) and central (f_β) Ly-β intensities, and central to total Ly-α and Ly-β intensities ratios. We have the following relationships:

$$f_\alpha = k_\alpha F_\alpha^{1.15\pm0.02}$$

$$f_\beta = k_\beta F_\beta^{1.03\pm0.02}$$

$$\frac{f_\beta}{F_\beta} = k \left(\frac{f_\alpha}{F_\alpha}\right)^{0.34\pm0.06}$$

It results that when the full range of solar structures is scanned (at low solar activity), relative variations of the central intensity are larger by about 15% than those of the total intensity in the Ly-α line. On the contrary, there is a rather good proportionality between the Ly-β central intensity and the total one. Nevertheless, there are some small relative variations of the Ly-β central to total intensity ratio and their amplitude is 3 times lower than in the Ly-α line.

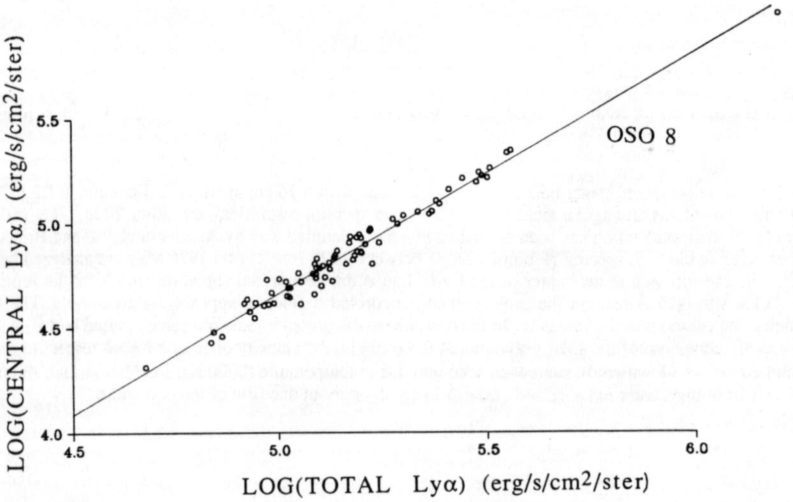

Fig. 10. Comparison in logarithmic scale of central and total Ly-α specific intensities calculated from OSO-8 spectra. The slope of the correlation straight line is 1.15. Note that the very high point on the right, which corresponds to an active region, is well aligned with other points.

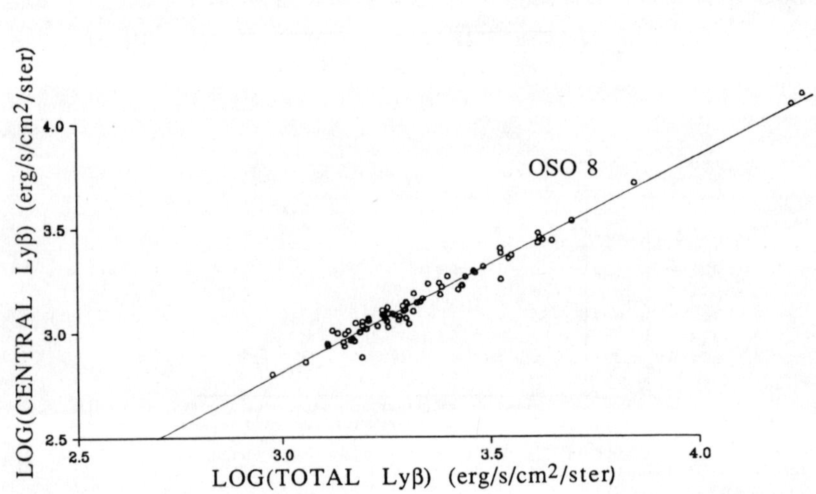

Fig. 11. Same as Fig. 10 for Ly-β. The slope of the correlation straight line is 1.03, which means that the intensity at line center is nearly proportional to the total intensity.

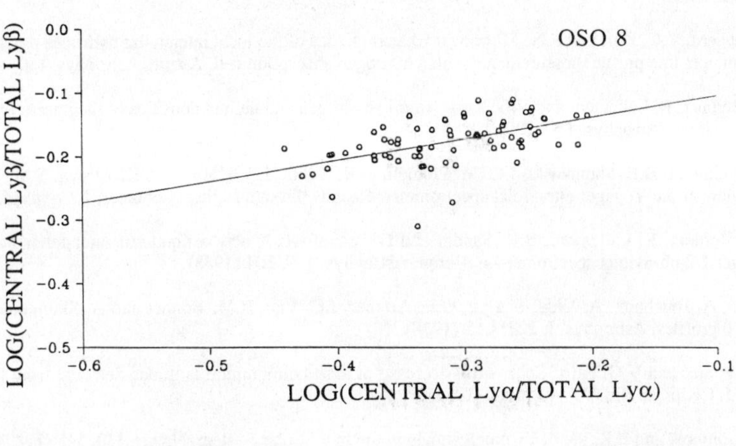

Fig. 12. Comparison in logarithmic scale of central to total Ly-α and Ly-β ratios. The slope of the correlation straight line is 0.34, which means that relative fluctuations of the central to total intensity ratio are three times lower in Ly-β than in Ly-α.

There is no simple rigorous way to convert spatial correlations between lines emitted by small regions of the Sun into temporal correlations of lines integrated over the whole solar disc. Nevertheless, we performed a simple computation of solar fluxes emitted in two situations corresponding to an increase of the total Ly-α flux by a factor 2 (see e.g. Vidal-Madjar and Phissamay /13/). The Ly-α and Ly-β fluxes were modelled with a two component emission (quiet and active) with respective areas varying from [0.9, 0.1] to [0.7, 0.3] and a Ly-α contrast of a factor 10 (see e.g. Lemaire et al /16/). Using correlation coefficients between central anf total fluxes derived from the present OSO-8 analysis, we show that the Ly-β flux (line center normalized to the total flux) varies by 4 % when the Ly-α flux (defined as the Ly-β one) varies by 9 %. Temporal variations of the central to total flux ratio are more than 2 times lower when using Ly-β instead of Ly-α, the factor 3 found from OSO-8 data analysis being somewhat reduced when passing on from small regions to the whole solar disk.

So, using a sharp temporal decrease of the central to total Ly-α flux ratio as observed by OSO-5 (Fig.6) and assuming, according to OSO-8 observations (Fig.12), a 2 or 3 times larger stability of the central to total Ly-β flux ratio (the central Ly-β flux being in fact nearly proportional to the total Ly-β flux), might lead to the large apparent excess of interplanetary Ly-α seen by V1 at the beginning of the cruise (Fig.5), no such an excess being seen in the Ly-β line.

But we must consider this explanation carefully:

- First, OSO-5 data of Fig.6 are obtained at the end of the operation time of the instrument, which could weaken their reliability. In particular, Fig. 1 from Vidal-Madjar and Phissamay /13/ shows at low solar activity a quite less good correlation than our Fig. 10.

- Second, V2 does not see such a large excess of interplanetary Ly-α intensity in the first weeks as V1 /8/, not more than an half (20%) of the excess seen by V1 (40%). But V2 measurements are made from Sep. 3 to Sep. 15, whereas V1 measurements are made from Sep. 14 to Oct. 6. So, the lack of overlapping prevents to compare in a significant way V1 and V2 data. Rapid variations of the line center solar Ly-α flux (with a comparatively much more stable line center Ly-β flux) can not be ruled out.

Concerning the feature observed later, when the spacecraft moves inside the ionization cavity, near the downwind axis, it is more difficult to attribute relative enhancement of Ly-α intensities with respect to those seen in the Ly-β line to solar effects. In this case, Ly-α intensities are well fitted by those computed from the model when using a lifetime of 1.5 10^6 s in the ecliptic plane at 1 AU from the Sun. Theoretical calculations of the lifetime indicate a more likely value of 1.2 10^6 s, which might be confirmed by the analysis of Voyager rotation maneuvers (Lallement, 1989, personal communication). When taking a shorter lifetime in the model, intensities decrease relatively more in the ionization cavity than in upwind and sidewind regions, which could enhance the OCVR inside the cavity (continuous curve of Fig.5) and more or less reconcile Ly-β intensities and modelled values, allowing to propose the same explanation, i.e. a higher Ly-α temporal variability. But, at the present time, to decrease the lifetime in the model leads to a general degradation of fits probably due to some unknown effect, not taken into account up to now.

CONCLUSION

We have compared the Ly-α and Ly-β interplanetary glows observed by the ultra-violet spectrometer on board Voyager 1 and Voyager 2 spacecrafts during the first year of their cruise. We find a large relative excess of Ly-α intensity as measured by Voyager 1 in september 1977 (≈40%), Ly-β data providing on the other hand a very good fit to the model. We propose the following explanation: There is a large variation of the ratio between central and total fluxes in the solar Ly-α line (as seen one year earlier in OSO-5 data) without correlative variation of the same ratio in the Ly-β line because of its better stability (as shown in the present paper from OSO-8 data). The possibility of an other source of interplanetary neutral hydrogen, like cometesimals, is excluded because of the very good agreement between Ly-β intensities and computed values during the same period.

The relative excess of Ly-α inside the ionisation cavity can not be explained by radiation transfer in the Ly-α interplanetary line because the increase of the Ly-α to Ly-β ratio is correlated to the position of the spacecraft instead to the direction of observation. Temporal variability of solar Ly-α flux in line center could be put forward but this explanation seems to us not satisfactory because of the coincidence between the bump of the ratio and the passage of the spacecraft inside the ionization cavity. We think that the present comparison points at a real and important phenomenon probably related to difficulties in interpreting Voyager maneuvers data and more generally systematic deviations from models when going from downwind to upwind regions.

REFERENCES

1. J.L. Bertaux, R. Lallement, V.G. Kurt and E.N. Mironova: Characteristics of the local interstellar hydrogen determined from Prognoz 5 and 6 interplanetary Lyman α line profile measurements with a hydrogen absorption cell, Astron. Astrophys. 150, 1 (1985)

2. E. Chassefière, J.L. Bertaux, R. Lallement and V.G. Kurt: Atomic hydrogen and helium densities of the interstellar medium measured in the vicinity of the Sun, Astron. Astrophys. 160, 229 (1986)

3. A.L. Broadfoot, B.R. Sandel, D.E. Shemanski, J.C. McConnell, G.R. Smith, J.B. Holberg, S.K. Atreya, T.M. Donahue, D.F. Strobel and J.L. Bertaux: Overview of the Voyager ultraviolet spectrometry: Results through Jupiter encounter, J. Geophys. Res. 86, 8259 (1981)

4. E. Chassefière, J.L. Bertaux, R. Lallement, B.R. Sandel and L. Broadfoot: A new estimate of interstellar helium parameters from measurements of Voyager 1/2 ultraviolet spectrometers, Astron. Astrophys. 199, 304 (1988)

5. P. Lemaire, J. Charra, A. Jouchoux, A. Vidal-Madjar, G.E. Artzner, J.C. Vial, R.M. Bonnet and A. Skumanich: Calibrated full disk solar H I Lyman-α and β profiles, Astrophys. J. 223, L55 (1978)

6. R. Lallement, J.L. Bertaux and V.G. Kurt: Solar wind decrease at high heliographic latitudes detected from Prognoz interplanetary Lyman alpha mapping, J. Geophys. Res. 90, A2, 1413 (1985)

7. T.M. Donahue, T.I. Gombosi and B.R. Sandel: Cometesimals in the inner Solar System, Nature 330, 548 (1987)

8. D.T. Hall and D.E. Shemanski: No cometesimals in the inner Solar System, Nature 335, 417 (1988)

9. H.E. Hinteregger: Representations of solar EUV fluxes for aeronomical applications, Adv. Space Res. 1, 39 (1981)

10. R. Lallement, J.L. Bertaux, E. Chassefière and B. Sandel: Ly-α observations by Voyager (1-18 AU), Adv. Space Res., this issue, in press (1990)

11. H.U. Keller, K. Richter and G.E. Thomas: Multiple scattering of solar radiation in the nearby interstellar medium. II, Astron. Astrophys. 102, 415 (1981)

12. A. Vidal-Madjar: Evolution of the solar Lyman alpha flux during four consecutive years, Solar Physics 40, 69 (1975)

13. A. Vidal-Madjar and B. Phissamay: The solar Lα flux near solar minimum, Solar Physics 66, 259 (1980)

14. G.E. Artzner, R.M. Bonnet, P. Lemaire, J.C. Vial, A. Jouchoux, J. Leibacher, A. Vidal-Madjar and M. Vite: The L.P.S.P. Experiment on OSO-8: I- Instrumentation, description of operations, laboratory calibration and prelaunch performances, Space Sci. Instr. 3, 131 (1977)

15. R.M. Bonnet, P.Lemaire, J.C. Vial, G. Artzner, P. Gouttebroze, A. Jouchoux, J. Leibacher, A. Skumanich and A. Vidal-Madjar: The L.P.S.P. instrument on OSO-8: II- In flight performances and preliminary results, Astrophys. J. 221, 1032 (1978)

16. P. Lemaire, P. Gouttebroze, J.C. Vial and G. Artzner: Optically thick lines: I. Observations, data reduction and modelling of an average plage, Astron. Astrophys. 103, 160 (1981)

LYMAN-ALPHA OBSERVATIONS FROM VOYAGER (1–18 AU)

R. Lallement,* J.L. Bertaux,* E. Chassefière* and B.R. Sandel**

**Service d'Aéronomie du CNRS, BP 3, 91371 — Verrières-le-Buisson, France*
***Lunar and Planetary Laboratory, University of Arixona, Tucson, AZ 85721, U.S.A.*

ABSTRACT

A fraction of Voyager 1 and 2 UVS measurements of the interplanetary Lyman-alpha background during the 1977-1983 period (radial distance 1-18 AU) is presented and compared with results from current models of the interaction between the sun and the neutral interstellar gas. It is shown that there are some systematic departures from the predicted pattern: 1) the position and brightness of the upwind maximum emission region on one part, and the downwind/upwind intensity ratio on the other part are contradictory and require two different values of the solar ionization: the downwind emission is most likely larger than predicted. 2) there is an excess of emission from the anti-solar directions as compared with regions close to the sun when moving outwards. Possible explanations include incorrect modeling (multiple scattering, non stationary processes), and inhomogeneous wind flow as a result of the heliospheric interface perturbations.

INTRODUCTION

A fraction of the Voyager 1 and 2 UVS measurements of the Lα Interplanetary emission between 1977 and 1983 (distance to the sun R between 1 and 18 AU) has been analysed. Data are compared with a model which was already

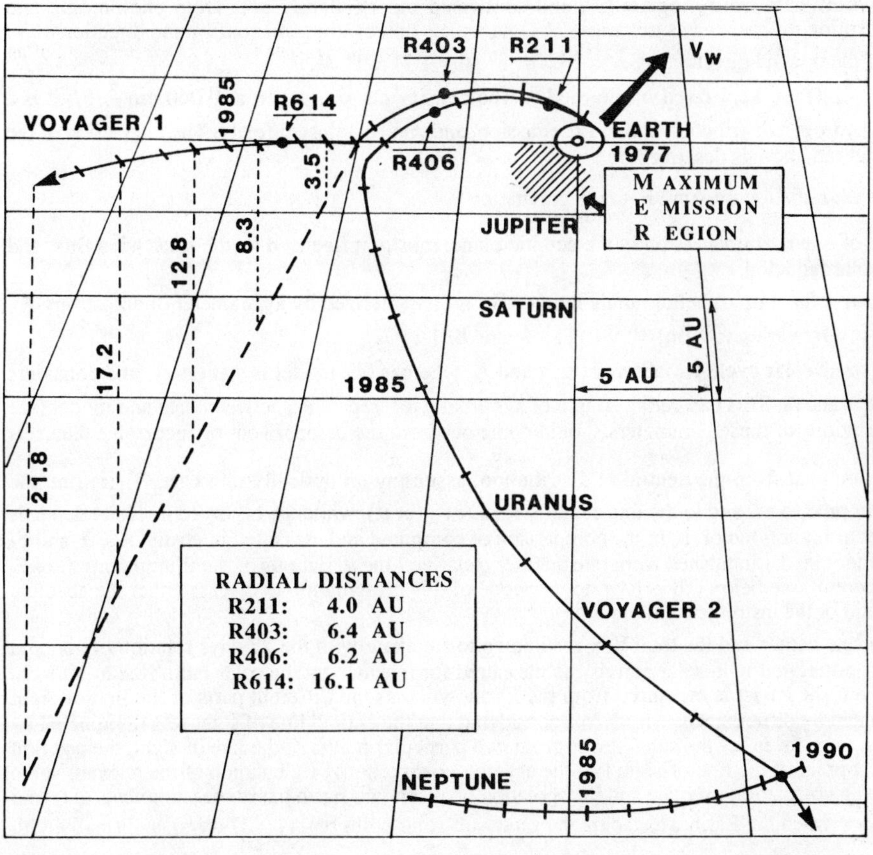

Figure 1: Voyager trajectories: the positions of the spacecraft during the roll maneuvers discussed in the text are indicated. The most precise location of the MER region (Maximum emissivity region) is obtained from VOY/1-R/403 or VOY/2-R/406 .

used to fit observations from the inner solar system only (.7 AU < R <1. AU : Mariner 10, Prognoz, Pioneer-Venus). Possible alternative outputs from this comparison could be :
- a carefully selected set of adjustable model parameters gives a good fit to the data (as good as close to the sun), showing that we understand pretty well the H distribution in the solar system
- no set of parameters gives a "perfect" fit to the data ("perfect" means that the deviations are only due to random data noise). In the second case, it would mean that some new ingredient would have to be added to the model, either in the H distribution, or in the way the model is computed.
Altough no full search for best-fit parameters has been done yet, it is already clear from the following analysis that the second conclusion is favoured, that is to say there are systematic departures from the "classical" model which require some explanations.

DATA

The trajectories of V1 and V2 are plotted on figure 1, together with the position of the Maximum Emissivity Region (hereafter MER, see Lallement, 1989 /1/. Two types of data set were examined: fixed lines of sight data (FL), and Roll Maneuvers data (RM). FL data were selected when there was no star in the UVS FOV, and when the scan platform stayed pointed in the same direction for some substantial time (hours), in order to allow a good signal to noise ratio. RM data were collected during maneuvers of the spacecraft deliberately performed in order to map the interplanetary L_α emission, at various places along the trajectory, marked on figure 1. Table 1 summarizes some data characterizing each Roll Maneuver. The spin axis was supposed to be oriented toward the sun, and the rotation rate was nominally 0.18° sec^{-1}. The scan platform was kept at a fixed cone angle to the spin axis during one complete rotation, and then moved at a larger cone angle. As the UVS line of sight was scanned across the sky, occasionally a star crossed the FOV as shown by a sharp peak of the signal. For the first maneuver that was examined, the direction of the line of sight at the time of these peaks, as given by the Jet Propulsion Laboratory, did not correspond to any star in the sky. In fact, we had to determine ourselves the real spin axis direction, accurate spin rate, and accurate timing of the rotation. This was done by a systematic search through all 4 rotation parameters and a star catalogue. In doing this excercise, we had to take into account the main FOV of UVS, and the auxiliary FOV port normally used during solar occultations, at 20° from the main FOV, between the spin axis and the main FOV. When computing the model Lα intensity, though, the Lα coming form this auxiliary port was ignored, since it accounts only for ≈ 5% of the signal at most.

MODEL

The flow of interstellar neutral H is characterized by a number density n_∞ (the "unpertubed" value), a temperature T and a flow velocity vector V. T = 8000 K± 1000, and V = 20 ± 1 km s^{-1} as measured by H cell absorption spectroscopy with Prognoz Lα instrument /2/. The direction of V was measured both from the axis of the Helium focusing cone on He 58.4 nm data from Prognoz /3/, and confirmed independently both from photometric and spectroscopic H cell absorption Prognoz measurements /2/. Other space observations do confirm the direction of V: the ecliptic coordinates of the upwind direction are $\lambda = 252° \pm 3°$, $\beta = +7° \pm 3°$ /4,5/.
These parameters (T, V, λ, β) are kept fixed in the model. The density n_∞ was taken at 0.060 cm^{-3} /6/; it is a scaling factor both in the hydrogen distribution and in Lα emission, and can be adjusted freely. The interaction of the H flow with solar wind and radiation is described by :
- the ratio $\mu = \frac{F_r}{F_g}$ of solar Lα radiation pressure F_r to the gravitation F_g.
- the ionization lifetime τ of one H atom at 1 AU in the ecliptic plane, mostly influenced by the solar wind flux, and more modestly by EUV solar radiation.
- the variation of lifetime as a function of heliographic latitude β_H is characterized by a parameter of anisotropy A : where the ionization rate $\nu = 1/\tau$ varies as $\nu(\beta_H) = \nu(0)\,[1 - A \sin^2 \beta_H]$
Obviously there may be some solar cycle variations of μ, τ and A, whereas our model is stationary, and considers only constant values of μ, τ and A. This can be the source of systematic discrepancies between data and model (see the discussion). Time variations of these parameters should come out from the comparison of successive data sets with models.
The model intensity is computed from the neutral H distribution, assuming an optically thin case. Therefore, the model intensity is proportionnal to n_∞ and to the line center solar Lα F_s (or μ), while the observed intensity depends on the chosen instrument calibration factor *a*.. In the comparison of computed and measured intensity, n_∞, *a* and F_s play a similar role and cannot be distinguished from one another (relative time variations of the illuminating flux F_s however are taken into account, see below). It is beyond the scope of this presentation to discuss either the absolute value of the data/model ratio or the instrument calibration.

Both the Lα solar flux at line center and the total flux are known to fluctuate with the 27 days rotation period and along the solar cycle. It is correlated with solar activity as measured for instance by the solar radio flux at 10.7 cm (the $F_{10.7}$ index). However, the $F_{10.7}$ is measured from the Earth, whereas the different parts of the line-of-sight from Voyager are illuminated by different solar disks. We considered that the entired line of sight was illuminated by the average value between two "faces" of the sun, viewed from two particular points of the line of sight, the position of Voyager and the farthest point of the line-of-sight (or the asymptotic direction) . Calculation of the relevant value of F_s at time t was done by a co-rotation exercise and interpolation between two Earth based measurements at t_1 and t_2,with $t_1 < t < t_2$, both done when the Earth was facing the same sub-solar point ($t_2 - t_1$ = 27 days), and finally with the statistical relationship with $F_{10.7}$ derived by Paxton et al (1988) (7), which is in good agreement with the analysis by Thomas and Anderson (1976) (8). While the 27 days modulation is of minor importance, the solar cycle variation is very large (50% increase between 1977 and 1982). This is the reason why those corrections have to be applied imperatively to the fixed line-of-sight data (FL), since one wants to use, all together, measurements made at different periods of the cycle. They have not yet been applied, however, to the second type of data (sky maps recorded within a few hours during roll maneuvers). As a result, the maps are used independently, and are not compared with one another from the point of view of the intensity level (this exercise would deserve further work).

For a variety of combinations of solar parameters μ, τ, A, a number of H density distributions were computed and stored in computer memory. Then, the intensity was computed for each measurement that we wanted to be compared to a model, specifying the location of the spacecraft and the direction of sight.

The Lα emissivity distribution presents a maximum region (MER) centered upwind (figure 1), at a distance $r_m \approx 2$ AU for $\tau = 2 \times 10^6$ s. If the ionization is increased (due to solar wind flux increase, for instance), this maximum region moves upwind (r_m increases), but the maximum emissivity decreases. At the same time, the cavity is enlarged. At large solar distances, (say >20 AU) the density distribution approaches n_∞ and is pretty uniform, so that it does not depend on the solar parameters (μ, τ, A). In contrast, in the inner solar system (say, inside $\approx$ 10 AU), the H distribution depends very much on these solar parameters. Schematically, the data-model comparison gives three ways of measuring the ionization (or τ): the location of the maximum emission region (MER), the intensity contrast between this region and the outer regions, and the upwind/downwind intensity ratio /1/.

COMPARISON DATA-TO-MODEL FOR FIXED DIRECTIONS OF SIGHT

On figure 2 are plotted the ratios between the measured Lα intensities (Voyager 1 and 2) up to $\approx$ 18 AU of radial distance and the computed intensities for a model with $\mu = 0.7$, $A = 0.50$, $\tau = 2.10^6$ s, corrected solar Lα. These model parameters are those which give the best fit for the 1977-78 points (1-5 AU) and are very similar to the values derived from contemporary Prognoz measurements in 1977 (see Fig 3). Three groups of points are circled and marked 1,2,3. The behaviour of group 1 (Voyager 1 points) is striking. Initially, we thought that the observed rapid decrease with time might be related to cometesimals. However, the measured Lβ signal shows no decrease during the same period of time, as it should if Voyager 1 crossed a shell of H (see ref /12/). One might think of a solar Lα variation without a corresponding solar Lβ variation /12/. However, simultaneous observations with Prognoz-6 Lα at the same location and approximately same direction show no sign of such a decrease with time. In addition, a 27 day modulation would give a complete cycle.

An other possible explanation would be that V-1 sensitivity changed a little bit at Lα during this $\approx$ 30 days period and then stabilized. Hereafter we will exclude group 1 from this analysis. Then the ratio is quite constant and the spread is quite small up to 4 AU (group 2 on Fig2) . At distances larger than $\approx$ 6 AU (group 3), the spread is much greater. The average value for the 1983 points is 5, to be compared to 3.5 in 1977-1978, for a model with $\tau = 2 \times 10^6$ s. In fact, given the known solar wind flux ionizing the flow of interstellar H, 2×10^6s is probably too large. A lower value $\tau = 1.0 \times 10^6$ (higher ionization) is used on figure 4 and 5 . Discarding the points of group 1, the ratio is now more constant throughout all radial distances. The average ratio value however has increased to about 6, while for points before 6 AU (Fig5) the ratio is decreasing with radial distance and the spread also is increased substantially . In conclusion of this section, we note the following points :

1. It is difficult to explain points of group 1 except from an instrumental effect or a solar line rapid variation /12/.
2. A model with $\tau = 2 \times 10^6$s fits rather well the distribution of H inside $\approx$ 5 AU, but not what is seen outside of 5 AU. Decreasing the lifetime τ, in accordance with solar wind measurements or even lower, improves the constancy of the ratio of the observed intensity to model values, but deteriorates the fit of data taken in the internal cavity (< 5 AU), as seen by the general decrease of the ratio and the increase of the spread for <5 AU points.

We have to be careful about the interpretation of these results: 1) The radial distance is increasing with time, and the detected effects could be due to time variations of the solar parameters : in a more realistic treatment, non-stationary models should be used, or at least different hydrogen distributions instead of assuming one for the whole 1977-1983 period. What is the expected trend of the data/model ratio variation between 1977 and 1981-83 as caused by μ or τ changes induced by solar flux variations ? Roughly the solar wind flux over this period decreased by at most 25% /9/, which could enhance the detected Lα emission by about 25-30 % from the filling of the cavity. At the same time, the solar Ly-α flux increased by about 25-40% /9/, which would, in principle, produce an inverse effect : depletion of the atoms by defocusing (increasd μ), enlargement of the cavity and decrease of the overall observed intensity everywhere by about 15-30 %. From these two opposite effects, one predicts an increase of less than 10%, or perhaps a decrease of the mean intensity level. This is significantly smaller than the observed increase when keeping constant the parameters, for $\tau = 2. \times 10^6$ s. Nevertheless these crude estimates give an idea of the difficulties in interpreting the FL data in the absence of non-stationary models. 2) It is difficult to estimate the errors on the solar Ly-α center-of-line calculations /12/. The correlation coefficients are subject to changes along the solar cycle and there are no direct measurements at the time observations were done. The Paxton et al (1988)(7) correlation was derived in 1985. As a comparison an other correlation was established by Thomas and Anderson (1976) (8)for the previous solar minimum which gives relative differences from the first one of the order of 25%.

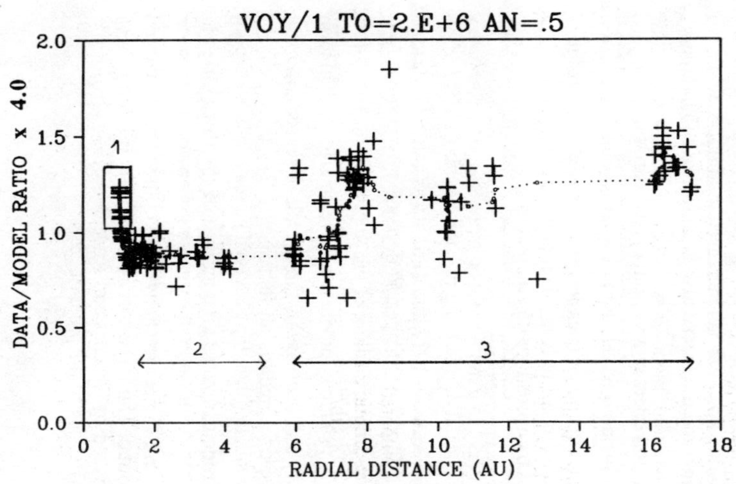

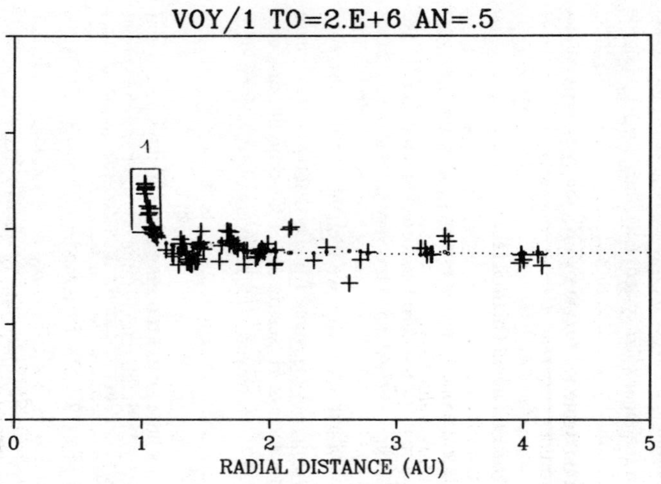

Fig 2: Data-to-model Ly-a intensity ratios are plotted as a function of spacecraft radial distance for the fixed directions of sight of Voyager 1, for μ=.7, A=.5 and τ = 2. 10^6s. These values provide the best fit for the 1977-1978 (R=1-5 AU) points. Points in group 1 have been excluded from the subsequent analysis (see text). There is a very large spread and the ratio increases at distances larger than 6 AU. The dotted line is a running average over 10 points.

Fig 3: Detail of Fig 2, for R < 5 AU. Discarding group 1, the data-to-model ratio is quite constant .

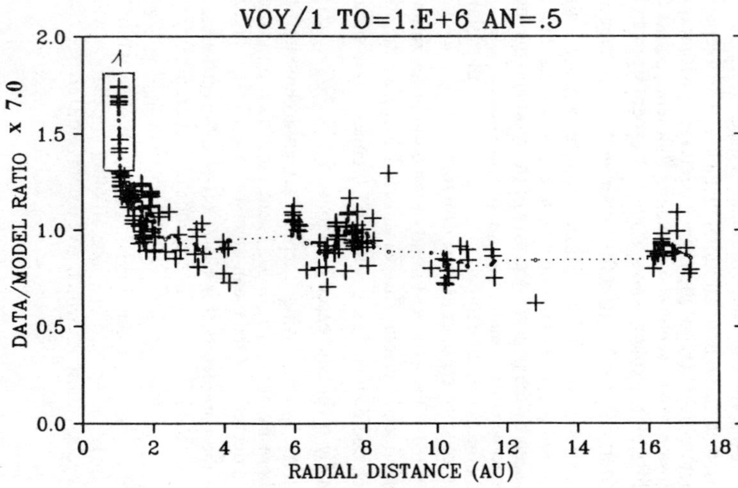

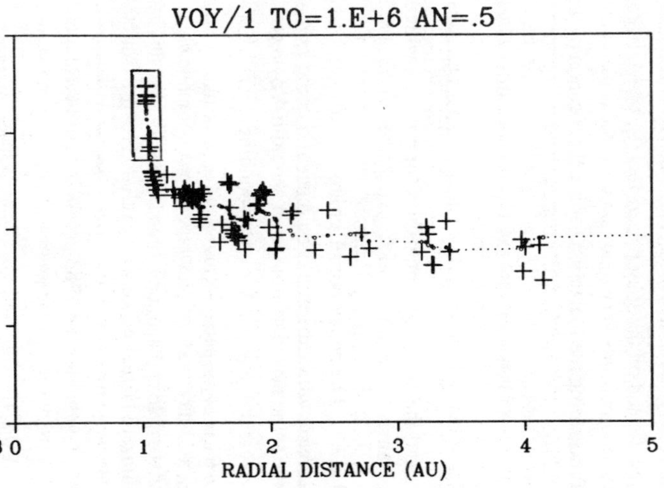

Fig 4: Same as Fig 2 for a double ionization rate ($\tau = 1.\ 10^6$ s). The ratio is nearly constant throughout all radial distances (discarding group 1), but the fit is not as good as for $\tau = 2.\ 10^6$ s for R< 6 AU.

Fig 5 : Detail of Fig 4, for R < 5 AU. Discarding group 1, the data-to-model ratio is decreasing between 1 and 4.5 AU and the spread is larger than for τ=2. 10^6s.

ROLL MANEUVERS

In this section, we study maps collected during Roll Maneuvers, each of them giving a snapshot of the whole sky (or a good fraction) at a given time. We comment on only 3 Roll Maneuvers, though there are some more that we have analyzed. They do however show the same pattern.
The Voyager 1 roll 403 and the Voyager 2 roll 406 correspond to nearly the same location and provide a good comparison between the two instruments. The results of the roll 403 are similar to those of roll 406 and are not shown here. Some data about the maneuvers are presented in Table 1 :

Table 1

Name	S/C	Date		Radial	Spacecraft position Ecliptic	Ecliptic
		Year	Day	distance (AU)	Longitude λ(°)	Latitude β(°)
B-211	V-2	1978	275	4.01	120.4	+2.2
B-406	V-2	1980	24	6.2	157.4	+3.6
A-614	V-1	1983	116	16.2	221.9	+23.6

The position of the S/C at the time of Roll Maneuvers is indicated on figure 1. Maps are shown where all points have been averaged in bins of 10 x 15° long./latitude, except at high latitudes where bins are larger in longitude (only one bin abovc 85° lat).
We recall here that we do not use the absolute value of the intensity, and we do not intercompare the maps from the point of view of the general intensity level. This would require a more detailed study of the solar flux variations. Our present purpose is to compare each data map with a choice of different models. In order to estimate the quality of fit of a model, the predicted intensities are first scaled to the best possible value using the scaling factor a_1, or the "best fit scaling factor" ,as marked on the figures. Then, the RMS residue in percent is :

$\sqrt{Q} = \sqrt{\frac{1}{n}\Sigma(\frac{I_{Data} - a_1 I_{model}}{I_{Data}})^2}$ where n is the number of points of measurements (about 1300 pts per roll).
We used a second scaling factor a_2 (quoted "used"), computed in such a way that the model is adjusted to the data in the region where the data/model ratio is minimum. In other words, by using this factor, a_2, all the differences (data)-(model) have positive values. The resulting relative difference gives an idea of the amplitude of the systematic departures, and can be simply derived from the data/model ratio as illustrated in Fig8. The relative difference map is identical to the data/model ratio map.

Absolute differences are defined as: $I_{data} - a_2 I_{model}$ in Rayleigh, while relative differences as $\frac{I_{data} - a_2 I_{model}}{I_{data}}$ in %. The absolute difference maps are more appropriate for search of an additional component, while relative difference maps are more useful for search of systematic discrepancies due to incorrect modeling for example.
It has been always possible to find sets of parameters giving a model with $\sqrt{Q} < 10\%$. However, even in such a case the data map departed from the model map, not randomly, but in a systematic fashion (some areas departing more than others). As we already said, we did not yet complete a full best-fit adjustment but fixed the parameters at a few discrete values in order to illustrate the general tendancies. The anisotropy degree A becomes a very minor parameter when going outwards on the upwind side and even for the first maneuver at 4 AU changes of A are unimportant effects as compared with the general systematic departures (note that there are always changes of the pattern induced by variations of A, but mainly because they are followed by a change of the average polar-equator lifetime: these changes are similar with lifetime changes and not anisotropy changes). The lifetime τ and the parameter μ have some similar influences such that to the first degree the whole emission pattern is sensitive to a combination of both of them. As a consequence for this short presentation, we mainly vary τ while keeping μ and A at fixed values.
It should be recalled, at this point, that it is not necessarily the model which gives the best overall fit to the data which corresponds to the real ISM, as far as systematic departures are present.

A) Roll maneuver B-211

Fig. 6 is a map of the data, where white indicates a region where no data were collected, in particular very near the sun. Fig. 7 is a map of a model with $\tau = 2.10^6$; the model is not computed where there are no data. It can be noted that the maximum region in the data map is more intense above the ecliptic plane, and less intense below the ecliptic plane in the minimum region, whereas the model map shows little North/South asymmetry. This might be due to a larger solar wind flux in the southern hemisphere at the time of these observations, or before.
With model $\tau = 2 \times 10^6$, $\mu = 0.7$, $A = 0.4$, figure 8 shows the data/model ratio map. The data is smaller than the model in the region around the sun and higher elsewhere. This is consistent with the idea that inner solar system observations at 1 AU, like the Prognoz data, agree well with 2×10^6 s, but the lifetime has to be decreased when looking outside. Decreasing the lifetime improves the RMS from 12.9 % for $\tau = 2 \times 10^6$s to 9.1 % ($\tau = 10^6$ s, A = 0.4 and $\tau = 10^6$s, A = O). A map of the relative differences is shown for the last case in fig 9. However, though the average RMS becomes rather low, systematic departures remain which amount to 40 - 50% in a significant part of the sky, but now they are displaced around the downwind direction. In other words, a small lifetime is more appropriate when fitting the upwind region, and the shape of the maximum emission region, but predicts an excess of emission downwind. Fig 10 shows the absolute differences map: absolute differences seem to be higher in the Northern hemisphere than in the Southern one, but there is no very clear tendency.

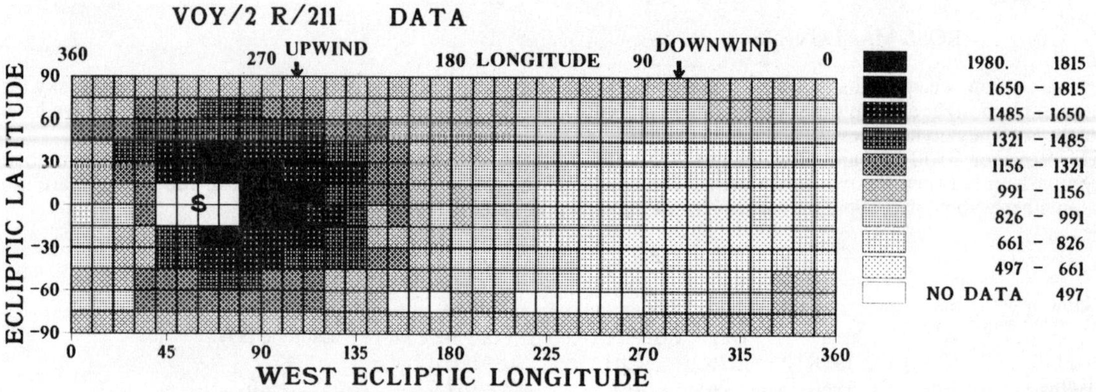

Fig 6 : Ly-alpha data map for Voyager 2 roll maneuver n° 211. Units of shading are Rayleighs, as deduced from the instrument calibration."S" marks the position of the sun.

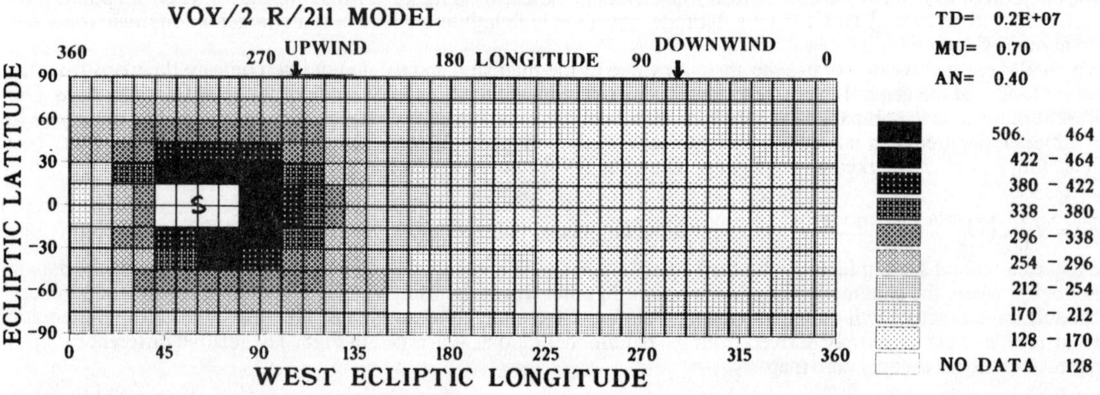

Fig 7: Model map in Rayleigh for the Voyager 2 roll maneuver 211 location. The density "at infinity" is 0.06 cm^{-3} and the illuminating flux, F_s, is 3.1 10^{11} photons cm^{-2} s^{-1} A^{-1}. Both act as scaling factors for the whole intensity map. The model parameters A (AN on the figure) = 0., μ (MU) = .7, τ (TD) = 2. 10^6 s. The model is not computed where there are no data.

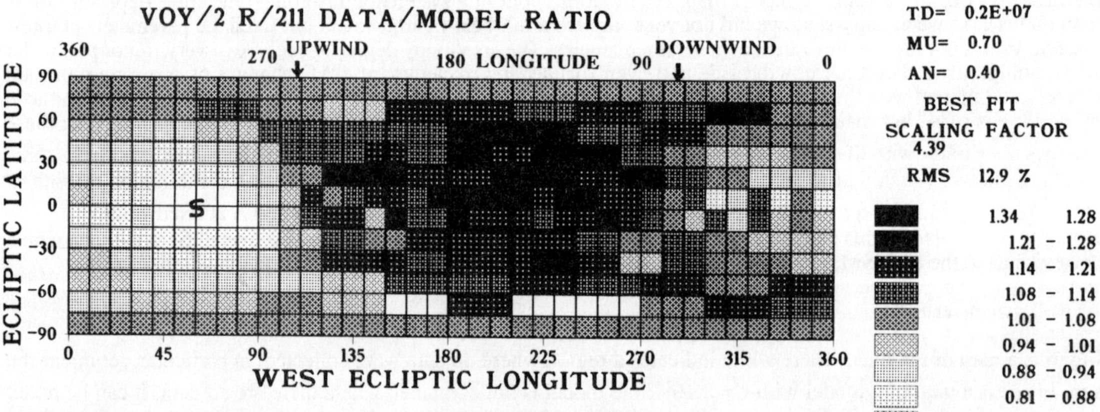

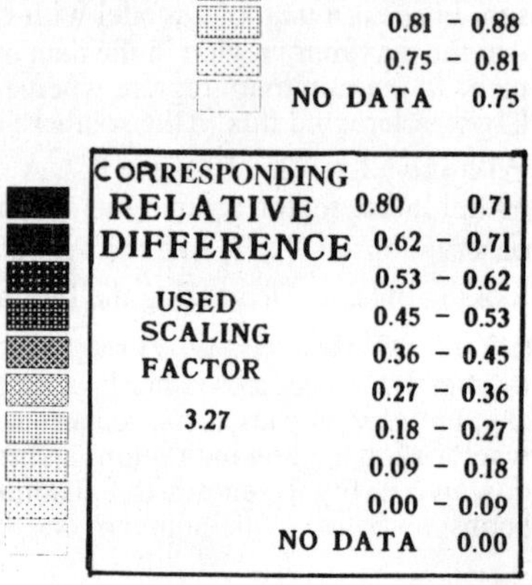

Fig 8: Data-to-model ratio map for the parameters of Fig 7: the model values have been all multiplied by the best fit scaling factor K indicated in the figure. The ratio varies between Rmin (.75) and Rmax (1.34) (left scale). Systematic departures are clearly shown : the model predicts too small intensities in the anti-solar directions, indicating that the lifetime is too large (see text) and the MER is not very well reproduced (strong gradient between 200 and 275 ° east longitude). If a smaller scaling factor K ' is choosed, in such a way that the model fits the data in the region where the ratio is minimum (K' = K* Rmin , the relative difference is positive everywhere (right scale). If one fits the region close to the sun, the anti-solar directions are up to 80% too bright as compared with model predictions.

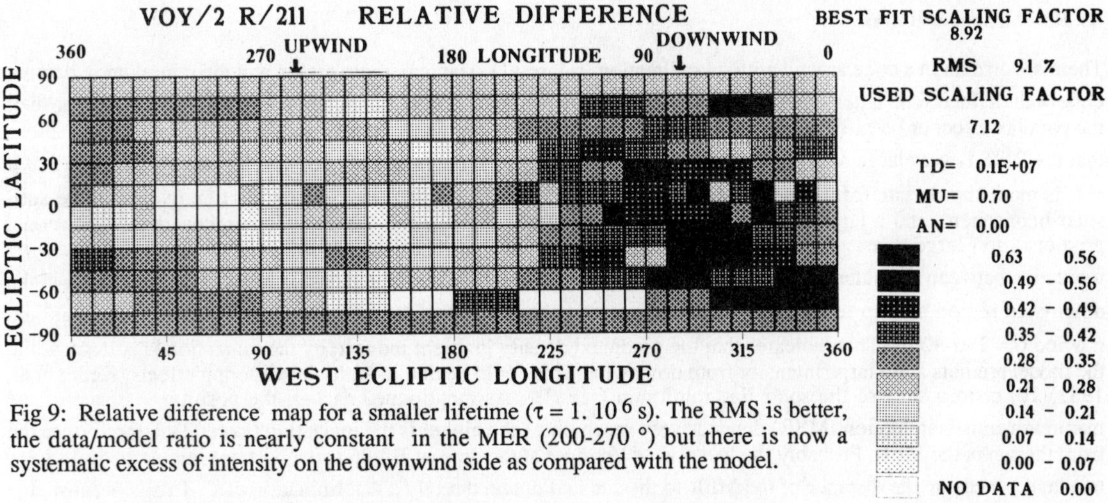

Fig 9: Relative difference map for a smaller lifetime ($\tau = 1.\ 10^6$ s). The RMS is better, the data/model ratio is nearly constant in the MER (200-270 °) but there is now a systematic excess of intensity on the downwind side as compared with the model.

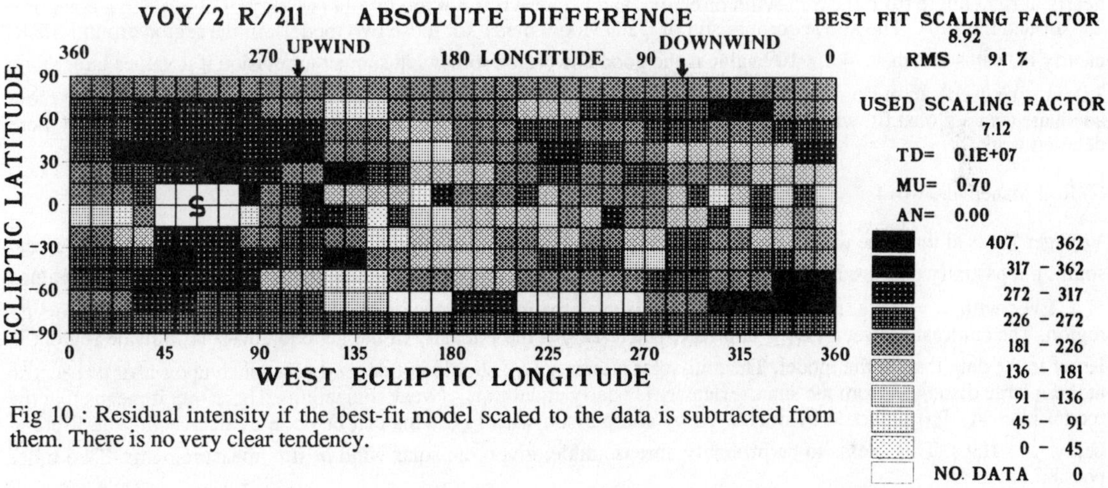

Fig 10 : Residual intensity if the best-fit model scaled to the data is subtracted from them. There is no very clear tendency.

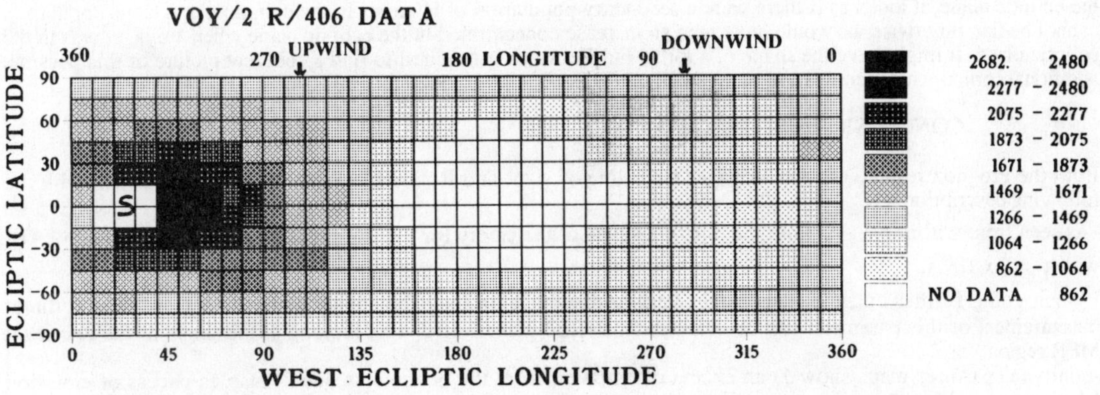

Fig 11 : Ly-alpha map for Voyager 2 roll maneuver n° 406. There is a large region void of data in the anti-solar direction.

B) Roll Maneuver B-406

There are no data in a cone around anti-solar direction (figure 11). The maximum region and the minimum region are quite well defined. The latter is not centered along the downwind direction, but rather at $\lambda \approx 50°$, probably because of the parallax effect or because of contamination by an extraneous source. A comparison with different models shows that the RMS is very large for $\tau = 2 \times 10^6$ (17%)(Fig 12), and decreases to 12 % for $\tau = 10^6$ s (fig 13). Note that $\mu = 1.$ is more appropriate for this period of time. For $\tau = 2 \times 10^6$ (Fig 13), there is an excess of emission in the anti-solar hemisphere, and a large gradient of the ratio in front of the sun and upwind. This indicates that the model predicts a too large density increase between the sun and the outer regions (or a too large density decrease by ionization between the outer regions and the sun). For $\tau = 1 \times 10^6$(Fig 14), the data is higher than the model in the downwind region and up to $\approx \lambda$=140° while the ratio is constant near the ecliptic plane in front of the Sun and upwind (λ= 230-320°). This indicates that the predicted density gradient induced by the ionization is correct, while the model predicts a too large intensity from downwind.This could be due to multiple scattering effects (Keller et al, 1981)(10) or to a shift of the solar line minimum (see /1/). As mentionned earlier, the position in space of the maximum emissivity region (MER) depends very much on τ ; the higher is the ionization (lower T_D), the further out from the sun is the MER. Probably the location of Voyager at the time of B-406 is the best one, along the trajectory, to actually measure the distance of the MER to the sun and obtain directly a determination of τ . This is because it is nearly at right angle from the Sun-wind direction, and because it is not too far out (we cannot observe in a cone of $\approx$ 25° around the sun). The above comparison of Data/Model maps for these two models, in the region around MER, clearly indicates that the $\tau = 1 \times 10^6$ value is the good one (value somewhat subject to revision if μ values larger than 1.0 must be used). In our opinion, determining τ from a limited observed region around the MER is certainly more adequate than a global fit, which can be biased by secondary effects. Certainly this type of maneuvers deserves more detailed work .

C) Roll Maneuver A-614

Voyager 1 was at that time well above the ecliptic plane. The data map (figure 14) shows accordingly the MER in the south, but partially obscured by the "no data" cone. The RMS fit is better for $\tau = 10^6$ (9.6%) than for larger values of τ. Even with $\tau = 10^6$ s, the data/model ratio is systematically larger in the whole anti-solar region, or outside region. The contrast between I_{MER} and I_{AS}, where I_{AS} is the intensity in the general anti-solar/outside region, is larger in the data than in the model. The anti-solar intensity I_{AS} should not depend very much upon τ or μ, because at these large distances from the sun the density is nearly equal to n_∞ (weak ionization). Therefore, it means that the model intensity I_{MER} should be decreased by comparison with I_{AS}. This can be done by decreasing the lifetime below $\tau = 10^6$ s. This seems to be probably unreasonable, given the solar wind *in situ* measurements. Two other possibilities are :

- radiative transfer smoothes out and decreases the intensity in respect to the optically thin approximation of the model in the MER region (a kind of "fogging" effect).

- along the line-of-sight the H density increases when going outwards from the Sun, faster than predicted by the model. This density gradient could be a consequence of the ionization of the incoming flow at the heliopause boundary, due to charge-exchange with the heated and compressed plasma.

When a model at $\tau = 10^6$ s is subtracted from the data, after proper scaling, the map of the absolute differences display a strong North South asymetry, even in the meridian of the antisolar direction (Fig 16). Since V-1 is above the ecliptic plane, it looks as if there were a secondary population of H atoms in the ecliptic plane. This population cannot be flat, otherwise we would have seen an increase concentrated in the ecliptic plane when Voyager was in the ecliptic plane. It might have the shape of a torus. Further work is required to find a coherent picture of this possible additional component by comparing between all maps.

CONCLUSION AND DISCUSSION

Both the Prognoz results, and the Voyager fixed line of sight and Roll maneuvers results are consistent with the following description :

- as seen from within 1-3 AU from the Sun, the shape of the cavity (or the Lα emission pattern) can be mimicked with $\tau = 2 \times 10^6$ s.

- when going further out, the fit becomes very poor with $\tau = 2 \times 10^6$ s, much better with $\tau = 10^6$ s; a direct measurement of this parameter can be obtained from Roll maneuver B-406, looking at the shape of the restricted MER region.

-relative departures maps show 1) an excess of emission from the downwind side 2) also an excess of emission from the anti-solar (external) regions when going outwards .

-absolute departures maps seem to indicate that there is an excess of intensity to the south as seen from V-1 when it is well above the ecliptic plane.

Possible explanation of data-model discrepancies are of three main types.

1) Shortcoming of the models

- radiative transfer neglected (this deserves further work).

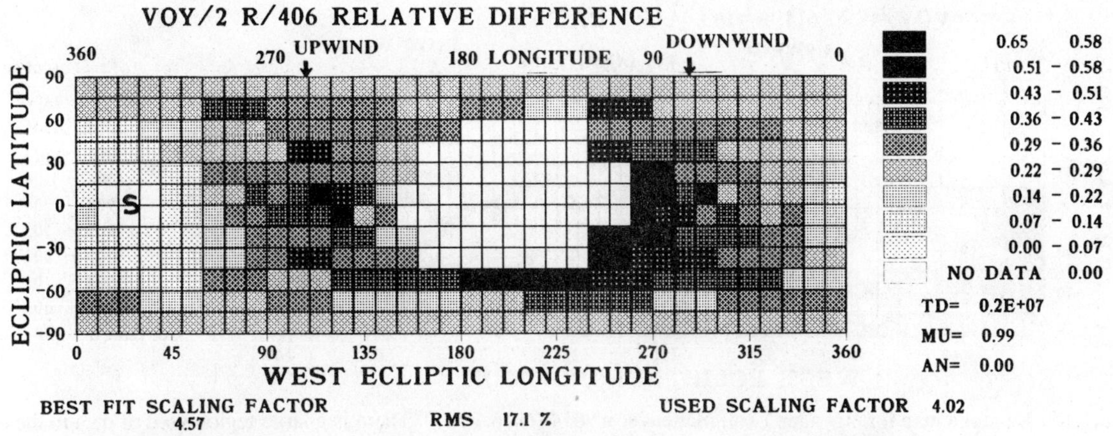

Fig 12: Relative difference map for a large lifetime (2 x10 ^{6}s) . There is a very large gradient between the sun and the anti-solar cone, around the MER. This indicates that the lifetime is too high.

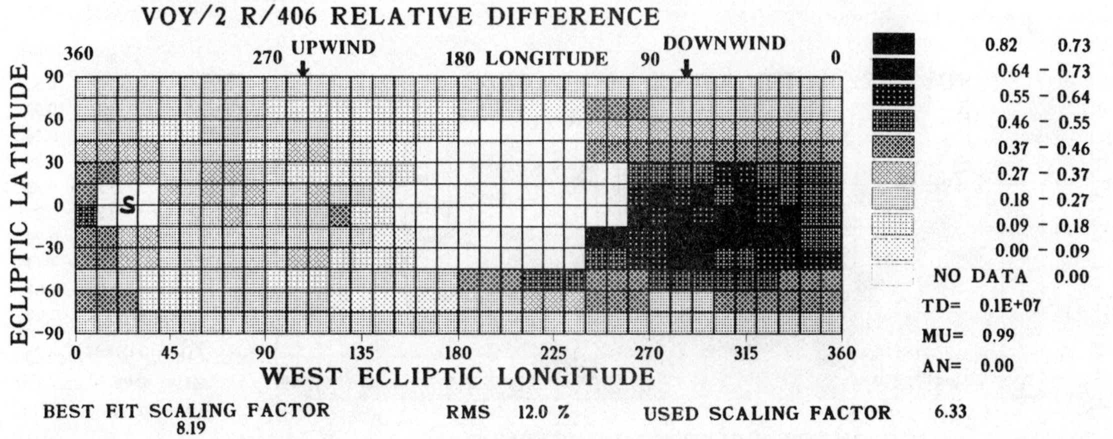

Fig 13: Same as fig 12, but for τ = 1. 10 ^{6} s. The MER is very well represented but there is a large excess of intensity on the downwind side.

- μ and τ assumed constant, whereas in fact they change as a function of solar cycle.
- influence of the shape of the solar Ly-alpha line neglected /12/.
- collisions with solar protons /11/.

2) The velocity and density distributions far from the sun are different from the supposed homogeneous simple Boltzmann distribution at velocity V, owing to perturbations and filtering at the heliospheric interface .

3) There are some additionnal sources of H in the solar system. Possible sources might be: H atoms evaporating from dust grains which are carried in the interstellar flow; H_2 or PAM molecules dissociated near the sun ; CHON particles (from "cometary " grains containing organic material), evaporating near the sun . These "cometary" grains could be either of interstellar origin, or solar system origin. The "torus" distribution (≈ 5 - to 20 AU or even further) may occur as a variant small cometesimals (!) at large distance from the sun.

References

1 - Lallement R.;1989 this issue.
2 - Bertaux J.L.,Lallement R.,Kurt V.G.,Mironova E.N.:1985, Astron. Astrophys. ,**150**,1
3 - Dalaudier F.,Bertaux J. L.,Kurt V. G.,Mironova E. N.: 1984,Astron. Astrophys. **134**, 184
4 - Weller C. S., Meier R. R., 1981 Astrophys. J., **246**, 386
5 - Broadfoot A. L., Kumar S, 1978 Astrophys. J., **222** , 1054
6 - Chassefière E., Bertaux J.L., Lallement R., Kurt V.G.: 1986, Astron. Astrophys. **160**,229.
7 - Paxton L. J., Anderson D.E., Stewart A.I.F., 1988, J. Geophys. Res., **93,** A3, 1766
8 - Thomas G.E.,Anderson D.E., 1976 , Planet. Space Sci., **24**, 303.
9 - Ajello J. M., Stewart A. I., Thomas G. E., Graps A.: 1987, Astrophys. J., **317**, 964
10- Keller H.U., Richter K., Thomas G.E.: 1981, Astron. Astrophys. **102,** 415
11- Chassefière E., Bertaux J.L.:1987, Astron. Astrophys. **174**, 239
12- Chassefiere E., Lallement R., Bertaux J.L., Vial J.C., Sandel B.R.: 1989, this issue.

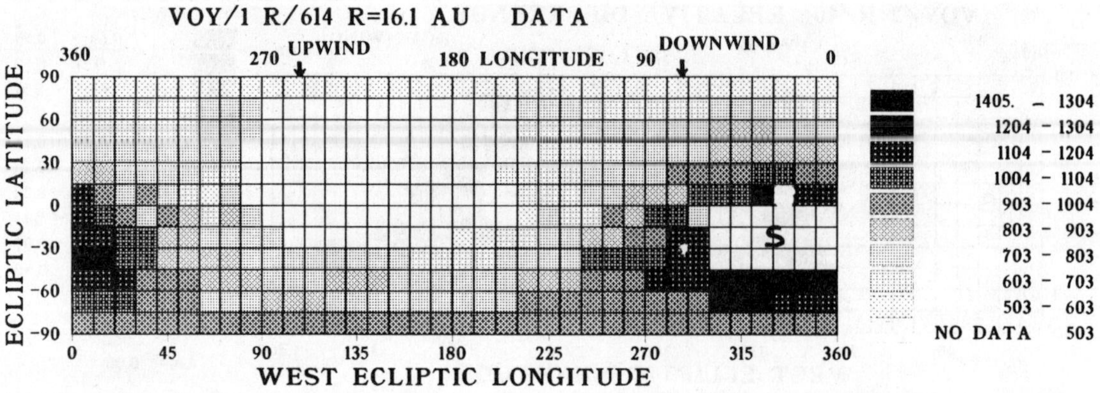

Fig 14 : Ly-alpha map for Voyager 1 roll maneuver n° 614 (r=16.1 AU). There is a large region void of data in the anti-solar direction.

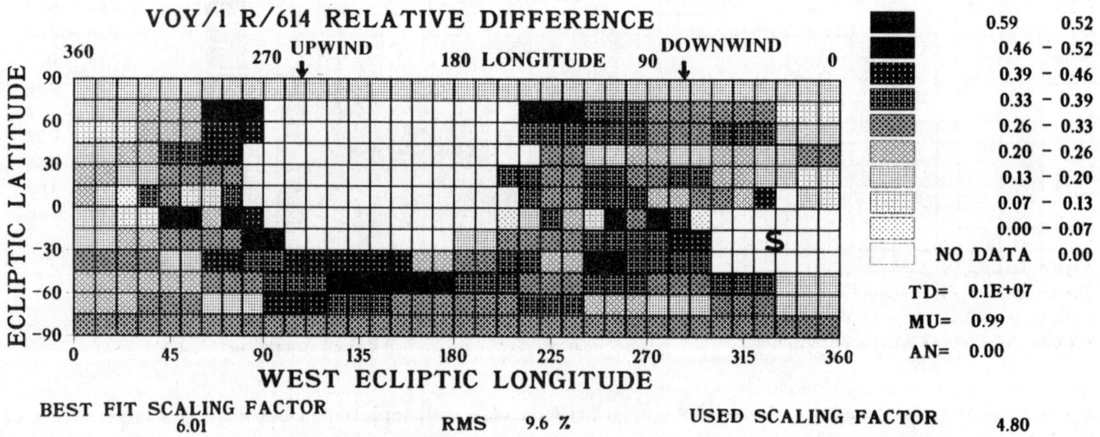

Fig 15 : Relative difference map for a small lifetime $\tau = 1.\ 10^6$ s. Even with this very low value, the anti-solar intensity is larger than predicted by the model. Since it is difficult to imagine that τ could be smaller, an explanation has to be found for this larger relative brigthness.

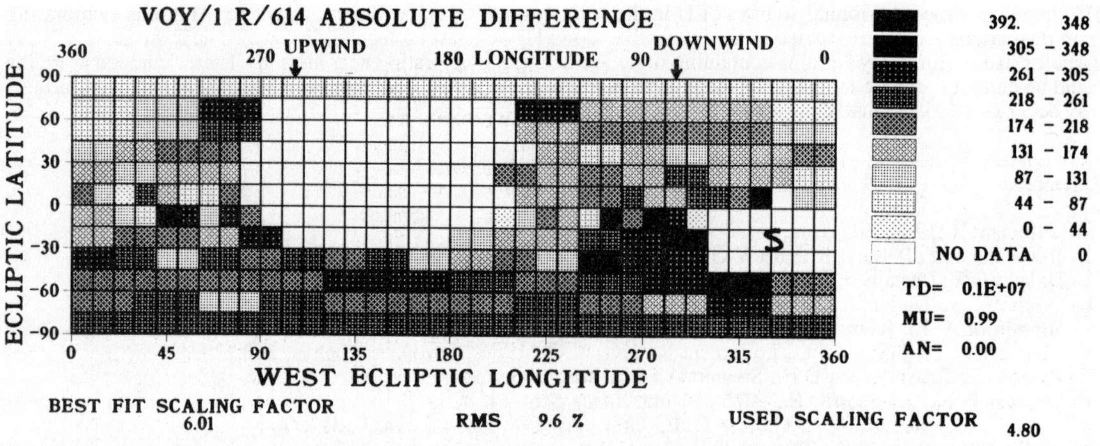

Fig 16: Absolute difference map for the same parameters: there is an excess of intensity in the south .

NEW CHANNEL FOR THE PHOTOIONIZATION OF HYDROGEN ATOMS IN THE SOLAR SYSTEM

M.A. Gruntman

Institute for Problems in Mechanics, Academy of Sciences, Prospekt Vernadskogo 101, Moscow 117526, U.S.S.R.

ABSTRACT

A new two step channel for the photoionization of hydrogen atoms in interplanetary space is proposed. Hydrogen atoms are excited by photons, then decay to the metastable H(2S) state, where they can be photoionized. Competing processes are considered, and the photoionization rate through the proposed indirect channel is calculated. This rate becomes higher than that of direct photoionization for the region closer than 0.4 a.u. to the Sun.

INTRODUCTION

The ionization rate is one of the key parameters in models describing neutral hydrogen atom populations inside of the heliosphere. The study of such populations provides an opportunity to get information on global heliospheric structure, and to derive physical parameters of interplanetary dust and interstellar gas, as well as planetary upper atmospheres and exospheres. Two major processes are believed to determine the ionization rate of the neutral atoms: i) ionization by solar photons, and ii) charge exchange with solar wind ions.

The Sun radiates intense Lyman series hydrogen lines. After photoexcitation to the state with the principal quantum number $n>2$, the hydrogen atom may emit a photon and transfer not to the ground state, but to the metastable H(2S) state. It can be shown that both the direct excitation to the metastable state and the ionization of excited atoms by collisions with solar wind plasma ions and electrons, can be neglected /1/. While being in this metastable state, a hydrogen atom may be ionized by solar photons with wavelength shorter than $\lambda = 3646$ Å (which are rather abundant in the solar spectrum). This way of photoionization was never, to my knowledge, taken into account earlier, and the aim of this work is to consider this indirect channel of ionization of hydrogen atoms in interplanetary space.

EXCITATION TO METASTABLE STATE

Intensities of solar Lyman lines decrease rapidly with the increase of photon energy, and, as will be shown, most of the indirect photoionization effect is due to the excitation channel through the H(3P) state. The total intensity of the solar Lyman-β ($\lambda = 1025.7$ Å) line is $E^{0}_{\beta} = 6\ 10^{-2}$ erg/(cm^{2}s) at Earth orbit /2/ (index "0" will correspond to value at Earth orbit). To assess the line width, let us assume that the same atoms that emit in Lyman-α ($\lambda = 1215.6$ Å) are responsible for the Lyman-β emission. The line width is determined by the Doppler shifts due to velocities of individual atoms, so that the ratio $\varkappa = \Delta\lambda/\lambda$ has to be constant for both lines. For Lyman-α, the line width is well known $-\Delta\lambda = 0.8$ Å (e.g. /3/), giving $\varkappa = 6.5\ 10^{-4}$. For the hydrogen atom, one can calculate exactly the probability of photoexcitation (e.g. /4/). Taking into account that the probability of H(3P) state transition to H(2S) is 0.118, one finally obtains the excitation rate to the H(2S) state $\beta^{2S}(R) = \beta^{2S}_{0}(R_{0}/R)^{2}$, where $\beta^{2S}_{0} = 4\ 10^{-7}\ s^{-1}$. The similar analysis for solar Lyman-γ line ($\lambda = 972.5$ Å) with energy flux $E^{0}_{\gamma} = 1.5\ 10^{-3}$ erg/(cm^2 s) /2/ shows that the excitation rate for H(2S) through the H(4P) state is $3.2\ 10^{-9}\ s^{-1}$ at Earth orbit, which is much lower than the rate through the H(3P) state. This channel can be therefore safely neglected, as can all other radiation channels.

FATE OF METASTABLE ATOM

Let us consider now what may happen with a metastable hydrogen atom in interplanetary space. First of all, without any external interactions this atom will decay to ground state by emission of two photons at a rate of $\delta^{d}=8.226\ s^{-1}$ /5/, i.e. with the life time $t_d=0.12$ s. Collisions with solar wind plasma electrons and ions, as well as interactions with the solar photons, may lead to either ionization of the atom or to a transition to another state which, in turn, may decay to the ground or a metastable state. Also, the decay of metastable atoms can be caused by the solar wind magnetic field.

It can be shown /1/ that the most important effect of the collisions with solar wind protons and electrons is the transition from $2S_{1/2}$ state to the $2P_{1/2}$ or $2P_{3/2}$ states which results in subsequent very fast (1.6 ns) emission of a Lyman-α photon leaving the hydrogen atom in the ground $1S_{1/2}$ state. The solution of the problem of collisionally induced transitions to 2P states of hydrogen atom was given in /6/. Calculations show that for solar wind plasma density $n_o=5\ cm^{-3}$, $V_{SW}=400$ km/s and $T_e^o=10^4$ K one obtaines $\delta_o^{d,e}=2.9\ 10^{-4} s^{-1}$ and $\delta_o^{d,p}=3.5\ 10^{-4} s^{-1}$ for H(2S) depopulation rates due to collisions with electrons and protons of solar wind respectively. Alpha particles constitute approximately 5% of solar wind number density, and the depopulation rate is proportional to the square of the ion charge. Therefore, the total effect of solar wind ions would be 20% more than presented by the value $\delta^{d,p}$.

A magnetic field is frozen into the solar wind plasma and moving relative to the metastable atom, which would then feel the electric field. For small electric fields, the H(2S) decay time, t_E, is well known (e.g. /4/), and $t_E > t_d$ everywhere in the solar wind /1/. Therefore the effect of a magnetic field can be neglected.

Interaction with solar photons may result either in photoionization, including the channel of indirect photoionization of the present work, or in excitation to higher levels of the atom with the subsequent transition to the ground state. The solar spectrum is assumed to be that of a blackbody with temperature $T_S=6000$ K. We assume here, also, that the solar radiation intensity is inversely proportional to the square of the distance from the Sun without change of spectral properties. Then the photoionization rate of H(2S) atoms would be $\delta^{i,ph}(R)=\delta_o^{i,ph}(R_o/R)^2$ and

$$\delta_o^{i,ph}=\mu c\int_{\omega_o}^{\infty}\sigma(\omega)\ dN_\omega$$

where $\omega_o=5.16\ 10^{15}\ s^{-1}$ is the threshold circular frequency for the photoionization, c is the velocity of the light, $\mu=I_o/(4\sigma_{SB}T_S^4)$, $I_o=1.39\ 10^6$ erg/(cm^2 s) is the solar constant (radiant flux density at 1 a.u.), σ_{SB} is the Stephan-Boltzman constant, $\sigma(\omega)$ is the photoionization cross section, and dN_ω is the photon number density distribution as a function of temperature T_S. For the H(2S) state, the ionization threshold corresponds to wavelength $\lambda=3646$ Å. This means that the major role in photoionization is played by the photons from the bulk of the distribution, and not from the far wing which is characterized by relatively large fluctuations in intensity. Therefore the blackbody approximation of the solar spectrum is adequate for the purpose. For the hydrogen atom, the photoionization cross section can be calculated exactly /7/, the value of μ is equal to $0.47\ 10^{-5}$, and one obtains $\delta_o^{i,ph}=1.06\ s^{-1}$.

For the excitation of the metastable atom to the H(3P) state, the photon wavelength is $\lambda=6562.8$ Å (Balmer Hα line). The solar irradiance in Hα is a Fraunhofer line with deep minimum down to 0.16 of the value of the continuum /8,9/. The continuum solar irradiance is equal to $1.6\ 10^2$ erg/(cm^2 s Å) /10/. The spectral radiation density responsible for the excitation of atoms depends then on the radial velocity of the atoms. For neutral hydrogen atoms originating in outgassing of interplanetary dust, the initial radial velocity is very small since dust particles are moving with almost circular orbits. The life time of atoms as neutrals is also small, which does not allow gravitation and radiation forces to transform the initial velocity to motion in radial direction. Therefore the spectral density for excitation would correspond to the minimum of the Fraunhofer Hα line. For interstellar hydrogen, the spectral density for excitation would be somewhat higher. We assume here that effective solar spectral irradiance is 0.2 of the level of continuum. The H(3P) state with the probability 0.118 returns back to the H(2S) state, and the depopulation rate by Balmer Hα photons is then $1.51\ s^{-1}$ at Earth orbit. A similar analysis shows that the depopulation rate through the excitation to H(4P) state is 0.19 s^{-1} at Earth orbit, verifying that this channel is relatively insignificant. The depopulation rate decreases rapidly with the increase of principal quantum number of excited state, and the excitation to the states with n>4 will be neglected. Then the total depopulation rate of H(2S) state by photoexcitation is $\delta^{d,ph}(R) = \delta_0^{d,ph}(R_0/R)^2$, where $\delta_0^{d,ph}=1.70\ s^{-1}$. This value is much higher than corresponding rates due to the interaction with solar wind plasma, and the effect of the latter can be totally neglected.

PHOTOIONIZATION RATE AND DISCUSSION

Let us now summarize the most important processes for indirect photoionization channel:

i) hydrogen atoms are excited to H(2S) state with the rate $\beta^{2S}(R)$;

ii) metastable atoms are spontaneously depopulated to ground state by two photon decay with the constant rate δ^d;

iii) metastable atoms are depopulated to ground state by interaction with solar photons at a rate $\delta^{d,ph}(R)$;

iv) metastable atoms are ionized by solar photons at rate $\delta^{i,ph}(R)$.

The total photoionization rate, $\eta(R)$, through the indirect channel is then

$$\eta(R) = \beta_0^{2S}\left(\frac{R_0}{R}\right)^2 \frac{\delta_0^{i,ph}}{\delta^d (R/R_0)^2 + \delta_0^{d,ph} + \delta_0^{i,ph}}$$

It is more convenient to consider the dependence on distance from the Sun not of the ionization rates, but the rates multiplied by $(R/R_0)^2$. For such quantities, the radial dependence vanishes when the ionization rates are inversely proportional to the square of the distance from the Sun, as is the case for ionization rates due to both direct photoionization and charge exchange on solar wind ions.

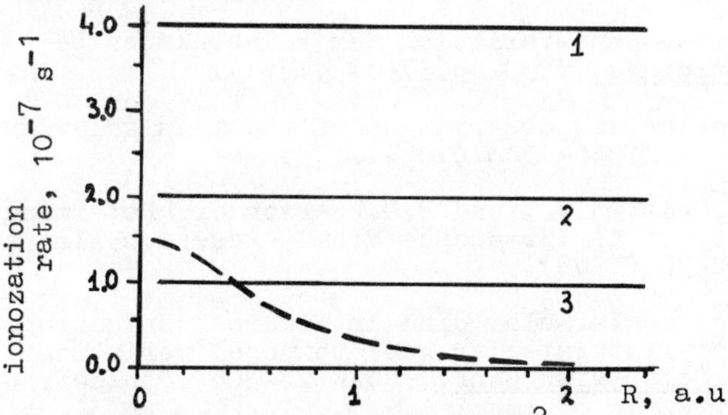

Fig.1. Normalized (multiplied by $(R/R_0)^2$) ionization rates dependences on distance from the Sun. Dashed line - indirect two step photoionization; solid lines - 1) due to charge exchange (at equator), 2) due to charge exchange (at poles), 3) direct photoionization.

The calculated dependence of $\eta^*(R) = \eta(R)\ (R/R_0)^2$ is shown in Figure 1. Also shown are ionization rates corresponding to direct photoionization and charge exchange with solar wind protons. The ionization rates due to charge exchange is believed now to depend strongly on heliographic latitude, its value at equator being twice as large as the value at solar poles /11/. At distances from the Sun of R 1-2 a.u., the ionization rate through the proposed two step channel is proportional to $1/R^4$, and at close approaches, R<0.5 a.u., to $1/R^2$. Here this effect is much more pronounced for neutral hydrogen atoms within Earth orbit, and it vanishes at large distances. At the distance of 1 a.u. from the Sun the proposed channel would add from 8% to 13% to the accepted values of total ionization rates depending on the heliographic latitude. At the distance of 0.42 a.u. from the Sun its effect is equal to the direct photoionization rate, and at closer approaches to the Sun it will add up to 30% at the equator and 50% in polar regions to the accepted values of total ionization rates of hydrogen atoms.

Although the considered effect seems to be too small to result in modifications of interstellar gas properties in Solar system, its influence on the neutral hydrogen atoms outgassing from interplanetary dust grains seems to be significant and must be taken into account. It should be noted that a relatively stronger effect due to the proposed channel can be expected in areas "protected" from solar wind, i.e. inside planetary magnetospheres, particularly for Earth and Venus. It is also interesting to mention that, as estimations show /1/, approximately one hundredth of the sky background far UV radiation is due to the two photon decay of metastable atoms in interplanetary space.

REFERENCES

1. M.A.Gruntman, Photoionization of hydrogen atoms in interplanetary space through excitation of metastable state, Preprint 418, Institute for Problems in Mechanics, The USSR Academy of Sciences, Moscow (1989).

2. J.G.Timothy, The solar spectrum between 300 and 1200 Å, In: The Solar output and its variations, ed. by O.R.White, Colorado Associated University Press, Boulder, Colorado (USA), 1977, p.237-251.

3. A.Vidal-Madjar, Solar spectrum at Lyman-alpha 1216 Å, In: The Solar output and its variations, ed. by O.R.White, Colorado Associated University Press, Boulder, Colorado (USA), 1977, p.213-236.

4. H.A.Bethe and E.E.Salpeter, Quantum Mechanics of One- and Two-electron atoms, Springer-Verlag, 1957.

5. J.Shapiro and G.Breit, Metastability of 2s States of Hydrogenic Atoms, Phys. Rev. v.113, p.179-181 (1959).

6. E.M.Purcell, The lifetime of the $2^2S_{1/2}$ State of Hydrogen in an Ionized Atmosphere, Astrophys. J. v. 116, p.457-462 (1952).

7. D.A.Frank-Kamenetsky, Physical processes in stars' interior (Fizicheskie protsessy vnutri zviozd), in Russian, Fizmatgiz, Moscow (1959).

8. K.-H.David, Die Mitte-Rand-Variation der Balmerlinien Hα-Hδ auf der Sonnenscheibe, Z. Astrophys. v.53, p.37-67 (1961).

9. O.R.White, A Photoelectric observation of the mean solar Hα-line profile, Astrophys. J. v.139, p.1340-1343 (1964).

10. J.C.Arvesen, R.N.Griffin Jr, and B.D.Pearson Jr, Determination of Extraterrestrial Solar Spectral Irradiance from a Research Aircraft, Appl. Optics v.8, p.2215-2232 (1969).

11. R.Lallement, Large scale Solar Wind latitudinal anisotropies as deducted from interplanetary/interstellar gas: temporal variations, In: Proceed. of the VI-th Intern. Solar Wind Conf., NCAR/TN-306, Boulder, Colorado, v.2, 1988, p.651-665.

GENERAL DESCRIPTION OF TIME-DEPENDENT DENSITY FLUCTUATIONS IN THE INTERPLANETARY NEUTRAL GAS DISTRIBUTION BY MEANS OF FOURIER TRANSFORMS

H.J. Fahr and K. Scherer

Institut für Astrophysik und Extraterrestrische Forschung der Universität Bonn, Auf dem Hügel 71, D-5300 Bonn 1, F.R.G.

In past years the intensity distribution $I_H(\vec{R},\vec{\Omega},t)$ of the interplanetary hydrogen-Lyman-α resonance glow was observed by several satellites. In all of these intensity measurements the corresponding photon-counting detectors integrate the resonance glow contributions of all sources along the line of sight according to

$$I_H(\vec{R},\vec{\Omega},t) \sim \int_0^\infty ds\, n_H(\vec{r},t)\, F_S(\vec{r},t) \tag{1}$$

where $n_H(\vec{r},t)$ is the density of the neutral H-atoms at space point $\vec{r}$ and time t, and F_S is the solar Lyman-α irradiation. $I_H(\vec{R},\vec{\Omega},t)$ is the H-Lyman-α resonance glow intensity seen at time t from a position $\vec{R}$ in interplanetary space along a line of sight with a unit vector $\vec{\Omega}$. The observations if carried out over a finite period of time and a variety of lines of sight $\vec{\Omega}$ contain information on the power spectrum of the intensity fluctuations as dependent on the wave vector $\vec{k}$ and frequency ω. In principle, one observes the Fourier backtransform of the power spectrum of the resonance glow intensity, $\mathcal{I}(\vec{k},\vec{\Omega},\omega)$. In what follows, we therefore will derive the relation between $I_H(\vec{R},\vec{\Omega},t)$ and $\mathcal{I}(\vec{k},\vec{\Omega},\omega)$.

For this purpose we start with the definition of the Fourier transform $\mathcal{g}(\vec{k})$ of an arbitrary function $g(\vec{r})$

$$\mathcal{g}(\vec{k}) = \frac{1}{(2\,\pi)^{3/2}} \int_{\mathbb{R}^3} g(\vec{r})\, \exp(-\,i\,\vec{k}\cdot\vec{r})\, d^3r \tag{2}$$

In the following text, the transform of any function g is denoted by corresponding script letters $\mathcal{g}$ as well as on purpose by a bar $\bar{g}$. The transform of the product of two functions n_H and F_S is given by the convolution of the two transformed factors

$$\overline{(n_H\, F_S)} = \int_{\omega'}\int_{k} \mathcal{n}_H(k-k',\omega-\omega')\, \mathcal{F}_S(k',\omega')\, d^3k'd\omega' = \mathcal{n}_H ** \mathcal{F}_S \tag{3}$$

The double asterisk symbol defines the operation of convoluting two functions. From equations (1) to (3) we derive the identity

$$I_H(\vec{R},\vec{\Omega},t) \sim \int_l dl \int_\omega d\omega \int_k d^3k \; \overline{(n_H\, F_S)}\; \exp(+i(\vec{k}\cdot\vec{r} + \omega t)) \tag{4}$$

If we can calculate the Fourier-transformed solar Lyman-α irradiation $\mathcal{F}_S(\vec{k},\omega)$ and the transformed density $n_H(\vec{k},\omega)$ of interplanetary neutral hydrogen, we can solve the integral equation (4) and thereby our problem of representing the intensity $I_H(\vec{R},\vec{\Omega},t)$ from known input functions.

The solar Lyman-α flux F_H may be obtained from corresponding observations at the orbit of the Earth. There are theoretical models available (Fahr and Rucinski, 1987; Fahr and Scherer, 1989) which describe its dependence on space-time coordinates.

The appropriate description of the density $n_H(\vec{r},t)$ is derivable from a nonlinear partial differential equation of the Boltzmann type for the velocity distribution function $f_H(\vec{r},\vec{v},t)$ of hydrogen atoms

$$\left[\frac{\partial}{\partial t} + \vec{v}\cdot\frac{\partial}{\partial\vec{r}} + \frac{\vec{K}(\vec{r},t)}{m}\cdot\frac{\partial}{\partial\vec{v}} + \nu(\vec{r},t)\right] f_H(\vec{r},\vec{v},t) = 0 \tag{5}$$

This differential equation is too difficult to be solved in general. But when we know the effective solar gravity force $\vec{K}(\vec{r},t)$ and the solar ionization field $\nu(\vec{r},t)$, e.g. as described in a model given by Fahr et al. (1987), formula (5) reduces to a linear partial differential equation.

We now evaluate the Fourier transform of Equation (5) to find an algebraic expression for $\mathcal{h}$. The product of the effective force $\vec{K}$ with the distribution function f is convoluted into two transformed factors, $\mathcal{K}$ and $\mathcal{h}$, as is the product of the ionization field ν with f. Thus one obtains (Fahr and Scherer, 1989)

$$i\,(\omega+\vec{k}\cdot\vec{v})\,\mathcal{h}(\vec{k},\vec{v},\omega) + \left[\frac{\vec{\mathcal{K}}(\vec{k},\omega)}{m}\cdot\frac{\partial}{\partial\vec{v}} + \overline{\nu(\vec{k},\omega)} + \frac{\vec{\mathcal{K}}(\vec{k},\omega)}{\varepsilon_\infty}(\vec{V}_\infty-\vec{v})\right] ** \,\mathcal{h}(\vec{k},\vec{v},\omega) = 0 \tag{6}$$

Here we have assumed that the distribution function is the product of an undisturbed function $f_\infty(\vec{v})$ and the fluctuation $h(\vec{r},\vec{v},t)$, i.e.

$$f(\vec{r},\vec{v},t) = h(\vec{r},\vec{v},t)\; f_\infty(\vec{v}) \tag{7}$$

where $f_\infty(\vec{v})$ is the distribution function outside the heliosphere depending only on the velocity $\vec{v}$ and is described here as a Maxwellian shifted by a wind velocity $\vec{v}_\infty$:

$$f_\infty(\vec{v}) = n_\infty \left(\frac{m}{2\pi\varepsilon_\infty}\right)^{3/2} \exp\left[-\frac{m}{2\varepsilon_\infty}(\vec{V}_\infty - \vec{v})^2\right] \tag{8}$$

with $\varepsilon_\infty = k\,T_\infty$ being the initial thermal energy at infinity.

In the next step we find the Fourier-transformed force and ionization fields. In configuration space these functions are given by (see Fahr et al. 1987)

$$\vec{K}(\vec{r},t) = K_E \vec{e}_r \frac{r_E^2}{r^2} T_K(t) = - K_E r_E^2 \; T_K(t) \; \nabla \frac{1}{r} \tag{9}$$

and

$$\nu(\vec{r},t) = \nu_E \frac{r_E^2}{r^2} T_\nu(t) \tag{10}$$

with

$$T_i = b_{0i} + b_{1i} \cos\{b_{2i}t\} \exp\{b_{3i}\cos\{b_{2i}t\}\} \tag{11}$$

($i \in \{K,\nu\}$). The explicit transformations of equations (9) - (11) are complicated; for more details see Fahr and Scherer, 1989. We give here only the results

$$\vec{\mathcal{K}}(\vec{k},\omega) = - r_E^2 K_E \mathcal{T}_K(\omega) \nabla \frac{\bar{1}}{r} = i\, r_E^2 \sqrt{\frac{2}{\pi}} \; \mathcal{T}_k(\omega) \frac{\vec{k}}{k^2} K_E \tag{12}$$

and

$$\overline{\nu(\vec{k},\omega)} = \sqrt{\frac{2}{\pi}} \; r_E^2 \; \mathcal{T}_\nu(\omega) \; \frac{1}{k} \; \nu_E \tag{13}$$

and $\mathcal{T}_i(\omega)$ are the transformations of time-dependencies of $\vec{K}$ and ν, which is given as an infinite sum of delta distributions. The convolution of an arbitrary function with a delta function reproduces the original function. Thus the convolution of the distribution function $\hbar(\vec{k},\vec{v},\omega)$ with the time-transformed force $\tau_K(\omega)$ and the ionization field $\tau_\nu(\omega)$ gives a constant M_i times the function $\hbar(\vec{k},\vec{v},\omega)$.

We now must find the convolution with respect to the space coordinates. We develop the function $\hbar(\vec{k}-\vec{k}',\vec{v},\omega)$ in a Taylor series around $\vec{k}$ and take only the first term into account. Then we have to find the following integrals:

$$F_1 = \left[\iiint \frac{\vec{k}'}{k'^2} \, d^3k' \right] \cdot \frac{\partial}{\partial \vec{v}} \hbar(\vec{k},\vec{v},\omega) \tag{14}$$

and, analogously, the second integral (for details see Fahr and Scherer, 1989).

We want to give the transformed Boltzmann equation in a dimensionless form and therefore have to introduce $\vec{\kappa} = \vec{k}/k_\odot$, $\mu = \omega/\omega_\odot$, and $\vec{w} = \vec{v}/w_\infty$. Then equation (6) yields

$$\left[\omega_\odot \mu + w_\infty k_\odot \vec{\kappa}\cdot\vec{w} + C_1\Gamma(\kappa)\, \kappa\cdot\frac{\partial}{\partial \vec{w}} + C_1\Gamma(\kappa)\, \vec{\kappa}\cdot(\vec{W}_\infty - \vec{w})\right] \hbar(\vec{\kappa},\vec{v},\omega) = i\, C_2 \Xi(\kappa)\; \hbar(\vec{\kappa},\vec{v},\omega) \tag{15}$$

with the shorthand notation

$$C_1 = r_E^2 \sqrt{\frac{2}{\pi}} \frac{k_\odot^2 M_K K_E}{m\, w_\infty} = - \sqrt{8}\,\pi \left[\frac{r_E}{r_\odot}\right]^2 M_K\, \omega_g \tag{16}$$

and

$$C_2 = r_E^2 \sqrt{\frac{2}{\pi}} \; k_\odot^2 M_\nu K_E = - \sqrt{8\,\pi} \left[\frac{r_E}{r_\odot}\right]^2 M_\nu\, \omega_E \tag{17}$$

With the ansatz

$$\hbar(\vec{\kappa}, \vec{w}, \omega) = \hbar_x(\vec{\kappa}, w_x, \omega)\, \hbar_y(\vec{\kappa}, w_y, \omega)\, \hbar_z(\vec{\kappa}, w_z, \omega) \tag{18}$$

we find the homogenous solution for one component, namely the z-axis

$$\hbar_z(\vec{\kappa}, \vec{w}, \omega) = h_{0z} \exp\left[\frac{w_z^2}{2} \left(1 - \frac{\omega_\infty}{C_1} \Gamma(\kappa)\right) - \left(w_{\infty z} + \frac{\Pi(\kappa, \mu)}{3\, \kappa_z}\right) w_z \right] \tag{19}$$

with

$$\Pi(\kappa, \mu) = \frac{\omega_\odot\, \mu - i\, C_2\, \Xi(\kappa)}{C_1\, \Gamma(\kappa)}, \quad \omega_\infty = 2\, \pi\, \frac{w_\infty}{r_s} \tag{20}$$

$\Xi(\kappa)$ and $\Gamma(\kappa)$ are, as the solution of the integrals F_1 and F_2, functions of κ only, namely

$$\Gamma(\kappa) = \frac{\pi}{\kappa^3} \left[\kappa + \kappa^3 + \frac{1}{2}(1 - \kappa^2)^2 \ln \frac{1-\kappa}{1+\kappa}\right] \tag{21}$$

and

$$\Xi(\kappa) = \frac{4\, \pi}{3\, \kappa} \left[1 - \sqrt{1 - \kappa^2}^{\,3} \right] \tag{22}$$

The density of the neutral particles is then given by integrating equation (19) over the velocity space.

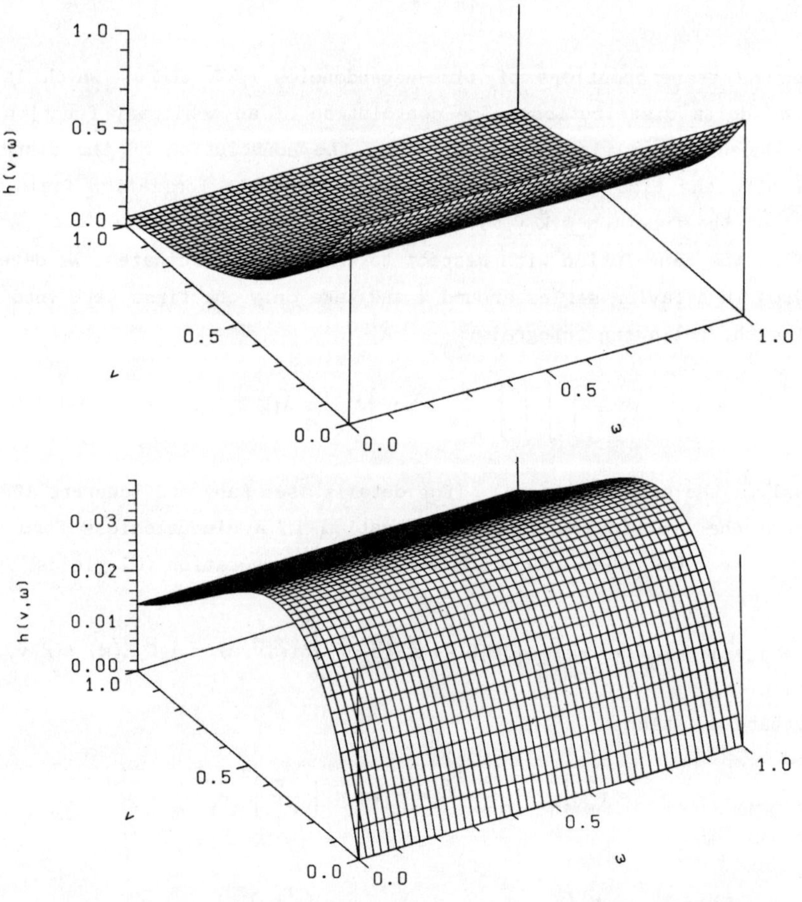

Figure 1 Real and imaginary part of the distribution function (19) for fixed $\vec{\kappa}$

$$n(\vec{\kappa},\mu) = \int_V f(\vec{\kappa},\vec{w},\mu)\, d^3w = \hbar_{0x}\hbar_{0y}\hbar_{0z} n_\infty \left[\frac{2\, C_1 \Gamma(\kappa)}{\omega_\infty}\right]^3 \exp\left[-\frac{W^2_{\infty z}}{2} + \frac{\left[\Pi(\kappa,\mu)\; C_1\Gamma(\kappa)\right]^2}{9\, C_1^2}\left(\frac{1}{\kappa_x^2}+\frac{1}{\kappa_y^2}+\frac{1}{\kappa_z^2}\right)\right] \tag{23}$$

Representing the solar irradiation $F_S(\vec{r},t)$ in an analogous way as the ionization field $\nu(\vec{r},t)$ in equation (10), we can evaluate the Fourier-transformed irradiation $\mathcal{F}(\vec{k},\omega)$ in the same manner. The convolution of the density with the solar irradiation is calculated in an analogous form in equation (14) and gives

$$\boxed{n_H\, F_S}\,(\vec{k},\omega) = r_E^2\, F_{SE} \sqrt{\frac{2}{\pi}}\; M_\nu\; \Xi(k)\; n(k,\omega) \tag{24}$$

where F_{SE} is a reference value for the solar irradiation at r_E.
Returning to equation (4) and exchanging the sequence of integrations, we find

$$I_H(\vec{R},\vec{\Omega},t) \sim \int_\omega d\omega\, e^{-i\omega t} \int_k n_H^{**}\, \mathcal{F}_S\, e^{-i(\vec{k}\cdot\vec{R})}\, d^3k \int_0^\infty e^{-i(k\cdot\Omega)l}\, dl \tag{25}$$

$$\sim \int e^{-i\omega t}\, \mathcal{I}_H(\vec{R},\vec{\Omega},\omega)\, d\omega$$

The last integral finally yields

$$\mathcal{I}_H(\vec{R},\vec{\Omega},\omega) = \int_k n_H^{**}\mathcal{F}_S\, e^{-i(\vec{k}\cdot\vec{R})} \left[\sqrt{\frac{2}{\pi}} - i\sqrt{\frac{\pi}{2}}\,\frac{1}{k}\,\delta(\vec{k}\cdot\vec{\Omega})\right] d^3k \tag{26}$$

This is the power amplitude of the hydrogen Lyman-α resonance glow intensity seen from $\vec{R}$ in a direction $\vec{\Omega}$, which is a directly accessible observable quantity. In the future we want to compare the observed data with numerical calculations based on formula (26). We have shown only the homogenous part of the differential equation (15).

VI REFERENCES

Bertaux, J.L., and Lallement, R.: (1984) Astron. Astrophys., **140**, 230

Bertaux, J.L., Lallement, R., Kurt, V.G., and Miranova, E.N.: (1985) Astron. Astrophys., **150**, 1

Broadfoot, A.L. and Kumar, S.: (1978) Astrophys. J., **222**, 1054

Chassefière, E., Bertaux, J.L., Lallement, R., and Kurt, V.G.: (1986), Astron. Astrophys., **160**, 229

Cook, J.W., Meier, R.R., Brueckner, G.E., and Van Hoosier, M.E.: (1981), Astron. Astrophys., **97**, 394

Fahr, H.J.: (1971) Astron. Astrophys., **14**, 263

Fahr, H.J.: (1978) Astron. Astrophys., **66**, 103

Fahr, H.J., Nass, H.U. and Rucinski, D.: (1985) Astron. Astrophys. **142**, 476

Fahr, H.J., Rucinski, D. and Nass, H.U.: (1987) Ann. Geophys., **A5**, 255

Fahr, H,J. and Rucinski, D.: (1989) Planet. Space Sci., **37**, 555

Fahr, H.J. and Scherer, K.: (1989) in press at Astron. Astrophys.

Lallement, R.: (1988) in "SOLAR WIND 6" Vol. II, pp. 651, Eds. Pizzo, Holzer, Sime
Lallement, R., Bertaux, J.L., and Dalaudier, F.: (1985) Astron. Astrophys., **150,** 21
Lallement, R. and Stewart, J.: (1989) in press at Astron. Astrophys.
Nass, H.U. and Fahr, H.J.: (1986) Advances in Space Research, **6(1)**, 365
Rucinski, D. and Fahr, H.J.: (1989) Astron. Astrophys., **974,** 234
Scherer, K.: (1988) Ph.D. Thesis, University of Bonn
Wallis, M.K.: (1975) Planet. Space Sci., **23,** 419
Witt, N., Ajello, J.M., and Blum, P.W.: (1979) Astron. Astrophys., **73,** 272
Wu, F.M. and Judge, D.L.: (1978) Astrophys. J., **225**, 1045
Wu, F.M. and Judge, D.L.: (1979) Astrophys. J., **231**, 594
Wu, F.M., Gangopadhyay, P., Ogawa, H.S., and Judge, D.L.: (1989) Astrophys. J., **336,** 10(
Gangopadhyay, P., Ogawa, H.S., and Judge, D.L.: (1989), Astrophys. J., **336,** 1012

THREE-DIMENSIONAL MODELS OF THE GLOBAL ZODIACAL DUST CLOUD — A BIMODAL MODEL OF THE SOLAR DUST HELIOSPHERE

B. Kneissel and I. Mann

Ruhr-Universität Bochum, D-4630 Bochum, F.R.G.

ABSTRACT

Several three-dimensional models of the interplanetary dust cloud or the dust heliosphere describing the number density of particles have been proposed according to infrared and optical measurements. Number densities close to the earth predicted by the different models converge sufficiently. But for regions close to the solar poles discrepancies between different models grow dramatically. Infrared observations demand lower particle densities, whereas most of the optical models suggest increasing particle densities. This problem may be solved under the assumption of different components within the global dust cloud with different particle properties. Finally the results from the dust experiment onboard the ULYSSES spacecraft will clarify whether this suggestion is reasonable.

INTRODUCTION

Deriving the three-dimensional particle number density distribution, $n(r,\beta_o)$ in helioecliptical coordinates /1/, of the zodiacal dust cloud from observational data bears several problems. The observed brightness in the visible wavelength range is not only related to the number density n of dust along the observer's line of sight (LOS) but also to nonisotropic scattering of sunlight at the particles. Determining the infrared brightness thermal emission of the grains, certainly isotropic, is governed by radiation balance temperature and thus unfortunately affected by scattering behaviour, too. Facing these ambiguities a variety of different models describing the distribution of interplanetary dust in terms of global properties has been proposed by researchers. Nevertheless these models have to match brightness observations.

COMPATIBILITY OF MODELS WITH OPTICAL AND INFRARED OBSERVATIONS

In detail, an observer located at solar distance R in the symmetry plane of the zodiacal dust cloud receives radiation scattered or emitted thermally from particles from each volume element P along the LOS at an angle Θ within his viewing cone. The scattering angle θ is the angle between the propagation directions of incident light and of light scattered at dust particles. The particles of volume element P are illuminated by the Sun at distance r. The angle between the observer's viewing direction (direction of LOS) and the direction towards the Sun is denoted by the elongation ε.

The discussed models about distribution of interplanetary particle number density are based on the assumption that the number density can be separated into one factor increasing with solar distance r according to a power law with exponent α and another one depending on the helioecliptical latitude β_o. Then the observed brightness in the optical and infrared wavelength range is calculated by integration along the LOS regarding the scattering or thermal properties of the grains in the volume elements (ref. /2/). Whereas the so called volume scattering function describes the average scattering pattern per volume element in the optical range, and thermal emission depends on the temperature distribution of the particles and their average emission cross-section in volume element.

According to the characteristics outlined above, Giese et al. /1,3/ carried out extensive investigations about the compatability of models with visible observations given by Dumont and Levasseur-Regourd /4/. Results show predictions by both optical and infrared models indicating a converging run of number densities close to Earth, significantly seen at large elongations, are in fair agreement with observational data. Towards the Sun the models, however, show contradicting results. Whereas most of the optical models suggest an increase of number density in alignment with visible observations, the infrared models demand a decrease there /5/. Due to these discrepancies special attention should be paid to particle properties in the infrared and optical range resulting from particles of different origin.

DIFFERENT COMPONENTS OF THE DUST CLOUD

In general, interplanetary dust originates from comets emitting dust during perihelion passage, or from mutual collision of asteroids, in order for the dust to penetrate the inner of Solar system by the decelarating forces of the Poynting-Robertson drag.

Cometary dust particles show different optical properties compared to particles of the global zodiacal dust cloud. Original cometary particles with mantles of cometary material /6,7/ are extremly dark /8/. Resulting from long period comets they move on high eccentric orbits beyond the asteroidal belt ($R>3.3$ AU). Particles of asteroidal origin show a higher reflectivity. According to Fechtig /9/ the asteroidal contribution makes up at least 2/3 of all ecliptic concentrated grains and should have an albedo between 10 to 20 %.

ASPECTS OF DYNAMICAL ANALYSIS

Dynamical investigation proposes a reasonable scenario of spatial distribution for these different components. The dynamics of the dust cloud were evaluated by deriving the distribution of the orbital elements from models describing spatial distribution of zodiacal dust /10,11/. The distribution of inclinations derived from the latitudinal part of the number density distribution reveals the existence of two components. The main contribution is concentrated towards the ecliptic plane with a maximum of inclination distribution between $\langle i\rangle=8°$ and $\langle i\rangle=15°$. An isotropic background with an inclination distribution proportional to sin (i), whith half of the particles in retrograde motion, is superimposed on the ecliptical component. Analyzing the different model predictions the isotropic component varies for the different models. According to models particles of the isotropic component may contribute up to 20% of the ecliptical number density. The isotropic component is suggested by research on dynamics to have its origin in long period comets.

LONG PERIOD COMETS A SOURCE OF THE ISOTROPIC COMPONENT

Analysis of collisional balance of the zodiacal dust cloud carried out by Grün et al. /12/ requires an additional source of dust particles. As recently found, however, the input of short period comets cannot refill the zodiacal dust cloud /13/. Then long period comets and related meteorite streams distributed isotropically in interplanetary space should be accounted as a potential source of zodiacal dust particles.

Pioneer 10/11 measurements give indications for an isotropic dust component. There was found a surprisingly high, nearly constant density of dust particles in a heliocentric distance from 1.1 AU to 20 AU in dust impact measurements /14/. This flux of particles was seen in at least two experiments. On the other hand the observed brightness from a photometric experiment /15/ for a great heliocentric distance ($R > 2.8$ AU) in the backscattering range ($\varepsilon>170°$) is very low (< 2 S10, 1 S10 $= 1.17 \cdot 10^{-9}$ erg cm^{-2} s^{-1} ster^{-1} Å^{-1} at $\lambda = 555$ nm /2/).

Despite these contradictory experimental results, Humes /14/ suggests an extended isotropic particle flux giving this nearly constant particle number density in the whole dust heliosphere. There the eccentricity of particle orbits is suggested to be $e = 0.99$. This isotropical component is not identical with the global zodiacal dust cloud which is obviously concentrated to the ecliptic plane. On top of that, zodiacal dust particles of this particle number density would produce a brightness much too high. Consequently, the particle densities measured in impact experiments of PIONEER cannot be explained as an effect of the global zodiacal dust cloud. Nevertheless the existence of a wide extensive isotropic dust cloud is not inconsistent with the distribution of long period comets. Finally a model, showing an ecliptic concentrated zodiacal component and an isotropic background, consistent with in-situ measurements and brightness observations, can be found from the proper choice of optical and thermal properties of particles in the isotropic background component.

PROPERTIES OF THE ISOTROPIC COMPONENT IN THE OPTICAL AND INFRARED RANGE

The volume scattering function for the isotropic component is derived from inversion of brightness integral in the backscattering domain from observations beyond the asteroid belt mentioned above. This was done under the assumption of constant particle density and a brightness with a magnitude of 2 S10 in 180° direction . Then this scattering function amounts to about 1/10 % of the one for the zodiacal light. Assuming the geometric albedo to be 3%, according to very dark material, we find the average cross-section per volume element. In contradiction to their optical darkness, these particles will give a high contribution to the infrared brightness. Consequently great discrepancies between infrared and optical brightness of the isotropic dust cloud will occur.

FIRST ESTIMATE FOR THE RESULTING BRIGHTNESS CONTRIBUTION

For a first test of these assumptions, we modelled the optical brightness of the isotropic component for viewing directions along the ecliptical plane and a perpendicular plane through the Sun (along the helioecliptical meridian). This brightness is found to be about 10% of the observed zodiacal light brightness. Its contribution to the optical brightness in the ecliptic plane is negligible. This is an acceptable result, regarding the deviations of different zodiacal dust models to the observed brightness.

For the infrared domain, the contribution of the isotropic dust cloud is only a small fraction of the observed brightness in the sunward direction, but contributes considerably to brightness in the gegenschein direction even for the results in the ecliptic plane. Finally, under the assumption of black body radiation, the resulting ecliptic brightness from the zodiacal dust cloud plus the contribution from the isotropic dust cloud gives a reasonable fit to the infrared data given by Murdock and Price /16/. A slight higher temperature would even improve the fit. These first estimates show that the combination of results from in-situ measurements and from remote sensing observations, as done here, should lead to very promising approaches. Of course the scenario developed has to stand the test from the dust experiment of the ULYSSES mission dedicated to the detection of the out-of-ecliptic distribution of interplanetary dust.

SUMMARY

The interplanetary dust cloud or dust heliosphere may be regarded as a superposition of one component mainly concentrated to the ecliptic plane with regular properties known from the zodiacal dust and a second one that is isotropical distributed and may result from long period comets. This approach promises to give a consistent and more physical explanation of results from in-situ measurements as well as optical and infrared brightness observations.

REFERENCES

1. R.H. Giese, B. Kneissel, and U. Rittich, Three-Dimensional Models of the Zodiacal Dust Cloud: A Comparative Study. ICARUS 68, 395-411 (1986)

2. C. Leinert, Zodiacal Light - A Measure of the Interplanetary Environment, Space Sci. Rev. 18, 281-339 (1975)

3. R.H. Giese and B. Kneißel, Threedimensional Models of the Zodiacal Dust Cloud: II. Compatibility of Proposed Infrarad Modells, ICARUS 81, 369-378 (1989)

4. R. Dumont, and A.-C. Levasseur-Regourd, Properties of Interplanetary Dust from Infrared and Optical Observations, Astron. Astrophys. 191, 154-160 (1988)

5. B. Kneißel, R.H. Giese, and I. Mann, The Three-Dimensional (3D-) - Distributions of Zodiacal Dust Derived from Infrared and Visual Measurements and their Compatibility including Dust Dynamics. Adv. Space Res. Vol. 10, No. 3-4, (3)181-(3)183 (1990)

6. J.M. Greenberg and R. Grim, The Origin and Evolution of Comet Nuclei In: Exploration of Halley's Comet (B. Battrick, E.J. Rolfe, and R. Reinhard, eds.), ESA SP-250, 255-264 (1986)

7. J.M. Greenberg, Dust in Diffuse Clouds: One Stage in a Cycle. In: Light on Dark Matter (F.P. Israel, ed.), Reidel, Dordrecht, 177-188 (1986)

8. M.S. Hanner and R.L. Newburn, Infrared Photometry of Comet Wilson (19861) at two Epochs, Astron. J. 97, 254-261 (1989)

9. H. Fechtig, Dust in the Solar System, Z. Naturforsch. 44a, 877-882 (1989)

10. B. Kneißel and R.H. Giese, The Dynamics of the Zodiacal Dust Cloud on account of Optical and Infrared Observations, Publ. Astr. Inst. Czech. Ac. Sci. 67, Proc. Vol. 2, 241-244 (1987)

11. B. Kneißel, The Global Distribution of Zodiacal Dust, Proc. IAU Symp. 139 in press (1990)

12. E. Grün, N. Pailer, H. Fechtig, and J. Kissel, Orbital and Physical Characteristics of Micrometeorids in the Inner Solar System as Observed by Helios 1. Planet. Space Sci. 28, 333-279 (1980)

13. Ch. Leinert and E. Grün, Interplanetary Dust, in Physics of the Inner Heliosphere (R. Schwenn and E. Marsch, eds.) Springer, in preparation

14. D.H. Humes, Results of Pioneer 10 and 11 Meteoroid Experiments: Interplanetary and Near-Saturn. J. Geophys. Res. 79, 5841- 5852 (1980)

15. D.W. Schuerman, J.L. Weinberg, and D.E. Beesson, The Decrease in Zodiacal Light with Heliocentric Distance during the Passage of Pioneer 10 through the Asteroid Belt, Bull. Amer. Astron. Soc. 9, 313 (1977)

16. T.L. Murdock and S.D. Price, Infrared Measurements of the Zodiacal Light. Astron. J. 90, 375-386 (1985)

ON THE POSSIBILITY OF DETECTION OF SMALL COMETS IN Ly-α

M. Banaszkiewicz, S. Grzedzielski, D. Ruciński and M.S. Staniucha

Space Research Centre, Ordona 21, 01–237 Warszawa, Poland

ABSTRACT

Possibility of detection of small comets in the heliosphere by simple monitoring technique of the Lyman-α interplanetary glow is discussed. Estimates of the detection possibility of water-type objects in the range of 0.1–1 km radius sublimating at a typical cometary rate are provided for both periodic and non-periodic populations of comets. A simple photometer with a 2° full FOV, scanning the sky by spacecraft rotation could detect up to 30 such small objects per year.

INTRODUCTION

Recently the hypothesis of a swarm of small (1-m to 100-m) cometesimals penetrating the inner Solar System has been introduced /1/. These icy objects, if numerous, should produce a great number of hydrogen atoms and give an enhancement of the Ly-α scattered light as compared with the background level. Such effect was first claimed to be seen in the Voyager-1 Ly-α spectrometer data measured shortly after launch /2/. However, the Voyager's observations were misinterpreted and the corrected values of the Ly-α intensities happened to be of the order of the background level /3/. The last result gives an upper limit on the population of cometesimals, but does not exclude the possibility of their existence.

In the size spectrum of small icy objects there is an interval between cometesimals and visible comets that extends from 100 m to about 1 km. The upper limit is given by the size of comets that still have high probability to be detected by classical astronomical observations. Smaller comets (fainter than 11 magnitude) are less likely to be discovered against the visible background of many stars with comparable magnitudes.

SUBLIMATION RATE

The insolated cometary nucleus absorbs a major part of the received energy. This energy is afterwards lost by the surface layer of the nucleus either by sublimation of volatiles from it, or by thermal radiation, or eventually by heat transport into deeper layers. Total sublimation rate is of course proportional to the ice covered area on the nucleus' surface. In the case of a uniform icy nucleus this area is equal to the surface of the insolated hemisphere, but some comets, like P/Halley, show only small parts of exposed ice on their surfaces, the rest being covered by the dust mantle. To find the sublimation rate one has to calculate the surface temperature of the nucleus. The simplest method is to use energy balance equation on the surface, but then the heat conduction term must be either neglected or assumed to have some *a priori* value. We use therefore another, more complicated but accurate method which consists of solving the heat conduction equation for the vertical section of the nucleus and requires knowledge of dynamical and physical parameters of the nucleus: spin axis orientation, rotation period, thermal inertia and other characteristics (cf. /4/). Dependence of the calculated sublimation rate on heliocentric distance, for several sets of nucleus parameters, is illustrated in Figure 1.

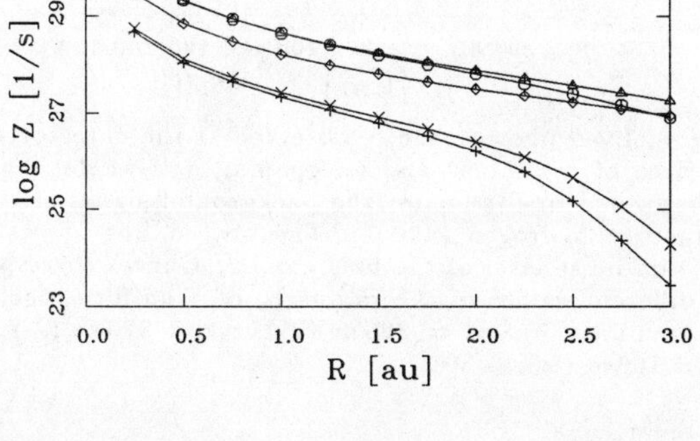

Fig. 1. Sublimation rate Z *versus* heliocentric distance R for a uniform nucleus [spin axis perpendicular (⊙) or parallel (△) to the orbital plane], and for a nucleus with 10% of its surface covered by an active region [inclination of the spin axis to the orbital plane I and latitude of the active region ϑ are: $I = 0°$, $\vartheta = 0°$ (+), $I = 0°$, $\vartheta = 45°$ (×), $I = 90°$, $\vartheta = 45°$ (◇)]. The nucleus radius = 1 km, the thermal inertia = $250\,\mathrm{W\,s^{1/2}m^{-2}K^{-1}}$, and the rotation period = 48 hours.

HYDROGEN CLOUD AND Ly-α SCATTERING

An excellent review of Ly-α scattering by hydrogen atoms forming the coma of a comet has been given by Keller /5/. H_2O molecules that sublimate from the surface leave the nucleus with an average thermal velocity of a few hundred $m\,s^{-1}$ for temperatures of 100–200 K which are typical for Sun-comet distances greater than 1 AU. They are subject to solar radiation field which photodissociates them in a time scale not exceeding some tens of hours. During that time an average molecule travels a distance of about 10^5 km from the nucleus. The first photodissociation process breaks H_2O molecules into H and OH. In the next, one oxygen and one hydrogen atom are produced from OH. From the observed shapes and sizes of the hydrogen coma around comets, it was found that the average velocity of the first H atom photodissociated from H_2O molecule is about $v_1 = 19\,\mathrm{km\,s^{-1}}$ while the second atom moves with the mean velocity of $v_2 = 8\,\mathrm{km\,s^{-1}}$ with respect to the nucleus. Hydrogen atoms move away from the nucleus till they are ionized by charge-exchange collision with solar wind or by solar photons. Ionization time for both processes acting together is 2.1×10^6 s at 1 AU. Hence the scale length of the cloud is 10^7 km. The dissociation region of H atoms (10^5 km) is then small as compared with the cloud size and can be regarded as a point source.

The radiation pressure in Ly-α deflects trajectories of atoms in the cometary system of coordinates and deforms spherical shells of hydrogen into ellipsoidal-like structures extending toward the comet's tail in the anti-Sun direction. Fortunately, within $\sim 5 \times 10^6$ km from the comet (from where comes the strongest signal in Ly-α) this deformation is negligibly small, which allows us to use the spherically symmetric density distribution, in which the hydrogen density is:

$$n_H = n_{H_1} + n_{H_2} = 0.5 \times \frac{Z}{4\pi d'^2}\left(\frac{1}{v_1}e^{-Bd'/v_1} + \frac{1}{v_2}e^{-Bd'/v_2}\right), \quad (1)$$

where n_{H_1}, n_{H_2} are densities of the two overlapping clouds of H atoms moving with velocities v_1 and v_2; d' is the distance from the nucleus, Z and B are the sublimation and ionization rates, respectively.

With reference to Figure 2, let O be a detector (placed at distance r from a nucleus) with the field of view given by a narrow cone with the half-angle ε; if the line-of-sight (axis of the cone) is at a distance $d \ll r$ from the center of the coma then the total number of atoms inside the column determined by the field of view is approximately equal to

$$N_H = \frac{Z\varepsilon^2}{2d}\left(r^2 + d^2\right)\left[\frac{\pi}{2}\left(\frac{1}{v_1} + \frac{1}{v_2}\right) + \frac{1}{v_1}\int_0^{Bd/v_1} K_o(y)\,dy + \frac{1}{v_2}\int_0^{Bd/v_2} K_o(y)\,dy\right], \quad (2)$$

where $K_o(y)$ is the modified Bessel function of order zero. For a comet at a heliocentric distance R, photons are received at the detector with area s at a rate $\dot{N} = N_H\,g(s/4\pi r^2)$, where the appriopriate g-factor (for details see /6/) is $g = 1.8 \times 10^{-3}(1\,\mathrm{AU}/R)^2\,\mathrm{s}^{-1}$.

This signal should be compared with the background level. The brightness of neutral interstellar hydrogen that flows through the Solar System and scatters photons in Ly-α depends on the comet position with respect to the apex direction and varies with the Solar activity; the lower and upper limits could be put at 200 and 500 Rayleighs, with mean 400 R. The background signal at the detector is: $\dot{N}_{bkg} = I\pi\varepsilon^2 s$, where I is the background Ly-α photon flux [$\mathrm{cm^{-2}\,s^{-1}\,sr^{-1}}$]. In Figure 3 the comet signal as a function of the distance from the nucleus is presented for some comet sizes. Also the background level and the statistical Poisson noise of the background is shown. For example, it can be seen that, if d is not larger than $\sim 1 \times 10^6$ km ($\sim 5 \times 10^6$ km) the signal from comets as small as 100 m (320 m) in radius could be discriminated against the statistical background

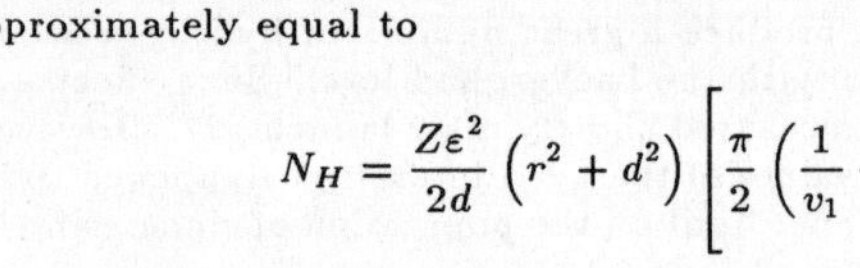

Fig. 2. Schematic representation of Ly-α scattering from cometary coma. Detector with the field of view given by a cone with a half-angle ε is at the point O and measures the light scattered by H atoms forming a coma around the nucleus.

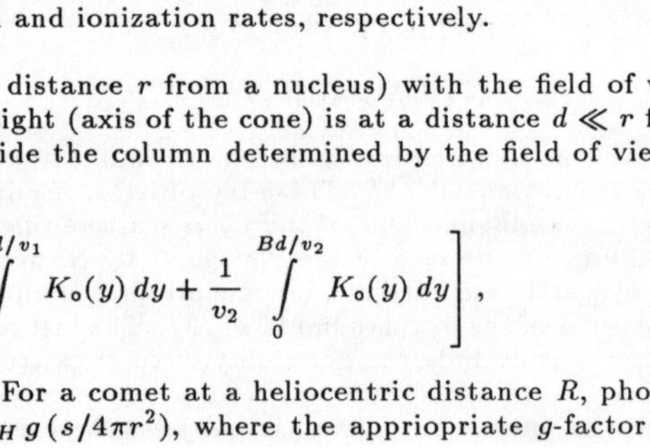

Fig. 3. Ly-α photon rate $\dot{N}$ observed at the detector with an area of $s = 1\,\mathrm{cm}^2$ and an opening half-angle $\varepsilon = 1°$. The dashed line indicates the background signal of interplanetary hydrogen (400 R). The dotted line shows the Poisson noise level of the background. Curves correspond to different values of the radius r_n of a uniform nucleus: $r_n = 1$ km ($\odot$), $r_n = 320$ m ($\triangle$), $r_n = 220$ m (+) and $r_n = 100$ m ($\times$).

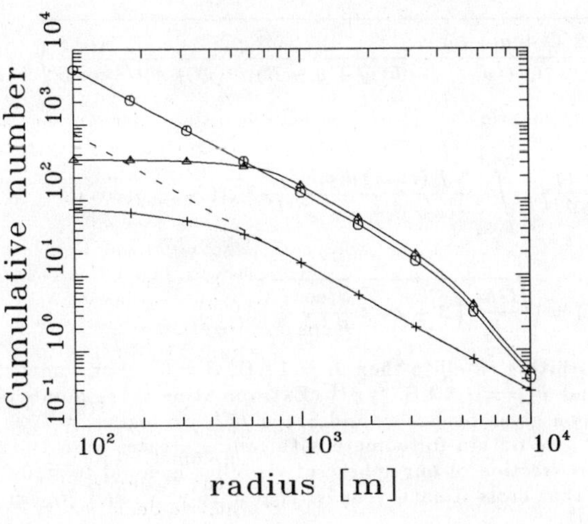

Fig. 4. Size statistics of comets. Curves denoted by ⊙ and Δ describe cumulative number of comets N_{cumul} with radius greater than a value of abscissa according to Everhart /10/ and Sekanina-Yeomans /8/, respectively. Short period comet statistics (+) is taken from Hughes /11/. Dashed line is an extrapolation of the Hughes function towards smaller sizes. All distributions are normalized to reproduce the observed number of objects brighter than 10.9 mag.

noise. When we look at a small comet from a large distance and the field of view ε is not small our equation (2) is not accurate. To be safe with our calculations we have to assume that the distance between observer and the smallest 100 m comet is not greater than 0.3 AU.

STATISTICS OF COMETS

There are two aspects in statistical approach to the comet population: size distribution and spatial distribution. These two distributions are assumed to be independent and there are no evidences that this assumption is invalid.

Size distribution is deduced from the observed magnitude distribution which is corrected for the observational selection effects and normalized according to the assumed dependence between sublimation rate and size of a nucleus. Many authors discussed possible form of the size distribution function (see /7/ for a review). For diameters greater than 1 km, corresponding to absolute magnitudes of less than 10.9, there is a high probability of detection in visual; here the statistics is relatively good. The size distribution for smaller comets is highly uncertain and varies from author to author. In case of long period (LP, $P > 200$ yr) almost parabolic comets two extreme hypotheses assume that either there are no small comets at all /8, 9/, or their distribution could be extrapolated from the curve describing large nuclei /10/; in the second case we can expect 300 comets with a radius greater than 100 m crossing the Jupiter's orbit each year. The statistics of short period (SP) comets was discussed by Hughes /11/. Selection effects can again decrease the number of observed comets fainter than 10.9 magnitude. Direct extrapolation gives 800–1000 comets with diameters greater than 100 m, but for this class of comets smaller objects are subject to faster destruction (evaporation) than comets on parabolic orbits. In Figure 4 different distributions for both populations of comets are shown.

The intrinsic distribution of orbital elements of SP comets is difficult to determine. From the distribution of the observed SP comets given by Marsden /12/ one can approximate the distribution in perihelion distance q, aphelion distance Q and orbital inclination i by $f_{SP}(q, Q, i) = \frac{1}{N_{SP}} f_1(q)\, f_2(Q)\, f_3(\cos i)$, with $N_{SP} = 120$, where $f_1(q) = 60\, q$, for $0 < q < 2$ AU, and

$$f_2(Q) = \begin{cases} 0 & Q < 4.0\,\mathrm{AU} \\ 22.5/(5.7 - Q) & 4.0\,\mathrm{AU} < Q < 5.3\,\mathrm{AU} \\ 30/(Q - 5.3) & 5.3\,\mathrm{AU} < Q \end{cases} \tag{3}$$

$$f_3(\cos i) = \begin{cases} 20.43 & 0^\circ \le i \le 18^\circ \\ 0 & 18^\circ < i \end{cases} \tag{4}$$

The remaining orbital elements (Ω, ω, M_o) are uniformely distributed. For the LP comets that all move approximately on parabolic orbits with $Q = \infty$, we obtain $f_{LP}(q, i) = f_4(q)\, f_5(\cos i)$, where $f_4(q) = 16.2\, q^{1/2}$ (from Kresák and Pittich /13/), and $f_5(\cos i) = 0.5$. The distributions for both populations were normalized to give the observed (for SP comets) or estimated (for LP comets) number of comets brighter than 10.9 mag in the vicinity of the Earth's orbit.

Now we can calculate the number of comets crossing a sphere centred at the detector at 1 AU from the Sun with a radius of 0.3 AU within which we can observe even the smallest comets. For the detection at a heliocentric distance R and an ecliptic latitude β the flux is equal to:

$$F_{SP}(R, \beta) = \frac{1}{R\pi^3} \int_0^R \frac{f_1(q)\, dq}{(R - q)^{1/2}} \int_{Q_{min}}^{Q_{max}} \frac{f_2(Q)\, dQ}{(Q - R)^{1/2}(Q + q)} \int_{\cos\beta_{max}}^{\cos\beta} \frac{f_3(\cos i)\, d(\cos i)}{(\cos^2\beta - \cos^2 i)^{1/2}}\, V_{rel}(R, \beta; q, Q, \cos i) \tag{5}$$

where $Q_{min} = 4$ AU, $Q_{max} = 15$ AU and $\beta_{max} = 18^\circ$. V_{rel} is the relative velocity between the comet and the detector:

$$V_{\rm rel}(R,\beta;\,q,Q,\cos i)=\sqrt{\frac{GM}{R}\left(1+\frac{2(Q+q-R)}{Q+q}-\frac{2\sqrt{2}\,qQ}{R(Q+q-R)^{1/2}(Q+q)^{1/2}}\frac{\cos i}{\cos\beta}\right)} \tag{6}$$

In the case of LP comets the corresponding formulae are:

$$F_{\rm LP}(R,\beta)=\frac{1}{2\pi R\sqrt{2GM}}\int_0^R\frac{f_4(q)\,dq}{(R-q)^{1/2}}\int_{\cos\beta_{\rm max}}^{\cos\beta}\frac{f_5(\cos i)\,d(\cos i)}{(\cos^2\beta-\cos^2 i)^{1/2}}\,V_{\rm rel}(R,\beta;\,q,\cos i), \tag{7}$$

and

$$V_{\rm rel}(R,\beta;\,q,\cos i)=\sqrt{\frac{GM}{R}\left(3-2\sqrt{2}\,\frac{q}{R}\frac{\cos i}{\cos\beta}\right)}. \tag{8}$$

If the detector is placed on board of an Earth orbiting satellite then $R = 1\,\mathrm{AU}$, $\beta = 0°$. For comets brighter than 10.9 mag we obtain $F_{\rm SP} = 4.8\,\mathrm{AU^{-2}yr^{-1}}$ and $F_{\rm LP} = 7.3\,\mathrm{AU^{-2}yr^{-1}}$. Extrapolating these numbers toward smaller objects (using the mass-magnitude relation given by Bailey and Stagg /7/, *viz.* $\log(m) = 16.9 - 0.5\,H$, and assuming the nucleus density of $0.25\,\mathrm{g\,cm^{-3}}$) we obtain for comets with radius greater than 100 m $F_{\rm SP} = 50\,\mathrm{AU^{-2}yr^{-1}}$ and $F_{\rm LP} = 110\,\mathrm{AU^{-2}yr^{-1}}$. The cross-section of our sphere of visibility is equal to $\pi\,(0.3\,\mathrm{AU})^2 \simeq 0.3\,\mathrm{AU}^2$. Therefore the number of small comets that cross it each year is: $\dot{N}_{\rm SP} = 15\ \mathrm{yr}^{-1}$, and $\dot{N}_{\rm LP} = 33\ \mathrm{yr}^{-1}$.

DISCUSSION AND CONCLUSIONS

A simple device to observe small comets in Ly-α is a channeltron with a baffle that limits the field of view. This instrument can be fixed to a spinning satellite and scan the sky making use of the rotational motion of the spacecraft.

Soviet Prognoz series satellites are good candidates to carry the Ly-α detector. The spin period of Prognoz-8 was 2 minutes. The satellite was oriented towards the Sun and possible excursions of spin vector away from this direction were not greater than 10°. If the detector is pointing perpendicularly to the spin axis, its field of view would cut out in space a disk broadening outwards with a thickness of about 0.1 AU at the distance of 0.3 AU from the detector. The disk is perpendicular to the ecliptic plane and tangent to the Earth's orbit. All comets that move perpendicularly to the disk not farther than 0.3 AU from the detector will cross the disk. Comets moving parallel to the disk have smaller probability to cross it. It is approximately equal to $2 \times 0.3 \times 0.1/(0.3)^2\pi = 0.2$. For the whole population of comets the probability lies somewhere in the range 0.4-0.6. With this correcting factor we can estimate the total number of small comets that might be detected to be 20 to 30 per year.

Several assumptions were made in the above considerations. First, we assumed the Everhart's size distribution of long period comets. Hence the large flux of these objects, which in fact could be much smaller if the Sekanina-Yeomans and Kresák hypothesis about deficiency of small long-period comets is valid. Small short-period comets should exist, at least because of continuous decay of larger objects due to their evaporation. On the other hand this process changes selectively the lifetime of comets according to their perihelion distance. Both size and elements distributions could be affected in that way and might even not be considered as independent. The number of short-period comets crossing the sphere of observation could also be overestimated by about 30% due to other approximations. We assumed that the flux of comets is constant for the whole cross-section and is equal to its value at $R = 1\,\mathrm{AU}$, $\beta = 0°$. In fact the flux decreases quickly with increasing β. There exists still one factor that could counterbalance the effect of the mentioned overestimate. We assumed that the distance of visibility is 0.3 AU for all comets. However objects greater than 100 m sublimate more intensively and can be seen from larger distances. Hence the flux of observed comets will increase.

In conclusion, it seems to be worth trying to set up a simple experiment to detect small comets in Ly-α. Even if the number of observed comets is less than estimated in this paper it will put a limit on the size distribution function and give some constraints on the processes of formation and death of comets.

This research was supported by the Polish Academy of Sciences within CPBP 01.20.

REFERENCES

1. L.A. Frank, J.B. Sigwarth and J.D. Craven, *Geophys. Res. Lett.* **13,** 307–310 (1986).
2. T.M. Donahue, T.I. Gombosi and B.R. Sandel, *Nature* **330,** 548–550 (1987).
3. D.T. Hall and D.E. Shemansky, *Nature* **335,** 417–419 (1988).
4. H. Rickman, Cl. Froeschlé, in: *Cometary Exploration I,* ed. T.I. Gombosi, Hungarian Acad. Sci. 1983, p. 75.
5. H.U. Keller, *Space Sci. Rev.* **18,** 641–684 (1976).
6. J.W. Chamberlain, *Theory of planetary atmospheres,* Academic Press, New York 1978, Sec. 6.2.1.
7. M.E. Bailey and C.R. Stagg, *Mon. Not. R. Astron. Soc.* **235,** 1–32 (1988).
8. Z. Sekanina and D.K. Yeomans, *Astron. J.* **89,** 154–161 (1984).
9. L. Kresák, *Bull. Astron. Inst. Czech.* **29,** 114–125 (1978).
10. E. Everhart, *Astron. J.* **72,** 1002–1011 (1967).
11. D.W. Hughes, in: *Asteroids, Comets, Meteors II,* eds. C.-I. Lagerkvist, B.A. Lindblad, H. Lundstedt and H. Rickman, Univ. Uppsala, Uppsala 1986, pp. 503–519.
12. B.G. Marsden, *Catalogue of Cometary Orbits,* fifth edition, Smithsonian Astrophys. Obs. (1986).
13. L. Kresák and E.M. Pittich, *Bull. Astron. Inst. Czech.* **29,** 299–309 (1978).

A LOWER LIMIT ON THE IONIZATION FRACTION OF THE VERY LOCAL INTERSTELLAR MEDIUM*

R.J. Reynolds

Department of Physics, The University of Wisconsin-Madison, 1150 University Avenue, Madison, WI 53706, U.S.A.

ABSTRACT

The ionization state of the interstellar gas immediately surrounding the Sun is very uncertain. Known sources of ionization--the EUV stars HZ 43, Feige 24, G191-B2B, and GD153, the portion of the soft x-ray background above 90 eV, and the cosmic ray flux--will produce an ionization ratio of hydrogen $n(HII)/n(HI) \approx 0.09$. However, this value is only a lower limit, since additional, as yet unidentified sources of ionization are probably present. There is some evidence that the hydrogen ionization ratio may even exceed unity, which would imply that this local gas cloud is part of a widespread, but little-understood, warm, ionized phase of the interstellar medium. If high ionization ratios are <u>not</u> present, then either current theories of white dwarf evolution or the generally adopted picture of the local interstellar medium may have to be significantly revised.

INTRODUCTION

Observations of Solar H Lα 1216 Å and He 584 Å emission backscattered from the interstellar H I and He I atoms flowing into the Solar system reveal that there is a relative velocity of about 20 km s^{-1} between the Sun and the very local interstellar medium (VLISM or local cloud) which has a density $\sim 10^{-1}$ cm^{-3} and a temperature $\sim 10^4$ K /1,2,3/. Lyman-α absorption line observations toward nearby stars also indicate that the extent of this local cloud is about 3 pc /4/. This cloud is surrounded by a much hotter ($\approx 10^6$ K) and lower density ($\approx 4 \times 10^{-3}$ cm^{-3}) medium that forms the interior of a large (≈ 100 pc radius), possibly supernova-created interstellar bubble. This bubble appears to be the source of much of the Galactic soft x-ray background /5/.

The degree of ionization of hydrogen and helium in the VLISM has important implications for the properties and recent history of the local interstellar medium and radiation field as well as the nature of the interaction between the VLISM and the heliosphere. Unfortunately, no accurate, direct measurements of the ionization state of the VLISM yet exist. The detection of emission lines from ions in this local cloud is ruled out by its exceedingly small emission measure. For example, an ionization ratio $n(HII)/n(HI) = 1$ would imply an emission measure $\sim 10^{-2}$ cm^{-6} pc, which is approximately one order of magnitude below the capabilities of current emission line detection techniques for warm interstellar clouds /6/. In principle, sensitive absorption line measurements toward nearby stars offer the best opportunity to obtain information on ionization ratios, although currently, the results of such absorption techniques are ambiguous because of large uncertainties in the derived column densities due to the effects of line saturation /7/ and insufficient sensitivity. The low column density of Mg I (from observations of the 2852 Å absorption line) toward α CMa (2.7 pc) has been interpreted recently by Frisch et al. /8/ as indicating an electron density in the range 5×10^{-5} to 3×10^{-3} cm^{-3} in the local cloud. However, the value of the Mg II recombination coefficient used to compute the electron density is very uncertain because of the dominance of dielectronic recombination and its strong dependence on temperature near the VLISM temperature /9/. The derived electron density also depends upon the (uncertain) gas phase abundance of magnesium relative to hydrogen. Because of these uncertainties, it appears that the available data confine the electron density of the VLISM cloud only to the range $5 \times 10^{-5} \lesssim n_e < 5 \times 10^{-1}$ cm^{-3} /8,10/.

A different approach is to calculate the ionization state of the gas from the flux of ambient EUV and x-ray radiation that has been detected by various satellite and rocket instruments. This directly observed flux provides a strong lower limit on the ionization fraction of hydrogen in the VLISM.

---

*This paper is adapted from an invited talk given at the Vth International Workshop on Cosmic Gas Dynamics held at the Institute for Problems in Mechanics, USSR Academy of Sciences, Moscow, October 10-14, 1988.

OBSERVED IONIZING RADIATION WITHIN THE VLISM

A significant flux of ionizing radiation from O and B stars apparently does not reach the VLISM /11/. However, Lyman continuum radiation from the nearby (63 pc), hot DA white dwarf HZ 43 has been detected by the Voyager spectrometer and by the Berkeley EUV spectrometer /12/. The measured flux F_λ ranges from about 0.02 photons $cm^{-2}\ s^{-1}\ Å^{-1}$ at 825 Å to about 0.15 photons $cm^{-2}\ s^{-1}\ Å^{-1}$ at about 350 Å, and the resulting ionization rate per hydrogen atom in the VLISM due to HZ 43 is $\Gamma(HZ\ 43) \approx 7.7 \times 10^{-17}\ s^{-1}$. According to Paresce /13/, three additional, hot white dwarfs--Feige 24, G191-B2B, and GD153--also have detectable EUV fluxes. The total flux observed from all four stars provides a $\Gamma(WD) \approx 1.5 \times 10^{-16}\ s^{-1}$.

Another observed source of ionization within the VLISM is the soft ($h\nu \gtrsim 90$ eV) x-ray background, which has a maximum flux of about 20 photons $cm^{-2}\ s^{-1}\ Å^{-1}$ at the lowest observed energies /14/. The hydrogen ionization rate produced by the observed portion of the background is $\Gamma(\text{x-rays}) \approx 3 \times 10^{-17}\ s^{-1}$. Direct observations of the total interstellar cosmic ray flux in the VLISM is prevented by the interaction of the lowest energy component of the cosmic rays with the Solar wind. However, outside the heliosphere the ionization rate due to cosmic rays should be comparable to that in the general interstellar medium, where a rate $\Gamma(CR) \approx 7 \times 10^{-17}\ s^{-1}$ is inferred /15/.

Therefore, the total hydrogen ionization rate within the VLISM from known sources is $\Gamma(\text{known}) \approx 2.5 \times 10^{-16}\ s^{-1}$. In steady state the hydrogen ionization ratio is given by

$$\frac{n(HII)}{n(HI)} \approx \left(\frac{\Gamma}{\alpha\ n(HI)}\right)^{1/2}, \qquad (1)$$

where α is the hydrogen recombination coefficient, which has an effective (case B) value of $3.1 \times 10^{-13}\ cm^3\ s^{-1}$ at 8000 K, the temperature of the VLISM hydrogen. Thus within the VLISM

$$\frac{n(HII)}{n(HI)} \gtrsim 0.09 \left[\frac{0.1}{n(HI)}\right]^{1/2}, \qquad (2)$$

or

$$n_e \gtrsim 0.009 \left[\frac{n(HI)}{0.1}\right]^{1/2}\ cm^{-3}. \qquad (3)$$

This result is expressed as a lower limit to emphasize the fact that any additional, as yet unobserved sources of ionization will increase n_e and the n(HII)/n(HI) ionization ratio. Since the Solar backscattering results indicate $n(HI) \sim 10^{-1}\ cm^{-3}$ in the VLISM, the electron density $n_e \gtrsim 10^{-2}\ cm^{-3}$. This limit significantly narrows the range of ionization allowed by the absorption line data toward nearby stars. Bruhweiler and Cheng /16/ have thoroughly examined both the H and He ionization rates in the VLISM due to the observed stellar EUV sources. Their results are consistent with those presented here.

A HIGH IONIZATION FRACTION IN THE VLISM?

Other Possible Sources

There probably are additional, as yet undetected sources of ionization for the VLISM that raise the ionization of hydrogen above the lower limit calculated above. For example, if the flux of the soft x-ray background is proportional to $E^{-2.5}$ photons $cm^{-2}\ s^{-1}\ eV^{-1}$ near 100 eV /14/ and extends below the 90 eV detection limit of present instruments to 40 or 30 eV, then the resulting ionization rate would be $\approx (2\text{-}10) \times 10^{-15}\ s^{-1}$, producing an ionization ratio $n(HII)/n(HI) \approx (0.25\text{-}0.57)\ [0.1/n(HI)]^{1/2}$, and thus an electron density $n_e \approx 0.02\text{-}0.05\ cm^{-3}$.

Slavin /17/ has examined the consequences of a conductive interface between the VLISM cloud and the surrounding lower density 10^6 K gas in the local cavity. His model, which includes for the first time the dynamical effects of a magnetic field on cloud evaporation, indicates that a significant amount of EUV radiation is produced by the interface, dominating all other known sources. The model predicts hydrogen and helium ionization ratios of n(HII)/n(HI) ≈ 0.25-1 and n(HeII)/n(HeI) ≈ 0.7-1.5 for the cloud with electron densities of 0.04-0.1 cm^{-3}.

Another potentially important source of ionization is the integrated EUV flux from the population of faint, hot white dwarf stars that must be present within the local cavity. According to Panagia and Terzian /18/, the local density of white dwarf stars brighter than absolute magnitude + 12.75 is approximately 1.4×10^{-3} stars pc^{-3}, which should be producing Lyman continuum radiation at the rate of about 4.6×10^{39} photons s^{-1} pc^{-3}. If the local cloud is a relatively rare and isolated object within the local cavity, as the H I column densities to stars in the Solar neighborhood imply, then the 6000 white dwarf stars within the approximately 100 pc radius cavity boundary /19/ will produce a local ionizing flux of about 5×10^4 photons cm^{-2} s^{-1}. This flux is more than 100 times that available from HZ 43 (before attenuation by the VLISM cloud) and would produce an ionization ratio n(HII)/n(HI) ≈ 1-2 within the VLISM ($\Gamma \sim 10^{-13}$ s^{-1}).

Finally, because the recombination time $\tau \approx 1 \times 10^6$ $(0.1/n_e)$ yrs within the VLISM is comparable to the main sequence lifetime of massive O stars and the mean interval between nearby (< 100 pc) supernovae, transient events may have strongly influenced its present ionization state. In particular, it has been suggested that a nearby supernova occurred within the last few hundred thousand years creating the local x-ray emitting cavity /20/, and that the local radiation field was higher a million years ago when Antares was an O star /21/.

Evidence for a High Ionization Fraction

A high degree of ionization within the VLISM does not appear to be inconsistent with any of the available data. It even has some observational support. A high fractional ionization of hydrogen along the line to Capella (14 pc), for example, has been proposed by Bobroff et al. /22/ in order to explain the absence of what was expected to be a measureable amount of He II 304 Å emission from the star. Consistency between their XUV observations of Capella and the H I column density to the star determined from Lα absorption appears to require that the ratio of He I to H I in the local cloud exceed 0.25, implying a hydrogen ionization ratio n(HII)/n(HI) > 1.5. Paresce /13/ has correctly cautioned that this interpretation depends very much on the accuracy of the stellar atmosphere model. However, a high He I to H I ratio in the local cloud also is supported by the EUV data toward HZ 43. There is an apparent discrepancy between the H I column density ($1.15 \pm 0.15 \times 10^{18}$ cm^{-2}) toward HZ 43 derived from the EXOSAT spectrum /23/, which is primarily shortward of the helium ionization edge at 504 Å and thus sensitive to He I absorption, and the H I column density ($4 \pm 2 \times 10^{17}$ cm^{-2}) derived from the Voyager spectrum /12/, which is primarily longward of 504 Å and thus most sensitive to H I absorption directly. Paerels et al. /23/ pointed out that this discrepency would be resolved if a large fraction of the hydrogen along the line of sight to HZ 43 were ionized.

A high electron density in the VLISM also has been indicated by a VLISM-heliosphere interaction model. Baranov /24,25/ has shown that, if the heliosphere bow shock is really as close as 30-45 AU from the Sun, as the Voyager radio data apparently suggest /26,27/, then an electron density of 0.3 cm^{-3} and a hydrogen ionization ratio near unity may be required for the inflowing VLISM gas. Moreover, the thermal pressure in a higher density, more highly ionized VLISM would be nearer the traditional value for the interstellar gas pressure (3,000-10,000 cm^{-3} K) than that (500-800 cm^{-3} K) provided by the observed neutral atoms alone.

Direct observational evidence for substantial ionization of hydrogen in the local interstellar medium is provided by the dispersion measure of a nearby pulsar. Gwinn et al. /28/ used very long baseline radio interferometric techniques to measure the parallax of PSR 0950+08, which has a dispersion measure of 2.97 cm^{-3} pc, corresponding to an electron column density $N_e = 9.16 \times 10^{18}$ cm^{-2}. The pulsar was found to have a distance of 127 ± 13 pc. Its location at galactic longitude 229°, latitude +44° appears to place it within the large, low H I column density region mapped by Frisch and York /29/ and Paresce /30/ and defined by H I column densities $< 5 \times 10^{18}$ cm^{-2}. The dispersion measure therefore suggests that N_e may be greater than N_{HI} along the line to the pulsar. Of course, the location of the ionized cloud(s) producing the dispersion measure is not known, but these data do indicate that within the local interstellar medium highly ionized clouds are present which have a mean electron density $n_e = 0.023/f$ cm^{-3}, where f is the fraction of the line of sight to the pulsar occupied by the clouds. This ionization does not appear to be produced by any nearby O, B, or hot white dwarf star /31/. Observations of faint Hα recombination emission toward a large number of much more distant pulsars indicates that these diffuse ionized clouds typically have an electron density of about 0.15 cm^{-3} and occupy a fraction f = 0.2 of the interstellar volume /32/.

SUMMARY AND CONCLUSIONS

The known sources of ionization imply a hydrogen ionization fraction $n(HII)/n(HI) \gtrsim 10^{-1}$ and an electron (HII) density $n_e \gtrsim 10^{-2}$ cm^{-3} in the VLISM. The degree of ionization is thus significantly higher than that suggested by some absorption line observations of the local cloud /8/, which probably reflects the large uncertainties involved in correctly interpreting the absorption line data. The electron (and proton) density in the VLISM therefore may be large enough for charge exchange processes to be important near the VLISM-heliosphere interface /33/.

There also are a number of reasons to expect the actual ionization fraction of the VLISM to be substantially larger than 10^{-1}. If the local cloud is highly ionized (i.e., $n(HII)/n(HI) > 1$ with $n_e \sim 10^{-1}$ cm^{-3}), then it could be part of the warm, diffuse, ionized, phase of the interstellar medium associated with the interstellar Hα background /34,35/. (It should be noted that, although hot white dwarf stars are expected to be a significant source of ionization for the VLISM, they are not sufficient to account for the interstellar ionization globally /37/). If the VLISM is not so highly ionized, then generally held ideas about white dwarf evolution or the structure of the local bubble may have to be substantially revised. An unambiguous measurement of the ionization state of the VLISM is therefore crucial.

REFERENCES

1. Dalaudier, F., Bertaux, J. L., Kurt, V. G., Mironova, E. N. 1984. Astron. Astrophys. 134, 171.
2. Bertaux, J. L., Lallement, R., Kurt, V. G., Mironova, E. N. 1985. Astron. Astrophys. 150, 1.
3. Chassefiere, E., Bertaux, J. L., Lallement, R., Kurt, V. G. 1986. Astron. Astrophys. 160, 229.
4. McClintock, W., Henry, R. C., Linsky, J. L., Moos, H. W., 1978. Astrophys. J. 225, 465.
5. Sanders, W. T., Kraushaar, W. L., Nousek, J. A., Fried, P. M. 1977. Astrophys. J. Lett. 217, L87.
6. Kutyrev, A. S., and Reynolds, R. J. 1989. Astrophys. J. (Letters) 344, L9.
7. Bruhweiler, F. C. 1984. In The Local. Interstellar Medium, IAU Colloq. No. 81, ed. Y. Kondo, F. C. Bruhweiler, B. D. Savage (NASA CP-2345), p. 39.
8. Frisch, P. C., York, D. G., Fowler, J. R. 1987. Astrophys. J. 320, 842.
9. Shull, J. M., Van Steenberg, M. 1982. Astrophys. J. Suppl. 48, 95.
10. Cox, D. P., Reynolds, R. J. 1987, Ann. Rev. Astron. Astrophys. 25, 303.
11. Reynolds, R. J. 1986. Astron. J. 92, 653.
12. Holberg, J. B., Sandel, B. R., Forrester, W. T., Broadfoot, A. L., Shipman, H.L., Barry, D. C. 1980. Astrophys. J. Lett. 242, L119.
13. Paresce, F. 1984a. In The Local Interstellar Medium, IAU Colloq. No. 81, ed., Y. Kondo, F. C. Bruhweiler, B. D. Savage (NASA CP-2345) p. 169.
14. Bloch, J. J., Jahoda, K., Juda, M., McCammon, D., Sanders, W. T., Snowden, S. L. 1986. Astrophys. J. Lett. 308, L59.
15. van Dishoeck, E. F., Black, J. N. 1986. Astrophys. J. Suppl. 62, 109.
16. Bruhweiler, F. C., and Cheng, K. P. 1988, Astrophys. J. 335, 188.
17. Slavin, J. D. 1989. Submitted to Astrophys. J.
18. Panagia, N., Terzian, Y. 1984. Astrophys. J. 287, 315.
19. Cox, D. P., Snowden, S. L. 1986. Adv. Space Res. 6, 97.
20. Cox, D. P., Anderson, P. R. 1982. Astrophys. J. 253, 268.
21. Bochkarev, N. G. 1987. Astrophys. Space Sci. 138, 229.
22. Bobroff, N., Nousek, J., Garmire, G. 1984. Astrophys. J., 277, 678.
23. Paerels, F. B. S., Bleeker, J. A. M., Brinkman, A. C., Gronenschild, E. H. B. M., Heise, J. 1986. Astrophys. J. 308, 190.
24. Baranov, V. B. 1986. Adv. Space Res. 6,5.
25. Baranov, V. B. 1987. Sov. Astron. Lett. 12, 300.
26. Kurth, W., Garnett, D., Scarf, E., Poynter, R. 1984. Nature 312, 27.
27. Suess, S., Dessler, A. 1985. Nature 317, 702.
28. Gwinn, C. R., Taylor, J. H., Weisberg, J. M., Rawlings, L. A. 1986.Astron. J. 91, 338.
29. Frisch, P. C., York, D. G. 1983. Astrophys. J. Lett. 271, L59.
30. Paresce, F. 1984b. Astron. J. 89, 1022.
31. Reynolds, R. J. 1990. Astrophys. J., (Jan. 1).
32. Reynolds, R. J. 1977. Astrophys. J. 216, 433.
33. Ripkin, H. W., Fahr, H. J. 1983. Astron. Astrophys. 122, 181.
34. Reynolds, R. J. 1987b. Astrophys. J. 323, 118.
35. Reynolds, R. J. 1989, in Galactic and Extragalactic Background Radiation,IAU Symposium No. 139, ed. S. Bowyer and C. Leinert (Dordrecht:Reidel), in press.
36. Kulkarni, S., Heiles, C. 1988. In Galactic and Extragalactic Radio Astronomy,ed. G. L. Verschuur, K. Kellerman (2nd ed., New York: Springer-Verlag), p. 95.
37. Reynolds, R. J. 1990. Astrophys. J. Lett. (January 10).

Session III

Entry and Dynamics of Galactic and Anomalous Cosmic Rays in the Heliosphere

(Curator — D. E. Page)

COSMIC RAYS IN THE LOCAL INTERSTELLAR MEDIUM

R.B. McKibben

Laboratory for Astrophysics and Space Research, Enrico Fermi Institute, University of Chicago, Chicago, IL 60637, U.S.A.

ABSTRACT

Determination of the local interstellar spectrum of galactic cosmic rays at low energies (E$\lesssim$1 GeV) has long been a major goal of cosmic ray astrophysics. The question is of interest because of the possible effects that a large population of low energy cosmic rays may have on the local interstellar medium, on interstellar clouds, and, perhaps, on the boundary of the heliosphere itself. Because of the solar modulation of low energy cosmic rays, no direct information concerning the low energy interstellar spectrum of galactic cosmic rays is available from measurements made near earth. However, several estimates of the form of the interstellar spectrum have been put forth by use of indirect arguments and models of the modulation process. The Pioneer and Voyager spacecraft are now at such large radial distances from the sun (>45 AU for Pioneer 10) that, even though all are still apparently deep in the modulation region, observations from these spacecraft are beginning to provide meaningful tests of some of the suggested forms for the local interstellar spectrum. We review estimates of this spectrum and compare them to current observations from the spacecraft in the outer heliosphere which place at least a lower bound on the cosmic ray spectrum in the nearby interstellar medium. Using a standard model for the modulation process to extrapolate to larger radii the observed fluxes of anomalous components, believed to be accelerated near the modulation boundary, suggests limits on the radial extent of the modulation region in the heliosphere.

INTRODUCTION

Cosmic rays were first discovered from observations near the base of Earth's atmosphere. It was soon realized that the radiation observed at the surface or at altitudes accessible by the balloons and aircraft available in the early 20th century was secondary rather than primary radiation, and that, while its intensity variations may have accurately reflected the intensity variations of the primary cosmic radiation at the top of the atmosphere, its nature bore only a distant relation to that of the primary radiation incident from space. Thus, one of the first goals of cosmic ray science was to determine the nature of the cosmic radiation at the top of the atmosphere. In the last half of the 20th century we have surmounted the obstacle of the atmosphere, and have made extensive observations of the primary cosmic rays that reach the Earth both in space and at the very top of the atmosphere. However, we still have not observed the primary spectrum of cosmic rays at low energies as it exists in interstellar space. Still another filter lies between us and the interstellar medium, one whose existence was unknown until the middle 1950's /1/, and whose serious investigation became possible only with the beginning of the space age in the late 1950's. That filter is the heliosphere, whose structure and behavior we are still only beginning to understand.

THE HELIOSPHERE AS SEEN BY COSMIC RAYS

While for very high energy cosmic rays the heliosphere has little effect, for cosmic rays with energies below about 1 GeV the heliosphere is at best semi-transparent, and, at energies below a few hundred MeV/nucleon, it is totally opaque as a result of the interaction of the charged cosmic rays with the interplanetary magnetic field (IMF) carried outwards from the sun by the solar wind. Figure 1 shows a schematic representation of the magnetic structure of the heliosphere as viewed from the north pole (Fig. 1A) and from a point in its equatorial plane (Fig. 1B). Also shown are the trajectories of the four spacecraft now exploring the outer heliosphere traced through the middle of 1989, corresponding roughly to the time of the most recent data that have been analyzed.

In the ecliptic plane, the dominant magnetic structure is the spiral IMF. For a typical solar wind velocity of 400 km/s and a nominal heliospheric radius of 100 AU to the termination shock in the solar apex direction, consistent, within wide uncertainties, with current observations and with pressure balance with the interstellar medium /2,3,4/, the spiral has approximately sixteen windings in the region of supersonic solar wind between the sun and the shock. Therefore the path length for a cosmic ray travelling inwards from the shock toward the earth along the field is several thousand AU. Beyond the shock, as discussed recently by, for example, Holzer /4/ and Lee /3/ the solar wind flow continues at a lower velocity as a subsonic incompressible flow. Because of the velocity decrease at the shock, the magnetic field becomes even more tightly wound and its strength increases significantly. Thus the region beyond the shock

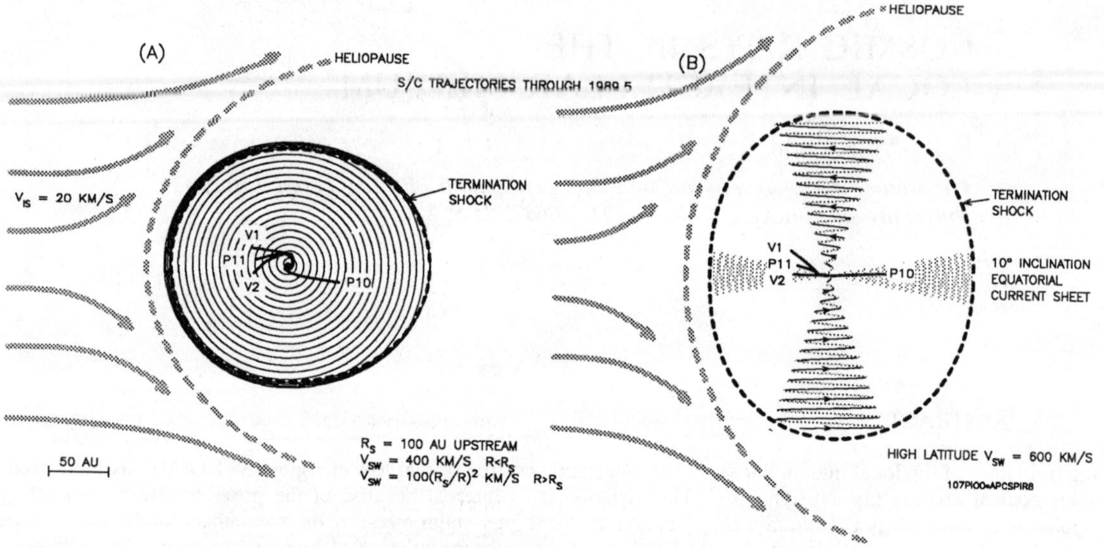

Fig. 1. Schematic of possible configurations of the heliosphere and the heliospheric magnetic field viewed from the pole (A) and from a point in the ecliptic (B). Shaded arrows represent flow of the interstellar medium around the heliopause. Magnetic field lines are shown only in regions of radial outflow of the solar wind. Spacecraft trajectories have been projected (A) or rotated (B) onto the plane of the figure.

may also be very important in controlling access of low energy cosmic rays to the inner heliosphere, as suggested recently on observational grounds by Webber /5/. Eventually, the flow is deflected by the pressure of the interstellar medium, and, as a result of the sun's motion through that medium, a heliospheric tail is formed. The nature of the flow in this region is very uncertain, and no attempt has been made in Figure 1 to show field lines in the region where the flow is no longer primarily radial. The pressure difference between the nose and tail of the heliosphere may also produce some asymmetry in the solar wind terminal shock (e.g. /4,6,7/), represented here by an arbitrary 20 AU extension of the shock in the tailward direction. A contact discontinuity, the heliopause, of uncertain shape and distance from the termination shock, separates plasma of solar and interstellar origin.

Away from the ecliptic, the field continues to describe a spiral on cones of constant latitude, such as the 70 degree latitude cone shown in Figure 1B. At solar minimum, the IMF is divided into hemispheres of opposite polarity by a current sheet rooted in the coronal magnetic field. The current sheet has a typical inclination of a few degrees to the sun's rotational equator at solar minimum, and, when carried out by the solar wind, this produces the wavy current sheet that plays a significant role in current theories of cosmic ray modulation (e.g. /8,9,10,11,12/). During the approach to solar maximum, this structure breaks down as the current sheet rises to very high latitudes, reflecting the increasing complexity of the solar coronal field. Detached current sheets may appear at high latitudes surrounding islands of polarity opposite to the dominant hemispheric polarity. Finally, near solar maximum the solar field is no longer predominantly dipolar, and the magnetic polarity of the solar polar regions reverses. Following solar maximum, the solar and interplanetary field structure relaxes back to the solar minimum configuration, but with the magnetic polarities of the hemispheres reversed. Although not drawn in Figure 1B because of the difficulty of representation, in the same way as in the ecliptic the component of the magnetic field perpendicular to the solar wind flow becomes stronger immediately outside the shock. However, since the high latitude field is more radial, the effect on the magnitude of the field, and thus perhaps on the cosmic rays as well, should be smaller.

The solar wind also shows significant variation with latitude, rising to an average speed in excess of 600 km/s over the polar regions at solar minimum /13/. Near solar maximum the high speed region of the solar wind retreats towards the pole. However the speed near the pole remains significantly higher than near the ecliptic. As a result of both the higher latitude and the higher solar wind speed, the length of the interplanetary field lines between the sun and the shock is considerably shorter at high latitudes than in the ecliptic, leading to the suggestion that the exclusion of cosmic rays over the polar regions may be less effective than in the ecliptic. There may also be significant departures from spherical symmetry for the heliosphere towards the poles as a result both of variations in the solar wind with latitude, and of changes in the hydrodynamic pressure of the interstellar medium as it flows around the heliosphere. The sketch in Figure 1B shows the effect of a change in dynamic pressure by a factor of two from equator to pole, as suggested, for example, by Suess /7/.

SOLAR MODULATION: BASIC OBSERVATIONS

The effect of the changing structure of the heliosphere on the cosmic radiation is shown clearly in Figure 2, where the intensity of cosmic-ray-produced secondary neutrons measured by the Climax neutron monitor at an altitude of

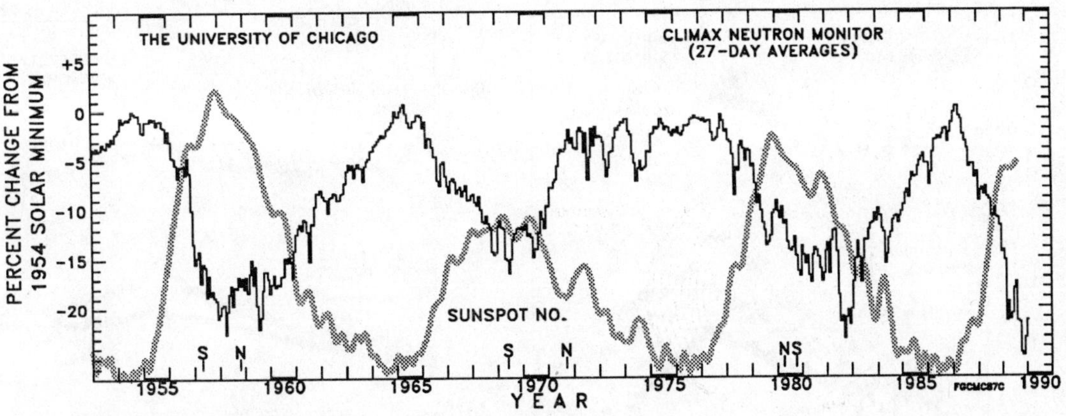

Fig. 2. 27-day average normalized intensity of ≳ 3 GV cosmic rays measured by the Climax neutron monitor and smoothed sunspot number.

~11,500 feet is plotted together with the smoothed sunspot number as an indicator of solar activity. The primary cosmic rays contributing to this counting rate have energies >2.2 GeV, so that even at these relatively high energies the effect of modulation by the solar wind and the variation from solar maximum to solar minimum are significant. Extending the observations to lower energies, Figure 3 shows the Climax counting rate and the intensities of primary cosmic rays measured above the atmosphere by the satellite IMP 8 down to energies as low as 11 MeV/nucleon. The data are plotted on the same scale and normalized to 1 at the solar minimum of 1977 to bring out the dependence of the strength of modulation on particle energy. While the maximum total variation in counting rate over a solar cycle for the Climax neutron monitor is only about 20% (corresponding to an intensity variation of ~44% at the top of the atmosphere after taking into account the specific yield function for Climax /55/), at energies below 100 MeV/nucleon, more than 90% of the flux observed at solar minimum was excluded at solar maximum.

Studies of the radial profile of the cosmic ray intensity demonstrate that significant modulation of low energy cosmic rays persists even at solar minimum. Figure 4, from Lopate et al. /14/, contains measurements of the intensity and radial gradients of low energy protons and helium nuclei from Pioneer 10, Pioneer 11, and IMP 8 extending from the most recent period of near-maximum modulation in 1983 through solar minimum in 1987 and up to the middle of 1989. The low energy helium shown in Figure 4 is dominated by the anomalous component, which will be discussed in more detail below. As indicated in Figure 1, by the middle of 1989 Pioneer 10 had reached a radius of more than 45 AU from the sun in the ecliptic towards the heliospheric tail, while Pioneer 11 was at a radius of about 30 AU and about 16° above the ecliptic in the direction of the solar apex.

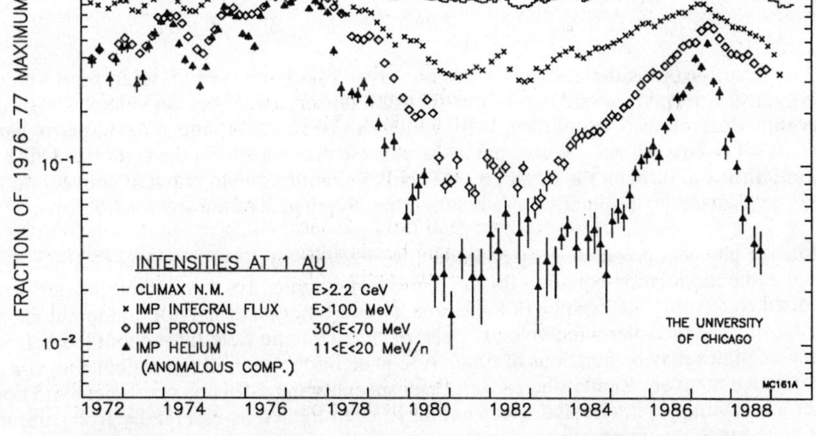

Fig. 3. Normalized intensities of galactic cosmic rays and anomalous helium. All fluxes have been normalized to 1.0 at the maximum flux observed during the 1977 solar minimum.

The data in Figure 4 show that in 1987, the helium flux at Pioneer 10 was approximately 30 times that measured at Earth, and the low energy proton flux, which may have included some contribution from anomalous protons at Pioneer 10 /14,15/ was about 6 times that measured at Earth. The near-exact alignment of the time of maximum flux, after taking into account the solar wind propagation delay, shows that both Pioneer 10 and IMP 8 were subject to the same process of solar modulation, although at significantly different levels. Thus, even beyond 45 AU, Pioneer 10 remained embedded in the solar modulation region of the heliosphere, and the interstellar spectrum at low energies remained hidden by the modulation process.

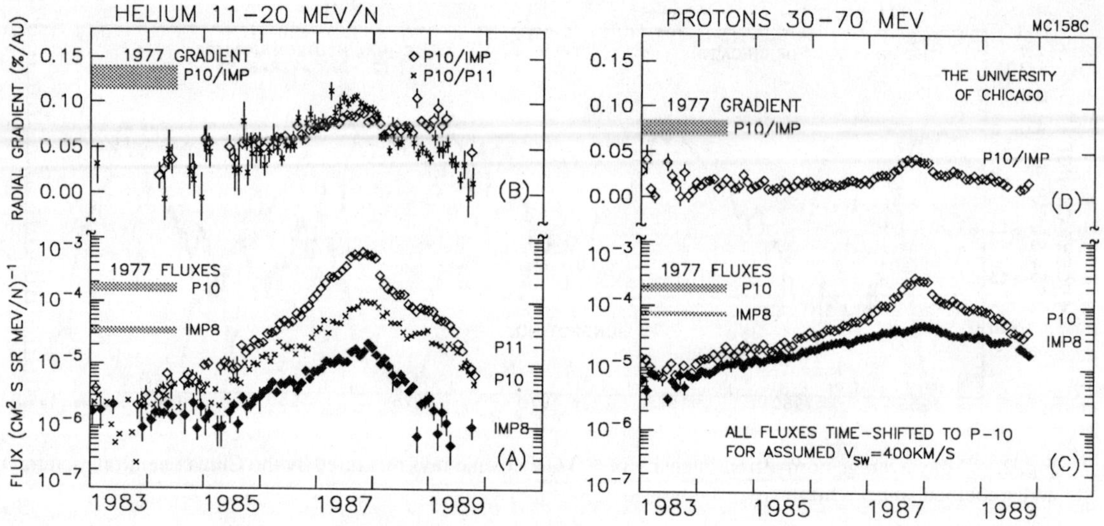

Fig. 4. 27-day average fluxes of 11-20 Mev/n helium (A) and 30-70 MeV protons (C) from Pioneer 10, Pioneer 11 and IMP8, together with radial gradients for the two particle species (B and D) measured through the period of the 1987 solar minimum. No measurements of protons are available from Pioneer 11 after 1980 because of a detector failure. The time axis is the time at Pioneer 10. Other data have been plotted at times shifted from the actual times of measurement by a time corresponding to the propagation delay for a constant solar wind velocity of 400 km/s from the point of observation to Pioneer 10.

SOLAR MODULATION: PROCESS AND MODELS

The physics of solar modulation is believed to be relatively well understood. In 1965 Parker /16/ wrote down the equation that is still the basis for investigations of modulation. The equation, shown in a slightly rewritten form as equation 1, contains terms describing all of the processes that we still today think contribute to the modulation.

$$\frac{\partial U}{\partial t} = \nabla \cdot (\kappa^{(s)} \cdot \nabla U) - (V_{sw} + V_d) \cdot \nabla U + \frac{1}{3} \nabla \cdot V_{sw} \frac{d(\alpha T U)}{dT} \qquad (1)$$

In this equation, U is the density of cosmic rays, V_{sw} is the vector solar wind velocity, T is the particle kinetic energy, $\kappa^{(s)}$ is the symmetric part of the diffusion tensor, and V_d is the velocity resulting from particle gradient and curvature drifts in the non-uniform IMF, which can be related to the antisymmetric part of the diffusion tensor, $\kappa^{(a)}$ /17/. $\alpha = (T+2m_0)/(T+m_0)$ where m_0 is the particle rest mass. On the right hand side the terms, in order, represent inward diffusion through the irregular IMF, bulk velocities due to outward convection by the solar wind and particle drifts, and adiabatic deceleration as a result of the divergence of the solar wind flow.

While the physical processes may be known, application to the real problem is considerably more difficult. In order to solve the modulation equation for the density of cosmic rays throughout the heliosphere, accurate information is required concerning the location of the boundary as a function of latitude and longitude and the values of the diffusion coefficients, of the solar wind velocity, and of the magnetic field throughout the heliosphere. Furthermore, each of these quantities may be functions of time. Adequate information is not available to permit the equation to be solved in a predictive manner. Rather, the parameters, especially the diffusion coefficient and boundary characteristics and, to a certain extent, the interstellar spectrum, have been treated as essentially free parameters and adjusted to provide adequate fits to the observations. Until very recently, no consideration was given to the possible modulating effects of the region between the solar wind termination shock and the heliopause because of the near-complete lack of information concerning this region.

Application of modulation models is saved from being a purely descriptive undertaking by the existence of an integral of the equation into which much of our ignorance can be condensed. It was first shown by Gleeson and Axford /18/, that, for a spherically symmetric heliosphere, if (1) $\partial U/\partial t \cong 0$, and (2) the cosmic ray flux within the heliosphere is approximately isotropic, the modulation can be parameterized by a quantity Φ, defined as

$$\Phi = \int_r^R \frac{V_{sw} dr}{3\, \kappa\, (P, r)}$$

where r is the radius of observation, R is the radius of the boundary outside which no modulation takes place, and κ, the diffusion coefficient, may depend upon rigidity, P, and radius. The importance and usefulness of Φ is that, within the constraints of spherically symmetric modulation models neglecting drifts, the same value of Φ describes modulation for all particle species with the same rigidity. This has been verified experimentally for a variety of nucleonic species by Garcia-Munoz et al. /19/. Physically, Φ is approximately equal to the mean energy lost by adiabatic deceleration in the course of propagation inward from the boundary, typically several hundred MeV for particles observed at 1 AU. As emphasized by Urch and Gleeson /20/, the particles in interstellar space with energies less than a few hundred MeV/nucleon are essentially completely excluded by the modulation process as a result of adiabatic deceleration. The particles observed at low energies near earth had much higher energies in interstellar space. Thus, as they demonstrated, the form of the spectrum observed at 1 AU at energies below a few hundred MeV/nucleon provides no information at all concerning the form of the spectrum at these energies in interstellar space.

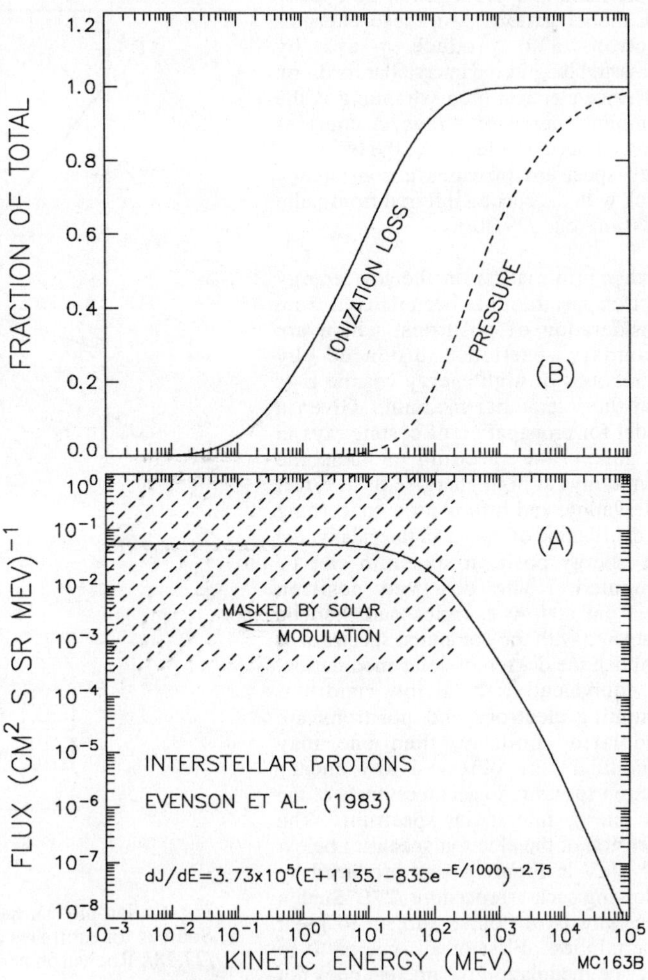

Fig. 5. (A) Sample interstellar proton spectrum from /39,53/ deduced by demodulation of spectra observed within the heliosphere. (B) Normalized pressure and energy deposit by ionization loss from the proton spectrum in (A).

It is a fair question to ask why the low energy form of the interstellar spectrum should be of particular interest. Figure 5A contains an example of one estimated form for the interstellar galactic cosmic ray proton spectrum used by Evenson et al. /39,53/. As shown in Figure 5B, for this choice of spectrum about one third of the pressure, or energy density, provided by the protons in the cosmic radiation is provided by particles with energies blocked out by the modulation process. Furthermore, almost all of the energy input to the interstellar medium by ionization loss is provided by cosmic rays with energies less than 100 MeV/n. Therefore, through their contribution to the interstellar pressure the low energy cosmic ray spectrum can have important effects on the dynamics of the interstellar medium /22/. Also, through their ionization loss they affect the thermal balance, and, through production of ions, the chemistry of the interstellar medium /23,24/. Furthermore, in the low energy spectra we should expect to find either evidence for local sources, or, since the low energy cosmic rays are near their end of range in the interstellar medium, possible significant differences in spectra of heavy elements as a result of their differing rates of ionization loss in the medium. Such differences could be important for deducing information concerning the history of cosmic rays in the interstellar medium.

LIMITS ON THE INTERSTELLAR COSMIC RAY SPECTRUM

Direct information concerning the spectrum of cosmic rays in interstellar space is basically unavailable except for rather high energy particles. Above a rigidity of a few 10's of GV, solar modulation has little effect on the cosmic rays, so the high energy form of the cosmic ray spectra for nuclei and electrons has been established by direct observations, with identification of chemical elements extending up to ~1 TeV/amu /25/, and separation of positrons and electrons to ~20-30 GeV /26/.

At lower rigidities the most direct information concerns the interstellar electron spectrum. Studies of electron synchrotron emission at radio wavelengths, taken together with reasonable assumptions concerning the galactic magnetic field, place limits on the cosmic ray spectrum for electrons. These limits are rather tight at energies above about 1 GeV, and are loose but still useful down to about 100 MeV. Figure 6 contains several estimates of the interstellar electron spectrum above a few hundred MeV based on the observed synchrotron emission as defined by a collection of early observations as analyzed by Cummings et al., /27,28/ and by more recent observations of radio emissions in the plane of the galaxy toward the anti-center considered by Rockstroh and Webber /29/, and toward the galactic poles by Webber et al. /30/. The various estimates are in reasonably good agreement, and are also in reasonable agreement with the spectral form deduced by Ip and Axford /31/ from theoretical considerations of shock acceleration and propagation in a leaky box model galaxy. At these energies electrons also produce γ-rays by bremsstrahlung in the interstellar medium, and since electron bremsstrahlung is the dominant source of γ-rays at energies below about 100 MeV /32/, the observed γ-ray spectrum provides a consistency check with the spectra inferred from radio emission (e.g. /29,30/).

Further information on the low energy electron spectrum has been derived from consideration of positrons, which are secondary particles produced by interactions of high energy cosmic rays with the interstellar medium. Given a model for propagation of cosmic rays in the galaxy, the spectrum of nucleonic cosmic rays at high energies from direct observation, and information concerning the distribution of matter in the galaxy, the low energy positron spectrum can be computed. The observed positron spectrum at low energies can then be compared with the computed spectrum to establish the degree of solar modulation, or, equivalently, Φ, at low rigidities. Assuming electrons and positrons are modulated similarly, then one may demodulate the observed low energy electron spectrum to get an estimate of the low energy interstellar spectrum. The extension of the electron spectrum below 100 MeV in Figure 6 was produced by following such a procedure /27/. Such a procedure is of course subject to great uncertainties. First of all, use of Φ to describe modulation parameters does not obviously include the effect of drifts. In modulation models including drifts, electrons and positrons are not necessarily modulated in a similar manner, since the pattern of drift velocities is dependent on the sign of the particle charge. Indeed, Garcia-Munoz et al. /33/ have found clear differences in the apparent modulation of helium nuclei and electrons of the same rigidity, leading to changes in the He/e ratio of up to a factor of 4 from 1978-1982, spanning the reversal of the solar magnetic dipole. Furthermore, all uncertainties concerning the nature and density of the interstellar medium affect the calculated positron spectrum, and, as emphasized by Müller and Tang /26/, the positron/electron ratio observed at low energies is not consistent with commonly used path-length distributions for cosmic rays in the interstellar medium. Therefore, extensions of the interstellar electron spectrum to low energies based on inferences from the positron spectrum should probably be considered uncertain at best.

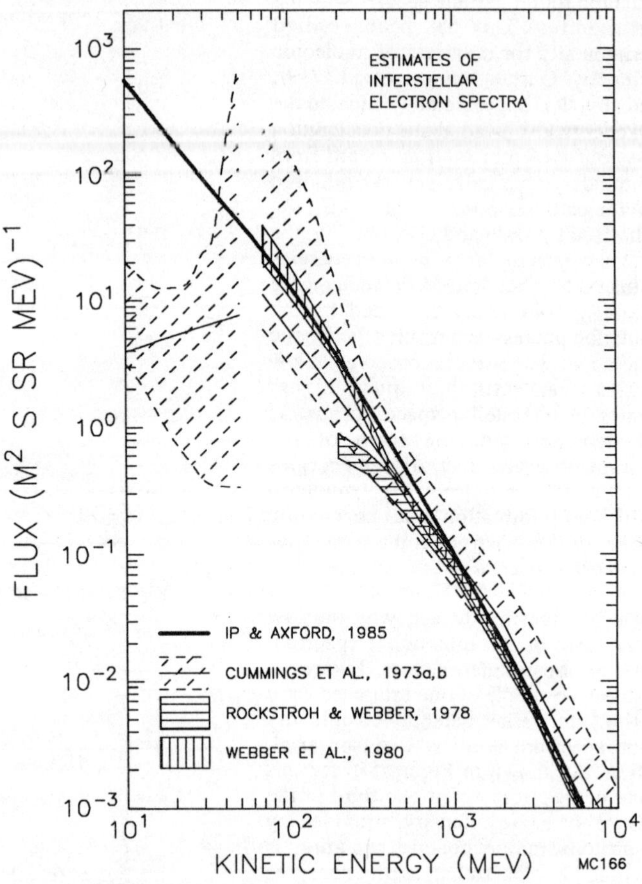

Fig. 6. Sample estimates of interstellar electron spectra. Sources for estimates are: Ip and Axford /31/; Cummings et al. /27,28/; Rockstroh and Webber /48/; Webber et al. /30/.

For nucleons below the energy for which the interstellar spectrum is directly observable, limits can be set from the observed intensity of γ-rays from π^0 decay, from the total pressure in the interstellar medium available for confinement of the cosmic rays, and from the ionization state of interstellar clouds as deduced from the observed chemistry in the clouds. This last limit is particularly interesting for low energy cosmic rays, since, as shown in Figure 5B, almost all of the ionization is produced by particles with energy below 100 MeV/n. From observations of three clouds Hartquist et al. /23/ have deduced an ion production rate of about 2×10^{-17} per atom per second in the clouds, although van Dishoeck and Black /24/ have argued that new information concerning reaction rates requires that the ion production rate for the various clouds be increased to between 1 and 8×10^{-16} per atom per second. For comparison, assuming a requirement of ~35 eV energy loss per ion produced /34/, the suggested interstellar proton spectrum shown in Figure 5A produces a rate of $\sim 2 \times 10^{-16}$ per atom per second, increasing to $\sim 5 \times 10^{-16}$ per atom per second if contributions from heavier elements in the cosmic rays are included in the abundances established by Garcia-Munoz and Simpson /35/. Thus, while the ionization rate for this sample spectrum is near the upper limit allowed, it is not in serious conflict with the observations. Of course neither this limit nor limits from γ-rays or the interstellar pressure provide specific information about the low energy spectrum, so that the form of the low energy nucleonic spectrum remains unknown. Most estimates currently in the literature depend upon demodulation of the modulated cosmic ray spectrum as observed in the solar system.

OBSERVATIONS VS. ESTIMATES OF THE INTERSTELLAR SPECTRA

For most of the history of the studies of cosmic rays, the only available observations of cosmic ray spectra were those made in the vicinity of earth. As noted above, for use with modulation models the interstellar spectrum was often treated as almost a free parameter to be adjusted together with the interplanetary diffusion coefficient to achieve fits to observed spectra. As a result, a number of different forms for the interstellar spectrum have been proposed, with varying degrees of justification.

Recently, instruments on the Pioneer 10 spacecraft have returned observations of low energy cosmic ray spectra from radii in excess of 40 AU from the sun made during the solar minimum of 1987. These spectra are the least modulated spectra ever obtained at low energies, and it is therefore of interest to compare them to interstellar spectra that have been suggested in the past. In Figure 7, spectra of protons and helium measured during the 1987 solar minimum and during the previous solar minimum in 1978 are compared to a number of suggested interstellar spectra that have been used in modulation studies.

The interstellar spectra shown in Figure 7 have been derived from a number of lines of argument. The spectra of Urch and Gleeson /36/ and Randall and Van Allen /37/ were chosen simply as reasonable estimates, correct at high energies and not in conflict with any known observations at lower energies. The estimates of Ip and Axford /31/ were derived from theoretical considerations of acceleration, propagation and escape from the galaxy to establish the shape, and then normalized to the high energy spectrum. The Ip and Axford spectra plotted here have indeed been further renormalized to agree with other spectra at high energies since, as originally published, they were found to lie significantly below the other spectra at energies of a few GeV. The spectra of Webber and Yushak /38/ were derived from measurements of the spectra of ^1H, ^2H, ^3He, and ^4He. For any given interstellar primary spectrum (^1H, ^4He), models of propagation through the galaxy allow prediction of the secondary spectrum (^2H, ^3He). The interstellar spectrum of the primary was then selected from the set of possible spectra by the requirement that a single set of modulation parameters operating on the selected primary and associated secondary spectrum describe the observations made at 1 AU. Unfortunately, it was not possible to find a single set of parameters that satisfactorily described both the helium and the hydrogen primary-secondary pairs. The spectra of Evenson et al. /39,53/, which are almost identical to as those used by Garcia-Munoz et al. in their subsequent work /21,41/ are derived from demodulation of the electron spectrum observed at 1 AU to match the interstellar spectrum of Cummings et al. /27,28/ shown in Figure 6, and then using the same parameters to demodulate the nucleonic spectra. As noted above, this procedure is open to question because of the neglect of possible effects, such as drifts, which depend on the sign of the particle charge. Therefore, of the suggested interstellar spectra shown in Figure 7, none can be assigned a high degree of confidence, and the ultimate test must be experimental.

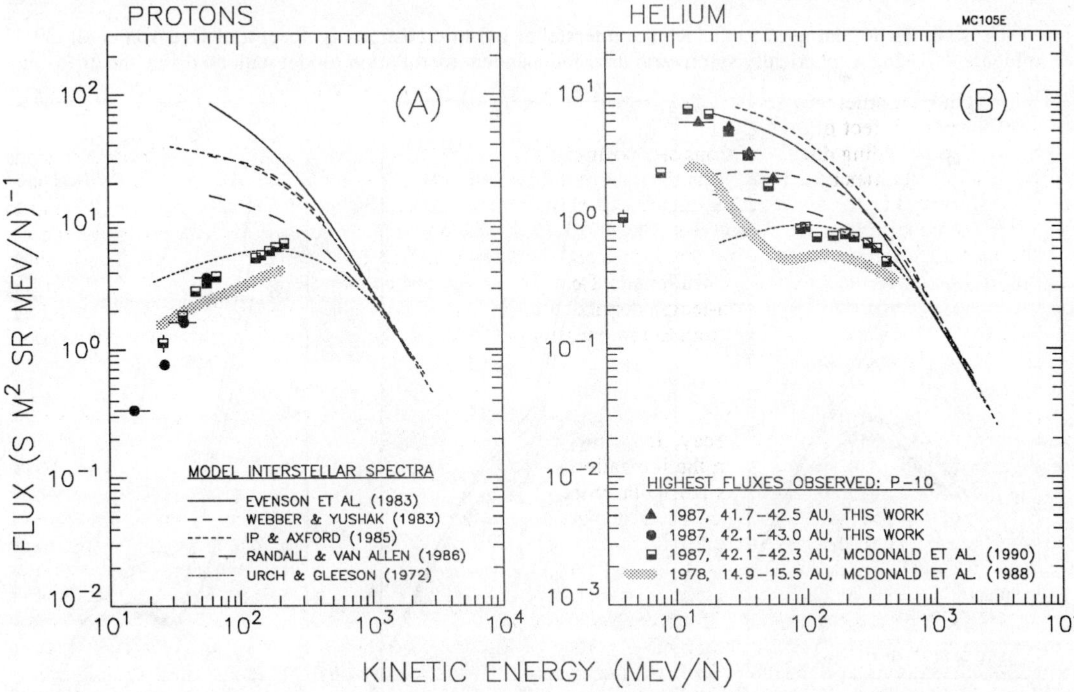

Fig. 7. Estimates of interstellar proton (A) and helium (B) spectra compared to observations during solar minimum at 42 AU prepared for this work and taken from /45/. Shaded lines represent maximum spectra observed in the previous solar minimum /46/. Sources for estimated spectra are Evenson et al. /39,53/; Webber and Yushak /38/; Ip and Axford /31/; Randall and Van Allen /37/; and Urch and Gleeson /36/.

For protons, as shown in Figure 7A, the 1987 Pioneer 10 observations clearly exclude the spectrum of Urch and Gleeson /36/, and appear also to be inconsistent with the spectrum suggested by Randall and Van Allen /37/. The remaining spectra are not at present seriously constrained by observations or theoretical considerations. The Evenson et al. spectrum /21,53/ is the highest at low energy and has been shown above to be consistent with available tests and limits from characteristics of interstellar clouds.

For helium, shown in Figure 7B, the situation is much more complicated as a result of the dominance of the anomalous helium at energies below about 100 MeV/n. Following the suggestion of Fisk et al. /40/, the anomalous helium, and the other observed anomalous components of oxygen, nitrogen, neon, argon, and, perhaps, hydrogen, are believed to consist of atoms that enter the heliosphere as interstellar neutrals, are ionized once by solar UV and convected to the solar wind termination shock where they are accelerated to energies typically of the order of a few (2-3) hundred MeV/charge /e.g. 43/. Once accelerated, these particles are subject to modulation by the solar wind in the same way as cosmic rays, except that, being only singly ionized, their magnetic rigidity is much higher than that of fully stripped cosmic ray particles of the same energy, so that they have easier access to the inner heliosphere. Their spectrum, however, is such that their intensity reduction as a result of adiabatic deceleration in the modulation process is much greater than that for low energy cosmic rays, so that they are very sensitive to changes in the level of modulation. Since they are subject to the same modulation process as are the cosmic rays, any model that describes modulation of cosmic rays may also be applied to the anomalous components.

Above 100 MeV/n, observations of helium spectra seem sufficient to exclude the interstellar spectra of Urch and Gleeson /20/ and Randall and Van Allen /37/, unless there is no modulation at 42 AU during solar minimum at these energies, an unlikely situation given the continued modulation of protons of comparable and slightly lower rigidities. The other spectra remain viable as far as direct observations can determine. Other tests offer no further constraints.

ANOMALOUS COMPONENTS AT THE BOUNDARY

Since anomalous components and galactic cosmic rays are subject to the same modulation process, it is interesting to apply demodulation techniques to the observed anomalous components to see if any constraints can be placed on the location of their acceleration region, presumably the solar wind termination shock. Recently, Garcia-Munoz et al. /21/ have undertaken such a program, using as a base observations of protons, helium, and oxygen made at 1 and 15 AU in 1977-78. The period is chosen as one in which the modulation in the heliosphere is most likely to be nearly in equilibrium, following the extended near-solar-minimum-like modulation conditions that were observed beginning in 1972. The period of minimum modulation in 1987 was so brief that equilibrium was not achieved, and only time-dependent modulation models can be expected to describe the behavior of low energy spectra during this period.

The approach is similar to that used to deduce the interstellar galactic cosmic ray spectra of Evenson et al. /39,53/ shown in Figure 7. Using a spherically symmetric time-independent modulation model with no drifts, the diffusion

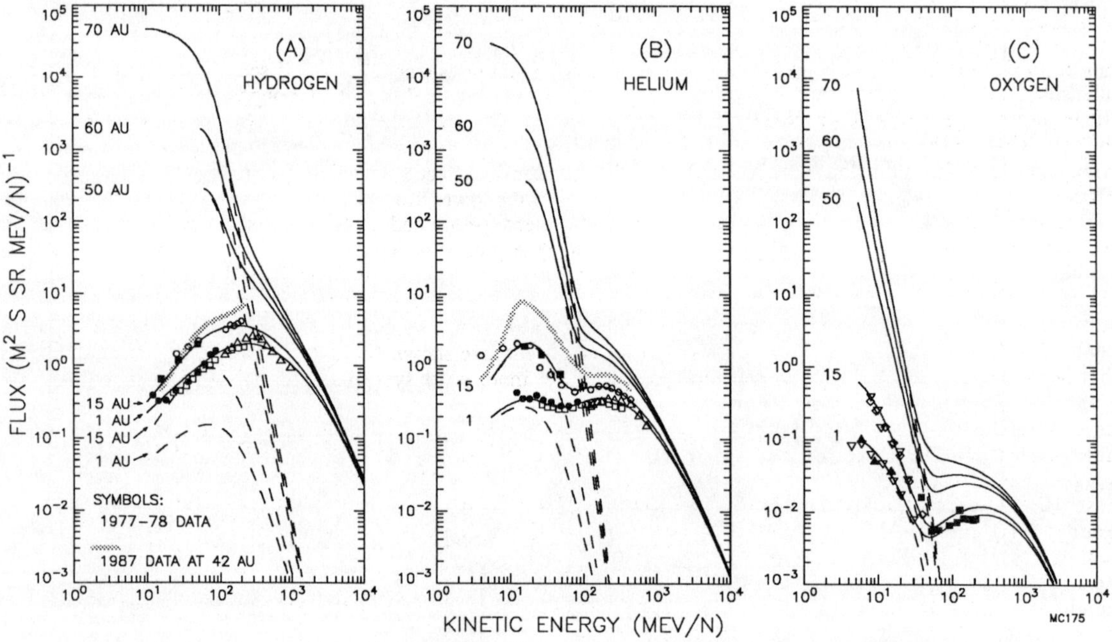

Fig. 8. Galactic cosmic ray plus anomalous component hydrogen, helium and oxygen domodulated to 50, 60, and 70 AU /21,53/. Solid lines are total spectra, anomalous component plus galactic cosmic rays. Dashed lines are anomalous component only. Shaded lines represent 1987 spectra from Figure7. Sources for data points are: filled circles /39/; open circles /47/; filled squares /21/; open squares /48/; open diamonds /49/; filled triangles /50/; open triangles /51/; inverted triangles /52/.

coefficient, or equivalently, Φ, and the interstellar spectrum necessary to satisfactorily fit the galactic helium spectrum at high energies at both 1 and 15 AU was determined. The interstellar spectrum so determined was very similar to that shown in Figure 7, and depended only weakly on the assumed location of the modulation boundary. Then, using the same diffusion coefficient, the spectra for singly ionized anomalous components required at the boundary to fit the observations at 1 and 15 AU were determined for assumed boundary locations of 50, 60, and 70 AU. Their results are shown in Figure 8 for hydrogen, helium, and oxygen. Also shown, for hydrogen and helium, are the spectra observed at 42 AU in 1987, which lie well below the spectra deduced for a boundary at 50 AU. While this may throw some question on the validity of the results, it is also plausible that it is simply the result of the far-from-equilibrium modulation conditions prevalent during the brief 1987 solar minimum.

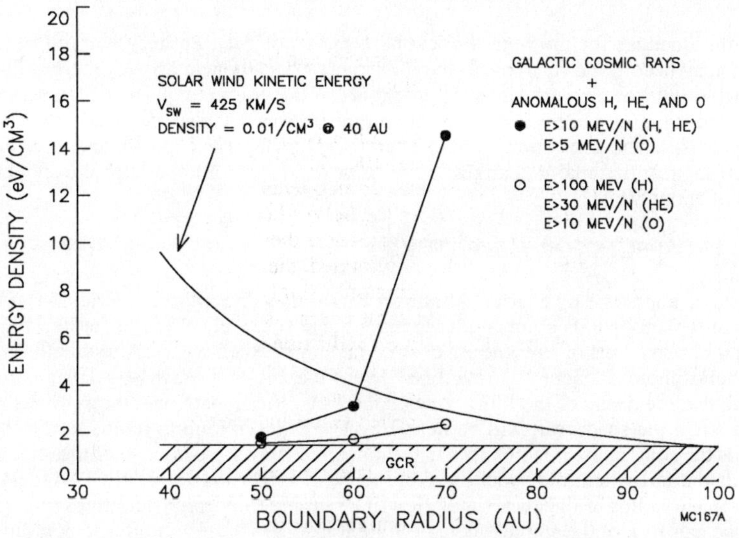

Fig. 9. Energy densities of total galactic cosmic rays ($1 \leq Z \leq 26$) + anomalous H, He, and O for the deduced boundary spectra shown in Figure 8, compared to the typical solar wind kinetic energy density as a function of boundary radius. The energy density of the galactic cosmic rays alone is shown as the shaded band.

For hydrogen, there was no clear evidence for an anomalous component at 1 AU in 1977-78. At 15 AU, however, the observed spectra were slightly higher than those expected for a modulation model that satisfactorily described other galactic cosmic ray components and the proton spectrum at 1 AU. Interpreting this excess flux, which is distinct from the so-called superfluxes observed even at 1 AU /41,42/, as due to a contribution from anomalous component hydrogen, the boundary spectra shown in Figure 8A result. As expected if the anomalous components are accelerated at the boundary, the spectra for protons and for other components as well are very soft, consistent with acceleration only up to energies of the order of a few hundred MeV per charge.

For all of the anomalous components, the very soft spectra and consequent enhanced sensitivity of the modulated intensity to adiabatic deceleration lead to the requirement for enormous fluxes at the boundary. Indeed, the deduced fluxes are so large that, as suggested by, for example, Lee /3/, they may have important dynamical consequences for the solar wind in the outer heliosphere. As a way of addressing this question, the energy density of the anomalous components may be compared with that of the flow energy density of the solar wind. Such a comparison is shown in Figure 9 for two cases as a function of boundary radius. For the lower limit case (open circles), the energy densities for the summed galactic cosmic ray and anomalous component spectra deduced for boundary radii of 50, 60, and 70 AU are calculated only for energies above the minimum energies at which changes in the boundary spectra have visible effects on the modulated spectra at 1 and 15 AU. For the higher case, meant to provide a "maxmum probable" estimate of the energy density, the lower energy bound for the calculation was lowered by up to a factor 10, corresponding to reasonable extrapolations of the deduced boundary spectra, as shown in Figure 8. The relative contributions of the hydrogen, helium, and oxygen components are shown in Table 1. In all cases, despite the fact that its contribution to the flux is far from obvious in the inner solar system, for the model considered here the anomalous hydrogen dominates the energy density in the outer heliosphere. The calculations suggest that, even for the lower limit case, the energy density of the energetic particles becomes comparable to the kinetic energy density of the solar wind in the neighborhood of 70-80 AU, and at even smaller radii for any reasonable extrapolation of the boundary spectra to lower energies. Using a different basic data set and a model including the acceleration at the boundary and the effects of drifts in modulation,

TABLE 1. Energy Density of Anomalous Components (eV/cm^3)
(Based on spectra from /21, 53/, cf. Fig. 8)

Species	Symbol (Fig. 9)	Min. Energy (MeV/n)	Radius (AU) 50	60	70
H	○	100	0.05	0.17	0.63
	●	10	0.17	1.01	9.82
He	○	30	0.03	0.09	0.26
	●	10	0.14	0.57	2.83
O	○	10	0.002	0.004	0.01
	●	5	0.015	0.079	0.43

Jokipii /43/ has come to similar conclusions.

The consequences for the outer heliosphere have not been entirely worked out. For example, while the energy density may become comparable to the flow energy density near 80 AU, there is little differential pressure acting on a solar wind volume element since the gradients of the energetic particles are relatively small. However, in that the solar wind imposes a convection velocity on the cosmic ray gas, one would expect some significant effects once the energy densities become comparable. Therefore, the observed spectra and intensities of the anomalous components seem to suggest that major changes in the solar wind flow, and perhaps the shock itself should be encountered by a radius of about 80 AU.

PROSPECTS FOR THE FUTURE

It would of course be of enormous interest to observe directly the interaction of the solar wind with the very high fluxes of the anomalous components expected near the boundary, and further, to penetrate the boundary and obtain our first observations of low energy cosmic rays unaffected by solar modulation. Of the four spacecraft now in the outer heliosphere, Pioneer 10, travelling toward the tail, will reach 80 AU soon after the year 2000, but at present it appears that the decay of the RTG power sources will terminate operations of the spacecraft's experiments late in the 1990's when the spacecraft will be near 70 AU. If the minimum radius to the shock should be as small as 70 AU towards the solar apex, it is possible that Pioneer 10 may see evidence of the shock's location even though it is going towards the tail where the shock is likely to be at larger radii. Figure 1 suggests that once Pioneer 10 passes the approximate radius of the apex shock, it will be on interplanetary field lines that connect directly to the shock in less than one half turn of the spiral, whereas while it is inside the minimum shock radius, the path length along the field to the shock is more than one complete turn. Therefore, when Pioneer 10 passes the minimum shock radius one might expect to observe very rapid changes in the modulation of cosmic rays and anomalous components that would have no counterparts on other spacecraft. The region beyond the minimum shock radius should be unlike any region we have so far observed in the heliosphere. Thus observations of cosmic rays and anomalous components on field lines directly connected to the shock would be of great interest in and of themselves, whether or not the unmodulated interstellar spectrum is observed.

Even though it is headed toward the solar apex, Pioneer 11 is less likely to see any evidence of the shock since it is at all times closer to the sun than Pioneer 10, and since it is expected to have adequate power to operate the scientific experiments only into the early 1990's. The Voyager 1 and 2 spacecraft will reach a radius of 80 AU early in the 21st century, at a time when they are still expected to be transmitting data to earth. Thus, if the minimum radius to the shock is 80 AU or more toward the solar apex, it is most likely to be the Voyager spacecraft that provide the first evidence for and direct observations of the shock.

Even if the Voyagers do encounter the shock, it is not clear that they will see the unmodulated interstellar cosmic ray spectrum, since little is known of the possible modulating effects of the low velocity, strong field region expected to exist between the shock and the heliopause. If the fields over the solar poles are nearly radial, it is possible that Ulysses, which will be over the poles in 1994-1995 will see a spectrum that is only weakly modulated. However, it has recently been pointed out /44/ that even small non-radial components in the polar field will most likely become dominant at large radii, giving rise to a significant barrier for penetration of low energy particles into the inner heliosphere. It seems most likely that for the direct observation of the true interstellar spectrum down to energies of the order of 1 MeV, a spacecraft dedicated to rapid escape from the solar system, such as the proposed Interstellar Probe /54/, will be required.

ACKNOWLEDGMENTS

This work would not have been possible without many helpful and informative discussions with M. Garcia-Munoz and J. A. Simpson. I also gratefully acknowledge the assistance of C. Lopate and A. Olson. This work was supported in part by NASA/Ames Grant NAG 2-380, by NSF Contract ATM 8908245, and by NASA Grant NAS 5-706.

REFERENCES

1. P. Meyer, E. N. Parker, and J. A. Simpson, The solar cosmic rays of February 1956 and their propagation through interplanetary space, Phys. Rev., 102, 1648, (1956).

2. W. I. Axford, The solar wind, Solar Phys. 100, 575, (1985).

3. M. A. Lee, The solar wind termination shock and the heliosphere beyond, Proc. Solar Wind 6, V. J. Pizzo, T. E. Holzer, and D. G. Sime (eds), NCAR Technical Note NCAR/TN-306+Proc, 2, 635, (1988).

4. T. E. Holzer, Interaction between the solar wind and the interstellar medium, Ann. Rev. Astron. Astrophys., 27, 199, (1989).

5. W. R. Webber, The interstellar cosmic ray spectrum and energy density. Interplanetary cosmic ray gradients and a new estimate of the boundary of the heliosphere, Astron. Astrophys., 179, 277, (1987).

6. V. B. Baranov, Private communication, (1989)

7. S. T. Suess, The Heliopause, submitted to Reviews of Geophysics, (1990).

8. R. A. Burger and M. S. Potgieter, The calculation of neutral sheet drift in two-dimensional cosmic ray modulation models, Astrophys. J., 339, 501, (1989).

9. M. S. Potgieter and R. A. Burger, Charge dependent modulation in the heliosphere as predicted by a wavy neutral sheet drift model, Proc. XXI Int'l Cosmic Ray Conf., (Adelaide), 6, 83, (1990).

10. J. Kota and J. R. Jokipii, Effects of drift on the transport of cosmic rays, VI, A three-dimensional model including diffusion, Astrophys. J., 265,573, (1983).

11. J. A. le Roux and M. S. Potgieter, The simulation of long-term modulation of cosmic rays with a time-dependent wavy neutral sheet drift model, Proc. XXI Int'l Cosmic Ray Conf., (Adelaide), 6, 87, (1990).

12. J. R. Jokipii and B. Thomas, Effects of drift on the transport of cosmic rays IV. Modulation by a wavy interplanetary current sheet, Astrophys. J., 243,1115, 1981.

13. T. Watanabe. Solar wind latitude/longitude variations derived from interplanetary scintillations, Adv. Space Res., 9, #4, 99, (1989).

14. C. Lopate, R. B. McKibben, K. R. Pyle, and J. A. Simpson, Pioneer 10 and 11 gradients of galactic cosmic ray nuclei and anomalous components through the period of solar minimum and to 46 AU from the sun, Proc. XXI Int'l Cosmic Ray Conf., (Adelaide), 6, 128, (1990).

15. R. A. Mewaldt, Temporal variations of anomalous cosmic rays and further evidence for anomalous cosmic ray hydrogen, Proc. XXI Int'l Cosmic Ray Conf., (Adelaide), 6, 160, (1990).

16. E. N. Parker, The passage of energetic charged particles through interplanetary space, Planet Space Sci, 13, 9, (1965).

17. J. R. Jokipii, E. H. Levy, and W. B. Hubbard, Effects of drift on cosmic ray transport I. General properties, Application to solar modulation, Astrophys. J., 213, 861. (1977).

18. L. J. Gleeson and W. I. Axford, Solar modulation of galactic cosmic rays, Astrophys. J., 154, 1011, (1968).

19. M. Garcia-Munoz, K. R. Pyle, and J. A. Simpson, The 1973-1984 solar modulation of cosmic ray nuclei, Proc. XIX Int'l Cosmic Ray Conf. (La Jolla), 4, 409, (1985).

20. I. H. Urch and L. J. Gleeson, Radial gradients and anisotropies due to galactic cosmic rays, Astrophys. Space Sci., 16, 55, (1972).

21. M. Garcia-Munoz, K. R. Pyle, and J. A. Simpson, Solar modulation in the heliosphere: time and space variations of anomalous helium and galactic cosmic rays, Proc. XXI Int'l Cosmic Ray Conf., (Adelaide), 6, 164, (1990).

22. D. Breitschwerdt, H. J. Völk, and J. F. McKenzie, Cosmic ray and wave driven galactic wind solutions, Proc. XX Int'l Cosmic Ray Conf., (Moscow), 2, 115, (1987).

23. T. W. Hartquist, J. H. Black, and A. Dalgarno, Cosmic ray ionization and the deuterium abundance, Mon. Not. R. Astr. Soc., 185, 643, (1978).

24. E. F. van Dishoeck and J. H. Black, Comprehensive models of diffuse interstellar clouds: physical conditions and molecular abundances, Astrophys. J. (Supp.), 62, 109, (1986).

25. J. M. Grunsfeld, J. L'Heureux, P. Meyer, D. Müller, and S. P. Swordy, Source spectra of primary cosmic rays, Proc. XXI Int'l Cosmic Ray Conf., (Adelaide), 3, 69, (1990).

26. D. Müller and K. K. Tang, The cosmic ray positron enigma, Proc. XXI Int'l Cosmic Ray Conf., (Adelaide), 3, 249, (1990).

27. A. C. Cummings, E. C. Stone, and R. E. Vogt, Evidence for primary interstellar cosmic-ray electrons, Proc. XIII Int'l Cosmic Ray Conf. (Denver), 1, 340, (1973).

28. A. C. Cummings, E. C. Stone, and R. E. Vogt, Interstellar electron spectrum from the galactic non-thermal radio emission, Proc. XIII Int'l Cosmic Ray Conf. (Denver), 1, 335, (1973).

29. J. M. Rockstroh and W. R. Webber, A new determination of the local interstellar electron spectrum from the radio background, Astrophys. J., 224, 677, 1978.

30. W. R. Webber, G. A. Simpson, and H. V. Cane, Radio emission, cosmic ray electrons and the production of γ-rays in the galaxy, Astrophys. J., 236, 448, (1980).

31. W. H. Ip and W. I. Axford, Estimates of galactic cosmic ray spectra at low energies, Astron. Astrophys., 149, 7, (1985).

32. J. B. G. M. Bloemen, K. Bennett, G. F. Bignami, L. Blitz, P. A. Caraveo, M. Gottwald, W. Hermsen, F. LeBrun, H. A. Mayer-Hasselwander, and A. S. Strong, The radial distribution of galactic gamma rays II. The distribution of cosmic ray electrons and nuclei in the outer galaxy, Astron. Astrophys., 135, 12, (1984).

33. M. Garcia-Munoz, P. Meyer, K. R. Pyle, J. A. Simpson, and P. Evenson, The dependence of solar modulation on the sign of the cosmic ray particle charge: Further study of a 22 year solar magnetic cycle, Proc.XX Int'l Cosmic Ray Conf. (Moscow), 3, 302, (1987).

34. S. P. Ahlen, Theoretical and experimental aspects of the energy loss of relativistic heavily ionizing particles, Rev. Mod. Physics, 52, 121, (1980).

35. M. Garcia-Munoz and J. A. Simpson, The elemental abundances of hydrogen through nickel in the low energy cosmic rays, Proc. XVI Int'l Cosmic Ray Conf. (Kyoto), 1, 270, (1979).

36. I. H. Urch and L. J. Gleeson, Galactic cosmic ray modulation from 1965-1970, Astrophys. Space Sci., 17, 426, (1972).

37. B. A. Randall and J. A. Van Allen, Heliocentric radius of the modulation boundary, Geophys. Res. Letters, 13, 628, (1986).

38. W. R. Webber and S. M. Yushak, A measurement of the energy spectra and relative abundance of the Cosmic ray H and He isotopes over a broad energy range, Astrophys. J., 275, 391, (1983).

39. P. Evenson, M. Garcia-Munoz, P. Meyer, K. R. Pyle, and J. A. Simpson, A quantitative test of solar modulation theory: The proton, helium, and electron spectra from 1965-1979, Astrophys. J. (Letters), 275, L15, (1983).

40. L. F. Fisk, B. Koslovsky, and R. Ramaty, An interpretation of the observed oxygen and nitrogen enhancements in low-energy cosmic rays, Astrophys. J. (Letters), 190, L35, (1974).

41. M. Garcia-Munoz, G. M. Mason, and J. A. Simpson, The appearance of super-fluxes of quiet time cosmic rays, Proc. XV Int'l Cosmic Ray Conf. (Plovdiv), 3, 209, (1977).

42. J. J. Beatty, M. Garcia-Munoz, and J. A. Simpson, The cosmic-ray spectra of ^{1}H, ^{2}H and ^{4}He as a test of the origin of the hydrogen superfluxes at solar minimum modulation, Astrophys. J., 294, 455, (1985).

43. J. R. Jokipii, The anomalous component of cosmic rays, Adv. Space. Res., this issue.

44. J. R. Jokipii and J. Kota, The polar heliospheric magnetic field, Geophys. Res. Letters, 16,1, (1989).

45. F. B. McDonald, N. Lal, R. McGuire, and T. T. von Rosenvinge, The modulation process: Comparison of successive solar minimum cosmic ray observations, Proc. XXI Int'l Cosmic Ray Conf., (Adelaide), 6, 144, (1990)

46. F. B. McDonald, T. T. von Rosenvinge, N. Lal, J. H. Trainor, and P. Schuster, The recovery phase of galactic cosmic ray modulation in the outer heliosphere, Proc. XX Int'l Cosmic Ray Conf., (Moscow), 3, 397, (1988).

47. F. B. McDonald, private communication, (1984).

48. T. T. von Rosenvinge, F. B. McDonald, F. B. Trainor, and W. R. Webber, The modulation of galactic cosmic rays between 1 and 17 AU, Proc. XVI Int'l Cosmic Ray Conf. (Kyoto), 12, 170, (1979).

49. A. C. Cummings, E. C. Stone, and W. R. Webber, Changes in the energy spectrum of anomalous oxygen during 1977-1985, J. Geophys. Res., 91, 2896, (1986).

50. D. Hovestadt, B. Klecker, G. Gloeckler, F. M. Ipavich, C. Y. Fan, and L. A. Fisk, Temporal variations of the anomalous oxygen (1974-1979) and disappearance in 1978, Proc. XVI Int'l Cosmic Ray Conf. (Kyoto), 3, 255, (1979).

51. W. R. Webber and S. M. Yushak, Measurements of cosmic ray 2H and 3He nuclei above 100 MeV/nuc using a balloon borne telescope, Proc. XVI Int'l Cosmic Ray Conf. (Kyoto), 1, 383, (1979).

52. W. R. Webber, F. B. McDonald, J. H. Trainor, and T. T. von Rosenvinge, A measurement of the radial gradient of the low energy anomalous oxygen component from 1-16 AU, Proc. XVI Int'l Cosmic Ray Conf. (Kyoto), 5, 353, (1979).

53. M. Garcia-Munoz, private communication, (1990).

54. Task Group on Solar and Space Physics (F. Scarf, Chrmn), Space Science Board, Nat'l Research Council, "Solar and Space Physics", Space Science in the Twenty-First Century: Imperatives for the Decade 1995-2015, National Academy Press, Washington, D. C., pp. 41-43, (1988).

55. J. A. Simpson and J. R. Wang, The eleven-year and residual modulation of cosmic rays (1952-1969), Astrophys. J., 161, 265, (1970).

DIFFUSION, DRIFTS, AND MODULATION OF GALACTIC COSMIC RAYS IN THE HELIOSPHERE

J. Kóta*

Lunar and Planetary Laboratory, The University of Arizona, Tucson, AZ 85721, U.S.A.

ABSTRACT

The role of particle drifts in the heliospheric transport of cosmic rays is reviewed. The basic transport-theory is outlined, the most profound effects of drifts are interpreted. Experimental material seems to be in favour of the drift concept. A new, more advanced 3-D model incorporating a non-uniform solar wind speed, rising away from the wavy neutral sheet is introduced and some preliminary results are presented. Among others, the model correctly predicts the negative latitudinal gradient of cosmic ray intensity that appeared during the last solar minimum, and disappeared as the tilt angle of the sheet increased. Also, radial gradients predicted for the two polarity cycles become closer to each other. Possible modulation effects beyond the solar wind termination shock are discussed. Preliminary results suggest that, at lower energies, a substantial part of modulation may occur beyond the shock.

INTRODUCTION

Entering the heliosphere, charged cosmic ray particles encounter magnetic and electric fields embedded in the outward moving solar wind. This interaction results in a modulation of cosmic ray flux inside the heliosphere: cosmic ray intensity is lowest during the periods of high solar activity and is less depressed (though still below the undisturbed interstellar level) during the years of quiet sun. The physics of modulation was first understood by Parker /1/ in terms of balance between the outward convection and inward diffusion. The appropriate transport-equations have been worked out by Gleeson and Axford /2/ and Jokipii and Parker /3/. In the late seventies, Jokipii, Levy, and Hubbard /4/ re-investigated the possible role of curvature and gradient drifts, and pointed out that inclusion of drifts may profoundly modify the modulation picture. Until now, drift still remained the most controversial subject of cosmic ray transport in the heliosphere. A most characteristic feature of drift is its sensitivity to the polarity of the magnetic field, drift will introduce a charge-asymmetry. Also, particle transport will be affected by the global structure of the heliospheric magnetic field and its neutral sheet. Owing to this, drift-models require 2-dimensional and ultimately 3-dimensional studies of particle transport. Recent reviews on the physics of cosmic ray modulation were written by McKibben /5/ and Jokipii /6/.

The present review will primarily be devoted to the discussion of drift-models, with special emphasis on 3-dimensional studies including a wavy heliospheric neutral sheet. After reviewing the transport-equation and the basic pattern of drift, we will outline the most profound predictions, then report on recent 3-dimensional numerical simulations. This paper considers galactic particles only, the physics of the anomalous component of cosmic rays is reviewed in another work in this volume by Jokipii /7/.

With the Pioneer and Voyager spacecraft traveling far out in the heliosphere, possible effects at and beyond the termination shock of the solar wind become increasingly important. In the last section, this topics will briefly be addressed.

TRANSPORT-EQUATION

Figure 1 (taken from /5/) shows a schematic picture of the heliosphere. The supersonic solar wind flows radially outward from the sun and carries the frozen-in magnetic field. The magnetic field consists of a regular spiral field and superimposed small-scale irregularities, waves traveling roughly together with the solar wind. Cosmic ray particles, which are typically far more energetic than solar wind protons, are scattered on these waves giving rise to diffusion and convection. The expansion of the solar wind plasma

* Permanent address: Central Research Institute of Physics, Budapest, Hungary

causes the cosmic radiation to cool adiabatically. The curvature and gradient of the large-scale magnetic field lead to particle drifts. Since particle speeds are much higher than solar wind speed, cosmic ray distribution will remain nearly isotropic. The appropriate equation for the isotropic part of the distribution function, $f(\mathbf{r},p,t)$ of cosmic rays at position, $\mathbf{r}$, momentum, p, and time, t, is

$$\frac{\partial f}{\partial t} = \nabla(\kappa^{(s)}\nabla f) - \mathbf{V}.\nabla f - \mathbf{V}_{dr}.\nabla f + \frac{\nabla.\mathbf{V}}{3}\frac{\partial f}{\partial lnp} \qquad (1)$$

with the consecutive terms of the R.H.S. accounting for diffusion, convection, drift, and adiabatic energy change, respectively. $\kappa^{(s)}$ stands for the symmetric diffusion tensor, $\mathbf{V}$ is the solar wind speed, and $\mathbf{V}_{dr}$ is drift velocity. The inclusion of dispersion due to forward and backward propagating waves would give rise to a stochastic, or 2nd order, Fermi acceleration. This term, though conceptually important, is expected to play little role for galactic cosmic rays, and is usually omitted.

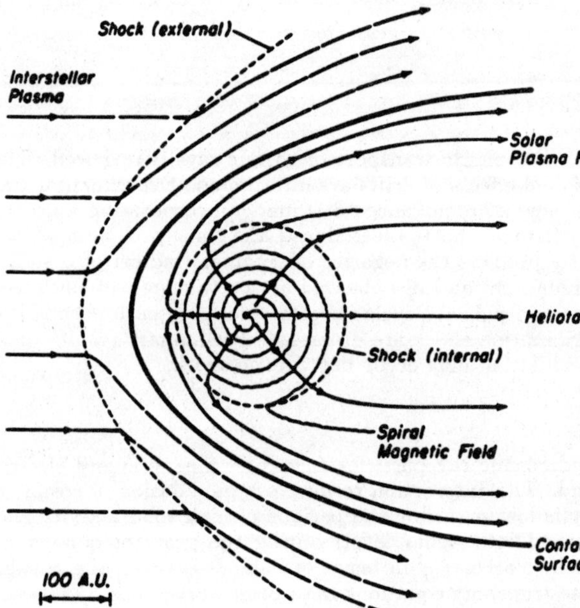

Fig. 1. Schematic view of a plausible model for the heliosphere (from /6/).

At some 50 to 150 AU from the sun, the solar wind must undergo a supersonic to subsonic transition, plausibly through a termination shock (marked as inner shock in Figure 1). Beyond this shock, the solar wind flow is subsonic, and deflected from the radial direction. In the simplest picture, a contact surface (heliopause) forms that divides the solar plasma from the interstellar medium. Outside the heliosphere, depending on the prevailing interstellar conditions, there may or may not be an external bow shock.

Though the scenario changes quite dramatically at crossing the terminal shock, the transport-equation (1) remains valid throughout the whole heliosphere. At the shock, particles gain energy through shock acceleration: this process is also accounted for by the energy-change term. In mathematical terms, the sign of $\nabla.\mathbf{V}$ determines if particles are accelerated or decelerated: positive values correspond to expansion and cooling, negative values imply compression and particle acceleration. Beyond the shock, in the subsonic region, the solar wind has a low Mach-number, and is nearly incompressible. Thus no significant energy change (neither acceleration nor deceleration) can be expected there.

Though energy loss and modulation are intimately related in the inner heliosphere, the lack of deceleration does not preclude a substantial modulation to occur between the termination shock and the heliopause. Because of the strong shock, convection speed should drop by about a factor of 4 at the shock. On the other hand, plasma density and the spiral field (which is nearly azimuthal at these distances) is expected to increase by the same factor. One can reasonably argue that diffusion should scale inversely with the magnetic field strength, so the net effect is that the ratio of convection to diffusion remains about the same on both sides of the shock.

Diffusion. In the presence of a large-scale regular magnetic field, diffusion becomes anisotropic and the diffusive streaming will be related to the density gradient by the full diffusion tensor, κ_{ij} /3/

$$\kappa_{ij} = \begin{pmatrix} \kappa_{\parallel} & 0 & 0 \\ 0 & \kappa_{\perp} & \kappa_T \\ 0 & -\kappa_T & \kappa_{\perp} \end{pmatrix} \tag{2}$$

with free, unrestricted diffusion ($\kappa_{\parallel}$) in the field's direction, and reduced rate of diffusion ($\kappa_{\perp}$) perpendicular to the magnetic field. The diffusion tensor will also contain an anti-symmetric term, κ_T, which changes sign with the reversal of magnetic polarity. This anti-symmetric diffusion is closely related, though not identical, with particle drifts. For conditions typical in the heliosphere, the scattering mean free path of cosmic rays well exceeds their gyro-radius. In this case, one expects κ_T to be much larger than $\kappa_{\perp}$, and

$$\kappa_T = \frac{v.p.c}{3qB} \tag{3}$$

with v being particle velocity, c the velocity of light, q the particle charge, and B the magnetic field strength. As soon as the scattering mean free paths and gyro-radii become comparable, particles cannot complete several cycles around the field before being scattered, and κ_T diminishes.

Assuming that scattering can be described by a relaxation time model (which mathematically corresponds to isotropic scattering), the components of the diffusion tensor can readily be evaluated. This idealized picture, however, ignores perpendicular diffusion due to the random walk of magnetic field lines which, in reality, may be a major part of perpendicular diffusion. The three diffusion coefficients should satisfy the relation:

$$\kappa_{\perp}(\kappa_{\parallel} - \kappa_{\perp}) \geq \kappa_T^2 \tag{4}$$

with the equality holding for the relaxation-time approximation.

Jump conditions. Plasma parameters, solar wind speed, and magnetic fields change abruptly at discontinuities. The two principal discontinuities of interest are shocks and the heliospheric neutral sheet where magnetic polarity changes sign. At these places the transport-equation (1) contains singularities, for instance drift becomes singular. To circumvent these singularities, jump conditions can be established, with the basic underlying physics that the f isotropic density and the s_n normal component of the net flux $\mathbf{s}$, measured in a fixed frame should be the same on both sides of a discontinuity surface, i.e.:

$$f\big|_1^2 = 0 \qquad and \qquad s_n\big|_1^2 = 0 \tag{5}$$

where the net streaming, $\mathbf{s}$, consists of diffusive and convective terms:

$$\mathbf{s} = -\kappa\nabla f - \frac{p}{3}\frac{\partial f}{dp}\mathbf{V} \tag{6}$$

In this expression, the diffusive streaming obviously includes the antisymmetric term. The above jump conditions are valid in general, as long as the second order 'stress-tensor' of the cosmic ray distribution is not considered. Recently, Jokipii, Morfill, and Kóta /8/ worked out a more general formula, including second-order terms.

Boundary Conditions. Since the external interstellar shock, if it exists at all, should be a weak one, and the interstellar flow-pattern is plausibly smooth and divergence-free, any modulation outside the heliopause should be negligible. Thus it seems reasonable to set f equal to the undisturbed interstellar level at the heliopause. Possible small galactic anisotropies (e.g. convection by the interstellar wind) can be ignored since these should completely be washed out in the heliosphere, at least for energies up to a few tens of GeV. On the other side, an absorbing boundary (f=0) is to be assumed close the sun. Results, except for very small radii, are fairly insensitive to the exact position of the inner boundary.

One should realize, that setting the boundary condition at the heliopause meets several difficulties. First, the subsonic region is largely unknown, and one has to rely upon a hypothetical solar wind flow pattern. Second, there will be some technical subtleties; for instance the geometry of the stream-lines are bound to be more complex than the simple radial lines inside the termination shock. Model calculations tend to take the shock as outer boundary, tacitly assuming that the effect of shock on modulation is negligible. Some recent works suggest, however, that this might not be so. Quantitative studies on this subject are, understandably, in an exploratory stage.

DRIFT EFFECTS

Drift Velocities in the Heliosphere.

The well known curvature and gradient drifts depend on the pitch angle of the gyrating particle. For a nearly isotropical distribution of particles, the average drift can be derived as the divergence of the anti-symmetric part of the diffusion tensor. This derivation implies that drift must be free of divergence, and drift solely would not lead to any modulation /4/.

In the interplanetary magnetic field, a wavy neutral sheet (Figure 2) divides the two 'hemispheres' of opposite magnetic polarity. The measure of waviness of the sheet is determined by the tilt angle of the solar dipole with respect to the rotation axis of the sun. The tilt angle is known to change during an 11-year cycle: the sheet is relatively flat around solar minimum, and becomes more wavy with increasing solar activity. The solar dipole reverses every 11 year, forming a 22-year solar magnetic cycle. In the years of 1970-1980, the interplanetary magnetic field pointed outward above the sheet and inward below the sheet. The opposite configuration prevailed in the sixties and eighties.

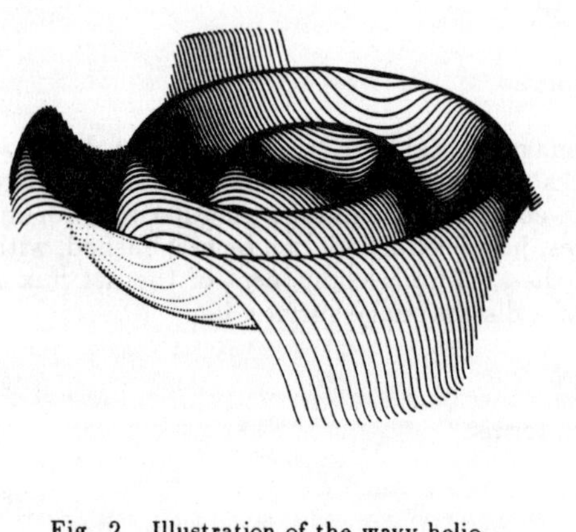

Fig. 2. Illustration of the wavy heliospheric neutral sheet (from /13/).

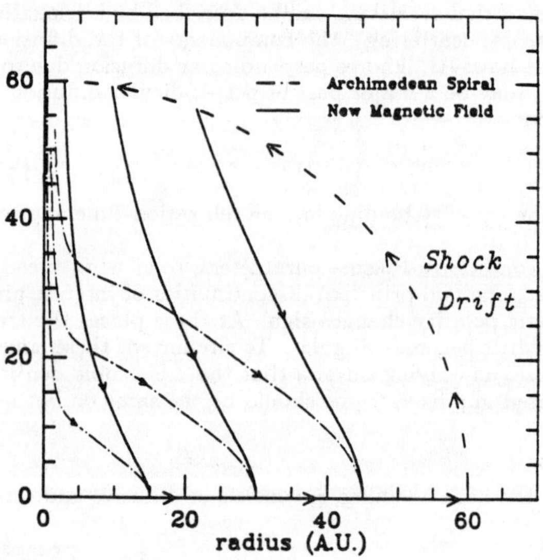

Fig. 3. Drift pattern for ~1 GeV protons in the heliosphere when the northern magnetic field is outward ($A > 0$). The sense of drift reverses with polarity reversal. Dashed lines are for the Archimedean spiral, and solid lines are for a modified field that is stronger in the polar regions (from /6/).

In Figure 3, dash-dotted lines indicate the drift pattern of ~1 GeV protons in the regular Parker spiral field of the configuration of the seventies (with usual notation $A > 0$), with a flat neutral sheet. Positively charged particles drift inward in the polar regions, downward to the neutral sheet, and outward along the neutral sheet. The sense of drift will be the opposite if either the magnetic polarity reverses ($A < 0$), or the particle charge is negative (electrons).

At the shock, there is a fast drift upward along the shock. In the subsonic regime, drift should be reduced by a factor of 4, because of the stronger magnetic field. If one simply continues the solar wind beyond the shock as a radial and incompressible flow then drift turns out to almost completely vanish everywhere, except for a very narrow polar cone and the neutral sheet.

The Archimedean spiral field falls off rapidly with heliocentric radius in the polar directions. As a result, drift becomes large at the poles posing some conceptual problems. What may be even more important for modulation, the weak field would imply that particles can very easily diffuse into the inner heliosphere through the polar regions. Recently Jokipii and Kóta /9/ proposed that, at large distances (>5 AU), the polar field is dominated by an irregular transverse component, which should both reduce polar drift, and impede the easy diffusive access. Drift velocities calculated from this assumption are shown by solid lines in Figure 3. The modified polar field will spread the drift streamlines over a larger area around the poles.

When can drift-effects be significant? Drift in itself would not produce any modulation, it acts through changing the path of particles into the heliosphere. Crudely, positively charged particles are arriving predominantly from the poles for the configuration $A > 0$, and are drifting in along the neutral sheet for $A < 0$. Typically, diffusion dominates drift on small time scale. However drift, unlike diffusion, is organized and accumulates on large time scales. Two conditions must be met before one can expect significant drift effects.

Noticeable drift effects can obviously be expected only if particles spend long enough time in the heliosphere to drift at least a significant portion of $\pi/2$ in latitude. The typical latitude drift is κ_T/r, combining this with the crude 'diffusive radial speed' κ_{rr}/r one requires :

$$\int_r^{R_{out}} \frac{\kappa_T}{\kappa_{rr}} \frac{dr'}{r'} > 1 \tag{7}$$

where κ_{rr} stands for the radial diffusion coefficient: $\kappa_{rr} = \kappa_{\|} cos^2\psi + \kappa_{\perp} sin^2\psi$, with the ψ garden hose angle being close to $\pi/2$ far from the sun.

Another condition is that latitudinal (i.e. perpendicular) diffusion should not be too strong to wash out the drift pattern. Particles will not be drifting completely confined to the neutral sheet or polar axis but will be spread in latitude. The extent of this latitude spread is determined by the ratio of $\kappa_{\perp}/\kappa_T$. Crudely, $\kappa_{\perp} \ll \kappa_T$ represents the case when drift is dominant, and one can simply think in terms of drift trajectories. The other extereme case is $\kappa_{\perp} \gg \kappa_T$ when perpendicular diffusion completely destroys the picture, and drift effects are negligible. In the intermediate case, when $\kappa_{\perp}$ and κ_T are comparable, drift effects must be present to a smaller or larger extent, depending on the value of $\kappa_{\perp}/\kappa_T$. In this case, however, there is already an interplay of drift and diffusion; results are less transparent, they cannot simply be interpreted by merely looking at drift-trajectories.

Drift Models, Predictions, Neutral Sheet Effects

In this section the most profound predictions of model calculations will be outlined. Some of these predictions can easily be interpreted as result of polar or neutral sheet drift into the heliosphere. Others are less transparent, and in some cases the dependence on the input parameters (for instance parallel and perpendicular diffusions as functions of position and momentum) is not yet thoroughly explored. Nevertheless, the remarkable consensus in the sense of drift effects obtained in different calculations indicate that these results remain valid (qualitatively at least) for a wide range of parameters.

Effects directly connected with the waviness of the neutral sheet represent a particularly intriguing subject, since waviness is known to vary during a solar cycle. The change in the extent of waviness, i.e. change of the tilt angle of the solar dipole, may constitute one of the most important factors in solar cycle variations. Throughout this paper, the tilt angle will be the only parameter to be changed during a cycle. Of course, there are other effects as well, which will not be covered in this work. For instance, more transient events occur at high solar activity. Their effect, the so called episodic modulation, should be super-imposed on the quasi-steady picture outlined here.

Jokipii and Kopriva /10/ were the first to carry out a numerical solution of the transport-equation (1) including drift for a 2-dimensional model with flat neutral sheet. The results verified that drift may indeed be important and can produce significant charge-asymmetries. Qualitative considerations /11,12/ and approximate calculations /13/ indicated that the waviness of the neutral sheet may play an important role. Roughly speaking, in the cycle when particles drift inward along the neutral sheet cosmic rays should be sensitive to the structure of the sheet; enhanced waviness will mean longer path along the sheet, which should result in larger modulation. In the opposite cycle, on the other hand, cosmic rays penetrate through the polar regions, thus they are insensitive to the neutral sheet and, in general, to conditions in the equatorial region.

A quantitative study involving a wavy neutral sheet requires a 3-dimensional numerical simulation. This was first done by Kóta and Jokipii /14/ in a fully 3-dimensional model. The model has subsequently been extended, more recent results will be reported on in the next section. Even this more advanced model is, however, still incapable to describe the period of solar maximum, when polarity reverses. It is not the drift concept that breaks down, it is the global field structure around reversal that is far too complex.

Potgieter and Moraal /15/ have devised a way to simulate a wavy sheet in a 2-dimensional model, retaining axial symmetry. These authors substitute the thin wavy sheet by a thick, symmetric transition region whose extent corresponds to the tilt angle of the sheet. The scheme is economic and has been succesful in reproducing some of the 'tilt-angle' effects.

In what follows, drift predictions will briefly be summarized and compared with experimental findings. Details of observations and the underlying physics of interpretation are illuminated in the surveys of McKibben /5/ and Jokipii /6/.

Shift of anisotropy. Historically the shift of the phase of the anisotropy was the first phenomenon explained by drift. Levy /16/ invoked drift to account for the shift toward earlier hours observed in the phase of the cosmic ray daily variation following the polarity reversal of the sun in 1969. The shift of anisotropy appears to be a natural and almost inescapable result in drift models, the phase shift (in the correct direction) is present in all drift calculations /15,17,18,19/, though details are still to be worked out /20/.

Polarity-dependent 11-year cycles. As mentioned earlier in this paper, in the cycle $A > 0$, positively charged particles drift inward through the polar regions, so they are expected to be insensitive to conditions in the equatorial region. This concept is supported by the observation of Shea and Smart /21/ who found a good correlation between cosmic ray intensity and turbulence in the ecliptic plane for $qA < 0$ and a poor correlation for $qA > 0$.

Correlation of modulation with tilt angle. It is conceivable that, in the case of $qA < 0$, cosmic rays need to drift inward along a longer path on the neutral sheet if the sheet is wavy. Larger tilt angle or more warps will mean longer time spent in the heliosphere, more deceleration, and lower cosmic ray flux. In the opposite cycle, $qA > 0$, on the other hand, particles come primarily from the poles, and have little chance to interact with the neutral sheet. Cosmic ray flux at the earth is thus expected to be rather insensitive to the waviness of the sheet for $qA > 0$. Numerical calculations verified this expectation: Kóta and Jokipii /14/ simulated a 22-year solar cycle with the tilt angle being the single parameter to change during the cycle, and found a close resemblence between predictions (Figure 5) and observations (Figure 4). The peaked profile of $qA < 0$ (1959-69) is followed by a plateau-type profile for $qA > 0$ (1970-80) similarly as observed in the Climax neutron monitor counting rates /5/. The polarity dependence of correlation between cosmic ray flux and the tilt angle was directly confirmed by Smith and Thomas /22/.

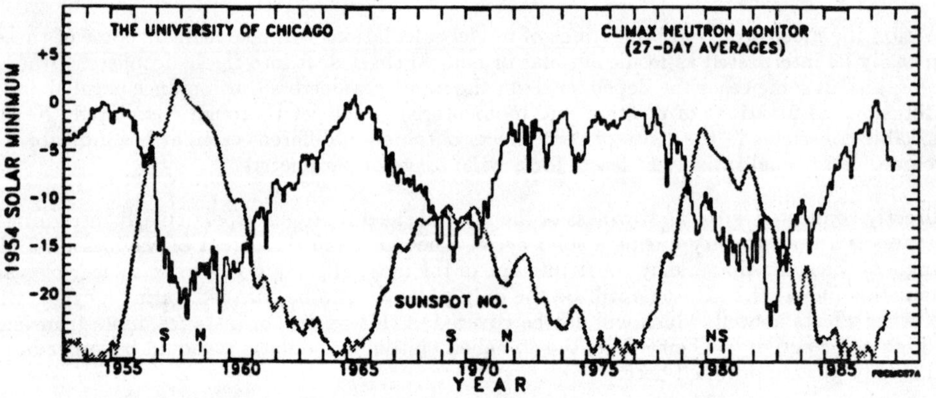

Fig. 4. 27-day average counting rates from the Climax neutron monitor (solid line) compared to the smoothed monthly average sunspot number (shaded line). **N**, **S** indicate times of magnetic polarity reversal of the sun's north and south poles (from /5/).

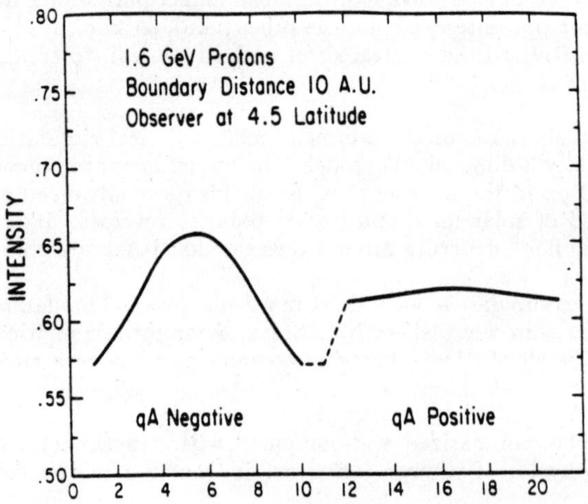

Fig. 5. Simulated 22-year cycle by changing the tilt angle of the neutral sheet (from /14/).

Charge asymmetry, Helium-electron ratio. Polarity dependence will obviously transform into charge asymmetry when cosmic ray species of opposite charges are compared. Garcia-Munoz et al /23/ found a clear evidence for charge asymmetry in the change of the electron/helium ratio in opposite 11-year cycles. At the same time, however, electrons failed to produce the predicted peaked-plateau difference shown in Figure 5. The situation concerning electrons still remains controversial. Some possibilities to resolve the discrepancies are discussed by Jokipii, Mewaldt, and Moraal /24/.

Radial gradients. Radial gradients appear to constitute another subject of controversy. First drift calculations /10,14,17/ predicted large differences for the radial gradients in the two cycles: markedly smaller ecliptic gradients were charactersitic for the solutions of $qA > 0$. In contrast to these theoretical results, observational gradients inferred from Voyager and Pioneer data /5,25,26,27/ indicate small or no polarity dependence at all. To make things even worse, the difference obtained by the Chicago group in the 30-70 MeV channel of Pioneer-10 turns out in the opposite direction /25/.

Early calculations turned out to seriously overestimate the asymmetry because of the overly simplified models they used (for instance, computational reasons dictated the use of a very small heliosphere). Later, more sophisticated drift calculations using more realistic parameters have considerably softened the discrepancy (see next paragraph) but the conflict has not yet been resolved in a reassuring way. The underlying physics has not yet been fully understood. For instance, the effect of perpendicular diffusion must be studied in more depth.

On one hand, a large perpendicular diffusion would obviously reduce the asymmetry of the predicted gradients. This would happen at the expense of the overall role of drift: large cross-field diffusion would reduce, though not entirely eliminate, all drift effects. At the other extreme, a complete lack of perpendicular diffusion would imply (via jump conditions (5) and (6)) zero gradient all along the sheet for $qA < 0$. Of course this latter case is non-physical, nevertheless it does indicate that gradients close to the sheet may depend on $\kappa_\perp$ in a subtle way. Taking the 'minimum $\kappa_\perp$' defined by the equal sign in condition (7), and considering a special case, Kóta /28/ deduced a 3-D force-field solution where drift effects were not supressed and yet radial gradients were identical for the two cycles.

It seems conceivable, that the smaller radial gradient for $qA > 0$ might be related to the easy diffusion in the polar regions. The modification of the polar field as suggested by Jokipii and Kóta /9/ does indeed increase primarily the gradient for $qA > 0$, and brings predictions closer to each other (see next paragraph).

It may be interesting to mention that, at the ~10 GeV range, the observed shift of the anisotropy can best be explained with a smaller radial gradient in the cycle of $qA > 0$ /16,17/. Indirect experimental results on radial gradients, on the other hand, do not show any noticeable difference between the two cycles /29/.

From experimental side, it should be borne in mind that radial gradients are inferred from few (two or three) points in space spanning over large distances thus the radial dependence of the gradient cannot be known accurately. If, for instance, the radial gradient happens to be greater at small heliocentric radii then the values for $qA > 0$ (derived from smaller distances) are relatively overestimated with respect to the values inferred for $qA < 0$ from larger distances.

Latitude gradients. Until now, latitudinal gradients have brought perhaps the strongest evidence in favour of drifts. Drift models are specific in that they predict opposite latitude gradients for the two cycles. In a 2-D model with a flat neutral sheet, cosmic ray flux should be maximum at the poles and decrease downward to reach minimum at the sheet for $qA > 0$. For $qA < 0$, high intensity is expected at both the neutral sheet and the poles because of the fast neutral sheet drift and polar diffusion, respectively. Minimum cosmic ray flux should be at medium latitudes, the exact location depending on the relative easiness of polar diffusion. The modified Parker-field /9/ impeding polar diffusion will shift the location of minimum cosmic ray flux toward higher latitudes. In any case, right at the sheet the latitude gradient is required to be negative for $qA < 0$, and positive for $qA > 0$.

A 3-D model including a wavy neutral sheet will give a more complex picture, as illustrated in Figure 6. Nevertheless, intensity minimum still remains close to the heliospheric equator for $qA > 0$, and shifts toward higher latitudes for $qA < 0$. Looking at the picture averaged over azimuth (i.e. over a full solar rotation) there are distinct differences from the 2-D picture. In a 3-D model, unlike in 2-D, the latitude gradient for $qA < 0$ need not apriori be negative at the equator.

The negative latitudinal gradient was first seen by the two Voyager spacecraft in the anomalous cosmic ray component /30,31/ then it has also been confirmed in galactic cosmic ray data /32/ during the present cycle (1980-90). In the preceding cycle positive gradients were observed in accord with predictions. The reversal of latitudinal gradient is a natural result of drift theories while it is hard to explain by any other current ideas. Besides drift, helicity-dependent transport /33/ is the only other process known to produce charge asymmetry. It is hard to see, however, how helicity could account for the reversal of latitude gradient.

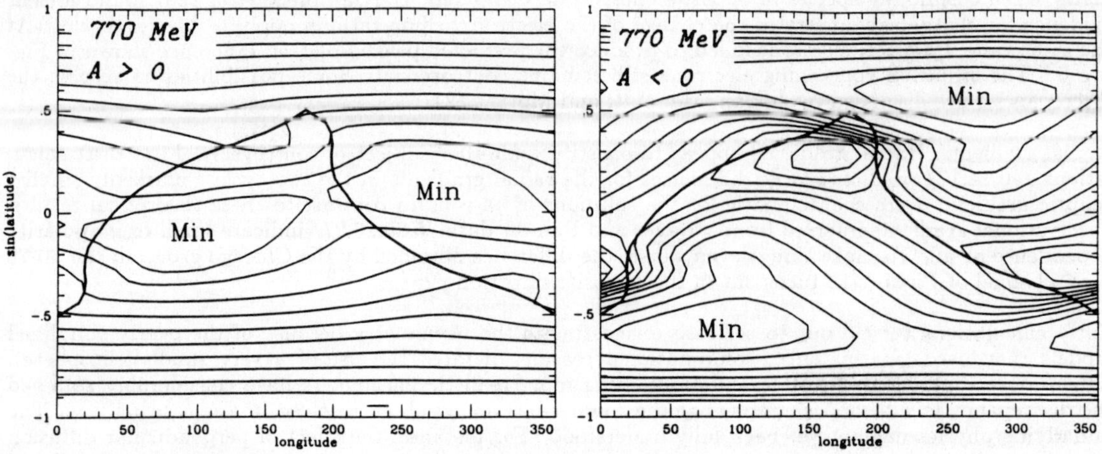

Fig. 6. Intensity contours of 770 MeV protons at a heliocentric distance of 10 a.u. obtained with a wavy neutral sheet (thick line). In both plots, intensity is high at the poles. Minimum (Min) occurs at low latitudes for $qA > 0$ (left panel) and at higher latitudes for $qA < 0$ (right panel).

Dependence of flux on magnetic latitude. During a solar rotation, the earth or a spacecraft gets closer to and farther from the warped neutral sheet. In other words, different heliomagnetic latitudes are scanned, where heliomagnetic latitiude refers to angular distance from the sheet. Since particle densities seem best organized by the distance from the sheet, it is tempting to predict cosmic ray flux from this single parameter. This approach is overly simplified that may lead to incorrect conclusions. 3-dimensional calculations /14/ show that cosmic ray intensity cannot be described by a single obvious parameter such as heliomagnetic latitude. For instance, 3-D models correctly reproduce the decrease of near earth cosmic ray intensity with increasing distance from the sheet, for both signs of qA (extrapolation from 2-D results would suggest the opposite for $qA > 0$). At the same time, Newkirk and Fisk /34/ pointed out a flaw in the predictions for the magnitude of the effect: model calculations gave a much larger effect for $qA < 0$, while experimental results did not indicate any significant polarity dependence.

Again, similarly as in the case of radial gradients, the discrepancy is reduced but not eliminated when model calculations use more realistic parameters /18/.

Newkirk and Fisk /34/ suggested a possible explanation without including drift. Instead, they ascribe the effect to the increase of solar wind speed away from the sheet. According to this, what we see on earth is an excess modulation due to the faster wind. It should be borne in mind that the effect of faster wind is not quite obvious: faster convection and the accompanying straightening of field lines will act in opposite directions, and the net effect could as well be less modulation. Quantitative studies require 3-D calculations.

NEW MODEL-CALCULATIONS

Recently, we (Kóta and Jokipii /35/) have developed a new fully 3-dimensional, time-dependent scheme for solving the transport-equation (1) including drift, diffusion, convection, and adiabatic cooling or acceleration. Compared to our earlier model /14/, the present one is more advanced in two ways:

i) The present scheme is equally capable to handle adiabatic cooling or acceleration. Thus the effects of a termination shock can be studied. Also acceleration inside the heliosphere (e.g. at corotating interaction regions) is automatically taken care of.

ii) The solar wind speed need not be uniform, it is allowed to increase away from the neutral sheet according to /34/. This variation of wind speed, combined with a wavy neutral sheet, will ultimately lead to dynamical effects in the expansion of the solar plasma. The radial evolution of the solar wind is calculated in a simple dynamical model: shocks, corotating interaction regions, merging interaction regions rise in a natural way. The model is steady-state in the corotating frame, the magnetic field is taken as completely frozen-in. The polar field is modified according to /9/.

Details of the model and numerical results will be given elsewhere /35/. Here I report only on a couple of interesting features of the results.

Figure 7 illustrates the typical radial dependence of the azimuthally averaged fluxes predicted for $qA < 0$, with flat and wavy neutral sheet, respectively. The only difference between Figures 7a and 7b is the tilt angle of the sheet. As seen in Figure 7a, ray intensities are predicted to increase toward the equator, corresponding to a negative latitudinal gradient for a flat sheet. On the other hand, Figure 7b demonstrates that near equatorial particle fluxes sharply decrease, and latitudinal gradient virtually disappears if the sheet is wavy. Our specific example shown in Figure 7 refers to 250 MeV protons, 30° tilt angle, and a calculation without including the termination shock. The kind of behaviour outlined above is, however, a very typical feature of the $qA < 0$ solutions that remains valid for a broad range of parameters.

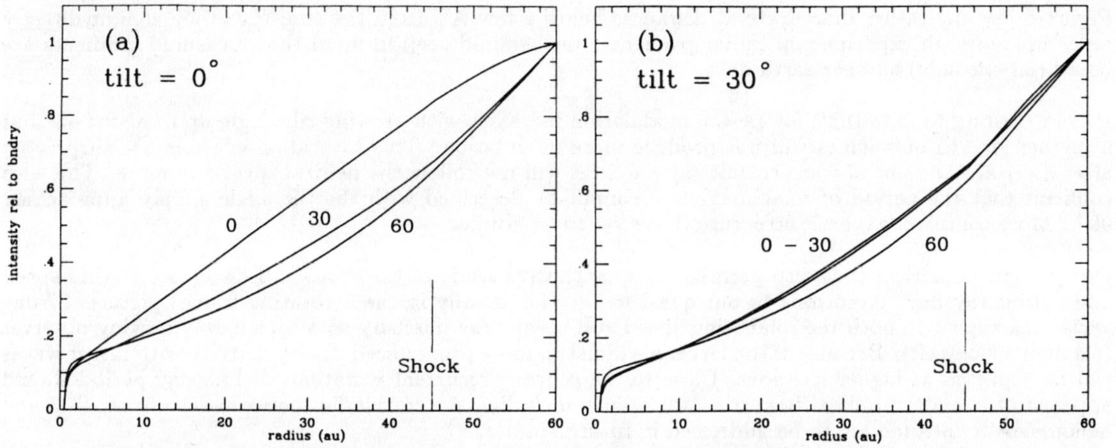

Fig. 7. Radial dependence of the azimuthally averaged fluxes of 250 MeV protons calculated for $A < 0$, with a flat sheet (a) and with a tilt angle of 30° (b). Numbers indicate latitudes of 0°, 30°, and 60°. Note the drop of equatorial flux and the disappearance of the negative latitudinal gradient on panel (b).

When comparing these theoretical results with experimental data, we find a remarkable agreement with the observations of the Voyager 1 and 2 spacecraft from 1984 to 1989 /36/. During this period Voyager 2 was in the ecliptic plane while Voyager 1 was about 30° above the ecliptic. Around solar minimum, when the neutral sheet was relatively flat, the counting rate of Voyager 2 increased to well above the Voyager 1 level, indicating a significant negative latitudinal gradient. As solar activity and the tilt angle of the sheet started to increase, cosmic ray intensity at Voyager 2 dropped, and the negative latitudinal gradient first diminished then entirely vanished.

A similar effect was even more evident in the anomalous cosmic ray measurements of the two Voyager spacecraft. Cummings, Stone, and Webber /37/ have found a direct correlation between the latitudinal gradient of anomalous cosmic rays and the tilt angle of the neutral sheet.

To illustrate the typical radial dependences of predicted cosmic ray fluxes, Figure 8 shows the azimuthally averaged intensities obtained for 250 MeV protons in a calculation including the termination shock. Again, for a broad range of parameters, solutions share the same characteristic features, at least inside the termination shock. Effects of the termination shock will be discussed in the next paragraph.

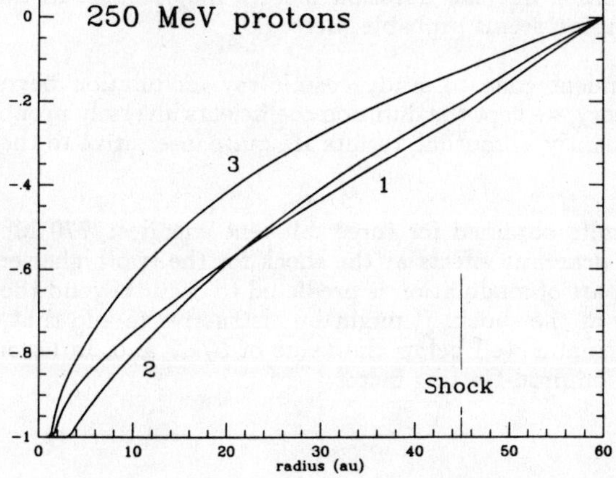

Fig. 8. Radial dependence of the azimuthally averaged fluxes of 250 MeV protons at 10° latitude, calculated for: $qA > 0$ with tilt angle 45° (1), and for $qA < 0$ with tilt angle 45° (2) and flat sheet (3).

In Figure 8 curves referring to a low latitude of 10° are presented for the two polarity cycles, with flat and wavy neutral sheet, respectively. For $qA > 0$, only the curve obtained for wavy sheet (45° tilt angle) is shown. In this cycle, near equatorial cosmic ray flux is insensitive to the neutral sheet, and flat sheet would yield slightly different result. For $qA < 0$, as expected, calculations give lower flux and higher radial gradient as the tilt angle is increased. Predicted radial gradients are still somewhat larger for the case of $qA < 0$, the difference, however, is not very significant.

In our calculations we did not attempt to achieve a best fit with observations but simply used reasonable parameters. It should also be mentioned that the primary purpose of our work was the study of the outer heliosphere, the spatial resolution and accuracy at small radii had to be sacrificed. The results shown in Figure 8 become rather inaccurate at distances below a few AU from the sun, thus they cannot directly be compared with experimental radial gradients. One should keep in mind that measured gradients are based (on one side) on near earth data.

It is interesting to note that, for $qA < 0$, modulation increases with growing tilt angle up to about 45° but a further growth of wavines will not produce more modulation. This saturation effect is not surprising: after a certain amount of warp cosmic ray particles will not follow the neutral sheet any more. This also confirms that the period of solar maximum cannot be described with the tilt angle simply approaching 90°. More complex magnetic structures have yet to be studied.

Our present numerical code also permits the quantitative study of the relation between solar wind speed and cosmic ray flux. According to our quasi-steady (i.e. steady in the corotating frame) picture, 27-day variations appear in both the solar wind speed and cosmic ray intensity seen by a slowly moving observer (earth or spacesraft). Because of the larger asymmetry, more pronounced 27-day activity with larger waves can be expected at higher latitudes. Close to the ecliptic, recurrent variations of 13.5 day period should appear in a greater number. Recurrent variations, including the correlation between cosmic ray flux and heliomagnetic latitude are to be addressed in future studies.

MODULATION BEYOND THE TERMINATION SHOCK

Webber and Lockwood /27/ have observed that radial cosmic ray gradients remain small and relatively constant during a solar cycle, thus a sustantial part of modulation should occur at very large distances from the sun, presumably near or beyond the solar wind termination shock. A similar argument was put forward by Lockwood and Quenby /38/.

Another indication for a possible modulation beyond the termination shock comes from the theory of anomalous cosmic rays. A substantial scattering is required beyond the shock to keep particles in the heliosphere long enough to get accelerated. This scattering should affect galactic cosmic rays, too. There is, however, a significant difference to keep in mind. Most of galactic particles in the inner heliosphere have already undergone a considerable deceleration, these particles were more energetic when they first entered the heliosphere. Anomalous cosmic rays, on the other hand, originate from low energy particles that become accelerated at the termination shock. Of course, anomalous cosmic rays are equally subject to cooling in the expanding solar wind, this, however, may be less effective because of the low relative charge of anomalous ions.

As already mentioned earlier in this paper, drifts and adiabatic cooling should be small in the subsonic, near-incpomressible wind. The wind speed must drop by a factor of about 4 with respect to its supersonic value. Diffusion is expected to decrease by the same factor, if one assumes that diffusion scales inversely with the magnetic field strength. The important factor will be perpendicular diffusion, since large scale fields must be almost completely azimuthal thus cosmic rays should diffuse accross the field to reach the termination shock. One should also bear in mind that the diffusion picture may change in the more turbulent subsonic regime, a more isotropic diffusion seems probable there.

We have applied our 3-dimensional time-dependent code to study cosmic ray modulation beyond the termination shock. In the absence of a better theory, we kept the diffusion coefficients inversely proportional to the magnetic field. Since the field is predominantly azimuthal, results are quite insensitive to the choice of parallel diffusion.

Figure 9 illustrates some typical numerical results obtained for three different energies: 770 MeV, 250 MeV, and 75 MeV. We find noticeable but not dramatic effects at the shock for the two higher energies. At lower energies, however, a relatively greater part of modulation is predicted to occur beyond the shock, and radial gradient is predicted to turn sharply at the shock. It might be instructive to add that, at low energies, the radial gradient inside the shock remains well below the value of V/κ_{rr} and, in accord with this, the spectrum hardens, yielding a smaller Compton-Getting factor.

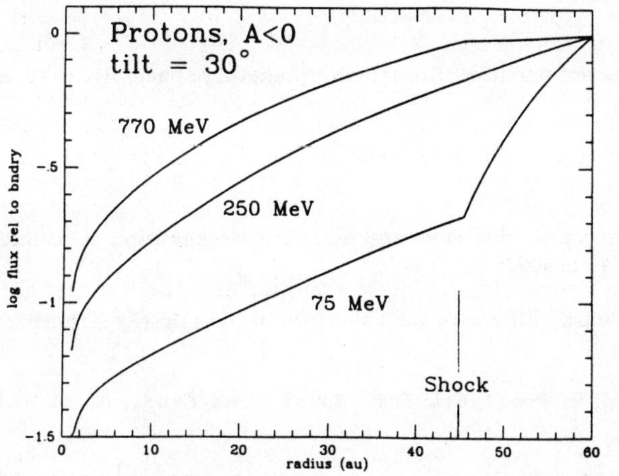

Fig. 9. Radial dependence of cosmic ray fluxes at 10° latitude calculated for energies 75 MeV, 250 MeV and 770 MeV. The shock is placed at 45 a.u. from the sun.

This kind of behaviour can be observed even in 1-dimensional solutions. It is interesting to mention that force-field solution does predict a diminsihing Compton-Getting factor at low energies, but does not correctly reproduce the decrease of the radial gradient. This effect seems to emerge at low energies where, due to the decrease of diffusion, $V.r/\kappa_{rr} > 1$. The force-field approximation breaks down for this case.

The most plausible interpretation of the results is that most of these low energy cosmic rays are particles that have spent long times in the heliosphere, and were cooled down at small heliocentric distances. These decelerated cosmic rays can then fill the heliosphere fairly uniformly.

In summary, our numerical results indicate that, at low energies, a substantial part of modulation may occur beyond the termination shock, and radial gradients, instead of disappearing, should increase beyond the shock. At higher energies, the effect of shock is noticeable but not dramatic.

SUMMARY

Though there is still some controversy on the role of drifts in the heliospheric transport of cosmic rays, experimental results are strongly in favour of the drift concept. The strongest evidence supporting drift is the observed reversal of the latitudinal gradient, which is a natural result of drift and is very hard to explain otherwise.

At the same time, there are also several problems that need attention. The magnitude of drift effects is uncertain, it remains to be seen if the transport-equation (1) should contain the full drift-term, or drift happens to be reduced as suggested by 2-D models /15/. More detailed, and methodical 3-D works should address this question. More important, the outstanding discrepancies, like the conflict of radial gradients, have to be critically examined. First, it should be clarified if the conflict does indeed exist or can be resolved in a *natural way* within the frame of the present model.

From experimental side, the Ulysses mission, together with more out of ecliptic data from the outer heliosphere should expand our horizon. From the side of theory, a better understanding of diffusion would be desirable. As we saw, the relative importance of drift crucially depends on cross-field diffusion. It is unfortunate that the perpendicular diffusion coefficient is still a poorly known, phenomenological parameter, leaving too much freedom for model calculations.

Preliminary results indicate that, at low energies, a considerable part of modulation may occur beyond the solar wind termination shock. Instead of reaching a plateau, counting rates are predicted to increase at a higher rate after crossing the shock. Smaller and less dramatic 'residual modulation' is expected at higher energies (above ~100 MeV/nucl.).

Cosmic ray transport during the period of solar maximum (i.e. polarity reversal) represents another intriguing and theoretically poorly understood subject. The heliospheric magnetic field must have a rather complex configuration around the time of reversal: the large-scale field will not be steady, and must also have a latidudinal component traveling outward with the solar wind. This transition period has not yet been addressed by models including drift.

Acknowledgement

Most of the results presented here represent a joint work with Professor J.R. Jokipii at the University of Arizona. The research was supported by the National Science Foundation under Grant ATM-8618260.

REFERENCES

1. E.N. Parker, The passage of energetic charged particles through interplanetary space, *Planet. Space Sci.* **13**, 9-49 (1965)

2. L.J. Gleeson and W.I. Axford, Cosmic rays in the interplanetary medium, *Astrophys. J.* **149**, L115-118 (1967)

3. J.R. Jokipii and E.N. Parker, On the convection, diffusion, and adiabatic deceleration of cosmic rays in the solar wind, *Astrophys. J.* **160**, 735-744 (1969)

4. J.R. Jokipii, E.H. Levy, and W.B. Hubbard, Effects of particle drift on cosmic ray Transport. I., *Astrophys. J.* **213**, 861-869 (1977)

5. R.B. McKibben, Cosmic ray modulation, in *Proc. 6th Int. Solar Wind Conf.*, ed. V.J. Pizzo, T.E. Holzer, and D. Sime, **II.**, 615-633 (1988)

6. J.R. Jokipii, The Physics of Cosmic Ray Modulation, *Adv. Space Res.* **9**, #12, 105-119 (1989)

7. J.R. Jokipii, The anomalous component of cosmic rays, this issue

8. J.R. Jokipii, G. Morfill, and J. Kóta, Jump conditions for cosmic rays at fluid discontinuities, *Astrophys. J.* **345**, L67-69 (1989)

9. J.R. Jokipii and J. Kóta, The polar heliospheric magnetic field, *Geophys. Res. Lett.* **16**, 1-4 (1989)

10. J.R. Jokipii and D.A. Kopriva, Effects of particle drift on the transport of cosmic rays. III. Numerical model of galactic cosmic ray modulation, *Astrophys. J.* **234**, 384-392 (1979)

11. L.R. Barnden and M. Bercovitch, Field gradient and curvature drifts and cosmic ray transport into the solar system, *Proc. 14th Int. Cosmic Ray Conf.*, Munich 1975, **3**, 875-880 (1975)

12. J. Kóta, Drift - the essential process in losing energy, *Proc. 16th Int. Cosmic Ray Conf.*, Kyoto 1979, **3**, 13-18 (1979)

13. J.R. Jokipii and B.T. Thomas, Effects of drift on the transport of cosmic rays. IV. Modulation by a wavy interplanetary current sheet, *Astrophys. J.* **243**, 1115-1122 (1981)

14. J. Kóta and J.R. Jokipii, The effect of drift on the transport of cosmic rays. VI. A three-dimensional model including diffusion, *Astrophys. J.* **265**, 573-581 (1983)

15. M.S. Potgieter and H. Moraal, A drift model for the modulation of galactic cosmic rays, *Astrophys. J.* **294**, 425-440 (1985)

16. E.H. Levy, Theory of solar magnetic cycle wave in the diurnal variation of energetic cosmic rays: physical bases of the anisotropy, *J. Geophys. Res.* **81**, 2082-2088 (1976)

17. A. Kadokura and A. Nishida, Numerical modeling of the 22 year variation of the cosmic ray intensity and anisotropy, *J. Geophys. Res.* **91**, 1-11 (1986)

18. J. Kóta and J.R. Jokipii, Effects of a wavy neutral sheet on cosmic ray anisotropies, *Proc. 19th Int. Cosmic Ray Conf.*, La Jolla 1985, **4**, 453-456 (1985)

19. G. Erdős and J. Kóta, Sector-structure of the interplanetary magnetic field and anisotropy of 50-1000 GV cosmic rays, *Astrophys. Space Sci.* **67**, 45-59 (1980)

20. H.S. Ahluwalia and J.F. Riker, The diurnal anisotropy of cosmic rays and the heliospheric transport parameters, *Planet. Space Sci.* **35**, 45-50 (1987)

21. M.A. Shea and D.F. Smart, Preliminary search for cosmic radiation and solar-terrestrial parameters correlated with the reversal of the solar magnetic field, *Adv. Space Phys.* **1**, #3, 147-150 (1981)

22. E.J Smith and B.T. Thomas, Latitudinal extent of the heliospheric current sheet and modulation of galactic cosmic rays, *J. Geophys. Res.* **91**, 2933- 2942 (1986)

23. M. Garcia-Munoz, P. Meyer, K.R. Pyle, and J.A. Simpson, The dependence of solar modulation on the sign of the cosmic ray particle charge, *J. Geophys. Res.* **91**, 2858-2866 (1986)

24. J.R. Jokipii, R.A. Mewaldt, and H. Moraal, in preparation (1990)

25. C.R. Lopate, R.B. McKibben, K.R. Pyle, and J.A. Simpson, Radial gradients of galactic cosmic rays to ~40 AU during declining solar activity, *Proc. 20th Int. Cosmic Ray Conf.*, Moscow 1987, **3**, 409-412 (1987)

26. F.B McDonald, T.T. von Rosenvinge, N. Lal, J.H. Trainor, and P. Schuster, The recovery phase of galactic cosmic ray modulation in the outer heliosphere, *Geophys. Res. Lett.* **13**, 785-788 (1986)

27. W.R. Webber and J.A. Lockwood, Interplanetary Radial cosmic ray gradients and their implication for a possible large modulation effect at the heliospheric boundary, *Astrophys. J.* **317** , 534-542 (1987)

28. J. Kóta, How does the global structure of the interplanetary magnetic field affect cosmic ray modulation, *Adv. Space Res.* **1**, #3, 135-138 (1981)

29. J.W. Bieber and M.A. Pomerantz, Cosmic ray North-South anisotropy in 1961-1983, *Proc. 19th Int. Cosmic Ray Conf.*, La Jolla 1985, **4**, 433-436 (1985)

30. A.C. Cummings, E.C. Stone, and W.R. Webber, Latitudinal and radial gradients of anomalous and galactic cosmic rays in the outer heliosphere, *Geophys. Res. Lett.* **14**, 174-177 (1987)

31. E.C. Stone, Cosmic ray studies out of ecliptic, *Proc. 20th Int. Cosmic Ray Conf.*, Moscow 1987, **7**, 105-114 (1987)

32. R.B. Decker, S.M. Krimigis, and D. Venkatesan, Latitudinal gradient of energetic particles in the outer heliosphere during 1985-86, *J. Geophys. Res.* **92**, 3375-3379 (1987)

33. J.W. Bieber, P. Evenson, and W.F. Mattheaus, Magnetic helicity of the IMF and the solar modulation of cosmic rays, *Geophys. Res. Lett.*, **14**, 864-867 (1987)

34. G. Newkirk and L.A. Fisk, Variation of cosmic rays and solar wind properties with respect to the heliospheric current sheet. 1. Five GeV protons and solar wind speed, *J. Geophys. Res.* **90**, 3391-3414 (1985)

35. J. Kóta and J.R. Jokipii, in preparation (1990)

36. R.B. Decker, S.M. Krimigis, and D. Venkatesan, Onset of cosmic ray modulation observed at Voyagers 1 and 2 during the early phase of solar cycle 22, *Proc. 21st Int. Cosmic Ray Conf.*, Adelaide 1990, **6**, 152-155 (1990)

37. A.C. Cummings, E.C. Stone, and W.R. Webber, Time variation of radial and latitudinal gradients of anomalous cosmic ray oxigen in the outer heliosphere, *Proc. 21st Int. Cosmic Ray Conf.*, Adelaide 1990, **6**, 190-193 (1990)

38. J.A. Lockwood and J.J. Quenby, The Pioneer-10 gradient measurements - the key to the modulation problem, *Proc. 20th Int. Cosmic Ray Conf.*, Moscow 1987, **3**, 406-408 (1987)

NEUTRON MONITOR INVESTIGATIONS RELATING MODULATED COSMIC RAY SPECTRA WITH HELIOSPHERIC MAGNETIC FIELD POLARITY REVERSALS

B. Popielawska* and J.A. Simpson**

*Space Research Center, Polish Academy of Sciences, ul. Ordana 21, 01–237 Warsaw, Poland
**The Department of Physics and Enrico Fermi Institute, The University of Chicago, Chicago, IL 60637, U.S.A.

ABSTRACT

The neutron monitors at Climax, Colorado (geomagnetic cutoff $\mathfrak{R}_c \sim 3$ GV) and Huancayo, Peru ($\mathfrak{R}_c \sim 13$ GV) are an important investigative tool since their more than 36 year continuous record now makes it possible to distinguish the long term, large scale dependence of solar modulation on the 11 year solar activity cycle from the dependence on the reversals of the solar polar and interplanetary magnetic field polarity. We analyse here the rigidity dependence of solar modulation during the increasing solar activity. After corrections for the progressive drifts of the geomagnetic cutoffs due to secular variation of the Earth's field, we find that - if we represent the magnetic rigidity dependence by $\mathfrak{R}^{-\alpha}$ - $\alpha \simeq 0.8$ for periods when the solar north pole magnetic field is positive. For periods when the solar north polar field is negative $\alpha \simeq 1.1$. We discuss some of the implications of these results for models of cosmic ray propagation in the heliosphere.

INTRODUCTION

Investigations of the long term, large scale changes in the modulated spectrum of galactic cosmic radiation and of their origin in interplanetary dynamical phenomena have made important contributions to the understanding of the heliosphere, whose existence was proven more than 33 years ago. With cosmic ray nuclei as test particles propagating through the heliosphere, it was shown - especially for spherically symmetric models of the heliosphere - that diffusion, convection and adiabatic deceleration could account for the principal large scale changes in modulated spectra over successive 11 year solar activity cycles. However, the experimental proof from spacecraft measurements (Smith and Thomas /1/) - that the heliospheric magnetic field is divided by a neutral current sheet with the magnetic field polarity of each "hemisphere" corresponding to the prevailing solar polar magnetic field polarity (which changes sign every ~11 years) - raised the question of how charged particles propagate through a heliosphere over this 22 year magnetic cycle.

Jokipii, Levy and Hubbard /2/ and Kota and Jokipii /3/ addressed this question by showing that curvature and gradient drifts in the heliospheric magnetic fields may make important and observable changes in the modulated spectra over the 22 year solar polar magnetic cycle. Later models included the waviness of the current sheet due to its overall inclination with respect to the solar rotation axis (e.g., Potgieter and Moraal /4/; Burger and Potgieter /5/) and possible contributions from the helicity of the interplanetary magnetic fields (e. g., Bieber et al. /6/). Experimental investigations over the last decade support these heliospheric models based on a wavy neutral current sheet and the 22 year solar magnetic cycle. However, all of the investigations do not agree on the importance of drifts (or the helicity of the magnetic field) relative to the demonstrated importance of diffusion, convection and adiabatic deceleration in solar modulation. The present report is another attempt to investigate this question.

Neutron monitor which record the changes in modulated spectra of relativistic nuclei in the cosmic rays are an important investigative tool, since their continuous record covers, as in the study reported here, more than 36 years or more than 1.5 solar magnetic cycles. Thus, it is possible to distinguish between the dependence of solar modulation on the 11 year activity cycle and the dependence on reversals of the solar polar magnetic field polarity.

Cooper and Simpson /7/ examined, with neutron monitor measurements over two solar activity cycles (No. 19 and 20), the origin of the apparent difference in the observed magnetic rigidity dependence of the modulated spectra between two successive cycles noting that the explanation was either due to heliospheric propagation differences between two solar activity cycles (e.g., Jokipii /8/), or due to the systematic drift of the geomagnetic cutoff for equatorial observations (Huancayo) which had been pointed out by Shea /9/ as a consequence of the secular variation in the Earth's magnetic field. Cooper and Simpson showed that this drift accounted for most, but not all, of the differences between two successive modulation cycles (see their Figures 1 and 2). They noted: "We have not yet determined whether the remaining difference between the two solar cycles can be accounted for by residual, systematic errors over the 26 year period ..." In the present paper we find, on the one hand, that residual systematic errors in cutoff or pressure correction drifts cannot account for the differences in the magnetic rigidity spectra of modulated cosmic radiation and, on the other hand, that the differences are correlated with the four polarity reversals of the solar magnetic fields which have occurred over the 36 year period of the measurements.

The analysis we report here was carried out in 1987 and later extended for an additional year. A preliminary report was presented in 1989 (Simpson /10/). Recently, Reinecke, Potgieter and Van Staden /11/ have reported a similar result based on neutron monitor measurements for the period 1976-87.

ANALYSIS OF THE NEUTRON MONITOR MEASUREMENTS

The Climax neutron monitor (at ~3400 m altitude) with a geomagnetic cutoff of $\Re_c \cong$ 3 GV and the Huancayo neutron monitor (also at ~3400 m altitude) with a geomagnetic cutoff $\Re_c \cong$ 13 GV provide the two sets of data over 36 years, as shown in Fig. 1 after correcting for drifts in both barometric pressure recorders and the geomagnetic cutoffs. (From Huancayo pressure corrected data we removed the positive trend in counting rate due to the decreasing geomagnetic cutoff with time at a rate of 3.5±0.2% per 30 years. This is consistent with the 2.4% per 20 years used by Cooper and Simpson /7/.)

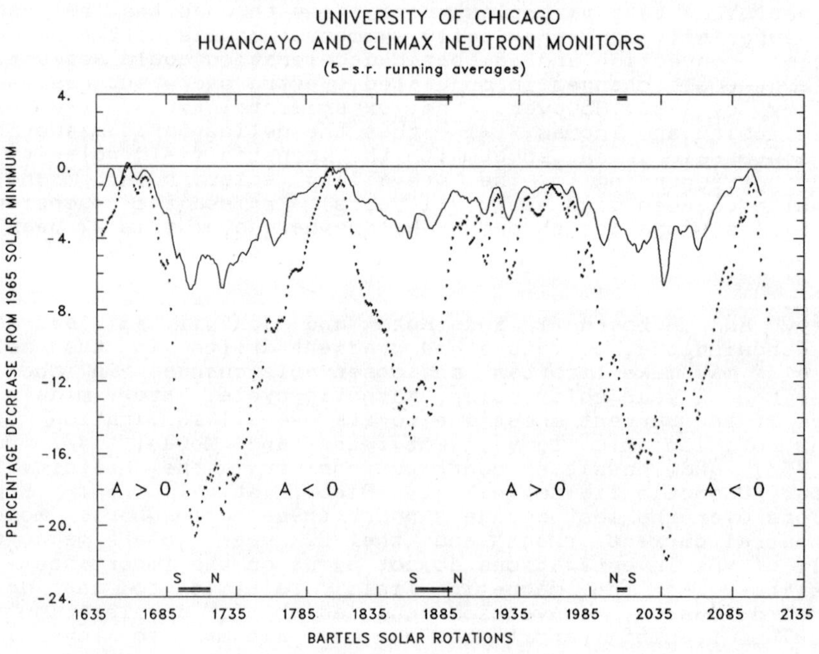

Fig. 1. Neutron monitor intensity measurements (5-solar rotation running averages) shown as percent changes from the normalized level of 100% at the 1965 level of solar minimum modulation. The dashed line is from Climax, and the solid line is from Huancayo (Jan. 1953 to Nov. 1987). Timing for solar polar magnetic field reversals is from /13/.

The data are shown as percent decreases of integral intensity above 3 GV and 13 GV, respectively, below the 1965 solar minimum selected as the 100% intensity level. The correlation of long term intensity changes at Climax and Huancayo are shown in Fig. 2 based on five solar rotation running averages in order to eliminate the effect on this correlation of short term changes in intensity, such as Forbush decreases.

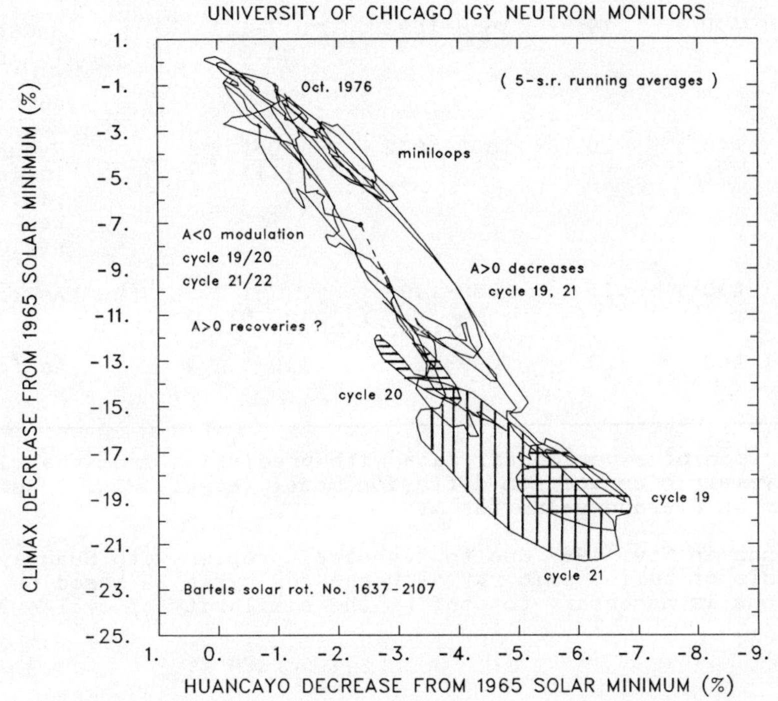

Fig. 2. Correlation plot of Climax and Huancayo neutron monitors from data derived from Fig. 1. Shaded areas are the hysteresis loops at the times near solar activity maxima.

We recognize in Fig. 2 the significant magnetic rigidity dependent hysteresis effect which occurs near the maximum of solar activity in successive solar activity cycles, as first reported by Simpson /12/. We further note that these loops essentially close at high levels of solar modulation. Limited space allowed for this short contributed paper does not give us a possibility to discuss here these transient effects. Further, we decided to restrict ourselves to the period of increasing solar modulation after the onset of new solar cycle. A complete report on results of analysis of the full 36 years period as shown in Figures 1 and 2 will be published elsewhere.

The times of solar polar north (N) and south (S) magnetic polarity reversals, as shown in Fig. 1 and Table 1, are from Webb, Davis and McIntosh /13/. The assigned character of the gradient drift for positively charged particles is inward from high latitude heliospheric regions when the north polar magnetic field is "away" from Sun (in another notation - "+" or A>0 in current drift models). With the north polar field reversed in polarity ("-" or A<0) positively charged particles drift inward along the wavy current sheet in the equatorial region of the heliosphere.

According to the above selection criteria we have extracted from the dataset a subset consisting of data starting at the cosmic ray maximum through the period of increasing modulation after the onset of new solar activity cycle up to either the solar polar magnetic field reversal (cycle 19) or the beginning of cosmic ray long-term recovery (cycle 20 and 21) whatever is earlier. This subset has been further divided into two groups of data with either A>0 or A<0 polarity of the solar polar field. The corresponding correlation plot is shown in Fig.3.

From Fig. 3 we conclude that for data from intervals separated by 1 or 3 solar cycles the slopes of correlation curves are significantly different,

TABLE 1 Cosmic ray nucleon component spectral changes at 1 A.U. during increasing solar activity over 4 solar cycles.

Date of Minimum Modulation	Polarity of Solar Polar Field	Solar Activity Cycle No.	Analysed Time Period B. s.rot.	Magnetic Rigidity Spectrum, $\mathfrak{R}^{-\alpha}$ α (e.g. /14/)[x]	C.R. Spectrum After Propagation to 1 A.U.
Aug.1954	+ (A>0)	19	1658–1692	~0.7–1.1 (~0.8)	Reduced modulation of the low rigidity particles
May 1965	– (A<0)	20	1803–1852	1.0–1.2 (1.1)	Reduced modulation of high rigidities with respect to low rigidities
Oct.1976	+ (A>0)	21	1958–1996	~0.7–1.1 (~0.8)	As for cycle 19
Feb.1987	– (A<0)	22	2099–2107	1.0–1.1 ?	As for cycle 20 ?

[x]– From comparison of experimental data with prediction from the classical spherically symmetric convection-diffusion model (e. g., /14/). Number in parenthesis is an average value for α.

?– Data set ends in Nov. 1987 due to technical problem with Huancayo station. Further analysis of full cosmic ray decrease for cycle 22 based on another pair of stations is necessary to confirm the similarity of cycles 20 and 22.

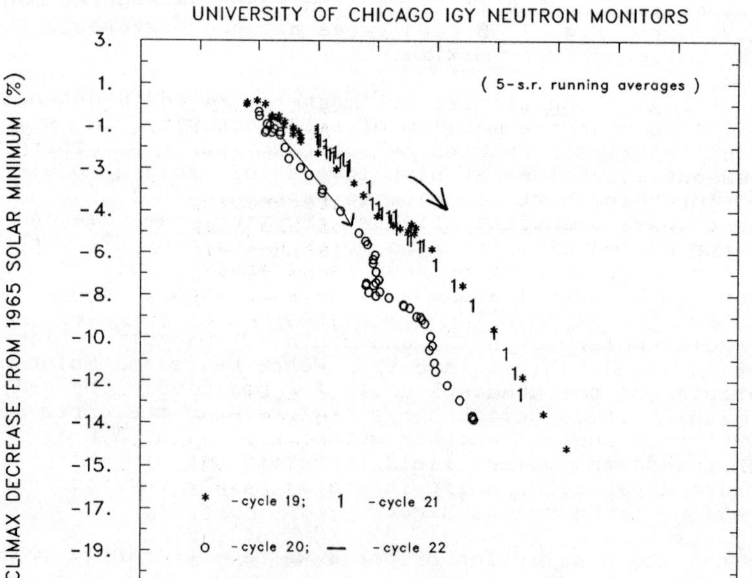

Fig. 3. The same as Fig.2, but only data listed in Table 1 are plotted to demonstrate the 22-year variation in rigidity dependence of cosmic ray modulation during the ascending phase of solar activity cycle. Cycle 19 and 21 cosmic ray decrease is during A>0 phase of solar magnetic cycle while cycle 20 and 22 decrease is during A<0 phase.

while for data separated by 2 solar cycles they are similar. Such a coincidence hardly could be the result of systematic, accumulative errors which we might have made in either the corrections for the barometric effect or the corrections for the continual decrease in the equatorial geomagnetic cutoff for the Huancayo neutron monitor. An immediate explanation of the observed effect is that the changes in magnetic rigidity spectra correlate with the solar polar magnetic field polarity, i. e., they are dependent on the sign of the polar fields, A>0 or A<0.

With only two integral cutoffs for our data sets (i.e., Climax and Huancayo) we are not justified in assuming a magnetic rigidity dependence for the modulated spectra more complex that $\exp(-\eta(t)*\Re^{-\alpha})$ (e. g., Mathews et al. /14/). Accordingly, for the case of A<0 a value of $\alpha \cong 1.1$ was found from comparison with predicted rigidity dependence. However, for A>0 there are two alternative approaches for determining α in view of the "curved" character of the correlation plot, namely:

1. by averaging the modulation over the entire range corresponding to 15% decrease for Climax, or alternatively;
2. by dividing the correlation curve into two linear intervals (i.e., for Climax 0-6% and 6-15%)

we find that the observed magnetic rigidity dependence can be fitted in the case 2 by $\Re^{-0.7}$ for modulation levels restricted to the range 0 to 6% at Climax up to $\Re^{-1.1}$ for higher modulation levels and by $\Re^{-0.8}$ in the case 1.

INTERPRETATION AND DISCUSSION

The two most important corrections which must be applied to the data from continually calibrated neutron intensity monitors for very long term investigations are:

1. barometric pressure indicator drifts which arise from aging components of aneroid or digital barometers, and;
2. systematic drift of the geomagnetic cutoff due to the secular variation of the geomagnetic field.

Although a best effort has been made over the more than 36 years of measurements, there always remain some uncertainties and potentially accumulative errors in making these corrections to the data. At present we place a tentative value of ±0.5 percent error on the data sets over the whole period. These results by themselves indicate that the correlation of cosmic ray magnetic rigidity spectral changes with solar polar magnetic field reversals reported here is not due to accumulative or repetitive errors in the corrections applied to the data.

In this paper we emphasized the independent check for the above conclusions, wherein the two data sets (for A>0 and A>0) would not be separable over 3 1/2 solar activity cycles if significant accumulative errors were present.

An equally difficult problem for interpreting long term, large scale modulation phenomena is the separation of the dependence on interplanetary dynamical effects of the *eleven* year solar activity cycle origin (diffusion, convection and adiabatic acceleration) from dependence on modulation factor such as drifts determined mainly by the magnetic polarity and neutral current sheet with its tilt angle all changing with the *twenty two* year solar magnetic cycle.

With a data set covering only 1 1/2 solar activity cycle in hand, Simpson and Wang /15/ showed that their investigation was consistent with the modulated spectrum of cosmic rays remaining essentially constant from the 1954 minimum through a complete solar cycle with $\Re^{-1.0\pm0.2}$. With the data set extended to more than two activity cycles (1953-1978) Cooper and Simpson /7/ were able to measure the effect of the geomagnetic cutoff drift and to obtain a residual difference in succesive solar cycles of modulation, but they could not distinguish between a residual error problem and a possible dependence of the modulation spectrum on interplanetary magnetic field reversals.

In this preliminary report we have shown that the magnetic rigidity spectra of the relativistic galactic cosmic ray nuclei during the increasing solar activity are dependent on the interplanetary magnetic field polarity. We conclude that cosmic ray propagation through the heliosphere when the north solar polar magnetic field is positive (A>0) leads to the reduced modulation of the low rigidity particles relative to high rigidity particles (a steeper spectrum) and *vice versa* for a negative north solar polar field (A<0). Independent reports demonstrating that there is an excellent correlation between the tilt angle of the neutral current sheet and the observed cosmic ray intensity , especially at low latitudes near and in the period of solar minimum (e.g., Webber and Lockwood /16/; Lopate and Simpson /17/) adds credibility to models that require curvature and gradient drift for the analysis of solar modulation. We point out, however, that our work does not exclude models based on magnetic helicity (e.g., Bieber et al. /6/). The analyses reported by Reinecke, Potgieter and Van Staden /11/ and Moraal et al. /18/ are consistent with the results reported here.

The origin of the difference in spectral shape observed during A<0 and A>0 periods is still an open question. On one hand Reinecke and Potgieter /19/ show that the modulated spectra depend on IMF polarity when wavy neutral sheet model is used even if the same rigidity dependence for the "classical" diffusion coefficient is assumed for both polarities. But they do not discuss the physical process responsible for that. On the other hand, Lopate and Simpson (private communication, 1990) point out that to account for their analysis of data near the solar minimum of 1987 there may be a significant difference in the magnitude of adiabatic deceleration at low magnetic rigidities between A<0 and A>0 periods (for latter periods it is smaller). Also, a rapid inward drift of high rigidity particles along the current sheet may explain the reduced modulation of these particles during A<0 periods. Both explanations are probably relevant to Reinecke and Potgieter results from numerical modelling.

One final remark is necessary: We included the measurements at the Huancayo, Peru station to the time of November 1987 which was the last visit by our technical staff. Since that time the terrorist group, Sendero Luminoso, has besieged the area making further calibration service checks impossible. Thus, we are withholding the release of Huancayo data beginning in 1988 pending further checks on the laboratory equipment at some future time when access is possible.

Acknowledgements. We are grateful to the many individuals who have contributed to the neutron monitor investigations, including: W. Dietrich for investigations on pressure drifts, S. Tejero for periodic maintenance of the mountain stations, C. Nickl and D Comcowich for operation of the stations in Huancayo and Climax, respectively. This research was supported in part by NSF Grant ATM 89-08245.

REFERENCES

1. E.J. Smith and B.J. Thomas, J.Geophys. Res. **91**, 2933 (1986)
2. J.R. Jokipii, E.H. Levy and W.B. Hubbard, Astrophys. J. **213**, 861 (1977)
3. J. Kota and J.R. Jokipii, Astrophys. J. **265**, 573 (1983)
4. M.S. Potgieter and H. Moraal, Astrophys. J. **294**, 425 (1985)
5. R.A, Burger and M.S. Potgieter, Astrophys. J. **339**, 501 (1989)
6. J.W. Bieber, P.A. Evenson and W.H. Matthaeus, Astrophys. J. **315**, 700 (1987)
7. J.F. Cooper and J.A. Simpson, Proc. 16th Intl Cosmic Ray Conf. (Kyoto, Japan), Vol.**12**, 176 (1979)
8. J. R. Jokipii, in: The Sun and the Heliosphere in Three Dimensions, ed. R.G. Marsden, D. Reidel Publ. Co., Dordrecht 1986, p. 375.
9. M.A. Shea, Proc. 12th Intl Cosmic Ray Conf. (Hobart, Australia), Vol. **3**, 859 (1971)
10. J.A. Simpson, Trans. Am. Geophys. U. **70**, 1256 (1989)
11. J.P.L. Reinecke, M.S. Potgieter and M.L. Van Staden, Proc. 21st Intl Cosmic Ray Conf. (Adelaide, Australia), Vol.**6**, 95 (1990)
12. J.A. Simpson, Proc. 8th Intl Cosmic Ray Conf. (Jaipur, India), Vol.**2**, 155 (1963)
13. D.F. Webb, J.M. Davis and P.S. McIntosh, Solar Phys. **92**, 109 (1984)
14. T. Mathews, P.H. Stoker and B.G. Wilson, Planet. Space Sci. **19**, 981, (1971)
15. J.A. Simpson and J.R. Wang, Astrophys. J. **161**, 265 (1970)
16. W.R. Webber and J.A. Lockwood, J. Geophys. Res. **93**, 8735 (1988)
17. C. Lopate and J.A. Simpson, Proc. 21st Intl Cosmic Ray Conf. (Adelaide, Australia), Vol.**6**, 9 (1990)
18. H. Moraal, M.S. Potgieter, P.H. Stoker and A.J. Van der Walt, J. Geophys. Res. **94**, 1459 (1989)
19. J.P.L. Reinecke and M.S. Potgieter, Proc. 21st Intl Cosmic Ray Conf. (Adelaide, Australia), Vol.**6**, 17 (1990).

THE PREDICTIONS OF A TIME-DEPENDENT DRIFT MODEL COMPARED WITH COSMIC-RAY INTENSITY OBSERVATIONS FROM 1976 to 1989

M.S. Potgieter and J.A. le Roux

Department of Physics, Potchefstroom University for C.H.E., Potchefstroom, South Africa

ABSTRACT

The recently developed time-dependent drift model with a simulated wavy neutral sheet in which changes in the waviness of the sheet are propagated radially outwards at the solar wind velocity is applied to the long-term modulation of cosmic rays. Actual neutral sheet tilt angles are incorporated in the model and the results are compared to the intensities ($E \geq 60$ MeV) observed on the IMP 8 spacecraft from 1976-1989.

INTRODUCTION

Correlation studies of calculated neutral sheet data /1/ with cosmic-ray intensities (positively charged particles) as observed by neutron monitors from 1976 - 1987 /2,3/ that there is a good correlation with a larger slope for regression lines when the northern-hemispheric interplanetary magnetic field (IMF) points towards the Sun ($A < 0$) compared to when $A > 0$. Quantatively this was in agreement with the prediction of steady drift drift models which employ a wavy neutral sheet (e.g. /4,5/) that positively charged particles drifting downwards from the polar regions to the equatorial plane of the heliosphere when $A > 0$, will be much less sensitive changes in the waviness of the neutral sheet compared to when these particles are drifting inward along the sheet from the heliospheric boundary when $A < 0$.

Recently, Webber, Potgieter and Burger /6/ went a step further with a more direct comparison of the long term variation of cosmic-ray intensities with tilt angle (α), as calculated with their newly developed steady state drift model, with the above-mentioned correlation of observed intensities with the real calculated tilt values at various energies form 1976 - 1987. In general the correlation between the model and the data when $A < 0$ (1980 - 1987) were found to be in support of a drift-dominated picture of the heliosphere at all energies considered. It must, however, be pointed out that some reduction in drift velocities, as much as a factor 2 at lower energies, were required for a continued good correlation at larger neutral sheet waviness. The total lack of sensitivity of the model intensities to changes in α when $A > 0$ were judged to mean that the decrease in observed intensities from 1976 - 1981 was not caused by the reaction of drift to changes in the waviness of the sheet, but by other means such as the propagation of systems of transient disturbances which were quite prominent during that period. It was concluded that an improvement can only be achieved by a time-dependent drift model which also includes the effect of propagating transients on the diffusion picture in the heliosphere.

Recently, it was also observed that comic-ray intensities are less responsive to changes in α after the solar minimum of 1987 /2/. This feature is beyond the scope of steady state drift models provided that all modulation parameters not associated with drift did not vary substantially over the modulation cycle. Although we do not exclude other possible explanations like the effect of a modulation barrier in the outer heliosphere /7,8/, or transient effects /9/, we have shown in a previous paper /10/ that a time-dependent drift model in which the changes in the waviness of the neutral sheet are propagated radially outwards at the solar wind speed can explain the observed asymmetry in the cosmic-ray intensity and neutral sheet tilt correlation before and after 1987. We consider the result of that paper as a major and important feature of time-dependent drift modulation not previously shown.

In this paper we shall attempt to establish to what extent a time-dependent drift model can simulate intensities compatible with what has been observed from 1976 to early 1989. This will hopefully shed light on the controversial issue of the relative magnitude and importance of gradient, curvature and neutral sheet drift in the solar modulation of galactic cosmic rays in the heliosphere. We will only consider cosmic-ray protons at one energy level (1 GeV), because a time-dependent approach to modulation requires a lot of computer power not available at this stage.

THE MODEL

Details of the model and the various parameters used for this study are given in paper /10/, except for the difference that we now use $K_{\perp} = 0.02K_{\parallel}$. Instead of using arbitrary values for the neutral sheet tilt angle, α, we now use actual values as calculated by Dr. T. Hoeksema. /1,11/. These tilt values are assumed to propagate from the Sun at the solar wind speed throughout the heliosphere, with an outer boundary at 60 AU, covering this distances in about 255 days. The temporal changes of the waviness of the neutral sheet are therefore properly and as realistically as possible incorporated into the time dependent code, making it possible to establish the general compatibility between the calculated long-term behaviour of cosmic-ray intensities with observations at Earth. For this purpose we use the IMP 8 counting rates as a function of tilt and time from 1976-1989.

RESULTS AND DISCUSSION

In Fig. 1 the intensity-tilt relationship of 1 GeV protons at Earth as generated by the time-dependent drift model with real tilt values is compared with the IMP 8 counting rates ($E \geq 60$ MeV) as a function of the real tilt angles at Earth. The period under consideration covers both polarities of the IMF; before 1980 the polarity epoch is indicated by $A > 0$ and after 1980 by $A < 0$. Both the calculated and observed intensities are normalized at the 1987 solar minimum. Comparison of these intensities shows that in general the model predicts the long-term behaviour of cosmic rays as a function of α rather well for both polarity epochs. Both the data and calculations show a moderate dependence on α when $A > 0$, but a strong α dependence when $A < 0$. The total modulation as a function of α from 1976 to 1980, when the polarity reversed with $\alpha = 72° \pm 3°$, and from 1980 to 1987, when $\alpha = 8° \pm 2°$, is also predicted well by the model. Even the relative short-term variations when $A < 0$ and $\alpha < 35° \pm 5°$ are reproduced well during the same period and can be interpreted that the neutral sheet, and therefore drift, play an important role when solar activity is low. On the other hand, note that with increasing solar activity, i.e., increasing α, the observed short-term modulation effects are not reproduced well by the model. This seems to happen over the total range in α when $A > 0$ (when the neutral sheet plays a less prominent role), indicating that other effects are also contributing significantly to modulation. This is true, especially for a period of about 2 years before and after the polarity reversal, when relative short-term variations in the intensity are evidently not caused by changes in α.

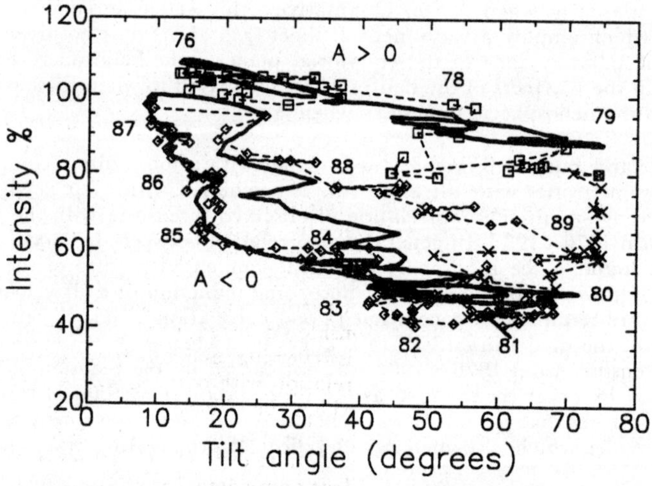

Fig. 1. Comparison of 1 GeV proton intensities (solid line) calculated with a time-dependent drift model using real neutral sheet tilt angles, α, and IMP 8 intensities ($E \geq 60$ MeV) as a function of α at Earth.

A very important and interesting feature in Fig. 1 is the split which occurs in both the observed and calculated intensities when the situation before and after solar minimum in 1987, with $A < 0$ and α a minimum, is compared. Our model with a time-dependent propagating wavy neutral sheet clearly produces the same kind of asymmetry in the intensity-tilt relationship present in the data before and after 1987. This gives credibility to our claim (see paper /10/) that changes in the wavy neutral sheet and the associated drift effects should be done with a time-dependent drift model. The reproduction of the split, together with the general good compatibility between the observed and calculated intensities, also give credibility to the claim that drift plays a role in modulation, at least when solar activity is low.

A further inspection of Fig. 1 shows that the model predicts the difference in the intensity-tilt relation before and after 1987 to diminish gradually with increasing tilt values after 1987. The predicted effect is much stronger than observed so that the good correlation between the predicted and observed intensities also diminishes with increasing α for both the long term and short term modulation. Again, whereas the model's predictions are reasonably good when α is small, the correlation with the IMP 8 data breaks down at larger tilt values when solar activity is high. Because α has increased almost dramatically after the solar minimum in 1987 - from 7°— 70° in just two years - the lack of correlation between the calculated and observed intensities occurs sooner in time but still at about the same tilt values, $\alpha > 30 - 35°$.

Another aspect of Fig. 1 which requires explanation is the absence of model results in our representation at large tilt values which includes the polarity change of the IMF. One of the reasons is that the model predicts an almost discontinuous jump in the intensity when the polarity is changed while it is obvious from the figure that the observed intensity actually changes gradually during times of high solar activity. This is despite the fact that we also let the new IMF polarity propagate radially outwards from the Sun at all latitudes at the solar wind speed, implying that a total change in the polarity throughout the heliosphere happens after ~ 255 days. Another reason is that the model also predicts a too large drop when the polarity of the IMF changes which means that the factor of modulation from minimum to maximum solar activity exceeds the factor of 3 present in the data. Since the model is not successful in explaining observations during times of high solar solar activity we were forced to treat the $A > 0$ and $A < 0$ epochs seperately in our computer runs and actually renormalise the $A > 0$ calculations to the normalised data of 1976.

To recap, Fig. 1 tells us that drift with the waviness of the neutral sheet as the modulation parameter is successful in explaining the long term modulation of cosmic rays at lower solar activity, but with increasing solar activity the drift effect proved to strong (e.g. the diminishing split effect, but especially the too large drop in intensity when the IMF polarity is switched). The short term modulation variation in the data is larger than the variation in the model calculated intensities for the whole period when $A > 0$ (1976 - 1981) as well as during large tilt values when $A < 0$ in comparison to the model intensities.

To address these problems it can be argued that all that is needed is a more realistic treatment of the neutral sheet and the polarity change of the IMF during high solar activity in the model. From observations during high solar activity it follow that the simple picture of a single highly warped neutral sheet is not valid any longer. The solar corona and magnetic field become very complex; the results of Hoeksema (1989) actually indicate the presence of multiple neutral sheets. The simple picture of particles drifting effectively towards Earth seems to become increasingly oversimplified with increasing solar activity. In addition to this observations also proved that the way the polarity of the IMF changes in the model is highly simplified compared to reality. In fact, the polarity reversal of the IMF was observed to start at mid latitudes, 40°— 50°, in the solar corona and migrated towards the poles in 1 — 1.5 years during solar cycle 20 /12/. Although modeling the neutral sheet and the polarity change in greater detail /13/ can probably remedy the almost discontinuous jump in the model calculated intensities during the polarity change of the IMF and even the too quick vanishing of the split effect, it is doubtful that the too large drift effect or the too small response of the model intensties to short term changes in tilt can be solved.

The significance of the too large drift effect also previously reported /14,15/ indicate that although the models are reasonably successful /5/, a critical look into the role of drift during larger solar activity has to be made. The IMF is obviously very turbulent during large solar activity so that it may well be that the weak scattering limit which the drift models are based on is not valid anymore. The point is that drift, whether it is time-dependent or in steady-state, can apparently not predict realistic modulation with increasing solar activity. Other modulating effects will have to be considered to explain, for instance, the short-term variations, especially during the period the IMF polarity is reversing, and which are beyond the scope of the present model. Perko and Burlaga /9/ achieved good correlation with data during the recovery phase from solar maximum (1981-1984) with a force-field model in which the diffusion is inversely proportional to the temporal fluctuations in the strength of the IMF. They will, however, have to illustrate that their assumption and model can also give the same reasonable agreement with data over 22 years. Nevertheless, it seems that what is really needed is a time-dependent model which is drift dominated during times of lower solar activity, but gradually becomes diffusion dominated as the Sun becomes more active. We have already started work in this direction by neglecting drift and incorporating the real IMF strength as a function of time in our two-dimensional time-dependent model. The purpose is to establish how successful such a model is compared to the presented drift model. The next step would then be to start working on a model which simulates a gradual transition from drift dominated to diffusion dominated modulation. This approach is supported by the observation that the IMF had increased from 4.5 nT in 1976 to 9.5 nT in 1980 /16/. Since $K_T \propto 1/B$, and therefore also drift, this observation implies that drift would decrease by at least a factor of 2 with increasing solar activity. This will, however, not solve the situation shown in Fig. 1 completely - test runs have indicated that even with drift reduced by a factor of 2 the model still is not able to predict the behaviour before and after a polarity reversal correctly.

SUMMARY AND CONCLUSIONS

The recently developed time-dependent drift model in which temporal changes in the neutral sheet waviness propagate at the solar wind speed proved capable of explaining the observed asymmetry in the cosmic ray intensity-tilt variation before and after 1987 /10/. In this paper, having incorporated the real neutral sheet tilt values in the model, we illustrate that the model is generally reasonably successful in explaining the intensity-tilt relationship of the IMP 8 counting rate against the real tilt values at Earth from 1976-1989.

From the short term intensity-tilt correlation as well as the split in the long term correlation before and after 1987, however, it appears that the model only gives good compatibility with the IMP 8 intensities at relatively low solar activity. The intensity-time correlation with the IMP 8 data shows a lack of response to temporal changes in tilt when it is large as well as an over large drift effect when the polarity of the IMF is switched. This has led us to the conclusion that a revision in the role of drift at larger solar activity is needed. It may be that what is really needed is a model in which there is a gradual transition from drift to diffusion dominated modulation with increasing solar activity where the diffusion is strongly effected by temporal variations in the strength of the IMF propagating at the solar wind speed.

ACKNOWLEDGEMENTS

We thank W.R. Webber for making the data available to us. MSP thanks W.R. Webber, J.A. Lockwood and F.B. McDonald for the interesting discussions and support during a short visit to the USA in May/June 1989.

REFERENCES

1. J.T. Hoeksema, J.M. Wilcox, and P.H. Scherrer, The structure of the heliospheric current sheet: 1978-1982, J. Geophys. Res. 88, # A12, 9910-9918 (1983).
2. E.J. Smith, and B.T. Thomas, Latitudinal extent of the heliospheric current sheet and modulation of galactic cosmic rays, J. Geophys. Res. 91, # A3, 2933-2942 (1986)
3. W.R. Webber, and J.A. Lockwood, Characteristics of the 22-year modulation of cosmic rays as seen by neutron monitors, J. Geophys. Res. 93, # A8, 8735-8740 (1988)
4. J. Kóta, and J.R. Jokipii, Astrophys. J. 265, 573 (1983)
5. M.S. Potgieter, and H. Moraal, A drift model for the modulation of galactic cosmic rays, Astrophys. J. 294, # 2, 425-440 (1985)
6. W.R. Webber, M.S. Potgieter and R.A. Burger, A comparison of predictions of a wavy neutral sheet drift model with cosmic ray data over a whole modulation cycle: 1976-1987, Astrophys. J. in press (1990)
7. J.A. Lockwood, and J.J. Quenby, The Pioneer 10 gradient measurements - the key to the modulation problem, Proc. 20th ICRC (Moscow) 3, 406-408, (1987)
8. M.S. Potgieter, and J.A. le Roux, A numerical model for a cosmic ray modulation barrier in the outer heliosphere, Astron. Astrophys. 209, 406-410 (1989)
9. J.S. Perko, and L.F. Burlaga, Cosmic ray variations and magnetic field fluctuations in the outer heliosphere, J. Geophys. Res. 92, # A6, 6127-6131 (1987)
10. J.A. le Roux, and M.S. Potgieter, A time-dependent drift model with a simulated wavy neutral sheet for the solar modulation of cosmic rays, this issue.
11. J.T. Hoeksema, Extending the sun's magnetic field through the three-dimensional heliosphere, Adv. Space Res. 9, # 4, 141-152 (1989)
12. D.F. Webb, J.M. Davis, and P.S. McIntosh, Observations of the reappearance of polar coronal holes and the reversal of the polar magnetic field, Solar Phys. 92, 109-132 (1984)
13. B.T. Thomas, B.E. Goldstein, and E.J. Smith, The effect of the heliospheric current sheet on the cosmic ray intensities at solar maximum: two alternative hypotheses, J. Geophys. Res. 91, # A3, 2889-2895 (1986)
14. M.S. Potgieter, J.A. le Roux, and R.A. Burger, Radial gradients and cosmic-ray modulation models, 20th ICRC (Moscow) 3, 287-290 (1987)
15. M.S. Potgieter, J.A. le Roux, and R.A. Burger, Interplanetary cosmic ray radial gradients with steady state modulation models, J. Geophys. Res. 94, # A3, 2323-2332 (1989)
16. E.J. Smith, D. Winterhalter, and J.A. Slavin, Recent Pioneer 10 observations of the distant heliospheric magnetic field, Proc. 6th SW Conf. 2, 581-585 (1987)

A TIME-DEPENDENT DRIFT MODEL WITH A SIMULATED WAVY NEUTRAL SHEET FOR THE SOLAR MODULATION OF COSMIC RAYS

J.A. le Roux and M.S. Potgieter

Department of Physics, Potchefstroom University of C.H.E., Potchefstroom, South Africa

ABSTRACT

We have developed a time-dependent drift model with a simulated wavy neutral sheet in which changes in the waviness of the sheet propagate radially outwards at the solar wind velocity. Using an asymmetric rate of change in the tilt angle , α, of the wavy neutral sheet (apparent in observations with respect to the solar minimum of 1987) in the model, we can explain the observed lesser response of the cosmic ray intensity to temporal changes in α after 1987 as well as the somewhat asymmetric behaviour of the intensity-time profile with respect to 1987.

INTRODUCTION

Recently, Webber, Potgieter and Burger /1/ have shown that a steady-state drift model can reproduce the observed long-term relation between cosmic ray intensities, at various energies, and the tilt angle, α, (Hoeksema et al. /2/) of the neutral sheet for the solar cycle 1976-1987. They conclude that this observation has a theoretical explanation in the drift picture of cosmic-ray modulation. Steady-state models are, however, only a first order approach to the complex process of solar modulation. During the past years there has been a steady increase of convincing evidence that long-term modulation can no longer be described satisfactory with steady-state models (e.g. /3,4,5,6/). New evidence in this regard is reported by Webber and Lockwood /7/, who find that the intensity of cosmic rays is less responsive to changes in α after the solar minimum in 1987 than before, whereas steady-state drift models predict a symmetrical response to changing α w.r.t. any solar minimum. Their observation could be interpreted as evidence against drift, and that it was caused by other effects, or it could be that it was the result of the time-dependent development of the wavy neutral sheet, or a combination of different effects.

In this paper we shall concentrate on the time-dependent drift option and shall illustrate that by considering changes in the waviness of the neutral sheet to propagate radially outwards at the solar wind speed in a time-dependent drift model, we can explain the mentioned asymmetry in the intensity-tilt correlation before and after 1987.

THE MODEL

We solve the time-dependent cosmic-ray transport equation numerically, in two spatial dimensions, using the alternating direction implicit method adapted to handle time-dependence (see also Le Roux and Potgieter /8/). The solar wind velocity is assumed to vary from 400 km/s in the equatorial plane to a maximum speed of 600 km/s at latitude above 45°. The diffusion coefficients parallel and perpendicular to the background IMF are respectively given by

$$K_{\parallel} = K_{o}\beta K_{p}(P)B_{e}/(3B), \quad \text{and} \quad K_{\perp} = 0.01K_{\parallel}. \tag{1a}$$

with the constant $K_o = 2.4x10^{21}$ cm^2 s^{-1}, which corresponds to low solar activity when α is small; β is the ratio of particle speed to the speed of light, and $K_p(P) = P/(1GV)$ when $P \geq 0.4$ GV, and $K_p(P) = 0.4$ when $P < 0.4$ GV. Here, B is the standard Parker spiral configuration of the interplanetary magnetic field (IMF) /9/ and $B_e = 5$ nT is the magnitude of B at Earth. For the asymmetric part of the diffusion tensor we used

$$K_T = \beta P/(3B), \tag{1b}$$

which describes curvature and gradient drift. The inner and outer boundary are taken at 0.1 AU and 60 AU respectively.

The waviness of the neutral sheet is simulated using the technique of Potgieter and Moraal /9/. Details of the different approaches to incorporating a wavy neutral sheet in numerical codes were given by Burger and Potgieter /10/. We start the code assuming solar minimum conditions, typically with $\alpha = 6° - 8°$, during an A>0 epoch (e.g., 1976), when the northern-hemispheric IMF points away from the Sun. The tilt angle is then systematically increased at the Sun in intervals of 27 day averages and propagated outwards with the solar wind speed until α reaches a maximum of 60° −70° within a specified period of time (4 - 6 years). It takes ~255 days for a change in the neutral sheet to propagate from the Earth to 60 AU. When the maximum waviness is reached throughout the heliosphere, the IMF polarity is changed within 255 days and α is then systematically decreased at the Sun and propagated outwards. During this A<0 period and before a solar minimum the waviness will be larger in the outer than in the inner heliosphere. When solar minimum (e.g., 1987) has been reached, α is increased again so that it may happen that while there is still large waviness in the outer heliosphere, there is almost no waviness in the middle with small but increasing waviness in the inner heliosphere. The spatial distribution of the neutral sheet will therefore, in principle, not be the same before and after a solar minimum.

RESULTS AND DISCUSSION

In Fig. 1 the calculated intensities for 1 GeV protons are shown as a function of the tilt angle α at Earth for the two polarity configurations, $A>0$ and $A<0$, of the IMF and for a steady-state and time-dependent model respectively. For both models we changed α linearly from 7°– 65°– 7° over a period of 6 years for each epoch. The rate of change was kept identical before and after each solar minimum to give a consistent picture of the differences between the two models.

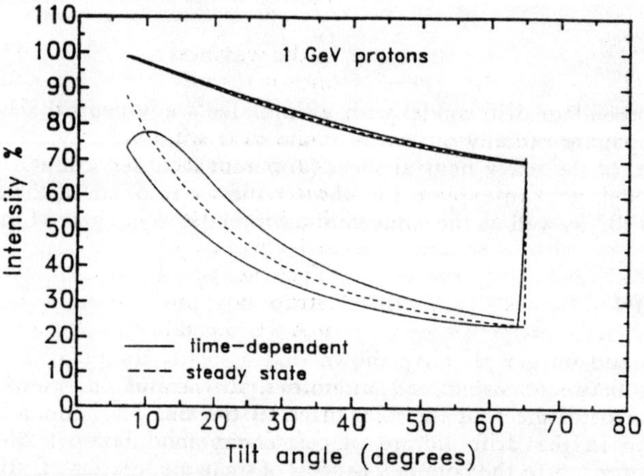

Fig. 1. Intensity at Earth as a function of the waviness of the neutral sheet for a steady-state and time-dependent drift model. The tilt angle was changed linearly from 7°– 65°– 7° over a period of 6 years for each epoch.

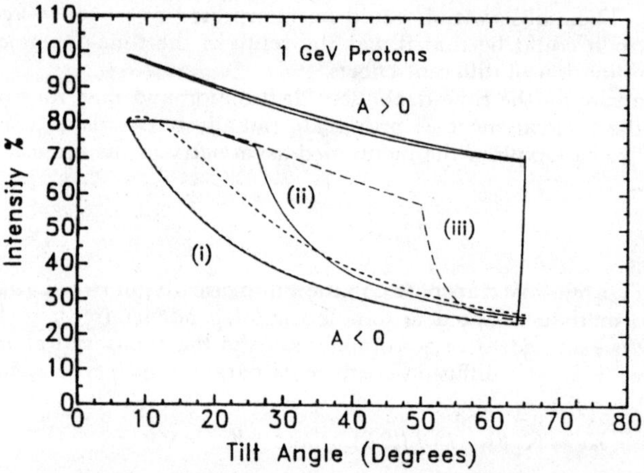

Fig. 2. Intensities at Earth as a function of the waviness of the neutral sheet for three different variations of α after an $A<0$ solar minimum. The tilt angle α was increased linearly (i) from 7°– 65° over a period of 2 years; (ii) from 7°– 26° in 1 solar rotation period and from 26°– 65° during the rest of the 2-year period, and (iii) from 7°– 50° in 2 solar rotation periods and from 50°– 65° during the rest of the 2 year period.

We first observe that in general the intensities are clearly more responsive to changes in α during the $A<0$ epoch than when $A>0$ (see also /1/). Second, the curve obtained with the time-dependent model shows a clear split around the solar minimum when $A<0$, but an almost negligible split when $A>0$. The steady-state profiles, on the other hand, show no split for both epochs and are centered between the time-dependent ones. The time-dependent curve lies below the steady-state curve with decreasing solar activity when $A<0$, because the waviness of the neutral sheet is larger in the outer than in the inner heliosphere compared to the steady state model where the waviness is the same at all radial distances. The path lengths which particles drift along the more realistic spatial distributed neutral sheet are longer in the time-dependent case resulting in somewhat lower intensities when plotted against α at Earth. After the $A<0$ solar minimum the time-dependent curve lies above the steady-state curve and reaches a maximum value sometime after the lowest α has been reached. In this case the waviness in the outer heliosphere is smaller than in the inner heliosphere so that the drift passage inwards toward Earth is easier resulting in a higher intensity for the same α at Earth.

The time-dependent and steady-state curves converge when α becomes large because the relative difference in the spatial distribution of the wavy neutral sheet for the two cases becomes increasingly less. Another way of looking at this effect is that for both polarity epochs, but especially when $A<0$, the longer drift path length which particles have to follow along the neutral sheet when α is large means an increase in the probability that particles will be scattered off the neutral sheet and therefore be less affected by temporal changes in the waviness of the sheet. When $A>0$ the effect of the propagating waviness is obviously not so prominent because particles drift downwards to the equatorial regions from higher latitudes and are nearly unaffected by changes in the waviness of the neutral sheet.

The predicted split in the intensity-tilt profiles when $A<0$ is supported by recent observations that the slopes of the intensity-tilt regression lines are flatter after than before 1987 (Webber and Lockwood /7,11/). We consider this effect as an indication of the role the wavy neutral sheet plays in modulation. The negligible split when $A>0$ provides another but as yet untested prediction of the time-dependent drift model.

In Fig. 2 we show what happens when the rate of change in the waviness is more rapid after than before solar minimum, when $A<0$. This feature was observed /11/ since it took 6-7 years for α to decrease from maximum (1980) to minimum (1986/7) but merely ~2 years after 1987 to increase to 70°. Without fitting data we want to demonstrate what can be expected when changes in α are made more realistic than in Fig. 1. In Fig. 2 we show three cases: First, where α was increased at a constant rate from 7°— 65° over a period of 2 years when $A<0$. Second, where α was increased linearly from 7°— 26° over a period of only ~1 solar rotation and linearly from 26°— 65° during the rest of the 2 year period. Such a rapid increase in α was observed in 1987 /7/. Third, we went further and increased α from 7°— 50° in just 2 solar rotation periods and from 50°— 65° in the rest of the 2 year while $A<0$. This case is, of course, an exaggeration in terms of what happened in the heliosphere after 1987, but will illustrate how powerful the time-dependent approach is. For all three cases the decrease of α from 65°— 7° when $A<0$ was done linearly over a period of 6 years.

Comparing the three cases mutually and with Fig. 1 one notes that when an asymmetry in the rate of change in α before and after an $A<0$ solar minimum is introduced in the time-dependent model the split in the intensity-tilt profiles around this solar minimum becomes significant; the more the asymmetry is made the larger the split becomes. A rapid increase of α at the Sun after solar minimum means that while large waviness has already reached the Earth it will take several months to reach the outer heliosphere. Until this larger waviness has reached larger radial distances cosmic rays will experience, in those regions, a waviness representing solar activity which occurred sometime before the solar minimum. The intensity therefore stays almost constant for some time after solar minimum despite the rapid increase in α at Earth.

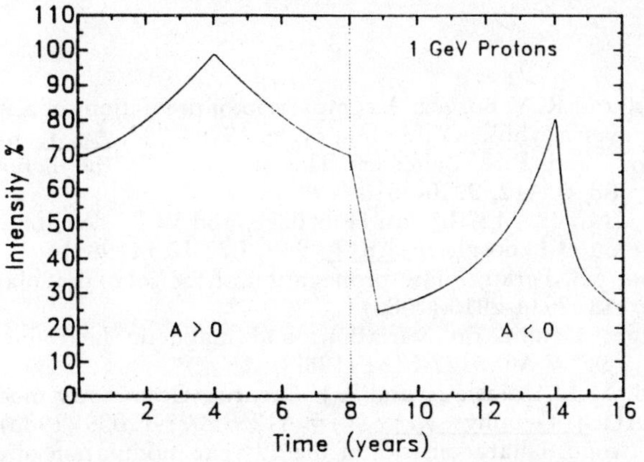

Fig. 3. Curve (ii) in Fig. 2 as a function of the corresponding time.

In Fig. 3 we show the 1 GeV intensities obtained with the second case in Fig. 2, but this time on a time scale rather than as a function of tilt angle. These calculated results show that while the intensity-time profile is plateau-like and symmetric around solar minimum when $A>0$ (e.g., 1976), it is peaked and asymmetric when $A<0$ (e.g., 1987). The intensity decrease is, in this case, faster after than before the solar minimum and is the consequence of the asymmetric rate of change in α before and after the solar minimum. These results are compatible with long-term observations made during 1972 - 1977 and after 1982. Note that the model predicts that the 1 GeV intensity is less for the $A<0$ (1987) compared with the $A>0$ (1976) solar minimum in accordance with observations/1/.

This study obviously emphasizes the role of the wavy neutral sheet in solar modulation. A further interesting implication of this work is that when the intensity at Earth is plotted as a function of the total length, L, of the wavy neutral sheet (i.e. between Earth and the outer boundary) rather than the tilt angle at Earth, a split should not occur with respect to the $A<0$ solar minimum. The reason is that in drift dominated modulation the effective length of the wavy neutral sheet, as a function of the tilt angle which depends on the activity of the Sun, determines the intensity at any observation point in the equatorial regions when $A<0$. Consequently, if drift dominates modulation and if L, which depends on solar activity and the position of the outer boundary, is calculated correctly, one should get symmetry with respect to any solar minimum when the intensity is plotted as a function of L. If drift is unimportant there should be no relation between the intensity and L. We consider this approach a strong method to establish the relative importance of drift in modulation.

Assuming that drift is important, a careful consideration of the mentioned effect will show that under certain conditions a split may also occur in the intensity-L relation with respect to 1987. Such a split should depend on how large the waviness of the neutral sheet is, at a given time before and after solar minimum, in the outer regions of the heliosphere, because with large waviness particles would tend to short-cut the wavy neutral sheet due to perpendicular diffusion described by $K_\perp$, making the effective L which particles would experience shorter than the actual length of the wavy neutral sheet. The very rapid increase in α after 1987 should enhance this effect, so that a split caused by the overestimation of L should occur soon after solar minimum - large waviness reached the outer heliosphere soon after solar minimum. This method can therefore also be used to establish the influence of $K_\perp$ on modulation. The relative contribution of the short-cut effect in the outer heliosphere to the overestimation of L will be less for an observer at Earth than further out so that at Pioneer 10, for instance, the predicted split should be more significant than at Earth. Another interesting implication of this intensity-L approach is that if the short-cut effect is known, it might be used to estimate the relative position of the heliospheric boundary before and after solar minima.

SUMMARY AND CONCLUSIONS

A newly developed time-dependent drift model was used to illustrate the modulating effect of a simulated changing wavy neutral sheet which evolved with time and was propagated outwards in the heliosphere at the solar wind speed. The primary effect was that the tilt angle dependence of the intensity at Earth was less after than before an $A<0$ (1987) solar minimum period, despite that α was changed linearly and symmetrically w.r.t. this minimum. When it was changed asymmetrically, as observations suggested, an even more significant effect was obtained - the intensity-tilt profile before solar minimum was notably above that for after the solar minimum. These results are compatible with observations before and after the solar minimum in 1987 /7/. With $A>0$ the split in the calculated curves was minimal. A further implication of the asymmetric rate of change in α is that the model also predicts a somewhat asymmetric intensity-time profile with respect to the solar minimum of 1987 in accordance with the effect of the rapid increase in solar activity after 1987.

In conclusion, our results support the role of drift, especially that of the wavy neutral sheet, in solar modulation. To explain the difference in the tilt-dependence of cosmic rays before and after 1987 it seems essential to use a time-dependent, instead of a steady-state, drift model.

REFERENCES

1. W.R. Webber, M.S. Potgieter, and R.A. Burger, A comparison of predictions of a wavy neutral sheet drift model with cosmic ray data over a whole modulation cycle: 1976-1987, Ap. J. in press (1990)
2. J.T. Hoeksema, J.M. Wilcox, and P.H. Scherrer, The structure of the heliospheric current sheet: 1978-1982, J. Geophys. Res. 88, # A12, 9910-9918 (1983)
3. F.B. McDonald, N. Lal, J.H. Trainor, M.A.I. Van Hollebeke, and W.R. Webber, The solar modulation of galactic cosmic rays in the outer heliosphere, Ap. J. 249, L71-L75 (1981)
4. M.A. Forman, F.C. Jones, and J.S. Perko, Phase propagation of the solar modulation of galactic cosmic rays, J. Geophys. Res. 91, # A3, 2914-2916 (1986)
5. J.S. Perko, and L.F. Burlaga, Cosmic ray variations and magnetic field fluctuations in the outer heliosphere, J. Geophys. Res. 92, # A6, 6127-6131 (1987)
6. L.F. Burlaga, F.B. McDonald, M.L. Goldstein, and A.J. Lazarus, Cosmic ray modulation and turbulent interaction regions near 11 AU, J. Geophys. Res. 90, # A12, 12027-12039 (1985)
7. W.R. Webber, and J.A. Lockwood, Characteristics of the 22-year modulation of cosmic rays as seen by neutron monitors, J. Geophys. Res. 93, # A8, 8735-8740 (1988)
8. J.A. le Roux, and M.S. Potgieter, Episodic cosmic-ray modulation in the heliosphere, Adv. Space Res. 9, # 4, 225 (1989)
9. M.S. Potgieter, and H. Moraal, A drift model for the modulation of galactic cosmic rays, Ap. J. 294, # 2, 425-440 (1985)
10. R.A. Burger, and M.S. Potgieter, The calculation of the neutral sheet drift in two-dimensional cosmic-ray modulation models, Ap. J. 339, # 1, 501-511 (1989)
11. J.A. Lockwood, and W.R. Webber, private communication (1989)

SOLAR EFFECTS ON UNDERGROUND MUONS AT 570 hg/cm^2

L. Bergamasco,* G. Cini Castagnoli,* M. Serio* and A. Provenzale**

*Istituto di Fisica Generale dell'Università — Via P. Giuria, 1 — 10125 Torino, Italy

**Istituto di Cosmo-Geofisica — Corso Fiume, 4 — 10133 Torino, Italy

ABSTRACT

This paper presents results on the intensity variations of underground muons originated by primary cosmic rays with mean rigidity about 1800 GV. We discuss their sensitivity to various solar effects, and the implications on the size of the solar cavity.

INTRODUCTION

Space-born instruments allow the direct study of the galactic cosmic radiation only up to rigidities of the order of a few GV (the rigidity is the energy per unit charge and is measured in Volts: 1 GV = 10^9 V). The effects of the heliosphere on cosmic rays of higher rigidity must be investigated indirectly through the secondary particles produced in the interactions of the primaries in the atmosphere, expecially muons which constitute the most penetrating component. The working hypothesis is that the imprint of solar modulation is transmitted by the primaries to the secondaries which are eventually detected on Earth. Of course in the various interaction and propagation processes the original information is somewhat altered by a succession of various spurious interplanetary, geomagnetic and atmospheric effects. The task of demodulating the earth-based measurements to local heliospheric values is therefore extremely difficult and risky. However since it is the only way to shed some light on a range of rigidities not directly explorable "in situ", the challenge is worthwhile.

Traditionally most of the studies on the solar effects on underground muons have been carried out at depths around 50 m.w.e. (the traditional units of underground depths are "meters of water equivalent": the rock burden, measured in meters, times the ratio of the density of the rock to the density of water) corresponding to primary rigidities of the order of a few hundreds of GV, where the presence of various manifestations of the solar influence has been thoroughly investigated and established.

However the existence of solar effects at rigidities higher than 500-1000 GV is still a controversial issue. The problem of the sensitivity of high rigidity primaries to solar effects is important because of its connection with the problem of the dimensions of the solar cavity. The important factor is the ratio of the heliospheric scale D_h to the scattering mean free path λ_s of the primaries: if D_h/λ_s is greater than 5-10, solar effects are to be reasonably expected. In this perspective ten years ago two of us proposed on the basis of our results on muons recorded at 60-70 m.w.e. a solar modulation region of at least 30 AU, against the conventionally accepted boundary value of 3 AU /1/. Our "bold" claim was then proven right by subsequent space explorations. Our present search for solar effects in muons detected at intermediate depths (100-1000 m.w.e) has again as ultimate end the assessment of the cut-off rigidity R_0 beyond which the influence of the Sun vanishes, and thus the placement of a lower limit on the heliospheric scale D_h.

In the present paper we present and discuss in this spirit results on the muons recorded in the station of Artyomovosk, at a vertical depth of 570 hg/cm^2. The sensitivity of the muon records to heliospheric influence will be examined under different aspects and analyzed with different techniques.

PRIMARY RIGIDITY RANGE

The response function of the muons recorded at 570 m.w.e. produced by primary particles of rigidity R is shown in fig.1 /2/. We show the calculations for the two viewing directions Q = 0° and Q = 30° inside which most of the arriving muon flux is included. Panel (a) shows the differential function F(R); panel (b) the percentage $I(\leq R)/I_{tot}$ of the primaries with rigidity $\leq$ R. We can see that roughly 75% of the muons have rigidity higher than 1000 GV, with average value around 1800 GV. These rigidities correspond to gyration radii around the local interplanetary magnetic field greater than 4 AU and to scattering mean free paths at least 4 or 5 times larger. Such values, when compared with the size of the heliosphere established by present Voyager observations ($\approx$ 50 A.U.) place these primaries on the borderline as far as the sensitivity to solar effects is concerned. Even if some influence is still to be

expected, its manifestations may be different than at lower rigidities and thus require different approaches and analysis techniques.
The obvious disadvantage in working at intermediate depths underground is the low counting rate, which requires large recording areas and long running periods in order to reach an acceptable statistical significance. For this reason we present here results of six years of measurements, relative to about 6 $\times 10^8$ muons. The advantage is given by the fact that the primary rigidities are high enough to ensure a relative insensitivity of the records to most geomagnetic and atmospheric effects.

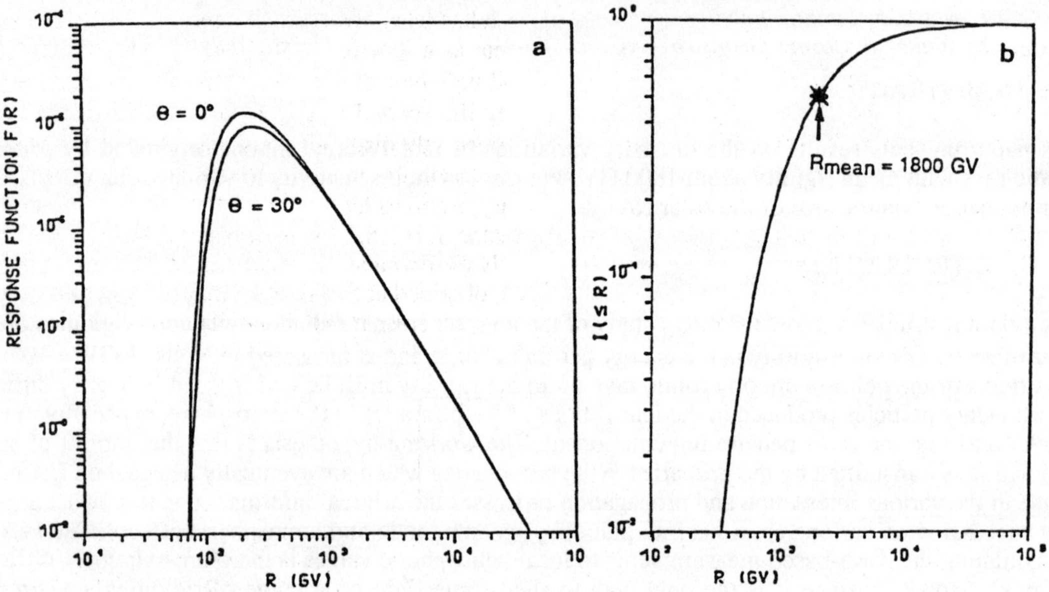

Fig. 1.Differential and integral response functions of muons originated by primaries of rigidity R which are recorded at a vertical depth of 570 m.w.e.

SOLAR TIME VARIATIONS

The study of muon intensity variations in solar time is a standard tool at shallow depths. The Fourier analysis of the 24-hour histograms of muon intensity measured over many years in different worldwide stations has made it possible to establish at R ≤ 300 GV the presence of a dominant first harmonic (so called "diurnal variation" with T = 24 hr) accompanied by a smaller second harmonic (T = 12 hr) and sometimes by a third harmonic (T = 8 hr) /3/.
The same kind of analysis performed on the muons recorded at 570 hg/cm^2 gives quite different results. Fig. 2 shows as a full line the 24-hour histogram of the muon intensity variations $\Delta N_\mu(t)/N_\mu$ obtained in six years of measurements (0<t<24 hr, Δt =0.25 hr); the dashed line is the fit to the histogram given by the superposition of the first three harmonics, whose parameters (amplitude and phase) are given.

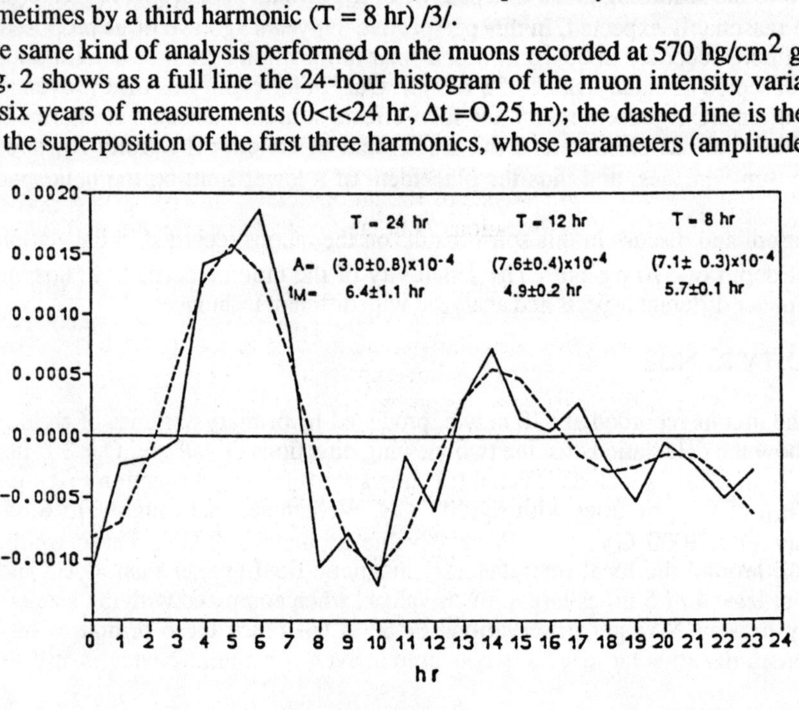

Fig.2. 24-hour histogram of the muon intensity variations in solar time (full line) and superposition of the first three harmonics (dotted line).

It may be seen that the fit reproduces only approximately the features of the experimental histogram, which implies the existence of higher order harmonics of not negligible amplitude; moreover the amplitude of the diurnal variation is hardly half the amplitude of the second and third harmonics.
The results of fig. 2 are long-time averages; particular heliospheric conditions may induce temporarily different behaviors and different relationships among the various harmonics/4/.

11-YEAR CYCLE

A second aspect of solar induced muon variations which has been investigated concerns the behavior of the counting rate N_m averaged over intervals of one year as a function of the corresponding number of sunspots R_Z. The number of sunspots is in fact a good indicator of the level of solar activity during a solar cycle. Fig. 3 shows the behavior of $N_{\mu av}$ during the second half of solar cycle 21, from 1981 (solar maximum) to 1987 (solar minimum). The data apparently display a consistent trend toward lower values of N_μ with increasing R_z, even if the inverse correlation may not be statistically significant; moreover as the mean annual value of the muon intensity becomes larger toward the solar minimum, the standard deviation of the data becomes smaller by a factor of ≈ 1.6. This behavior is reminiscent of the recovery phase of the 11-year modulation cycle normally observed at low rigidities, in neutron monitors or shallow depth installations data. The results of fig. 3, obtained at 570 m.w.e., may then be interpreted as an indication that at least some mechanisms of the 11-year solar modulation retain their efficiency up to rigidities of a few thousand GV.

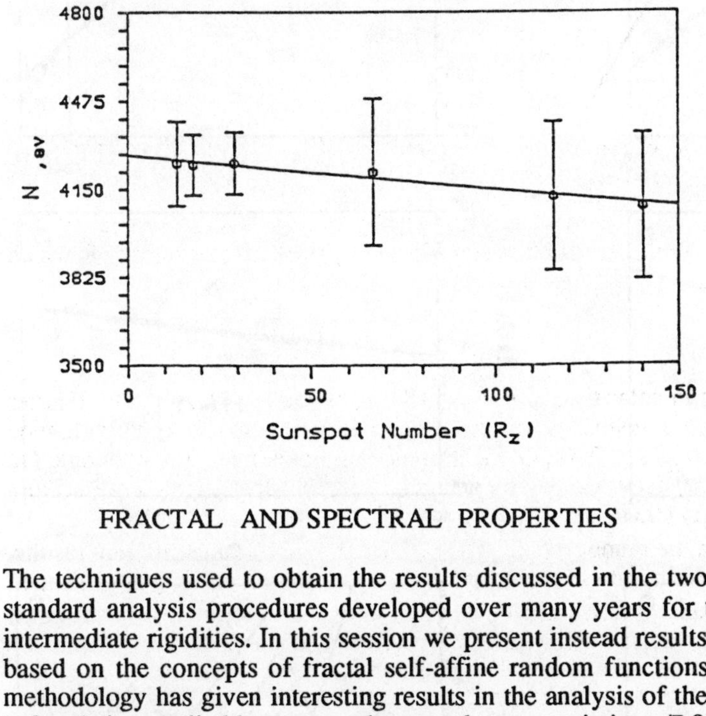

Fig.3. Mean muon counting rate during the second half of solar cycle 21, from 1981 (solar maximum) to 1987 (solar minimum).

FRACTAL AND SPECTRAL PROPERTIES

The techniques used to obtain the results discussed in the two previous sections are taken from the standard analysis procedures developed over many years for the study of cosmic rays at low and intermediate rigidities. In this session we present instead results obtained by means of new techniques based on the concepts of fractal self-affine random functions and of scaling exponents. This new methodology has given interesting results in the analysis of the properties of the IMF variations /5,6/ before being applied by us to underground muon variations /7,8/, and is becoming a valid complement to traditional spectral analysis procedures.
Self-affinity is connected with the fractal nature of the signal considered, and constitutes a particular case of the more general notion of scaling. From a physical point of view self-affinity means that the signal fluctuations at different scales are not completely irregular but are instead related to each other by a general (statistical) recursive relation characterized by a "scaling exponent" $0 \leq H \leq 1$.
To search for statistical self-affinity in the muon time series and to compute the scaling exponent H and the extent of the scaling range, we follow the two methods discussed in details in our papers /7,8/, that is: 1) the "yardstick length" procedure, which computes the curve length $L(n)= \Sigma_i |x(t_i + n\Delta t)- x(t_i)|$, where i = 1, 2n, 3n, and so on: in the small n limit, if the signal has self-affine properties, L(n) behaves as n^{H-1}, where H is the scaling exponent; 2) the " average absolute jump" procedure, which computes the differences $\Delta X(\lambda) = \langle|x(t_j + \lambda\Delta t) - x(t_j)|\rangle$, where j now spans all the points in the time series: if the series has self-affine properties, in the small λ limit, $\Delta X(\lambda)$ behaves as λ^H. If the self-affine behavior of the signal is physical, the values for the scaling exponents and the extent of the scaling

regions obtained by the two different procedures should agree with each other within the errors. The outcome of the fractal analysis is then related to the information given by traditional spectral analysis by the fact that the self-affine nature of a time series with scaling exponent H implies the existence of a power spectrum which behaves as $f^{-\alpha}$, with $\alpha = (2H + 1)$ /9/. For details see the references in /7,8/.
In this analysis we have considered again the one-year records extending from 1981 to 1987, obtaining for all of them a good agreement between the two values for the scaling exponent given by the two procedures. This result indicates that the underground muon variations are consistently self-affine in character over a scaling range which extends from a few hours to about two months (the presence of a lower and an upper limit is typical of natural fractals), independently of the level of solar activity.

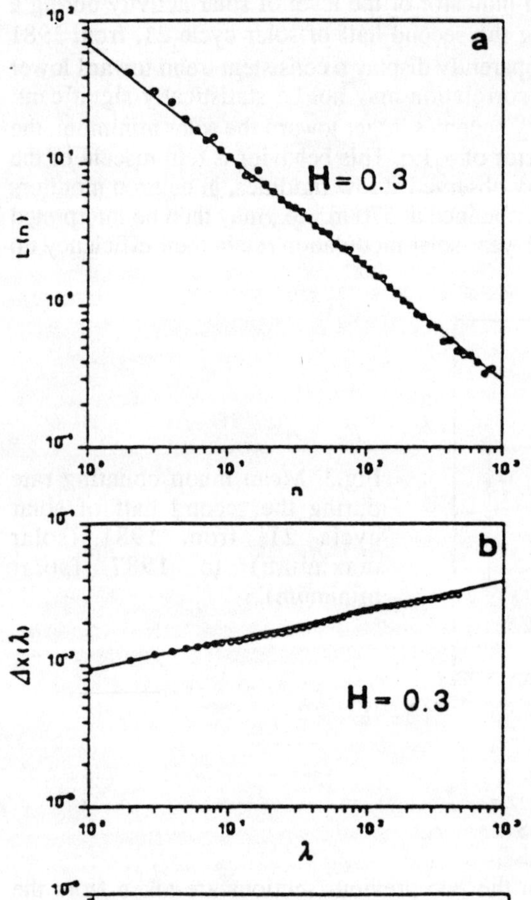

Fig. 4. Results of the fractal analysis for the year 1981 (solar maximum).

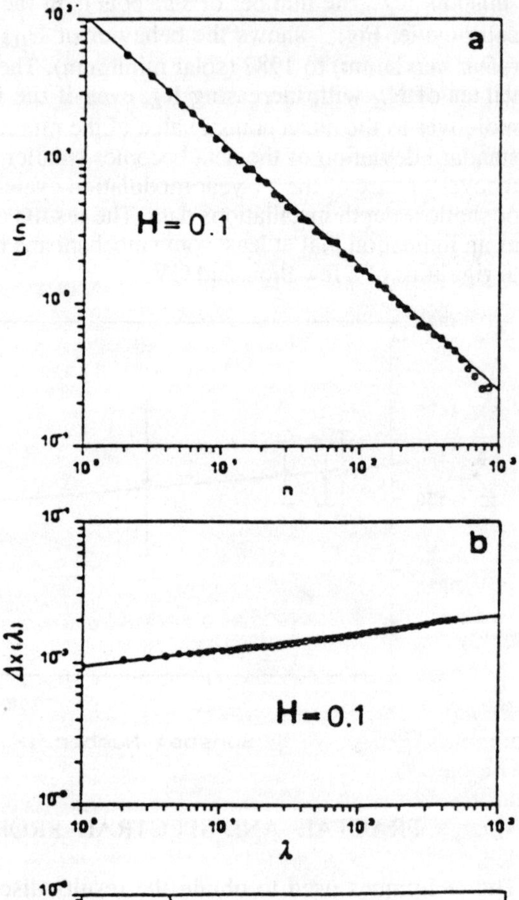

Fig. 5. Results of the fractal analysis for the year 1987 (solar minimum).

As examples we show in figs. 4 and 5 the results relative to the two extreme years 1981 and 1987. In each figure panels (a) and (b) display the yardstick length L(n) and the average absolute jumps $\Delta X(\lambda)$ from which the value of the scaling exponent is derived; panel (c) shows the experimental power spectrum and its power-law prediction on the basis of the indications of the two previous panels.
It is evident from the results of the two figs. 4 and 5 that the changes in the heliosphere conditions which occurred over the last half of solar cycle 21 affected the value of the scaling exponent H and thus that of the spectral exponent a, which become larger near solar maximum and then decrease with the reduction in the level of solar activity. The correlation between the parameters H and αderived for the different years and the corresponding mean annual sunspot number R_Z is shown in fig. 6. It will be important to check in the present solar cycle, which seems to be particularly active, whether the linear direct correlation of H and a with the level of solar activity as indicated by the annual sunspot number continues to hold. It seems anyway plausible to interpret the transition from the "equilibrium values" $H \approx 0.1$, $\alpha \approx 1.2$ at solar minimum to the values $H \approx 0.3$, $\alpha \approx 1.6$ at solar maximum as a reflection of the enhanced solar effects on the 1800 GV primaries.

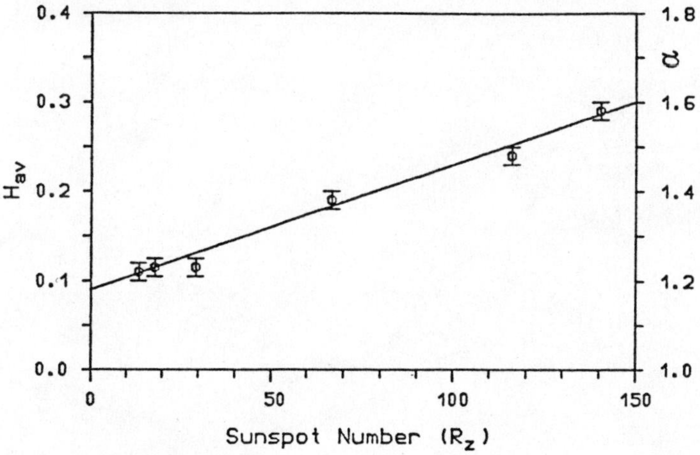

Fig. 6. Scaling exponent and power law exponent as a function of the corresponding mean annual sunspot number, from 1981 to 1987.

CONCLUSIONS

We have analyzed different types of variations in the intensity records of muons at 570 m.w.e. with the aim of investigating the sensitivity of primaries of mean rigidity ≈ 1800 GV to solar effects.
The outcomes of three different analysis procedures are consistent in indicating that these primaries during their stay in the heliosphere are affected in different ways by the changes and the phenomena which occur within it. This conclusion suggests that the solar modulation region may extend beyond 50 AU, even if the last word on this issue must as always be left to future space observations.

REFERENCES

1. L. Bergamasco, S.Alessio, G. Cini and A.R. Osborne, in: Proc. Intern. Cosmic Ray Conf., IUPAP, Kyoto 1979,4, p. 163.
2. K. Fujimoto, K. Murakami, I. Kondo and K. Nagashima, in: Proc. of Intern. Cosmic Ray Symp. on High Energy Cosmic Ray Modulation, ed. C.R. Laboratory, University of Tokio, Japan 1976, p.50.
3. S.Kudo, S. Mori, D.B. Swinson, S. Yasue and Y. Rue, in: Proc.20th Intern. Cosmic Ray Conference, ed. IUPAP, Moscow1987, 4, p.140.
4. L. Bergamasco, G. Cini, A. Osborne, A. Provenzale, V. Kudryavtsev, V. Kuznetsov and O. Ryazhkaya, J. Geophys. Res., 94, in press (1989).
5. L.F. Burlaga and L.W. Klein, J. Geophys. Res., 91, 347,1986.
6. L.F. Burlaga and W.H. Mish, J. Geophys. Res., 92, 1261,1987.
7. L. Bergamasco, A. Provenzale, A. Osborne, G. Cini,V. Kudryavtsev, V. Kuznetsov and O. Ryazhkaya, J. Geophys. Res., 94, 2667,1989.
8. L. Bergamasco, A. Provenzale, G. Cini, M. Serio, V. Kudryavtsev, V. Kuznetsov and O. Ryazhkaya, J. Geophys. Res., in press (1989).
9. S.Panchev, Random Functions and Turbulence, Pergamon, New York, 1971.

THE ADDITIONAL FLUXES OF COSMIC RAYS IN THE STRATOSPHERE IN THE VARIOUS HALF-PERIODS OF THE 22-YEAR SOLAR MAGNETIC CYCLE

A.K. Svirzhevskaya, N.S. Svirzhevsky and Yu.I. Stozhkov

P.N. Lebedev Physical Institute of the Academy of Sciences of the U.S.S.R., 53 Leninsky Prospect, Moscow, 117924, U.S.S.R.

ABSTRACT

The cosmic ray intensity measurements in the stratosphere made during the last three decades cover a period lasting one and a half times the 22-year solar magnetic cycle. Throughout the 1971-1979 half-period of the solar magnetic cycle we have observed in the energy range 100-400 MeV/n additional fluxes of cosmic rays which were absent in the previous 1959-1970 half-period. In 1975-1976 solar activity minimum the excess fluxes were evaluated as 300 $m^{-2}s^{-1}sr^{-1}$ on the assumption the additional particles were protons. The time variations of the stratospheric additional fluxes are in good correlation with an anomalous He component outside the Earth's magnetosphere. We expected the additional fluxes would not appear after the 1980 solar magnetic field reversal. However, they have reappeared in 1985 with the same delay after the reversal as the anomalous He component. The value of the additional fluxes in the 1986-1987 solar minimum was 230 $m^{-2}s^{-1}sr^{-1}$.

INTRODUCTION

The idea of the additional fluxes of low energy particles in the stratosphere arose from comparison of experimental data on the cosmic ray intensity during two time periods with the opposite polarity of the interplanetary magnetic field (IMF). We have found that at the same cosmic ray intensity level in the middle stratosphere (80-100 g/cm^2) the intensity in the upper stratosphere (5-25 g/cm^2) in 1971-1979 was obviously higher than that in 1959-1970 /1, 2/. In this paper we display what were the changes of the intensity dependence on atmospheric pressure for the mentioned periods.

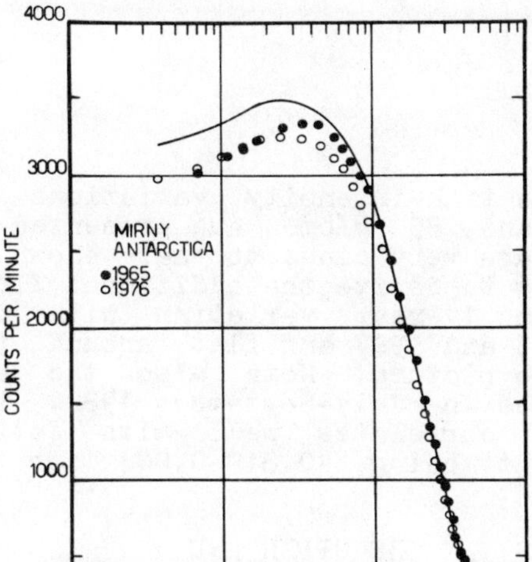

Fig. 1. Count rates versus atmospheric depth during the two solar minima in 1965 and 1976 at the Soviet antarctic station Mirny. Solid line represents the 1976 count rate normalized to the 1965 data at 100 g/cm^2.

Fig. 1 shows counting rate versus the atmospheric pressure at the Soviet antarctic station Mirny during consecutive solar minima in January-June 1965 and July-December 1976. A solid line represents the intensity of the secondaries in 1976 normalized to that in 1965 at the atmospheric depth 100 g/cm^2. The normalized 1976 intensity coincides with the 1965 intensity at pressures above 100 g/cm^2 and exceeds the 1965 intensity at lower pressures. In the upper stratosphere the excess amounts to about 7% of the total intensity. In 1976 the same additional fluxes were recorded in the northern hemisphere at the high latitude Murmansk station. To explain this change of the cosmic ray intensity vertical profiles in the atmosphere we have to assume a softening of the primary cosmic ray spectra at energies of a few hundreds MeV/n in 1976. Excess fluxes of low energy protons and He nuclei observed in 1976-1977 relative to 1965 were reported by Garcia-Munoz et al. /3/ and Webber et al. /4/. The experimental data obtained with the balloon born telescope in /4/ showed that in 1977 the fluxes of protons and He nuclei at the energy below 100 MeV/n were a factor 1.4-1.5 higher than those in 1965. According to our measurements the additional fluxes have existed over the whole of the 1972-1979 period as well as since 1985 up to now.

STRATOSPHERIC COSMIC RAY MEASUREMENTS

The stratospheric measurements have been carried on in the USSR in Moscow and Murmansk since 1957 and in Alma-Ata since 1962. Since 1963 the measurements have also been carried on in Antarctica at Mirny station. Once a day at 0630 UT radiosondes are launched to record the vertical intensity profiles from ground level up to altitudes of 30-35 km. The charged particles detectors used in measurements consist of two Geiger-Müller counters arranged as telescope and interlayed by 7 mm Al absorber. The effective area of a single counter is 16.4 cm^2, thresholds are 0.2 MeV for electrons and 5 MeV for protons. To maintain the homogeneity of data sets an elaborate procedure of detector calibration before flight was developed to make a correction for its registration efficiency, size, wall thickness, etc.

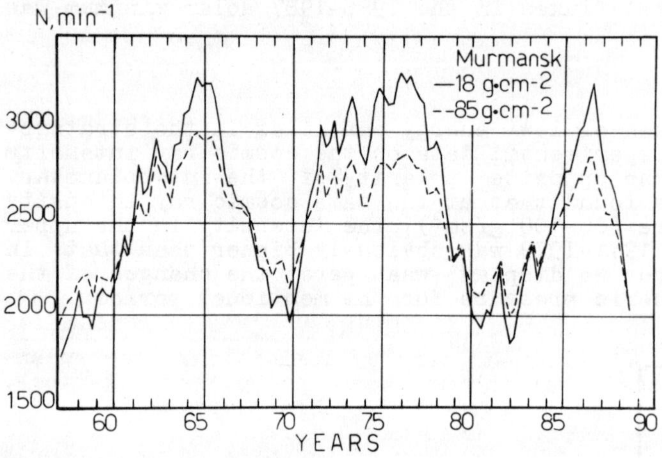

Fig. 2. Three month average values of counting rates for two atmospheric depths of 18 and 85 g/cm^2 in Murmansk in 1957-1989.

As an example of the stratospheric data the time-intensity variations in Murmansk at two atmospheric depths of 18 and 85 g/cm^2 are presented in Fig. 2. The intensity variations at Mirny are very close to that shown in Fig. 2. It is these data that are used below to derive the additional fluxes in the upper stratosphere. The long-term 11-year variation with the sharp peaks around the solar minima of 1965 and 1987 and flat around the 1976 solar minimum is clearly seen in the picture. Note also the deep intensity decrease during the recovery phase in July-September 1982. The cosmic ray intensity in the stratosphere correlates well with Wolf's sunspot numbers, the correlation coefficient being -0.819 ∓ 0.004 for the years 1957-1983.

CORRELATION BETWEEN THE INTENSITIES IN THE UPPER AND MIDDLE STRATOSPHERE

Fig. 3 shows the correlation between counting rates at the pressure 18 and 85 g/cm^2 at high latitude stations Murmansk (cutoff rigidity 0.60 GV) and Mirny (cutoff rigidity 0.02 GV). The 1959-1970 data are plotted as dark points, the 1971-1980 data as open circles and the 1981-1988 data as open triangles for both Murmansk and Mirny. A difference between these periods is the polarity of the interplanetary magnetic field. In the years 1971-1980 the north solar polar magnetic field pointed away from the Sun while during the previous and subsequent periods the magnetic field had the opposite direction. The points in Fig. 3 tend to cluster along the two separate lines so that the secondaries intensity at 18 g/cm^2 in 1971-1979 and 1985-1988 is higher than that in 1959-1970 for the same intensity level at 85 g/cm^2. Within the time periods between the reversals of the solar magnetic field the relation between the intensities at various atmospheric depths is quite definite and independent of the solar modulation level. A simple equation was found to express this relation:

$$N_{18} = a\cdot N_{85}^2 + b\cdot N_{85},$$

where N_{18} and N_{85} are the counting rates at 18 and 85 g/cm^2, respectively. Constants a and b calculated by a least square analysis for the years 1959-1968 are equal $(1.78\mp0.09)\cdot 10^{-4}$ and 0.56 ∓ 0.03, respectively. Using the equation written above and constants a and b we can calculate an expected intensity at 18 g/cm^2 for an arbitrary given N_{85}. Now we define the additional flux ΔN as a difference between the observed intensity N_{18} and the calculated one for the years 1959-1989.

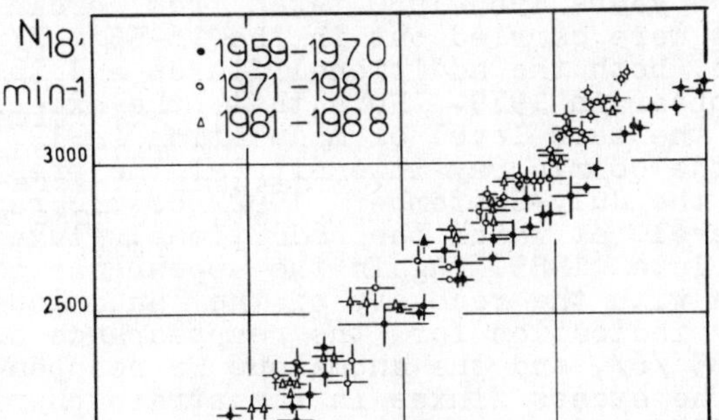

Fig. 3. Correlation of cosmic ray intensity in the upper stratosphere with the intensity in the middle stratosphere for the three successive half-periods of the 22-year solar magnetic cycle. The same symbols are used at both Murmansk and Mirny stations for each of the time intervals.

RESULTS AND DISCUSSION

The additional fluxes in Murmansk and Mirny versus time are plotted in Fig. 4. Their values in 1959-1970 are near zero by definition. The enhancement began in 1971 simultaneously at the northern and southern stations, and the maximum fluxes were observed in 1976-1977. The reduction was noticed in 1979 and in 1980 the additional fluxes disappeared completely. In 1985 around the next solar minimum they recovered and existed up to early 1989. The detectors used in this experiment are not capable of identifying the kind of particles of which the excess fluxes consist. Taking into account the measurements carried out in the stratosphere by Webber et al. /4/ we suggest that in our experiment protons with a small contamination of He nuclei are detected. The 85 g/cm^2 maximum range in air and the cutoffs for the launch sites set the upper and lower bounds on the energy of the additional flux particles. That gives the energy interval of 180-400 MeV for protons and of 45-400 MeV/n for He nuclei. As the ceiling altitude for balloon flights is about 4 g/cm^2 we have to set the lower bound on the value of energy for He nuclei at 70 MeV/n. Appropriate rigidity intervals are 0.6-0.9 GV for protons and 0.7-1.8 GV for He nuclei. For the 1976 solar minimum the integral energy spectrum of the additional flux, under the assumption that the additional particles are protons, was obtained from the residual range of particles. The value of the excess proton flux was 300 $m^{-2}s^{-1}sr^{-1}$ in the 100-400 MeV/n energy range which was in a good agreement with the results reported by Webber et al. /4/. A somewhat lower value of 230 $m^{-2}s^{-1}sr^{-1}$ was estimated for the additional fluxes in the 1987 sunspot minimum.

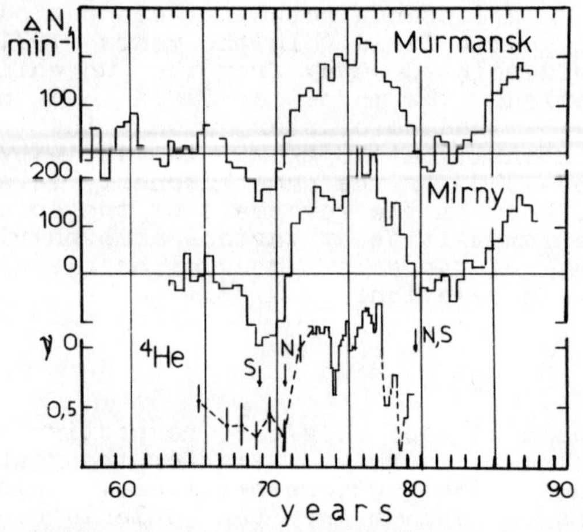

Fig. 4. The time dependence of the additional particle fluxes in the stratosphere at high latitude stations and of the energy spectrum of 16-55 MeV/n He nuclei according to /5/. The arrows indicate the times of the solar polar magnetic field reversals.

The time dependence of the stratospheric additional flux correlates with the anomalous He component at the orbit of the Earth. In Fig. 4 the spectral index γ of the anomalous He in the years 1965-1980 taken from Garcia-Munoz et al. /5/ is shown. Measurements were carried out in the 16-55 MeV/n energy range. As seen from this Figure, both the additional fluxes and the anomalous He appear in 1971 and disappear in 1979. Note that the excess fluxes appear and disappear at nearly the same level of modulation in 1971 and 1979 which we suggest is given by the cosmic ray intensity at the pressure 85 g/cm^2 (see Fig. 2). Due to the July-September 1982 cosmic ray intensity depression the modulation level at which the additional fluxes could recover was reached only in the late 1985. Again the appearance of the additional fluxes coincides in time with the recovery of the anomalous component. It is known that the first indication for the reappearance of anomalous component was seen in May 1985 /6/, and the anomalous He reappeared at the Earth's orbit in 1986 /7/. The excess fluxes in the stratosphere reappeared during the second half of 1985 and reached their maximum value in July-December 1987. In spite of their noticeable reduction in 1988 they were clearly present in January-April 1989.

The existence of the additional fluxes of low energy particles in 1972-1979 (and their absence in 1960-1970) seemed a natural consequence of the drift effect on cosmic ray propagation in the IMF. The modulation models in which drifts play an important role /8, 9/ predict the increase of proton and He fluxes in the heliosphere near the Earth during the time periods when the north solar magnetic field points away from the Sun, as it was in the years 1971-1980. However, the presence of the additional fluxes in 1985-1989 at the opposite polarity of the IMF shows that the relation between the additional fluxes and the IMF polarity is more complex. We must stress a distinct difference between the cosmic ray spectra at energies below 400 MeV/n in the years 1959-1970 and 1985-1989 when the polarity of the interplanetary magnetic field was the same. The cosmic ray spectra in 1959-1970 were obviously harder than that in 1971-1980 as well as in 1981-1989.

REFERENCES

1. T.N. Charakhchyan, Geomagnetizm i Aeronomia. 18, 385 (1978).
2. A.K. Svirzhevskaya, N.S. Svirzhevsky, Yu.I. Stozhkov, and T.N. Charakhchyan, Izv. Akad. Nauk SSSR, ser. fiz. 51, 1809 (1987).
3. M. Garcia-Munoz, G.M. Mason, and J.A. Simpson, in 15th Int. Cosmic Ray Conf. 3, 208 (1977).
4. W.R. Webber, J.C. Kish, and D.A. Schrier, in 18th Int. Cosmic Ray Conf. 3, 35 (1983).
5. M. Garcia-Munoz, G.M. Mason, and J.A. Simpson, in 17th Int. Cosmic Ray Conf. 3, 270 (1981).
6. N.L. Grigorov, N.V. Alekseev, V.V. Bobrovskaya et al., in 20th Int. Cosmic Ray Conf. 3, 264 (1987).
7. M. Garcia-Munoz, K.R. Pyle, and J.A. Simpson, in 20th Int. Cosmic Ray Conf. 3, 438 (1987).
8. J.R. Jokipii, and J.M. Davila, Astrophys. J. 248, 1156 (1981).
9. M.S. Potgieter, in 19th Int. Cosmic Ray Conf. 4, 425 (1985).

ION ACCELERATION TO COSMIC RAY ENERGIES

Martin A. Lee

Space Science Center, Institute for the Study of Earth, Oceans and Space, University of New Hampshire, Durham, NH 03824, U.S.A.

ABSTRACT

The acceleration and transport environment of the outer heliosphere is first described schematically. Acceleration occurs where the divergence of the solar wind flow is negative, that is at shocks, and where second-order Fermi acceleration is possible in the solar wind turbulence. Acceleration at the solar wind termination shock is presented by reviewing the spherically-symmetric calculation of Webb et al. /43/. Reacceleration of galactic cosmic rays at the termination shock is not expected to be important in modifying the cosmic ray spectrum, but acceleration of ions injected at the shock up to energies ≤ 300 MeV/charge is expected to occur and to create the anomalous cosmic ray component. Acceleration of energetic particles by solar wind turbulence is expected to play almost no role in the outer heliosphere. The one exception is the energization of interstellar pickup ions beyond the threshold for acceleration at the quasi-perpendicular termination shock.

INTRODUCTION

Since the early to mid 1970's the acceleration of energetic particles has been a major focus of heliospheric studies. Prior to that time the solar wind was viewed as a relatively passive medium with regard to energetic particles, merely providing an outward flow with magnetic fluctuations to disperse the ejections of solar flare particles on the Sun and inhibit the access of cosmic rays from interstellar space. Indeed the dominant process of energy exchange perceived between these particles and the solar wind was the distinctly enervating process of adiabatic deceleration of these solar and galactic cosmic rays in the diverging solar wind.

However, in the 1970's studies of newly discovered energetic particle populations began in earnest and they generally required heliospheric origins. Fisk /1/ developed the forerunner of the current theory of diffusive shock acceleration /2/ to understand the origin of energetic storm particles (ESP) at traveling interplanetary shocks. Sarris and Van Allen /3/ argued that ESP enhancements could be produced by single energizing encounters of the ions with these shocks in a process known now as shock drift acceleration /4,5/.

In 1973, 1974 the anomalous cosmic ray component was discovered by Garcia-Munoz et al. /6/, Hovestadt et al. /7/, and McDonald et al. /8/. It soon became clear that these ions could not be of galactic origin, and once Fisk et al. /9/ realized that their composition immediately followed from an origin as interstellar pickup ions, a heliospheric acceleration site had to be found. Fisk /10/ and Klecker /ll/ initially proposed acceleration by transit-time damping of magnetosonic waves in the outer heliosphere. The currently accepted mechanism and site of acceleration is diffusive shock acceleration at the solar wind termination shock /12,13,14,15/.

The corotating ion events were first discovered in the 1960's but were thought to be of solar origin. However, Pioneer 10 observations in 1975, 1976 /16,17/ revealed that their intensity increased with heliocentric radial distance r, and once again a heliospheric acceleration site had to be found. When Pioneer 10 reached r ~ 4-6 AU it discovered corotating interaction regions (CIR) /18/ bounded by a forward and a reverse shock, and observed that the corotating ion events peaked at the shocks. A shock origin for these events was soon developed by Palmer and Gosling /19/, Fisk and Lee /20/, and Pesses et al. /21/.

Further populations of energetic particles including the "diffuse" and "reflected" ions at Earth's bow shock /22,23/ and energized pickup ions at comets /24,25/ (although these ions with speeds ≤ 1000 km/s should hardly be dignified as energetic particles!) have been discovered and their heliospheric origins studied. The inner heliosphere has in fact become a veritable laboratory for energetic ion acceleration. Electrons fare less well: apart from streams of keV electrons from the region of Earth's bow shock tangent to the interplanetary magnetic field /26,27/ and efficient electron acceleration in Jupiter's magnetosphere (not truly a heliospheric source), electrons appear to be passive spectators in the acceleration processes.

We know less about the outer heliosphere. However, as Voyager 1,2 and Pioneer 10,11 penetrate ever further the far reaches of the heliosphere, it is appropriate to ask: Are ions accelerated to cosmic ray energies in the outer heliosphere?

ACCELERATION AND TRANSPORT IN THE OUTER HELIOSPHERE

The theoretical basis for the acceleration and transport of energetic particles in cosmic plasmas is generally the transport equation due originally to Parker /28/ but refined and extended in important ways by Gleeson and Axford /29/, Jokipii et al. /30/, and Luhmann /31/

$$\frac{\partial f}{\partial t} + (\mathbf{V} + \mathbf{V}_D) \cdot \nabla f - \nabla \cdot \mathbf{K} \cdot \nabla f - \frac{1}{3} \nabla \cdot \mathbf{V}\, p \frac{\partial f}{\partial p} - p^{-2} \frac{\partial}{\partial p}\left(p^2 D \frac{\partial f}{\partial p}\right) = Q \qquad (1)$$

Here f(p,**x**,t) is the particle omnidirectional distribution function, p is the magnitude of particle momentum, **V** is the background plasma bulk velocity (more precisely the average velocity of the scattering irregularities), **K** is the symmetric part of the diffusion tensor due to scattering on magnetic field fluctuations, D is the momentum diffusion coefficient (generally proportional to the intensity of scattering irregularities /32/ or the square of the velocity shear /33/), and Q is a source term. Also $\mathbf{V}_D = pvc(3e)^{-1} \nabla \times (B^{-2}\mathbf{B})$ is the drift velocity arising from a combination of gradient, curvature and magnetization drift, where v is particle speed, e is particle charge, and **B** is the average magnetic field.

Equation (1) is based on the assumptions that $V/v \ll 1$ and that particle anisotropy is at most of order V/v. It clearly does not describe anisotropic particle populations or processes which generate large anisotropies ("reflected" ions or "shock drift" acceleration, respectively), nor does it describe reliably, for example, the transport of cometary pickup ions in the frame of the comet for which $V/v \sim 1$. Nevertheless, for most energetic particle ($V \ll v$) applications in the outer heliosphere, where temporal and spatial scales are sufficiently large that irregularities tend to isotropize the velocity-space distributions, equation (1) is valid. It has a 25-year history of successful applications in the inner heliosphere, and has been used to study energetic particle transport in more distant astrophysical sites.

Acceleration arises from the last two terms on the left-hand side of equation (1): If $\nabla \cdot \mathbf{V} < 0$ the first term describes energization due to coupling of the particles to fluid compression. The second term describes stochastic or second-order Fermi acceleration.

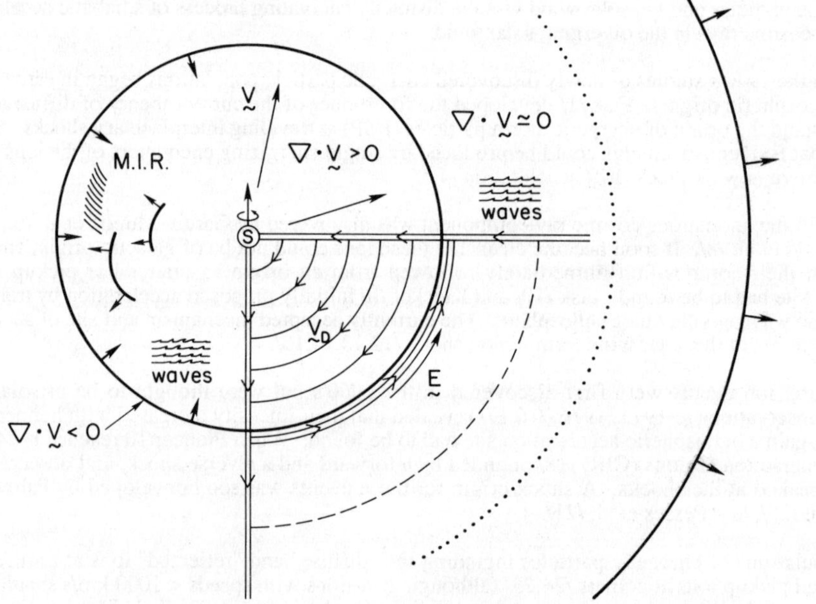

Fig. 1. The acceleration/transport environment for energetic particles in the outer heliosphere. A meridional slice through the heliosphere including the Sun's axis of rotation is shown with the Sun at the center. Thick solid lines denote shocks, the dotted line denotes the heliopause, and thin solid and dashed lines denote particle drift streamlines for an inward-oriented magnetic field in the upper hemisphere. Waves and merged interaction regions (MIR) are included schematically. The vector **E** gives the direction of the motional electric field near the solar wind termination shock. See text for a detailed description.

Figure 1 shows schematically the acceleration/transport environment in the outer heliosphere. It shows a meridional slice through the approximately spherical heliosphere. The Sun is at the center and the solar wind emanates radially. The dark lines represent shocks and the arrows denote the shock normal directed into the upstream flow. The circle is the solar wind termination shock, expected at r ~ 60-70 AU /15/. The arcs to the left of the Sun represent large traveling interplanetary shocks which may originate from a coalescence of smaller shocks and occasionally penetrate the outer heliosphere /34/. The large arc to the right shows the possible forward shock in the interstellar medium due to the Sun's relative motion of 20 km/s /35/ through the "very local interstellar medium" (VLISM) /36/. However, the most recent estimates of the interstellar ionized mass density, magnetic field strength, and temperature indicate that the relative speed is sub-fast-magnetosonic so that this shock should not exist. The dotted line represents the heliopause separating solar and interstellar ionized gases. Also shown explicitly is the sign of $\nabla \cdot \mathbf{V}$, which is positive in the solar wind, resulting in adiabatic deceleration of the energetic particles, approximately zero beyond the termination shock, and only negative at the shocks. The wiggly lines indicate waves in the solar wind, particularly downstream of the termination shock, which contribute to D. The "merged interaction regions" (MIR) in the outer heliosphere, studied intensively by Burlaga and colleagues /37/, are characterized by large magnetic field fluctuations, which also contribute to D.

The light solid lines in Quadrant IV of Figure 1 indicate the streamlines of $\mathbf{V}_D$, whose role in acceleration we return to in the next section. In the outer solar wind inside the termination shock the average magnetic field approximately satisfies $\mathbf{B} \propto r^{-1} \sin\theta\, \hat{e}_\phi$ /38/, which yields $\mathbf{V}_D \propto (\sin\theta)^{-1}\, \hat{e}_\theta$. Beyond the termination shock in a spherical geometry $\mathbf{B} \propto r \sin\theta\, \hat{e}_\phi$ /39/, which yields $\mathbf{V}_D = 0$. Eventually the spherical symmetry is compromised by the deflection of the flow due either to the Sun's motion through the VLISM or to the relative increase in magnetic pressure at mid-latitudes. Accordingly a small drift, denoted by the dashed line, occurs in the subsonic flow. The double light lines indicate large (singular in this idealized geometry) drift along the solar poles and the equatorial current sheet. In this schematic diagram we neglect the warp of the current sheet arising in the more general case in which the axes of magnetic symmetry and solar rotation are not aligned /40, 41/. The arrows give the direction of $\mathbf{V}_D$ for the case in which $e > 0$ and the magnetic field at the Sun is oriented inwards (outwards) in the upper (lower) hemisphere. Finally, the direction of the electric field, $\mathbf{E} = -c^{-1}\, \mathbf{V} \times \mathbf{B}$, at the termination shock is indicated.

Acceleration due to compression ($\nabla \cdot \mathbf{V} < 0$), predominantly at shocks, must compete with deceleration due to expansion ($\nabla \cdot \mathbf{V} > 0$). For particles which sample the entire heliosphere out to the subsonic flow beyond the termination shock, such as high energy cosmic rays with energies $\geq$ 10 GeV/nucleon, deceleration dominates acceleration since $\int_{Vol} d^3\mathbf{x}\, \nabla \cdot \mathbf{V} = \int_S d\mathbf{A} \cdot \mathbf{V} > 0$, where "Vol" is the volume of the heliosphere within a surface S lying outside the termination shock. This statement neglects the weighting of acceleration and deceleration due to **K** which controls the lifetime of particles in various regions but it is essentially correct. In order to have efficient compressive (shock) acceleration the particles must be localized near the shocks, which effectively limits their energies to at most a few GeV/nucleon.

Shocks in the outer heliosphere which are potential sites of heliospheric acceleration are the forward shock in the interstellar medium, traveling shocks formed by a coalescence of flare-associated shocks, the solar wind termination shock (all three depicted schematically in Figure 1), and a combination of forward and reverse shocks produced by interactions between solar wind streams. These latter shocks are generally limited in latitudinal extent and have criss-crossed a few times in the outer heliosphere, leaving behind the merged interaction regions mentioned previously and very weak shocks if they exist at all. As noted earlier the interstellar forward shock probably does not exist, and is in any case weak and spatially removed from the heliosphere.

Traveling shocks in the outer heliosphere capable of effective acceleration are not common. Figure 2 reproduces the figure from Pyle et al. /34/ which shows Pioneer 10 data near solar maximum in the outer heliosphere. It is clear that the large spatial and temporal scales of the outer heliosphere contribute to a simplified structure in energetic particles. Individual solar flare particle bursts coalesce to form single enhancements with durations of 1-4 months arising from separate periods of enhanced solar activity. These are seen as periods A,B and C in the 11-20 MeV proton profile with an additional small enhancement between periods B and C. Incidentally the approximately 1/2-year period of these recurring enhancements is a clear signature of the 154-day periodicity in solar activity discovered in the emission of solar gamma rays by Rieger et al. /42/. During this 2-year interval only periods A and C have an embedded shock, and the top panel shows that only the second shock is a strong shock (based on a characteristic fast magnetosonic speed in the outer heliosphere of 30 km/s). These two shocks are responsible for the larger enhancements of energetic particles in periods A and C which peak at the shocks. However, it is clear that even at these distances and correspondingly large acceleration timescales the enhancements do not extend in energy beyond ~ 100 MeV/nucleon and, as shown in the last panel, do not contribute to the galactic cosmic ray spectrum.

The termination shock, on the other hand, is a strong shock (its speed relative to the upstream flow is V) and it is always present. We turn to acceleration at the termination shock in the next section.

Finally, we list the seed particles available for shock acceleration in the outer heliosphere: adiabatically cooled solar flare particles, particles previously accelerated at traveling shocks (ESP events) or CIR shocks, interstellar pickup ions, galactic cosmic rays, and the solar wind plasma. Although the solar wind dominates the number density of available seed particles, injection threshold speeds may favor the other seed particles since shocks in the outer heliosphere tend to be quasi-perpendicular, (i.e. pickup ions are presumably accelerated at the termination shock to form the anomalous cosmic ray component).

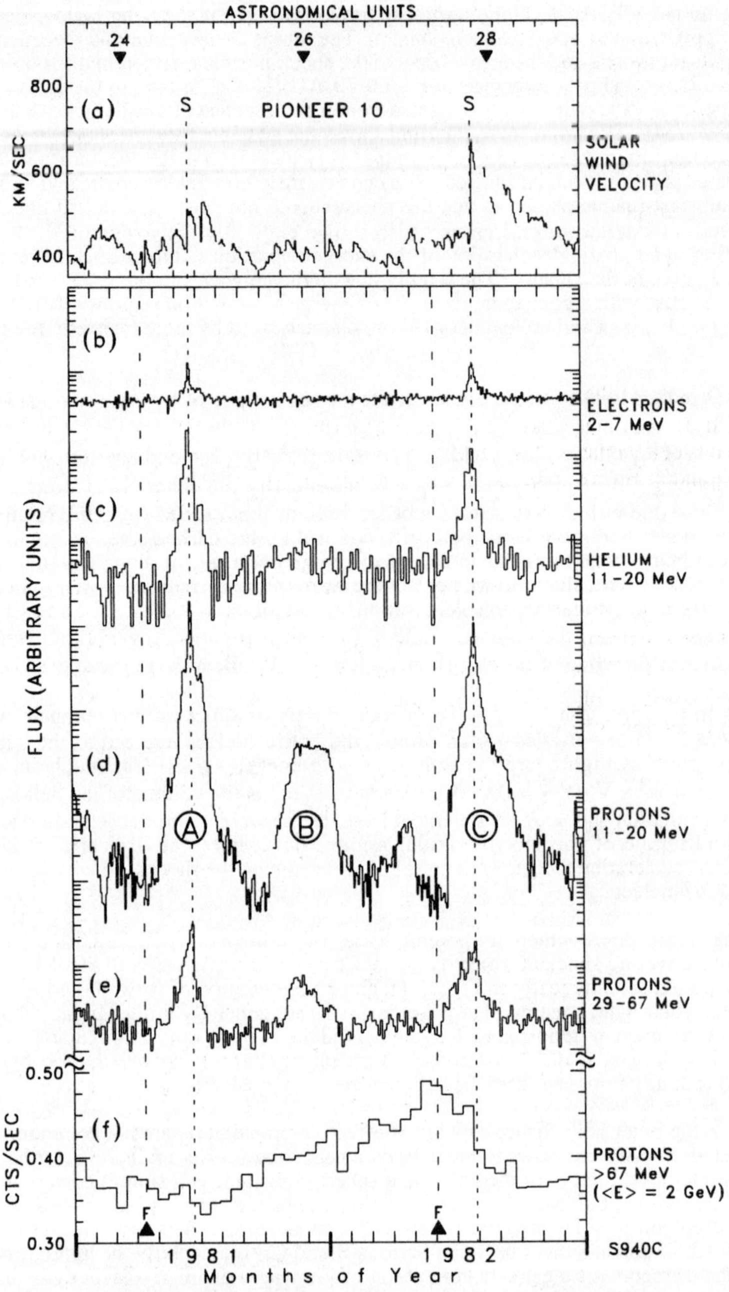

Fig. 2. An overview of a 2-year period of Pioneer 10 data near 26 AU (Figure 1 from Pyle et al. /34/). Energetic particle enhancements, particularly ions, in the energy range 10-70 MeV occur following periods of solar activity, the largest of which are denoted by solid triangles. The enhancements are particularly large and peaked at the passage of traveling interplanetary shocks.

ACCELERATION AT THE TERMINATION SHOCK: WEBB, FORMAN AND AXFORD

Webb et al. /43/ considered particle acceleration at the solar wind termination shock for an idealized configuration with spherical symmetry in which drift transport and second-order Fermi acceleration are neglected. The termination shock was placed at $r = R$ and $V\hat{e}_r$ and K_{rr} were chosen to be $V(r < R) = V_1$, $V(r > R) = V_2(r/R)^{-2}$, $K_{rr}(r<R) = K_1(r/R)$, and $K_{rr}(r>R) = K_2(r/R)$. Note that **K** is taken to be independent of p. Under these assumptions stationary solutions to equation (1) were found for various choices of $f_o(p)$ and $Q(r,p)$ subject to the boundary conditions (1) $f(r \rightarrow \infty, p) \rightarrow f_o(p)$, (2) $f(r \rightarrow 0, p)$ is finite, and (3) continuity of $f(r,p)$ and the differential flux $S(r,p) = - K_{rr}\, \partial f/\partial r - 1/3Vp\, \partial f/\partial p$ at $r = R$. The solutions are derived by Fourier transform in variable ln p, solution of the resulting equidimensional system in r, and Fourier inversion of the resulting expressions.

Although this configuration is highly idealized it includes most essential features of acceleration at the termination shock and yet is still amenable to analytical techniques. We shall return to some of its limitations at the end of this section. Numerical solutions to equation (1) describing acceleration and transport in the outer heliosphere including drift transport and more general choices of **K** (r,θ,p) are presented by Jokipii /15/ and Kota /44/ in this issue. The original work on acceleration of particles at the termination shock was by Jokipii /45/.

If $f_o(p)$ or $Q(r,p) \propto \delta(p-p_o)$, then the fundamental spectrum produced for $p > p_o$ is a power-law spectrum, $f(r,p) \propto (p/p_o)^{-\mu}$, with

$$\mu = \frac{3}{1-x}\left[\frac{1}{2} + \frac{x}{\eta_1(1-x)} + \frac{x\beta}{1-\beta} + \left\{\left[\frac{x}{\eta_1\,(1-x)} + \frac{1}{2}\right]^2 + \frac{2x\beta}{\eta_1\,(1-x)(1-\beta)}\right\}^{1/2}\right] \tag{2}$$

where $x^{-1} = r_c = V_1/V_2$ is the shock compression ratio, $\eta_1 = V_1R/K_1$, $\eta_2 = V_2R/K_2$, and $\beta = \exp(-1/2\,\eta_2)$. If either $\eta_1 \to 0$ or $\eta_2 \to 0$, $\mu \to \infty$, which implies no acceleration. In this limit particle lifetime either upstream or downstream of the shock vanishes; particles are then not coupled to the compression, and therefore not accelerated. In the limit that η_1 and/or η_2 are large, we obtain with the leading correction terms

$$\mu \sim \frac{3}{1-x}\left[1 + \frac{x\beta}{1-\beta} + \frac{2x}{\eta_1(1-x)}\right] \tag{3}$$

which shows that the spectrum softens with decreasing η_1 and η_2. The hardest spectrum, $\mu \to \mu_o = 3\,(1 - x)^{-1}$, obtains in the limit $\eta_1, \eta_2 \to \infty$, which corresponds to either $R \to \infty$ or $K_1, K_2 \to 0$. In this limit the standard spectral index for stationary planar shock acceleration, μ_o, is obtained /46,47,48/. At momenta $p < p_o$ adiabatic deceleration in the diverging solar wind dominates acceleration at the shock.

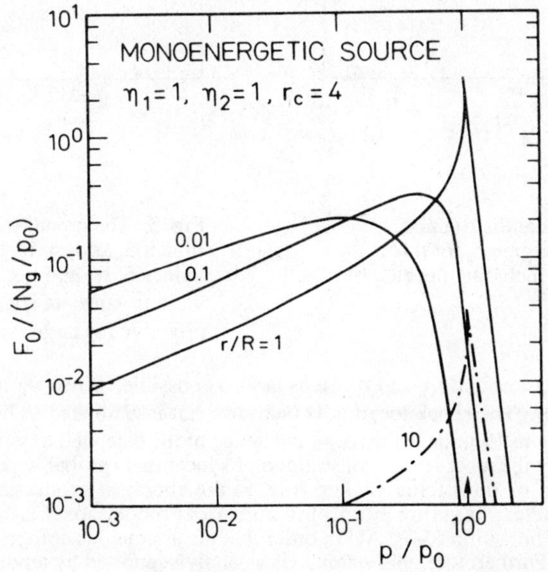

Fig. 3. The omnidirectional distribution function, $f(r,p)$ [$\equiv F_o(r,p)$] for the case $f\,(r \to \infty, p) = N_g\delta(p-p_o)$ and the parameters indicated. The adiabatic deceleration for $p < p_o$ and the soft power law spectrum for $p > p_o$ are apparent.

Figures 3-5 display $f(r,p)$ for the case in which $Q = 0$ and $f_o(p) = N_g\,\delta(p-p_o)$ [in the notation of Webb et al. /43/ $F_o(r,p) = f(r,p)$]. Figure 3 shows the momentum spectra for a strong shock at various heliocentric radial distances; the power law spectra for $p > p_o$ and the dominance of deceleration in the inner heliosphere are evident. Figure 4 shows the spectra at the shock for three compression ratios; $r_c = 1$ corresponds to no shock at all, and therefore no acceleration. Figure 5 shows that a larger value of η_1 suppresses the particle intensity in the inner heliosphere and yields more efficient acceleration. Webb et al. /43/ have also derived the solution in the case $f_o = 0$ and $Q \propto \delta(r-r_o)\,\delta(p-p_o)$, but the essential features of the acceleration are apparent in the case portrayed in Figures 3-5.

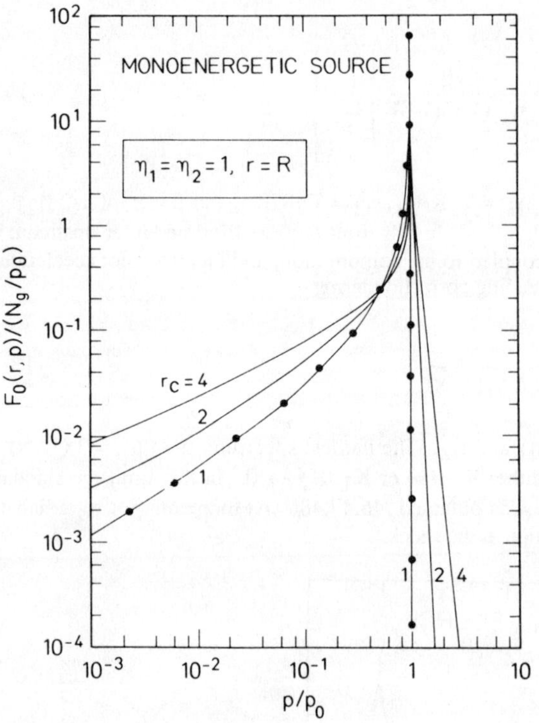

Fig. 4. The omnidirectional distribution function as described in Figure 3, for three shock compressionratios including the case of no shock: $r_c = 1$ (dots).

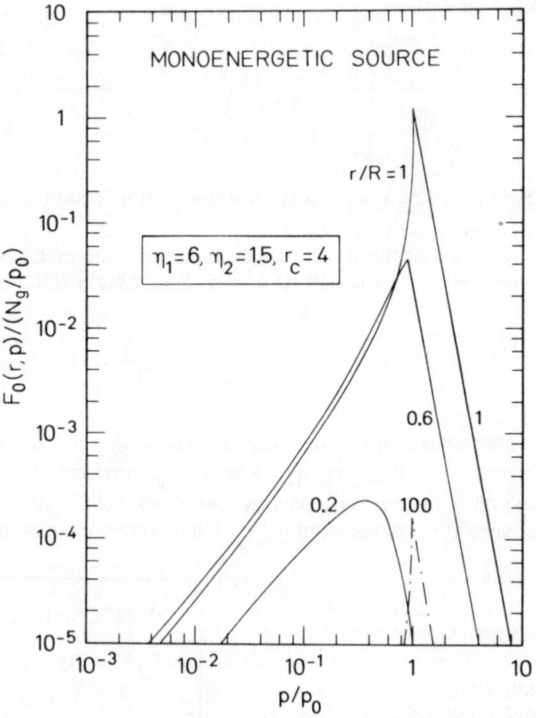

Fig. 5. The omnidirectional distribution function, as described in Figure 3, for larger values of η_1 and η_2, as indicated, and for various values of r/R. Acceleration is more effective for larger values of η_1 and η_2.

Clearly efficient acceleration requires that η_1 and η_2 be as large as possible. Since V_1 and V_2 are essentially fixed, we concentrate on R and K_1, K_2. The shock location is unknown but is estimated to lie in the range 50 AU < R < 150 AU /49/. Indeed determining its location is the major challenge of the extended missions of Voyagers 1 and 2 and Pioneers 10 and 11. Interestingly the most severe constraint on its location is probably provided by calculations such as this one for the acceleration of interstellar pickup ions at the shock to produce the anomalous cosmic ray component. Jokipii /15/ estimates, based on fitting his numerical model spectra to observations in the inner heliosphere, that the shock must lie within 65-70 AU in order that the anomalous component pressure not exceed the ram pressure of the solar wind. Furthermore, this estimate is strongly supported by a poll of eminent scientists! The results of the poll taken at the New Hampshire Symposium on the Outer Heliosphere are shown in Figure 6. Clearly 61 AU is an excellent estimate!

A characteristic scattering mean free path at cosmic ray energies (~ 1 GeV/nucleon) in the inner heliosphere is 1 AU /50/ which yields $K_{rr} \sim 10^{23}$ cm^2 s^{-1}. With $V_1 = 400$ km s^{-1} and R = 60 AU we then have $\eta_1 \sim 0.4$. At about the same energy Morfill et al. /51/ predict $K_{rr} \simeq 2 \times 10^{22}$ cm^2 s^{-1} within the inner heliosphere based on scattering by solar wind Alfvén waves. Theoretical estimates of K_{rr} in the outer heliosphere vary, and depend crucially on the role of perpendicular diffusion in the outer heliosphere. Actually the observed gradient of ~ 2%/AU, independent of heliocentric radial distance, indicates $K_{rr} \simeq 3 \times 10^{22}$ cm^2 s^{-1}. The diffusion coefficient in the heliosphere beyond Voyager and Pioneer could of course differ from this value. Drury and Falle /52/, for example, have investigated instabilities in the outer heliosphere driven by the cosmic ray gradient itself, which could increase (decrease) K if perpendicular (parallel) diffusion dominates radial transport. However, independent of all these detailed considerations, the largest possible values of η_1 and η_2 for the bulk of cosmic rays are in the range 0.5 - 1; otherwise the extrapolated pressure of galactic cosmic rays in interstellar space is unacceptably large /49/. We take $\eta_1 = \eta_2 = 1$ as the largest possible values at cosmic ray energies.

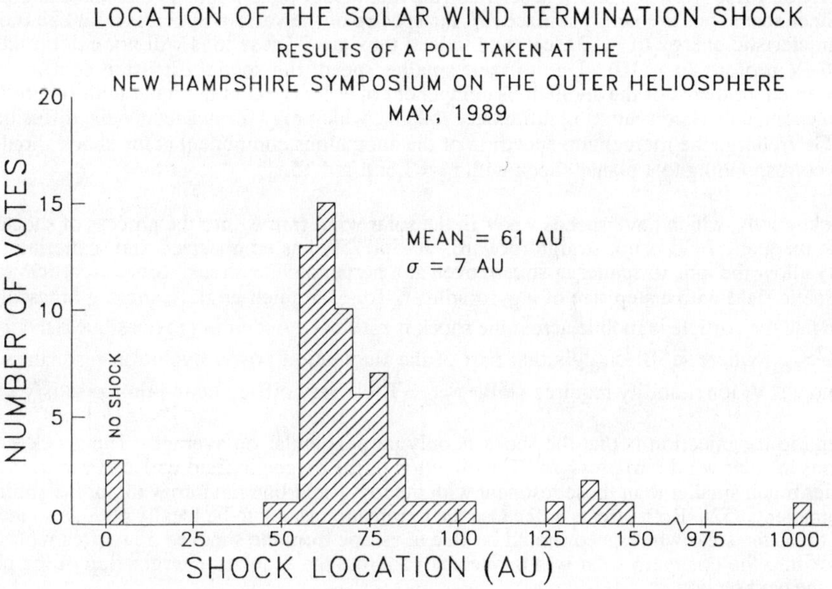

Fig. 6. Results of a poll on the location of the solar wind termination shock taken with eminent informed specialists on the outer heliosphere!

The compression ratio r_c of the termination shock is often taken to be 4, consistent with an estimated Mach number of 16 based on a solar wind speed of 400 km s^{-1} and an Alfvén speed of 25 km s^{-1}. However, interstellar pickup hydrogen should substantially decrease the Mach number in the outer heliosphere. Reasonable parameters at 60 AU in the solar wind are a solar wind proton density of 2×10^{-3} cm^{-3}, a magnetic field strength of 9×10^{-7} G, a pickup hydrogren number density of 2×10^{-4} cm^{-3} /35,53/, and a pickup hydrogen pressure of 7.6×10^{-14} dynes cm^{-2}/54/. Neglecting solar wind plasma pressure we then predict a compression ratio of $r_c = 3.6$ for the perpendicular termination shock.

These values yield a spectral index $\mu = 11.3$, which is large in comparison with the galactic cosmic ray spectral index $\mu_o \simeq 4.7$ at high energies.

Accordingly, acceleration of particles with energies of ~ 1 GeV/nucleon at the solar wind termination shock is not particularly effective. If the solution satisfying $f(r \to \infty, p) = f_o(p) \propto \delta(p - p_o)$ is convoluted with a galactic cosmic ray spectrum of the form $f_o(p) \propto p^{-\mu_o}$, then the resulting spectrum at energies ≥ 1 GeV/nucleon will also be approximately of the form $f(r,p) \propto p^{-\mu_o}$, with the result that the shock acceleration contributes only to a small overall enhancement in cosmic ray intensity.

In fact, acceleration of ions to GeV/nucleon energies is less efficient than described by the spherically-symmetric solutions of Webb et al. /43/. The spatial diffusion coefficient, K_{rr}, is inferred to increase with p, resulting in more rapid escape of high energy particles from the vicinity of the shock. With $K_{rr}(r,p) \propto p^{\alpha}$, balancing the third (diffusion) and fourth ($\propto \nabla \cdot V$) terms in equation (1), which dominate at high energy, yields for monoenergetic injection an exponential spectrum of the form, $f \sim \exp[-a(r)p^{\alpha}]$ /20,55/, where a(r) is a function of r to be determined.

Particle drift will also limit acceleration. As shown schematically in Figure 1 particles accelerated at the termination shock also drift in latitude along the shock front. If diffusive transport normal to the ambient magnetic field is negligible, Jokipii /56/ has shown within the theory of diffusive shock acceleration that the energy gain ΔE of a particle at the shock is proportional to the distance Δd the particle has drifted. For a quasi-perpendicular shock such as the termination shock, that relation is $\Delta E \cong e\mathbf{E} \cdot \Delta \mathbf{d}$, where **E** is the motional electric field shown schematically in Figure 1. Therefore, the energy gain of the particle arises primarily from drift through the heliospheric potential. An additional energy gain due to compression is only comparable for those few particles whose diffusive transport may compensate their drift motion. According to Figure 1 energy gains at the termination shock are effectively limited to $\Delta E = e\int_0^{\pi/2} d\theta \, Rc^{-1} \, VB(\theta{=}\pi/2) \sin\theta = e\,R\,c^{-1}\,VB(\theta{=}\pi/2)$. Using the same parameters quoted above we find $\Delta E \cong$ 0.3 GeV/charge. After gaining energy at the shock particles drift in along the current sheet and out along the heliospheric poles as shown in Figure 1 (or the reverse for the opposite orientation of **B**) where adiabatic deceleration depletes previous energy gains. Drifts circulate particles throughout the heliosphere and reduce the efficiency of shock acceleration.

Therefore, although the solar wind termination shock is not effective at reaccelerating galactic cosmic rays or accelerating any seed population to energies of ~ 1 GeV/nucleon, it can be efficient at energies ≤ 0.3 GeV/charge. In this issue Jokipii /15/ describes how the shock accelerates the interstellar pickup ions to create the anomalous cosmic ray component. Since these ions are singly ionized /9/, the spectra of the various species should be ordered by total energy with a characteristic energy of ~0.3 GeV, and indeed they are. These ions will not exhibit the soft spectra characteristic of GeV protons ($\mu \sim 10$). Due to their smaller speed, the spatial diffusion coefficient is smaller, consistent with observed gradients of the anomalous component of ~ 10%/AU in the inner heliosphere/57/. Near the shock the ions will excite enhanced scattering turbulence /52,58/, which may further reduce the diffusion coefficient. At energies ≤ 0.3 GeV/charge the momentum spectrum of the anomalous component at the shock should be a power law with an index corresponding to a planar shock with $r_c = 3.6$: $\mu \cong 4.15$.

Injection of the pickup ions, which have speeds $v \leq V$ in the solar wind frame, into the process of shock acceleration against the oblique magnetic field is not straightforward. Jokipii /59/ has emphasized that scattering normal to the magnetic field may allow the ions to scatter upstream even at a perpendicular shock. Since a particle random walks normal to the magnetic field with a step size of a gyroradius r_g for each pitch-angle scattering timescale τ along the field, the condition that the particle is mobile across the shock is $r_g/V > \tau$. According to quasilinear theory /60/ $\tau \cong \Omega^{-1} B^2/\langle|\delta \mathbf{B}|^2\rangle_{res}$, where $\langle|\delta \mathbf{B}|^2\rangle_{res}$ is that part of the fluctuation power cyclotron resonant with the ions. Since $\Omega r_g = v_\perp$ and $v \leq V$, ion mobility requires $\langle|\delta \mathbf{B}|^2\rangle_{res} > B^2$, a very difficult condition to satisfy.

A more likely scenario for injection is that the shock is only perpendicular on average. The shock surface will be warped by variations in solar wind ram pressure. The direction of the magnetic field will also vary due to fluctuation power at frequencies much smaller than those resonant with the ions, field line random walk, or instabilities driven by energetic particle gradients /52/. Both effects will cause the termination shock to be locally oblique so that pickup ions with speeds a few times the solar wind speed should be able to escape from the shocked downstream plasma along the magnetic field back into the upstream solar wind. We shall address the required energization of the pickup ions to accomplish this in the next section.

ACCELERATION BY TURBULENCE IN THE OUTER HELIOSPHERE

The heliospheric magnetic field and plasma are turbulent on virtually all spatial scales down to the thermal proton gyroradius. Some of the turbulent power originates at the Sun inside the Alfvén critical point and enters the solar wind as a spectrum of outward propagating waves. Turbulent processes soon modify the spectral shape and excite inward propagating waves /61,62/. However, additional turbulent power originates in the solar wind due to shock dissipation, instabilities driven by energetic particles /52,58/, isotropization of interstellar pickup ions /60/, and stream interactions /37/.

Near 1 AU the total power in magnetic field fluctuations is observed to be nearly saturated with $\langle|\delta \mathbf{B}|^2\rangle \cong 1/2\, B^2$. Beyond 1 AU, $B \propto r^{-1}$ and, in the absence of additional excitation, $\langle|\delta \mathbf{B}|^2\rangle \propto r^{-3}$ according to WKB theory /63/. Thus, even with additional wave excitation, the total wave power is limited to a fraction of B^2 in the outer heliosphere. With $B \cong 7 \times 10^{-5}$ G at 1 AU, $B^2/8\pi \cong 0.07$ eV cm^{-3} at 30 AU, for example, which implies that the total available power in magnetic and velocity fluctuations is about 0.03 eV cm^{-3}.

Energetic particle acceleration by this turbulence is not efficient. The power levels are small compared with energy densities in cosmic rays (~ 1eV cm^{-3} /49/). Although the wave power levels at 30 AU are larger than those inferred recently for anomalous component hydrogen (~10^{-3} eV cm^{-3} /64/), they become comparable at larger distances. Also the energetic particles can only interact readily with resonant turbulence, which limits the frequency range of power which can in principle be transferred to the particles.

The mechanisms for acceleration are not efficient. Transit-time damping of magnetosonic waves, as originally proposed by Fisk /10/, hinges on the compressive component of the turbulence, which is not large: observed power in the compressive component is usually smaller by an order of magnitude. Furthermore, transit-time damping requires that the Landau resonance condition be satisfied: $k_{||}/k \cong V_A/v_{||}$, where $\mathbf{k}$ is the wavevector and $||$ denotes components parallel to $\mathbf{B}$. Effective energetic particle acceleration therefore requires waves nearly normal to $\mathbf{B}$. Although the WKB evolution of wavevectors favors the radial direction /65/, which is nearly normal to $\mathbf{B}$ in the outer heliosphere, it appears that acceleration by transit-time damping is limited by the rate at which turbulent processes can replenish waves in a very small portion of wavevector space.

Particles can also be accelerated by second-order Fermi acceleration in the turbulent wave field. But again the acceleration rate is small: as stated in the last section, the pitch-angle scattering timescale is $\tau^{-1} \cong \Omega\langle|\delta \mathbf{B}|^2\rangle_{res} B^{-2}$. For saturated wave power with a k-dependence of k^{-2} for $k > 2\pi/L$, where L is the correlation length, the timescale becomes $\tau^{-1} = 2\pi v/L$. The acceleration timescale, τ_a, is larger by a factor $v^2/V_A{}^2$ so that $\tau_a \cong vL(2\pi V_A{}^2)^{-1}$. Even giving the waves the "benefit of the doubt" by taking a small correlation length of $L = r/100$, with $V_A = 0.1$ V we find that particles can only be accelerated to a speed of 6V in the transit time of the solar wind to radial distance r. Thus, second-order Fermi acceleration does not appear to be effective for energetic ions. It can be effective for interstellar pickup ions to compensate their adiabatic deceleration and accelerate them to speeds of several times V. A linear calculation of this process has been performed by Isenberg /66/ based on a fixed wave power. The additional acceleration is probably sufficient to provide the seed population for injection and acceleration at the quasi-perpendicular solar wind termination shock to produce the anomalous cosmic ray component.

However, at 30 AU pickup hydrogen is expected to have a number density about 5% that of solar wind hydrogen. The energy density of the pickup hydrogen is then about 0.15 eV cm^{-3} /54/, which is larger than the total power in the ambient field, let alone the fluctuations. Thus a self-consistent theory for the second-order Fermi acceleration of pickup hydrogen is required, which accounts for wave damping. A formalism based on quasilinear theory has been presented by Lee et al. /67/, and application of the formalism to interstellar pickup ions is underway. Wave damping should not affect the interstellar pickup minor ions (e.g. He, Ne, O, N) which resonate with waves at much lower frequencies, but it should make acceleration of pickup hydrogen much less efficient. This distinction between hydrogen and the minor ions would mean a reduced relative injection rate for hydrogen at the termination shock and could account for an inferred underabundance of hydrogen in the anomalous cosmic ray component /15, 68/.

CONCLUSIONS

Although the heliosphere is a veritable laboratory for energetic particle acceleration processes, and the outer heliosphere is no exception, the spatial extent of the heliosphere generally limits energy gains to substantially less than a GeV/nucleon. Acceleration occurs at traveling shocks in the outer heliosphere, but these are relatively weak and uncommon. Nevertheless, they create large enhancements in $\leq$ 100 MeV/nucleon ions. The most impressive site of acceleration in the outer heliosphere is undoubtedly the solar wind termination shock. It is steady and expected to be strong. Although, as we have shown, its finite size does limit the extent to which galactic cosmic rays can be reaccelerated, ions which can be injected at the quasi-perpendicular shock can experience energy gains up to ~ 300 MeV/charge. Interstellar pickup ions are apparently injected and accelerated at the termination shock to create the anomalous cosmic ray component as described by Jokipii /15/.

The solar wind turbulence in the outer heliosphere can also accelerate ions by transit-time damping or second-order Fermi acceleration, but these processes are far less efficient than shock acceleration and far less energy is available. Second-order Fermi acceleration, however, is able to accelerate interstellar pickup ions in the outer heliosphere to speeds several times the solar wind speed, which is apparently sufficient to overcome the injection energy threshold at the termination shock.

Acknowledgements

The author wishes first and foremost to thank Professor S. Grzedzielski and the entire local organizing committee for a very rewarding Colloquium and a wonderful week in Poland. Special thanks are due to Hanna and Tomasz Zarnowiecka, who helped him with last-minute changes in travel plans. The author also wishes to acknowledge useful discussions with J.R. Jokipii during the weekend in Warsaw prior to the Colloquium. This work was supported, in part, by NASA Space Physics Theory Program Grant NAGW-76, NASA Grant NAGW-1510, and NSF Grant ATM-8903703.

REFERENCES

1. L.A. Fisk, Increases in the low energy cosmic ray intensity at the front of propagating interplanetary shock waves, J. Geophys. Res. 76, 1662 (1971)

2. L. O'C. Drury, An introduction to the theory of diffusive shock acceleration, Rep. Prog. Phys. 46, 973 (1983)

3. E.T. Sarris and J.A. Van Allen, Effects of interplanetary shock waves on energetic charged particles, J. Geophys. Res. 79, 4157 (1974)

4. T.P. Armstrong, G. Chen, E.T. Sarris, and S.M. Krimigis, Acceleration and modulation of electrons and ions by propagating interplanetary shocks, in: Study of Travelling Interplanetary Phenomena, ed. M.A. Shea, D.F. Smart, and S.T. Wu, D. Reidel, Dordrecht 1977, p.367.

5. R.B. Decker, The modulation of low-energy proton distributions by propagating interplanetary shock waves: A numerical simulation, J. Geophys. Res. 86, 4537 (1981)

6. M. Garcia-Munoz, G.M. Mason, and J.A. Simpson, A new test for solar modulation theory: The 1972 May-July low-energy galactic cosmic-ray proton and helium spectra, Astrophys. J. 182, L81 (1973)

7. D. Hovestadt, O. Vollmer, G. Gloeckler, and C.Y. Fan, Differential energy spectra of low-energy (<8.5 MeV per nucleon) heavy cosmic rays during solar quiet times, Phys. Rev. Lett. 31, 650 (1973)

8. F.B. McDonald, B.J. Teegarden, J.H. Trainor, and W.R. Webber, The anomalous abundance of cosmic ray nitrogen and oxygen nuclei at low energies, Astrophys. J. 185, L105 (1974)

9. L.A. Fisk, B. Kozlovsky, and R. Ramaty, An interpretation of the observed oxygen and nitrogen enhancements in low-energy cosmic rays, Astrophys. J. 190, L35 (1974)

10. L.A. Fisk, The acceleration of energetic particles in the interplanetary medium by transit time damping, J. Geophys. Res. 81, 4633 (1976)

11. B. Klecker, The anomalous component of low-energy cosmic rays: a comparison of observed spectra with model calculations, J. Geophys. Res. 82, 5287 (1977)

12. M.E. Pesses, J.R. Jokipii, and D. Eichler, Cosmic ray drift, shock wave acceleration, and the anomalous component of cosmic rays, Astrophys. J. 246, L85 (1981)

13. L.A. Fisk, The anomalous component, its variation with latitude and related aspects of modulation, in: The Sun and the Heliosphere in Three Dimensions, ed. R.G. Marsden, D. Reidel, Dordrecht 1986, p. 401.

14. J.R. Jokipii, Particle acceleration at a termination shock, 1, Application to the solar wind and the anomalous component, J. Geophys. Res. 91, 2929 (1986)

15. J.R. Jokipii, this issue.

16. F.B. McDonald, B.J. Teegarden, J.H. Trainor, T.T. von Rosenvinge, and W.R. Webber, The interplanetary acceleration of energetic nucleons, Astrophys. J. 203, L149 (1976)

17. C.W. Barnes and J.A. Simpson, Evidence for interplanetary acceleration of nucleons in corotating interaction regions, Astrophys. J. 210, L91 (1976)

18. J.T. Gosling, A.J. Hundhausen, and S.J. Bame, Solar wind stream evolution at large heliocentric distances: Experimental demonstration and the test of a model, J. Geophys. Res. 81, 2111 (1976)

19. I.D. Palmer and J.T. Gosling, Shock-associated energetic proton events at large heliocentric distances, J. Geophys. Res. 83, 2037 (1978)

20. L.A. Fisk and M.A. Lee, Shock acceleration of energetic particles in corotating interaction regions in the solar wind, Astrophys. J. 237, 620 (1980)

21. M.E. Pesses, B.T. Tsurutani, J.A. Van Allen, and E.J. Smith, Acceleration of energetic protons by interplanetary shocks, J. Geophys. Res. 84, 7297 (1979)

22. J.T. Gosling, J.R. Asbridge, S.J. Bame, G. Paschmann, and N. Sckopke, Observations of two distinct populations of bow shock ions, Geophys. Res. Lett. 5, 957 (1978)

23. F.M. Ipavich, G. Gloeckler, C.Y. Fan, L.A. Fisk, D. Hovestadt, B. Klecker, J.J. O'Gallagher, and M. Scholer, Initial observations of low energy charged particles near the Earth's bow shock on ISEE 1, Space Sci. Rev. 23, 93 (1979)

24. G. Gloeckler, D. Hovestadt, F.M. Ipavich, M. Scholer, B. Klecker, and A.B. Galvin, Cometary pick-up ions observed near Giacobini-Zinner, Geophys. Res. Lett. 13, 251 (1986)

25. A.J. Coates, A.D. Johnstone, B. Wilken, K. Jockers, and K.-H. Glassmeier, Velocity space diffusion of pickup ions from the water group at comet Halley, J. Geophys. Res. 94, 9983 (1989)

26. K.A. Anderson, Measurements of bow shock particles far upstream from the Earth, J. Geophys. Res. 86, 4445 (1981)

27. C.S. Wu, A fast Fermi process: Energetic electrons accelerated by a nearly perpendicular bow shock, J. Geophys. Res. 89, 8857 (1984)

28. E.N. Parker, The passage of energetic charged particles through interplanetary space, Planet Space Sci. 13, 9 (1965)

29. L.J. Gleeson and W.I. Axford, Cosmic rays in the interplanetary medium, Astrophys. J. 149, L115 (1967)

30. J.R. Jokipii, E.H. Levy, and W.B. Hubbard, Effects of particle drift on cosmic-ray transport. I. General properties, application to solar modulation, Astrophys. J. 213, 861 (1977)

31. J.G. Luhmann, A quasi-linear kinetic equation for cosmic rays in the interplanetary medium, J. Geophys. Res. 81, 2089 (1976)

32. J. Skilling, Cosmic ray streaming - I: Effect of Alfvén waves on particles, Mon. Not. R. astr. Soc. 172, 557 (1975)

33. J.A. Earl, J.R. Jokipii, and G. Morfill, Cosmic-ray viscosity, Astrophys. J. 331, L91 (1988)

34. K.R. Pyle, J.A. Simpson, A. Barnes, and J.D. Mihalov, Shock acceleration of nuclei and electrons in the heliosphere beyond 24 AU, Astrophys. J. 282, L107 (1984)

35. E. Chassefiere, J.L. Bertaux, R. Lallement, and V.G. Kurt, Atomic hydrogen and helium densities of interstellar medium measured in the vicinity of the Sun, Astron. Astrophys. 160, 229 (1986)

36. V.B. Baranov, this issue.

37. L.F. Burlaga, F.B. McDonald, M.L. Goldstein, and A.J. Lazarus, Cosmic ray modulation and turbulent interaction regions near 11 AU, J. Geophys. Res. 90, 12027 (1985)

38. E.J. Weber and L. Davis, Jr., The angular momentum of the solar wind, Astrophys. J. 148, 217 (1967)

39. T.E. Holzer, Interaction of the solar wind with the neutral component of the interstellar gas, J. Geophys. Res. 77, 5407 (1972)

40. J.R. Jokipii and B. Thomas, Effects of drift on the transport of cosmic rays IV. Modulation by a wavy interplanetary current sheet, Astrophys. J. 243, 1115 (1981)

41. J.R. Jokipii and J. Kota, Cosmic rays near the heliospheric current sheet 2. An ensemble approach to comparing theory and observation, J. Geophys. Res. 91, 2885 (1986)

42. E. Rieger, G.H. Share, D.J. Forrest, G. Kanbach, C. Reppin, and E.L. Chupp, Evidence for 154 day periodicity in the occurence of hard solar flares, Nature 312, 623 (1984)

43. G.M. Webb, M.A. Forman, and W.I. Axford, Cosmic-ray acceleration at stellar wind terminal shocks, Astrophys. J. 298, 684 (1985)

44. J. Kota, this issue.

45. J.R. Jokipii, Acceleration of cosmic rays at the solar-wind boundary, Astrophys. J. 152, 799 (1968)

46. A.R. Bell, The acceleration of cosmic rays in shock fronts, 1, Mon. Not. R. astron. Soc. 182, 147 (1978)

47. R.D. Blandford and J.P. Ostriker, Particle acceleration by astrophysical shocks, Astrophys. J. 221, L29 (1978)

48. W.I. Axford, E. Leer, and G. Skadron, The acceleration of cosmic rays by shock waves, Proc. Int. Conf. Cosmic Rays 15th 11, 132 (1977)

49. M.A. Lee, The solar wind termination shock and the heliosphere beyond, in: Proceedings of the Sixth International Solar Wind Conference, ed. V.J. Pizzo, T.E. Holzer, and D.G. Sime, NCAR/TN-306 + Proc, NCAR, Boulder 1988, p. 635.

50. L.A. Fisk, The interactions of energetic particles with the solar wind, in: Solar System Plasma Physics, Volume 1, ed. E.N. Parker, C.F. Kennel, and L.J. Lanzerotti, North-Holland, Amsterdam 1979, p. 177.

51. G.E. Morfill, H.J. Völk, and M.A. Lee, On the effect of directional medium-scale interplanetary variations on the diffusion of galactic cosmic rays and their solar cycle variation, J. Geophys. Res. 81, 5841 (1976)

52. L. O'C. Drury and S.A.E.G. Falle, On the stability of shocks modified by particle acceleration, Mon. Not. R. astr. Soc. 223, 353 (1986)

53. V.M. Vasyliunas and G.L. Siscoe, On the flux and the energy spectrum of interstellar ions in the solar system, J. Geophys. Res. 81, 1247 (1976)

54. M.A. Lee and W.I. Axford, Model structure of a cosmic-ray mediated stellar or solar wind, Astron. Astrophys. 194, 297 (1988)

55. M.A. Lee and L.A. Fisk, Shock acceleration of energetic particles in the heliosphere, Space Sci. Rev. 32, 205 (1982)

56. J.R. Jokipii, Particle drift, diffusion, and acceleration at shocks, Astrophys. J. 255, 716 (1982)

57. A.C. Cummings and E.C. Stone, Composition, gradients, and temporal variations of the anomalous cosmic-ray component, in: Proceedings of the Sixth International Solar Wind Conference, ed. V.J. Pizzo, T.E. Holzer, and D.G. Sime, NCAR/TN-306 + Proc, NCAR, Boulder 1988, p. 599.

58. M.A. Lee, Coupled hydromagnetic wave excitation and ion acceleration upstream of the Earth's bow shock, J. Geophys. Res. 87, 5063 (1982)

59. J.R. Jokipii, Rate of energy gain and maximum energy in diffusive shock acceleration, Astrophys. J. 313, 842 (1987)

60. M.A. Lee and W.-H. Ip, Hydromagnetic wave excitation by ionized interstellar hydrogen and helium in the solar wind, J. Geophys. Res. 92, 11041 (1987)

61. D.A. Roberts, L.W. Klein, M.L. Goldstein, W.H. Matthaeus, The nature and evolution of magnetohydrodynamic fluctuations in the solar wind: Voyager observations, J. Geophys. Res. 92, 11021 (1987)

62. E. Marsch and C.-Y. Tu, On the radial evolution of MHD turbulence in the inner heliosphere, J. Geophys. Res. submitted (1989)

63. J.V. Hollweg, Transverse Alfvén waves in the solar wind: Arbitrary **k**, $\mathbf{v}_O$, $\mathbf{B}_O$, and $|\delta\mathbf{B}|$, J. Geophys. Res. 79, 1539 (1974)

64. E.R. Christian, A.C. Cummings, and E.C. Stone, Evidence for anomalous cosmic-ray hydrogen, Astrophys. J. 334, L77 (1988)

65. H.J. Völk and W. Alpers, The propagation of Alfvén waves and their directional anisotropy in the solar wind, Astrophys. Space Sci. 20, 267 (1973)

66. P.A. Isenberg, Evolution of interstellar pickup ions in the solar wind, J. Geophys. Res. 92, 1067 (1987)

67. M.A. Lee, T.J. Bogdan, and P. Schneider, Coupled stochastic Fermi acceleration and wave damping in space plasmas, Proc. Int. Conf. Cosmic Rays 21st 5, 348 (1990)

68. R.B. McKibben, this issue.

THE ANOMALOUS COMPONENT OF COSMIC RAYS

J.R. Jokipii

University of Arizona, Space Sciences 411, Tucson, AZ 85721, U.S.A.

ABSTRACT

The anomalous component, observed first in the early 1970's, poses a significant challenge to our understanding of cosmic-ray origin and transport. It appears at present that most of its observed properties, including composition, time-dependence, energy spectrum and spatial gradients, can be understood as a natural consequence of the acceleration of freshly-ionized interstellar neutral atoms at the termination shock of the solar wind. It therefore provides a valuable probe of the outer heliosphere. The models show significantly-improved agreement with the data if the polar heliospheric magnetic field is modified to be larger and more transverse, as suggest recently. The models also suggest that the energy density of the anomalous component may modify the solar wind flow and shock if the shock is at a heliospheric distance which is significantly greater than 80-100 A.U.

INTRODUCTION.

The anomalous component observed in the quiet time cosmic-ray flux has been the subject of considerable discussion and speculation since its discovery in the early 1970's /1,2,3/. In the anomalous component the fluxes of helium, nitrogen, oxygen, neon, and more recently protons and low levels of carbon /4,5/, are observed to be enhanced in a region of the energy spectrum ranging from a kinetic energy of 20 MeV to perhaps 300 MeV. The observed radial intensity gradient is positive out to beyond the maximum distance reached by current spacecraft, indicating that this component is not of solar origin, and probably originates in the outer solar system. The spectra of anomalous cosmic-ray oxygen, helium and protons are shown in figure 1.

This anomalous component was not seen during the 1965 sunspot minimum, but was very much evident in the 1975 minimum. For this reason it was initially thought that there might be a large 22- year variation in the intensity, and that we would perhaps not see the anomalous component in the 1986 minimum. However, this was not to be the case, and the anomalous component was indeed seen again around the last sunspot minimum.

The first significant advance in our understanding of the anomalous component was due to Fisk, Koslovsky and Ramaty /6/ (here FKR). They suggested that the anomalous component was the result of heliospheric acceleration (by some unspecified mechanism) of freshly ionized interstellar particles. Interstellar neutral atoms stream into the solar system as a consequence of the motion of the solar system relative to the interstellar medium. These neutral atoms have a probability of being ionized, either by solar ultraviolet radiation, or by charge exchange with the solar wind. The newly-ionized, singly-charged particles are then swept out of the inner solar system by the solar wind and subsequently accelerated. This suggestion explains very nicely the composition of the anomalous component since only initially neutral particles can participate in the process. Species such as carbon, which are nearly completely ionized in the interstellar medium, are not expected to enter the solar system and, indeed, are observed to be significantly depleted. This process remains the working hypothesis for the source of particles to be accelerated.

Subsequently, Fisk /7/ and Klecker /8/ worked out models in which the particles are accelerated by plasma waves in association with co-rotating interaction regions, at distances of the order of 5-10 A.U. However, since the observed intensity continues to increase far beyond this region, the anomalous component must originate even further out.

Pesses, Jokipii and Eichler /9/ (here PJE) pointed out that many features of the anomalous component could be explained if the acceleration of the freshly ionized interstellar particles occurs at the termination

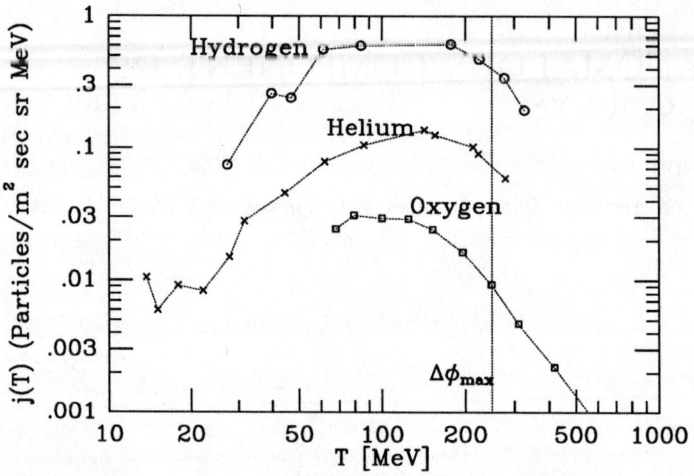

Fig. 1. Energy spectra of anomalous oxygen, helium and helium observed near 21 A.U. in 1985. The anomalous hydrogen from /5/, and the helium and oxygen were computed from digitized values provided by A. C. Cummings and correspond to the values in /4/. The vertical line indicates the electrostatic potential difference between the equator and pole of the heliosphere.

shock of the solar wind, by the mechanism of diffusive shock acceleration. In this picture, the successful idea of FKR for explaining the composition is retained. Clearly, in this picture, the intensity should continue to increase with radius out to the termination shock, in agreement with observation.

In 1986, Jokipii /10/ presented results from a quantitative two- dimensional numerical simulation of the PJE model, in which the full transport equation was solved. In this model, it was clear that a) the freshly-ionized, singly charged particles could be accelerated very efficiently at the termination shock of the solar wind and, b) the essential features of the spectrum and the gradients could explained very naturally in terms of this model, with reasonable values for the parameters. Because of the high magnetic rigidity which is a consequence of the small electric charge, the gradient and curvature drifts of the particles, both along the face of the shock and in the solar wind play a major role in producing the observed spectrum and intensity gradients.

In the rest of this paper numerical simulations and available observations are discussed in the context of this model.

TRANSPORT THEORY

In order to model the anomalous component quantitatively, we must consider energetic charged-particle transport in a magnetic field, which is a function of position and time, and which is frozen into the solar wind which is moving outward at a velocity $\mathbf{U} = V_w \hat{\mathbf{e}}_\mathbf{r}$. The particles are subjected to four distinct transport effects. Since they tend to stay on a given field line, they are convected with the fluid flow. In addition, because the magnetic field varies systematically over large scales there are curvature and gradient drifts which are coherent over large distances. The $\mathbf{U} \times \mathbf{B}$ electric field of the wind (or, equivalently, the expansion of the wind) causes associated energy changes. Finally, there is a random walk or spatial diffusion which is caused by scattering due to random magnetic irregularities carried with the flow, which also maintains a nearly-isotropic angular distribution.

The resulting superposition of coherent and random effects were combined first by Parker /11/ to obtain the generally accepted transport equation, which may be written for the quasi-isotropic distribution function $f(\mathbf{r}, p, t)$ of cosmic rays of momentum p at position $\mathbf{r}$ and time t:

$$
\begin{aligned}
\frac{\partial f}{\partial t} = & \frac{\partial}{\partial x_i}\left[\kappa_{ij}\frac{\partial f}{\partial x_j}\right] && (\textit{diffusion}) \\
& - U_i\frac{\partial f}{\partial x_i} && (\textit{convection}) \\
& - V_{d_i}\frac{\partial f}{\partial x_i} && (\textit{guiding-center drift}) \\
& + \frac{1}{3}\frac{\partial U_i}{\partial x_i}\left[\frac{\partial f}{\partial \ell np}\right] && (\textit{energy change}) \\
& + Q(x_i, t, p) && (\textit{source})
\end{aligned}
\tag{1}
$$

The labels next to each of the various terms indicate the associated physical effect. The guiding-center drift velocity is given in terms of the local magnetic field **B** and the particle charge q by $\mathbf{V}_d = (pcw/3q)\ \nabla \times (\mathbf{B}/B^2)$. This transport equation is remarkably general, and has been used in most discussions of cosmic-ray transport and acceleration over more than two decades. It appears to be a good approximation if there is enough scattering by the magnetic irregularities to keep the distribution function nearly isotropic, and if the particles have random speeds substantially larger than the background fluid convection speed. In particular, since the fluid velocity need not be a continuous function of position, shocks can be discussed within the framework of this equation. In fact, **all** of the standard theory of diffusive shock acceleration is contained in this equation.

HELIOSPHERIC CONFIGURATION

The general configuration of the inner heliosphere is reasonably well understood, and extrapolation to the termination shock probably does not introduce significant uncertainties. The configuration beyond the shock is not known, and we adopt a simple configuration which contains the basic features expected.

The solar wind velocity is taken to be constant and radial with magnitude $V_w = 400$ km/sec, out to a spherical shock at a radius R_{sh} at which it drops by a factor of four (strong shock) and then decreases as $1/r^2$ (small Mach number flow) out to an outer boundary R_b, where the energetic particle are presumed to escape. Typically, R_{sh} is taken to be some 60-90 A.U. and R_b some 30%-50% larger.

We next specify the ambient magnetic-field configuration. The large-scale structure of the magnetic field was clarified considerably by observations carried out on the Pioneer 10 spacecraft /12/, and the inferred relationship to observed coronal structure. During the years around each solar sunspot minimum, the field is generally organized into two hemispheres separated by a thin, nearly equatorial current sheet, across which the field reverses direction. In each hemisphere the field is generally assumed to be a classical Parker Archimedean spiral, with the sense of the field being outward in one hemisphere and inward in the other. At solar sunspot minimum, the current sheet is nearly equatorial. The field direction alternates with each 11-year sunspot cycle, so that during the 1975 sunspot minimum, the northern field was directed outward from the sun (conventionally denoted as $A > 0$), but in 1965 the northern field pointed inward ($A < 0$). Observations indicate that the inclination, or waviness, of the current sheet relative to the equator increases as a function of time away from sunspot minimum. The structure for the years near sunspot maximum is not simple. In the rest of the discussion, it will be generally assumed that the overall magnetic structure is given by this model, recognizing that it may not be a particularly good representation in the few years around maximum sunspot activity. This field is assumed to continue beyond the termination shock, with the spiral angle reflecting the local V_w.

It should also be noted here that a recent analysis by Jokipii and Kota /13/ suggests that the unobserved **polar** magnetic field may differ considerably from the classical Parker model at all times. This will be discussed further below.

ACCELERATION AT THE TERMINATION SHOCK

With the heliospheric model structure established, we consider the transport and acceleration of charged particles. We find that all four terms in equation (1) - convection, diffusion, energy change and drifts, are comparable in magnitude. This mandates a numerical solution of the equations. The numerical model follows the acceleration of low-energy, singly-charged particles injected into the solar wind *a la* FKR and then accelerated at the termination shock. Hence, the source function Q is set to correspond to the injection of low- energy particles (presumed to be singly-charged ions of oxygen, helium or hydrogen) at the termination shock, although the results would be similar for injection at any point interior to the shock. The method of solution is then to follow these particles in time as they are accelerated at the shock and propagate throughout the heliosphere. The characteristic time for these processes to approach a steady state is found to be 2 - 3 years.

General Considerations

The proposed model combines diffusive shock acceleration, (which has been studied extensively in the past decade, starting with the important papers of Bell, /14/, Blandford and Ostriker, /15/, and Axford, Leer and Scadron, /16/) with propagation in the solar wind. The acceleration process can be viewed in many ways. As was pointed out in the previous section, it is contained within the cosmic-ray transport equation (1), if one allows the fluid flow to have a compressive discontinuity corresponding to the shock wave. Most work on diffusive shock acceleration has concentrated on quasi-parallel shocks, where the major energy gain comes from compression along the field lines.

However, as was pointed out in Jokipii /17,18,19/, diffusive shock acceleration can occur for any angle between the shock normal and the magnetic field and is in general due to both compression and drift. The process at quasi-perpendicular shocks is more closely related to drift in the $\mathbf{U} \times \mathbf{B}$ electric field than to compressive Fermi acceleration. In fact, it is readily demonstrated that if the scattering frequency is significantly less than the gyro-frequency, the energy ΔT gained by a particle having electric charge Ze, at a quasi-perpendicular shock, is approximately the electrostatic potential $\Delta\phi$ gained in drifting along the shock face:

$$\Delta T \approx Ze\Delta\phi, \tag{2}$$

It follows, then, that there is in general a characteristic energy $T_c \approx Ze(\Delta\phi)_{max}$ above which the spectrum begins to decrease rapidly. Particles will "drift off" of the shock before gaining more energy. The cutoff need not be steep, since some particles will be scattered back and forth, receiving more or less energy than T_c. The above result is a consequence of the general relation between energy change and drift for energetic charged particles in hydromagnetic flows /20/,

$$\left[\frac{dT}{dt}\right]_{drift} = \nabla \cdot \mathbf{U}_\perp, \tag{3}$$

where $\mathbf{U}_\perp$ is that component of $\mathbf{U}$ which is normal to the local magnetic field.

Now, the termination shock of the solar wind is, for the most part a nearly-perpendicular shock. Hence, it is of interest to consider the electrostatic potential difference between the heliospheric equator and pole, which is readily shown to be independent of the wind velocity and heliocentric radius. We may write

$$\begin{aligned}(\Delta\phi)_{max} &= \frac{eB_r r^2 \Omega_\odot}{c} \\ &\approx 240 MeV\end{aligned} \tag{4}$$

where c is the speed of light and the numerical value results from using a radial magnetic field of 3.5 γ at a radius of 1 A. U. and a solar rotational angular velocity $\Omega_\odot = 2.9 \times 10^{-6}$ (corresponding to a magnetic field magnitude of 5 γ at 1 A.U.).

Singly-charged particles accelerated at the termination shock would then have a spectrum which exhibits a decrease above an energy between 200 and 300 MeV. Clearly, the spectra in Figure 1 are quite similar to each other in shape and exhibit a decrease beyond about 250-300 Mev, giving support to the idea that they are accelerated at the termination shock.

Acceleration Rate

In addition to the above, we must consider the acceleration **rate**. It seems reasonable to require that the termination shock be capable of accelerating the anomalous component in a time somewhat less than a sunspot cycle, or 11 years, since the entire magnetic structure reverses on this time scale. In order to study acceleration rate, we must go to time-dependent solutions. The particles gain only a small amount of energy (a fraction of order V_w/w) each time they cross the shock. The time for acceleration of particles from an initial momentum p_o to a momentum p may be written as /21,22/

$$\tau_a \simeq \frac{3}{(V_1 - V_2)} \int_{p_o}^{p} \{\frac{\kappa_1}{V_1} + \frac{\kappa_2}{V_2}\} \frac{dp}{p} \quad \text{or}$$
$$\frac{1}{p}\frac{dp}{d\tau_a} \simeq \frac{V_{sh}^2 (r-1)}{3r[\kappa_1 + r\kappa_2]}. \tag{5}$$

where, if θ is the angle between the shock normal and the magnetic field, κ_1 $(\kappa_2) = \kappa_{xx} = \kappa_{\|} sin^2(\theta) + \kappa_{\perp} cos^2(\theta)$ as a function of momentum p in the upstream (downstream) region, and V_1 is the shock velocity with respect to the upstream gas. The solution at any time is a power law up to a gradual cutoff at a momentum near that defined in equation (5). Lagage and Cesarsky /23/, considering quasi-parallel shocks only, showed that since the parallel scattering mean free path must be at least equal to the gyro-radius r_g (the so-called Bohm limit), for these shocks there was a maximum rate of energy gain given by setting κ_{xx} in equation (5) equal to $\frac{1}{3} r_g w$. Application of this Bohm limit to a solar-wind termination shock at 100 A.U., using nominal values of the magnetic field and solar- wind velocity at the termination shock, and setting $\lambda_{\|} = 3r_g$, leads to a characteristic time for acceleration at a quasi- parallel shock of about 20 years, which is probably too large. However, these considerations do **not** apply to quasi- perpendicular shocks, such as the termination shock /18,19/, for which the acceleration can be much more rapid. In fact, we find in our simulations that the characteristic acceleration time is of the order of one year. This can happen because the relevant κ in equation (5) is then $\kappa_{\perp}$, which is not restricted by the Bohm limit. In fact one can show that for simple scattering laws the Bohm value is an *upper* limit on the value of $\kappa_{\perp}$ /19/.

For such nearly perpendicular shocks the inequality $\lambda_{\|}/r_g << w/V_{sh}$ must be satisfied/19/. This condition is well- satisfied for the simulations discussed below.

SIMULATION RESULTS

PJE pointed out that if the polar field is given by the Parker spiral, the termination shock is essentially a perpendicular shock everywhere but near the poles, where it is quasiparallel. The injection of low-energy particles is preferred there. In this first version of the model, the energy spectra calculated near the heliospheric equatorial plane, at 1 A.U. show a peak in intensity at an energy in the range 1-10-MeV and the energy of the peak for $A < 0$ (northern solar field in) is approximately a factor of 3-5 higher than for $A > 0$. Although this agreed with data up until 1985, more recent observations showed that the energy spectrum continued to evolve as we approached the last sunspot minimum, and the difference between the spectra became much smaller /4/. This cannot be accommodated within the early form of the model.

However, Jokipii and Kota have shown /13/ that the interplanetary magnetic field in the polar regions may **not** be well-approximated by the Parker spiral. In this proposed magnetic field configuration, the

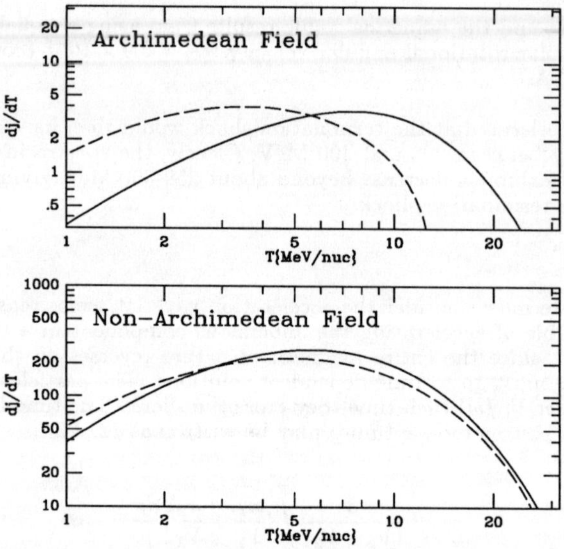

Fig. 2. Computed energy spectra of singly-charged oxygen at 20 A.U. Top curve was computed using the standard Archimedean magnetic field, whereas the bottom was computed using a field in which the polar magnetic field was enhanced as suggested by Jokipii and Kota /13/, equation (4), with $\delta = 0.003$.

magnetic field is more intense and effectively perpendicular at all latitudes, most of the time, so there is no reason to assume that the polar regions are more preferred as the injection site for the particles. Instead, injection would be expected to be more nearly independent of the polar angle. One would then expect the theory to produce spectra for the two signs of the magnetic field which are more nearly equal. The theoretical spectra from a sample simulation, in which the injection was uniform over the shock, are shown in figure (1), where the kinetic energy of peak for $A > 0$ is at an energy only slightly smaller than that for $A < 0$. Note that the relative normalization (amplitude) of the two spectra are not significant and may be changed by small variation of the parameters. The fact that the modified field gives better agreement provides some support for the existence of such a field.

In fact, the computed energy spectra of anomalous protons, helium, neon, etc are in excellent agreement with the observed spectra.

Radial and Latitudinal Gradients of the Anomalous Component

Figure 3 illustrates typical contour plots of the intensity of the modelled anomalous oxygen as a function of heliocentric radius and polar angle, in a solar meridional plane. Similar plots are obtained for the other anomalous elements as well. We see that the intensity increases with radius, much as does that of the galactic cosmic rays, out to the termination shock (at a radius of 60 A.U.). However, beyond the termination shock, which is where the particles are accelerated, the intensity decreases out to the outer boundary of the heliosphere, where the intensity approaches 0. This is a characteristic form of the intensity for this particular model and we would anticipate that when a spacecraft passes through the vicinity of the heliospheric termination shock, the intensity of the anomalous component should begin to decrease. This would provide strong support for the model. The **maximum** intensity occurs along the shock at a latitude which shifts as the sign of the magnetic field changes. If $A > 0$, the particles drift toward the pole along the shock face and then inward and down from the poles to the current sheet, the intensity maximum is near the poles for A positive. On the other hand, when $A < 0$ positively-charged particles at the shock drift toward the equator along the shock and then inward along the current sheet, executing a clock-wise

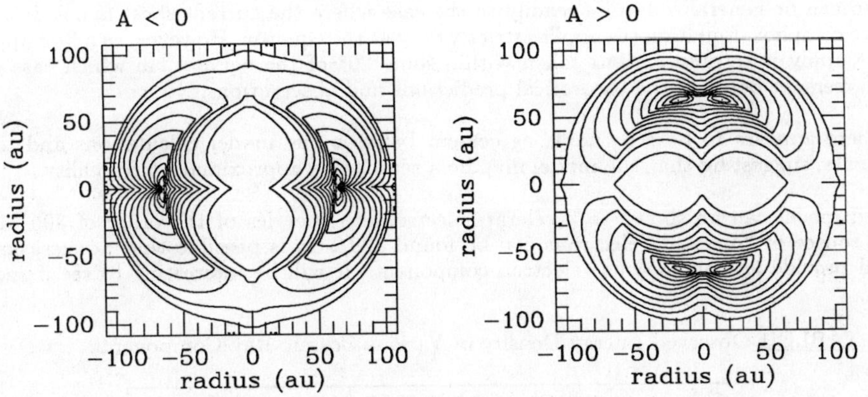

Fig. 3. Computed contours of equal intensity of singly- charged oxygen at an energy of about 5 Mev/nucleon, for the two signs of the interplanetary magnetic field. The minima occur at the center of each plot and at the outer edge.

cycle, shifting the maximum at the shock toward the equator. This is a property of essentially all of our simulations.

The radial and latitudinal gradients of the anomalous component have been measured by a number of observers /24,25,26/. In particular it is found that in the inner solar system the radial gradient of the anomalous oxygen is as high at 16% per astronomical unit, whereas in the outer solar system it is 3% per astronomical unit. Careful analysis /26/ indicates that this is indeed a radial variation of the gradient and not a temporal change. The magnitude of the radial gradient computed in the model is quite consistent with that observed, both in absolute magnitude and in radial variation.

An absolutely unique prediction of models of cosmic ray modulation and the anomalous component which involve drift is that the latitudinal gradient near the current sheet, near sun spot minimum, should change sign between sun spot cycles. In particular it is predicted that during the last sun spot minimum, near 1975, the intensity of both galactic cosmic rays and the anomalous component should **increase** away from the current sheet, whereas during the 1986 sun spot minimum the sign should change and the cosmic rays would **decrease** away from the current sheet. The reason for these predictions is that when the sign of the magnetic field of the sun is such that it is out in the northern hemisphere, positively-charged particles accessing the inner heliosphere must do so by drifting in through the poles. As they drift, they lose energy in the $\mathbf{U} \times \mathbf{B}$ electric field, and the intensity will decrease as one approaches the current sheet. On the other hand, during the alternate sun spot minima, when the magnetic field is inward in the northern hemisphere of the sun, the particles will drift inward along the current sheet and then out toward the poles, resulting in a decrease in intensity away from the current sheet. Near sun spot minimum, when the current sheet is nearly flat, the effects of these drifts are expected to be the most important. In 1977, Pioneer observed a positive gradient of anomalous helium away from the current sheet /24/, as predicted. Observations carried out during the last sun spot minimum during a period in 1984 and 1985, when the current sheet tilt went below the latitude of the Voyager I spacecraft, Cummings and Stone /26/ reported that the sign of the latitudinal gradient changed from being positive to negative, just as predicted by the theory.

One may further check the model by comparing the magnitude of the latitudinal gradient with the predictions of the model. In particular, if the current sheet is flat there is a very simple relationship between the latitudinal gradient near the current sheet and the radial gradient. The relation, written in terms of the angle ψ between the radius vector and the magnetic field, and the antisymmetric part of the diffusion tensor κ_A, is

$$\frac{1}{r}\frac{\partial f}{\partial \theta}\bigg|_{\theta=\frac{\pi}{2}} = -\frac{\kappa_A}{\kappa_\perp} sin(\psi)\bigg|_{\theta=\frac{\pi}{2}} \quad sgn(B_r|_{\theta<\frac{\pi}{2}}) \tag{6}$$

This equation can be generalized quite readily to the case where the current sheet is not flat, although it becomes quite complex. Equation (6) applies strictly only at the equator. However, as a first approximation one may just apply it to observations taken within some 20° of the equator, in which case one obtains reasonable agreement between the theoretical predictions and observations.

All in all, there appears to be reasonable agreement between the model calculations and the observed spatial gradients, suggesting that the model may be a reasonable approximation to reality.

The proposed model can be shown to accelerate electrons to energies of the order of 300 MeV as well, if a suitable source of "injected" electrons can be found /29/. It is possible that acceleration of Jovian electrons will provide an "anomalous" electron component. It will be interesting to see if such electrons are observed.

TABLE 1 Observed Energy Density of Various Cosmic-Ray Components

Galactic Cosmic Rays @ 1 A.U.	0.5 ev/cm^3
Anomalous Hydrogen @ 21 A.U.	0.0018 ev/cm^3
Anomalous Helium @ 21 A.U.	0.00037 ev/cm^3
Anomalous Oxygen @ 21 A.U.	0.00018 ev/cm^3

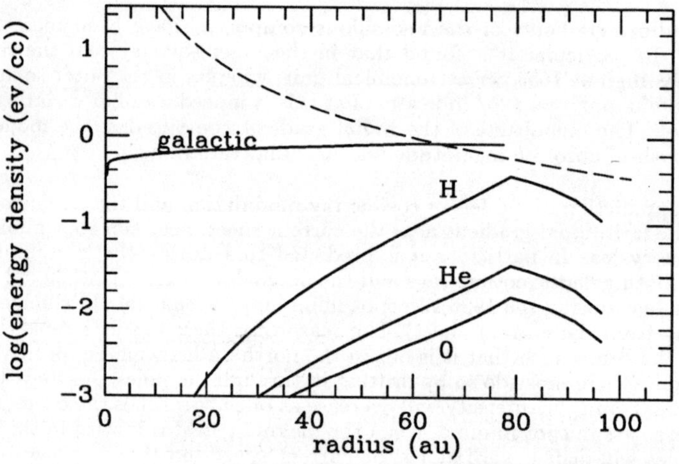

Fig. 4. The solid lines show the computed energy density as a function of radius, scaled to give the observed value at 21 A.U., from the spectra shown in figure 1. The solar wind value corresponds to the slow solar wind parameters given by Hundhausen /28/. scaled as $1/r^2$. Each energetic particle curve has been normalized to give the energy density from Table 1 at the proper heliocentric radius.

THE ENERGY DENSITY OF THE ANOMALOUS COMPONENT

The energy density of the anomalous component is surprisingly large and increases with radius. Table 1 lists the observed energy densities of the various components as computed from the data shown in figure 1. The galactic cosmic-ray data are from Gloeckler and Jokipii, /27/.

Figure 4 illustrates the energy density of these components and that of the solar wind as a function of radius, as obtained from a sample simulation. Very similar curves are obtained for a variety of parameters.

It is clear that the solar wind energy density is much larger than that of the energetic particles within about 60 A.U., but that the anomalous hydrogen becomes important beyond some 70 A.U. and peaks at the radius of the termination shock, here taken to be 79 A. U. Because of the large gradient of the anomalous component, it should be dynamically more important than the galactic cosmic rays. In any event, for these simulations, if the termination shock is significantly beyond 80 A.U., the solar-wind energy density would be comparable to that of energetic particles, and the dynamical effects would be significant. The radius at which this occurs is parameter-dependent to some extent. Nontheless, it is expected that the termination shock cannot be significantly beyond 80 A.U. without modification of this model.

SUMMARY AND CONCLUSIONS

The anomalous component provides a challenge to theories of cosmic ray transport and origin. It appears at present that most of its observed properties, including composition, time- dependence, energy spectrum and spatial gradients, can be understood as a natural consequence of the acceleration of freshly-ionized interstellar neutral atoms at the termination shock of the solar wind. Because of its relationship to large-scale features of the solar wind and to the local interstellar medium and its interaction with the solar wind, it will serve as a valuable probe of the plasma and magnetic field of the heliospheric boundary regions. For example, the fact that a modified polar magnetic field in the models /13/ gives significantly improved agreement with observations provides some evidence that the polar field is indeed larger and more transverse than previously thought. It is expected that the energy density of the anomalous component may approach that of the solar wind if the termination shock is at a heliospheric radius which is significantly greater than some 80 A.U., providing us with the opportunity to study plasma dynamics in a new regime where the cosmic rays affect the dynamics. In addition, the anomalous component should help us to better understand the general problem of cosmic-ray acceleration in astrophysics.

ACKNOWLEDGEMENTS

This research was supported, in part, by the National Science Foundation under Grant ATM-8618260 and by the National Aeronautics and Space Administration under Grant NAGW-1931.

REFERENCES

1. M. Garcia-Munoz, G.M. Mason, and J.A. Simpson, A new test for solar modulation theory: the 1972 May-June low energy galactic cosmic-ray proton and helium spectra, Astrophys. J. (Letters), 182, L81, (1973).

2. D. Hovestadt, O. Vollmer, G. Gloeckler, and C.Y. Fan, Differential energy spectra of low-energy (¡8.5 MeV per nucleon) heavy cosmic rays during solar quiet times, Phys. Rev. Letters, 31, 650, (1973).

3. F.M. McDonald, B.J. Teegarden, J.H. Trainor, and W.R. Webber, The anomalous abundance of cosmic-ray nitrogen and oxygen nuclei at low energies, Astrophys. J. (Letters), 187, L105, (1974).

4. A.C. Cummings, and E.C. Stone, Elemental composition of the anomalous cosmic ray component, Proc. 20th Intl. Cosmic Ray Conf., Moscow, 3, 421, (1987).

5. A.C. Cummings, and E.C. Stone, Evidence for anomalous cosmic-ray hydrogen, Astrophys. J., 334, L77, (1988).

6. L.A. Fisk, B. Kozlovsky, and R. Ramaty, An interpretation of the observed oxygen and nitrogen enhancements in low energy cosmic rays, Astrophys. J. (Letters), 190, L35, (1974).

7. L.A. Fisk, Increases in the low-energy cosmic ray intensity at the front of propagating interplanetary shock waves, J. Geophys. Res., 76, 1662, (1971).

8. B. Klecker, The anomalous component of low energy cosmic rays: a comparison of observed spectra with model calculations, J. Geophys. Res., 82, 5287, (1977).

9. M.E. Pesses, J.R. Jokipii, and D. Eichler, Cosmic ray drift, shock wave acceleration, and the anomalous component of cosmic rays, Astrophys. J. (Letters), 246, L85, (1981).

10. J.R. Jokipii, Particle acceleration at a termination shock 1., Application to the solar wind and the anomalous component, J. Geophys. Res., 91, 2929, (1986).

11. E.N. Parker, The passage of energetic charged particles through interplanetary space, Planet. Space Sci., 13, 9, (1965).

12. E.J. Smith, and B.T. Thomas, Latitudinal extent of the heliospheric current sheet and modulation of galactic cosmic rays, J. Geophys. Res., 91, 2933, (1986).

13. J.R. Jokipii and J. Kota, The polar heliospheric magnetic field, Geophys. Res. Lett., 16, 1, (1989).

14. A.R. Bell, The acceleration of cosmic rays in shock fronts - I, Mon. Not. Roy. Astr. Soc. 182, 147, (1978a).

15. R.D. Blandford, and J.P. Ostriker, Particle acceleration by astrophysical shocks, Ap. J., 221, L29, (1978).

16. W.I. Axford, E. Leer, and G. Skadron, Acceleration of Cosmic Rays at Shock Fronts Proc. 15th Int. Cosmic Ray Conf., Plovdiv, 11, 132, (1978).

17. J.R. Jokipii, Particle drift, diffusion and acceleration at shocks, Ap. J., 255, 716, (1982).

18. J.R. Jokipii, Rate of energy gain and maximum energy in diffusive shock acceleration,, Astrophys. J., 313, 842 (1987).

19. J.R. Jokipii, Mechanism of particle acceleration at shocks: the effect of magnetic field geometry, Proceedings of the 6th Intl. Solar Wind Conf., Volume II, pg 481, Estes Park, Colorado, (1987).

20. J.R. Jokipii, Rapporteur paper: Modulation theory, Proceedings of the 16th Intl. Cosmic-Ray Conf., Kyoto, August, (1979).

21. M.A. Forman, and G. Morfill, Time-dependent acceleration of solar wind plasma to Mev energies at co-rotating interplanetary shocks, in Proc. 16th Intl. Cosmic Ray Conf. Kyoto, 5, 328, (1979).

22. L. Drury, An introduction to the theory of diffusive shock acceleration of energetic particles in tenuous plasmas, Rpt. Prog. Phys, 46, 973, (1983).

23. P.O. Lagage, and C.J. Cesarsky, Cosmic-ray shock acceleration in the presence of self-excited waves, Astron. Astrophys., 118, 223, (1983).

24. R.B., McKibben, K.R. Pyle, and J.A. Simpson, The solar latitude and radial dependence of the anomalous cosmic ray helium component, Astrophys. J. (Letters), 227, L147, (1979).

25. Frank B. McDonald and Nand Lal, Variations of galactic cosmic rays with heliolatitude in the outer heliosphere, Geophys. Re. Letters, 13, 781, (1986).

26. A.C. Cummings, E.C. Stone, and W.R. Webber, Latitudinal and radial gradients of anomalous and galactic cosmic rays in the other heliosphere, Geophys. Res. Letters, 14, 174, (1987).

27. G. Gloeckler and J. R. Jokipii, Solar modulation and the energy density of cosmic rays, Ap. J., L41, (1967).

28. A. J. Hundhausen, Coronal expansion and solar wind , Springer-Verlag, New York, (1972).

29. J. R. Jokipii, H. Moraal and R. Mewaldt, in preparation.

EFFECT OF SMALL AND LARGE SCALE LENGTH PERTURBATIONS ON PARTICLE DRIFTS

R.A. Burger

Department of Physics, Potchefstroom University for C.H.E., 2520 Potchfstroom, South Africa

ABSTRACT

Drift moderated modulation models for galactic cosmic rays, which employ the weak-scattering form of the transverse diffusion coefficient in describing charged particle drifts, have been shown to predict too large charge-dependent effects. In this paper two mechanisms which may both reduce drift effects, and also influence its rigidity dependence, are discussed. First the effect of scattering (i.e. the resonant effect of small scale perturbations) and the random walk of magnetic field lines (nonresonant effects) on drift, as embodied in a general diffusion tensor, is considered. The second case concerns scale lengths longer than the resonant wavelength. It is shown that these medium to long scale length perturbations may cause significant deviations from expected local drift velocities. However, the spatial average of drift in the presence of these perturbations is shown to be smaller than that for a smooth field. Seen as a whole, the results of this paper is an indication that perturbations in a magnetic field, regardless of their scale lengths, may influence both the magnitude and the rigidity dependence of drift effects.

INTRODUCTION

Since the first modulation model which included the effects of particle drifts was published more than a decade ago /1/, various studies /2,3,4/ have shown that while drift effects may be significant, they appear to be too large when compared to observations. Furthermore, the recent model calculations performed by Webber et al. /3/. suggest that the need for reduced drift effects at lower energies (corresponding to $P \approx 0.8$GV) may be stronger than at higher energies ($P \gtrsim 1.5$GV). One can argue on the one hand that some other mechanism such as magnetic helicity /5,6/ may oppose drift, thereby decreasing its effectiveness. On the other hand, one should also consider means by which drift in itself can be made less effective. In this paper the latter approach will be used, and the influence of cross-field transport due to scattering and the random walk of magnetic field lines (diffusive effects) on drifts, as well as the "coherent" effects of magnetic field perturbations with scale lengths greater than the resonant wavelength will be discussed. The aim of this study is to provide some theoretical justification for reducing drift effects in numerical computations, and it will be shown that these diffusive and/or "coherent" effects may change both the magnitude and rigidity dependence of drift effects.

DIFFUSIVE EFFECTS

In most current drift models the assumption of weak scattering leads to the use of diffusion coefficients parallel and perpendicular to the mean magnetic field ($\kappa_\parallel$ and $\kappa_\perp$ respectively) whose spatial dependence and rigidity dependence are usually chosen to fit data on scattering mean free paths and (selected) cosmic ray measurements. While these two coefficients might differ from one model to the other, the same form for the transverse coefficient κ_T (which describes drift effects due to gradients and curvatures in the magnetic field), viz. that prescribed by the weak scattering assumption, is used. However, as was pointed out in the Introduction, the magnitude of drift effects in these models need to be reduced, and furthermore drift might not in actual fact be proportional to $P^{1.0}$ at all rigidities /7,3/. One way to modify drift effects is to employ the general diffusion tensor derived by Forman et al. /8/, and not only its weak scattering approximation. In the general case, the elements of the tensor are

$$\kappa_\parallel = \frac{v\lambda_\parallel}{3} \tag{1a}$$

$$\kappa_\perp = \frac{vR_L}{3}\frac{\lambda_\perp/R_L}{1+(\lambda_\perp/R_L)^2} \tag{1b}$$

$$\kappa_T = \text{sign}(qB)\,\kappa_\perp\frac{\lambda_\perp}{R_L} \tag{1c}$$

where v is the particle speed, q and B the signed particle charge and magnetic field respectively, $\lambda_\parallel$ ($\lambda_\perp$) the scattering mean free path parallel (perpendicular) to the mean magnetic field, and R_L the magnitude of the particle's Larmor radius. In the limit of weak scattering, $\lambda_\perp \gg R_L$, equations (1b) and (1c) reduce to

$$\kappa_\perp = \frac{vR_L}{3}\frac{R_L}{\lambda_\perp}, \tag{2a}$$

$$\kappa_T = \text{sign}(qB)\frac{vR_L}{3} \equiv \text{sign}(qB)\frac{\beta P}{3B}. \tag{2b}$$

In most current drift models, the form (2b) for κ_T is used, while empirical forms for $\kappa_\parallel$ and $\kappa_\perp$ are chosen to fit data. It was pointed out by Kadokura and Nishida /9/ that this procedure can easily lead to discrepancies if a form for $\kappa_\perp$ is chosen such that, for example, the condition for weak scattering required for (2b) is violated in the choice for $\kappa_\perp$. To clarify this point, note that from equation (1b), the maximum value of $\kappa_\perp$ in this scenario is $vR_L/6$ when $\lambda_\perp = R_L$, and that $\kappa_\perp$ decreases from this value if the ratio $\lambda_\perp/R_L$ is increased or decreased: for a particular value of this ratio, say a, $\kappa_\perp$ has the same value as for $1/a$. Thus, if one chooses a value for $\kappa_\perp$ that exceeds the maximum value allowed by equation (1b), it is not immediately obvious that the form (2b) for κ_T may be used in conjunction with that particular $\kappa_\perp$. An interesting property of the diffusion coefficients given by equation (1) is that for $\lambda_\perp/R_L \lesssim 1$, κ_T is smaller than the weak scattering limit given in (2b). Furthermore, if the ratio $\lambda_\perp/R_L$ in equation (1b) is not a constant (in contrast to the assumption of Kadokura and Nishida /8/), but depends in some way on particle rigidity, this may result in a rigidity dependence for κ_T which is different from $P^{1.0}$, prescribed by the result for weak scattering given in equation (2b).

In principle, therefore, diffusive effects may influence both the magnitude and the rigidity dependence of κ_T. However, the forms given in (1) apply only for rigidities above 0.8 GV /8/, and given the problems experienced by quasi-linear theory in general /10/, employing a diffusion tensor such as that given in equation (1) may not be the appropriate approach, and one should consider some alternatives, such as coherent effects.

COHERENT EFFECTS

The effect on diffusion of magnetic field perturbations with scale lengths longer than the resonant wavelength, but shorter than that of the background field, has been investigated by Morfill et al. /11/, but this detailed analysis has not yet been extended to include particle drifts. In what follows, simple geometric arguments will be used to show that this type of perturbations may in fact play a crucial part in determining the effectiveness of particle drifts.

Moussas et al. /12/ have shown that the standard expressions for gradient and curvature drift remain valid even in a highly perturbed magnetic field, and concluded that in the interplanetary medium the *local* gradients and curvatures determine the drift velocity, rather than the background Parker spiral field configuration. These authors calculated a drift velocity form *in situ* Pioneer 10 magnetic field measurements obtained at 5 AU, which turned out to be some four times larger than the drift velocity calculated from the Parker spiral field. At first glance this result appears to be in conflict with the by now established need for reduced drift velocities /2,3,13/. It will now be shown that this result can be accommodated in the same framework as that which leads to smaller overall drift effects. Consider a magnetic field

$$\mathbf{B} = B_0[(1 + \alpha y)\mathbf{e}_x + f(x)\mathbf{e}_y], \tag{3}$$

with α a constant, which models a smooth background field in the x-direction, with some arbitrary transverse disturbance $f(x)$ superimposed on it. This magnetic field is obviously divergence-free, and the well-known drift velocity for a near-isotropic particle distribution /14,15,16/,

$$\mathbf{v}_D = \frac{pv}{3q}\nabla\times\frac{\mathbf{B}}{B^2}, \tag{4}$$

only has a component in the z-direction. For simplicity it is assumed that $\alpha y \ll 1$, in which case the drift velocity due to the background field (i.e. with $f(x) = 0$) is independent of position. It is straight forward to show that the magnitude of the drift velocity in the disturbed field, normalized with respect to that in the background field, is given by

$$D = \frac{1 - f^2}{(1 + f^2)^2} + \frac{1}{\alpha}\frac{d}{dx}\left(\frac{f}{1 + f^2}\right), \tag{5}$$

which will be called the drift fraction. Note that a prerequisite for equation (5) to be physically significant is that the gyroradii of the particles must be much smaller than the scale length of the disturbance. By its definition, if D is greater than 1 the drift due to the perturbed field exceeds that due to the background field, and vice versa for $D < 1$. It is clear that if the disturbance $f(x)$ oscillates as function of position, D will also oscillate, and the second term in equation (5) can change its sign, which (depending on the relative magnitude of the two terms) could even cause the drift in the disturbed field to be oppositely directed to that in the background field. Equation (5) may therefore explain the result of Moussas et al. /12/.

To proceed with the analysis, it should be kept in mind that as the particles drift in the z-direction, they will typically move with respect to the disturbance in the x- and y-directions as well, thus "sampling" the disturbance across an area perpendicular to the direction of drift. To accommodate this effect, one can calculate the spatial average of the drift velocity (4) across an area A,

$$\tilde{v}_D = \frac{pv}{3q}\frac{1}{A}\int_A\!\!\int \nabla\times\left(\frac{\mathbf{B}}{B^2}\right)\cdot d\mathbf{A} = \frac{pv}{3q}\frac{1}{A}\oint_C\left\{\frac{B_x}{B^2}dx + \frac{B_y}{B^2}dy\right\}. \tag{6}$$

The choice of the closed integration path C will be determined by the nature of the disturbance under consideration. For the generic disturbance $f(x)$, equation (3) may be substituted into equation (6), yielding

$$\tilde{v}_D = \frac{pv}{3q}\frac{1}{A}\oint_C\left\{\frac{1 + \alpha y}{(1 + \alpha y)^2 + f^2(x)}dx + \frac{f(x)}{(1 + \alpha y)^2 + f^2(x)}dy\right\}. \tag{7}$$

In discussing the behavior of this spatially averaged drift velocity, it is useful to define the spatially averaged drift fraction not simply as an average of the quantity D, but rather as

$$\tilde{D} = \frac{\tilde{v}_D}{\left[\tilde{v}_D\right]_{f(x)=0}}. \tag{8}$$

To gain some insight into the significance of this equation, two idealized disturbances will be considered. The first is an infinitely long sinusoidal wave, whose amplitude remains constant throughout space. The second disturbance models a spectrum of waves with equal amplitudes but with different wavenumbers, resulting in a wavetrain whose amplitude diminishes in two directions along one axis, and remains constant along the other two axes.

For the case of a single sinusoidal wave, we consider a disturbance

$$f(x) = e\sin(kx), \tag{9}$$

where, from equation (3), e is the amplitude of the disturbance expressed as a fraction or multiple of B_0, and k the wavenumber. The closed integration path C of equation (7) is chosen such that the integration over x is performed at $y = \pm \Upsilon$, the only constraint on Υ being $\alpha\Upsilon \ll 1$. The integration path for the integration over y is chosen to be at $x = 0$ and $x = 2\pi/k$, which means that the integrals over y in equation (7) drops out. The remaining two integrals over x can be combined, and to first order in αy equation (8) becomes

$$\tilde{D} = \frac{k}{2\pi}\int_0^{2\pi/k} \frac{1 - e^2\sin^2 kx}{\left(1 + e^2\sin^2 kx\right)^2}\,dx = \frac{1}{(1+e^2)^{3/2}}. \tag{10}$$

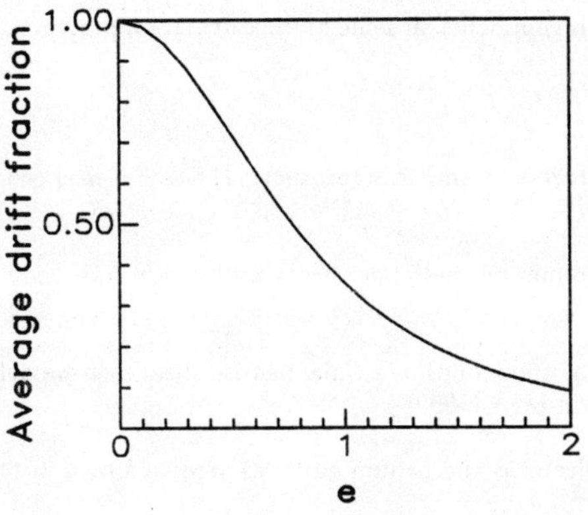

Fig. 1. Average drift fraction as a function of the relative amplitude e of a sinusoidal disturbance.

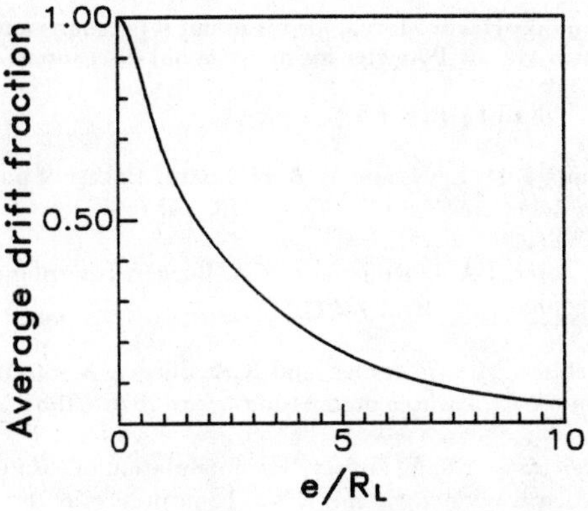

Fig. 2. Average drift fraction for a spectrum of waves as function of the ratio of relative amplitude to particle gyroradius.

This result, shown graphically in Figure 1, predicts a 50% decrease in the average drift velocity for $e \approx 0.76$, and if $e = 1.0$, the drift is reduced to 35% of the unperturbed value.

Turning now to the second case, we model a spectrum of waves with equal amplitudes by choosing $f(x)$ as

$$f(x) = \int_0^{k_{max}} e\cos(kx)\,dk. \tag{11}$$

To ensure that the gyroradius R_L of the particles under consideration is smaller than the shortest wavelength in the disturbance, the wavelength of the latter is chosen somewhat arbitrarily to be $2\pi R_L$. Equation (11) then becomes

$$f(x) = \frac{e}{R_L}\,\frac{\sin(x/R_L)}{x/R_L}. \tag{12}$$

Following the procedure outlined to calculate $\tilde{D}$ for the case of a single sinusoidal wave, but setting the limits of integration along the x-axis at $\pm 2\pi R_L$ so that the integrals over y again drop out, results in

$$\tilde{D} = \frac{1}{4\pi R_L}\int_{-2\pi R_L}^{2\pi R_L} \frac{1 - (e/R_L)^2\sin^2(x/R_L)/(x/R_L)^2}{\left[1 + (e/R_L)^2\sin^2(x/R_L)/(x/R_L)^2\right]^2}\,dx. \tag{13}$$

This integral was evaluated numerically, and the behavior of the averaged drift fraction is shown in Figure 2 as a function of the ratio e/R_L. As was the case for a single sinusoidal wave, it is again the amplitude of the waves constituting the disturbance which determines the degree to which drift is reduced. For this particular disturbance, however, the reduction also depends on the particles' gyroradius, i.e. on their rigidity. For a given value of the parameter e, this figure shows that in the presence of this disturbance, drift is less effective at low rigidities (e/R_L large) than it is at high rigidities (e/R_L small). Note that this behavior is in qualitative agreement with the findings of Webber et al. /3/ that the need for reduced drift effects at smaller rigidities may be stronger than at higher rigidities.

DISCUSSION AND CONCLUSIONS

In this paper it was shown that the much-needed reduction of drift effects can in principle be justified by considering diffusive and/or coherent effects. The validity of the former effect has not yet been proved beyond any doubt, and one must look toward quasi-linear theory to answer some of the remaining questions regarding diffusion coefficients for strong scattering.

The main result of this paper is that perturbations with scale lengths greater than the resonant wavelength may cause large fluctuations in local drift velocities, but will on the average lead to reduced drift effects. It was also shown that the reduction may depend on particle rigidity, and that particles with smaller rigidities may experience less drift than those with greater rigidities. Although this analysis is based on two rather idealized cases, the physical reason for this reduced drift effects is quite clear: A transverse oscillation on a field line increases the (average) scalar magnitude of the field, without affecting the (average) vector magnitude. From the expression for drift velocity given in equation (4), it therefore follows that the average direction of the drift velocity can be expected to remain unchanged, but that its average magnitude is decreased. Jokipii and

Kota (1989) recently gave an important physical content to this idea. They surmised that due to granulations on the solar surface, the polar magnetic fields may contain a transverse, turbulent contribution which falls off inversely proportional to radial distance ($1/r$). On the other hand, the undisturbed Parker spiral field falls off as $1/r^2$. In the nomenclature of the current paper, the Jokipii and Kota model essentially corresponds to the case $e \gg 1$ in the outer heliosphere, which shows that the drift effects may be effectively quenched there, as was suggested by them.

It is clear from the results presented in the present paper that drift effects can in principle be influenced by all fluctuations with scales equal to or greater than the resonant wavelength. Obviously, the weak-scattering approximation for κ_T, as used in most current drift models, may therefore not be the appropriate form for this coefficient. If its magnitude and its rigidity dependence are determined by all of these fluctuations, it does not seem unreasonable that (due to our rather limited knowledge of conditions throughout the heliosphere) future drift models may employ empirical forms for κ_T.

ACKNOWLEDGEMENTS

The author thanks Harm Moraal for his useful suggestions and for his comments on some of the calculation presented in this paper, and also Marius Potgieter for many useful discussions.

REFERENCES

1. J.R. Jokipii, E.H. Levy, and W.B. Hubbard, Effects of particle drift on cosmic-ray transport. I. General properties, application to solar modulation , Ap. J. 213, 861 (1977)

2. M.S. Potgieter, J.A. le Roux, and R.A. Burger, Interplanetary cosmic-ray radial gradients with steady-state modulation models, J. Geophys. Res. 94, 2323 (1989)

3. W.R. Webber, M.S. Potgieter, and R.A. Burger, A comparison of predictions of a wavy neutral sheet drift model with cosmic ray data over a whole modulation cycle: 1976 - 1987, Ap. J., in press (1990)

4. M.S. Potgieter and R.A. Burger, The modulation of cosmic-ray electrons and helium nuclei as predicted by a drift model with a simulated wavy neutral sheet, Ap. J., in press (1990)

5. J.W. Bieber, P.A. Evenson, and W.H. Matthaeus, Magnetic helicity of the Parker Field, Ap. J., 315, 700 (1988)

6. J.W. Bieber and R.A. Burger, Cosmic ray streaming in the Born approximation, Ap. J., in press (1990)

7. M.A. Lee and L.A. Fisk, The role of particle drifts in solar modulation, Ap. J., 248, 836 (1981)

8. M.A. Forman, J.R. Jokipii, and A.J. Owens, Cosmic ray streaming perpendicular to the mean magnetic field, Ap. J., 192, 535 (1974)

9. A. Kadokura, and A. Nishida, A., Numerical modeling of the 22-year variation of the cosmic ray intensity and anisotropy, J. Geophys. Res., 91, 1 (1986)

10. J.W. Bieber, C.W. Smith, and W.H. Matthaeus, Cosmic-ray pitch-angle scattering in isotropic turbulence, Ap. J., 334, 470 (1988)

11. G.E. Morfill, H.J. Volk, and M.A. Lee, On the effect of directional medium-scale interplanetary variations on the diffusion of galactic cosmic rays and their solar cycle variation, J. Geophys. Res., 81, 5841 (1976)

12. X. Moussas, J.J. Quenby, and J.F. Valdes-Galicia, Drift motion and perpendicular diffusion of energetic particles in interplanetary space based on spacecraft data, Astrophys. Space Sci., 86, 197, (1982)

13. M.A. Forman, Solar modulation of galactic cosmic rays, Proc. 20th Int. Conf. Cos. Rays, 8, 165 (1987)

14. B. Rossi S. Olbert, Introduction to the physics of space, McGraw-Hill, New York, 1970

15. P.A. Isenberg and J.R. Jokipii, Gradient and curvature drifts in magnetic fields with arbitrary spatial variations, Ap. J., 234, 746, (1979)

16. R.A. Burger, H. Moraal, and G.M. Webb, Drift theory of charged particles in electric and magnetic fields, Astrophys. Space Sci., 116, 107, (1985)

17. J.R. Jokipii and J. Kóta, The polar heliospheric magnetic field, Geophys. Res. Letters, 16, 1, (1989)

CHARACTERISTICS OF LARGE FORBUSH DECREASES ASSOCIATED WITH INTERPLANETARY MAGNETIC CLOUDS

Badruddin

Department of Physics, S.P. Degree College, Shohratgarh, Siddharth Nagar — 272 205 (U.P.), India

ABSTRACT

Three categories of magnetic clouds, i.e. clouds associated with shocks, with stream interfaces and with cold magnetic enhancements, have been utilized. Differences in amplitudes and profiles of cosmic ray decreases in association with these three types of clouds are found. Large Forbush-type decreases are associated with clouds preceded by a shock, whereas clouds followed by an interaction region produce smaller decreases in cosmic ray intensity. Those clouds associated with cold magnetic enhancement may produce a decrease of still smaller amplitude. Forbush-type decreases seem to be due to shocked plasma ahead of the magnetic cloud; the tangential discontinuity and magnetic clouds may not be the main causes of these decreases.

Using magnetic clouds identified during 1978-82 and their associated Forbush decreases we confirm our earlier result that the recovery time of Forbush decreases increases with the duration for which the magnetic cloud engulfs the earth and a simple empirical relation has also been obtained between these parameters. This relation may be used for computing the recovery time of Forbush decreases in the outer heliosphere as they propagate away from the earth if we know the expansion rate of the clouds and acceleration/deceleration of these disturbances as they move away and as long as they retain their identity.

INTRODUCTION

Since their discovery in 1938, the cause of Forbush decreases has been searched for in the fields and flows from the sun. It has been generally accepted that the structures of high magnetic field strength in the heliosphere are responsible for Forbush decreases. However, these regions may be the interplanetary magnetic clouds/loops of ordered field topology ejected from active regions on the sun, interplanetary shocks having relatively ordered field structure, the turbulent fields in the environment of shocks or tangential discontinuities /1/. The various proposed mechanisms of major importance responsible for these decreases are the gradient drifts in magnetic 'blob'/2/, gradient drifts in the environment of shock of relatively ordered structure /3/, reflection/screening of particles by smooth field lines in magnetic loops/clouds/4,5/, scattering due to turbulent fields between the shock front and the magnetic loop/cloud/6-10/, extra adiabatic cooling behind the shock front /11/.

In order to identify the interplanetary configuration and physical process mainly responsible for Forbush decreases, we have used the magnetic clouds identified in the plasma and field observations at 1 AU. Sixty four magnetic clouds were identified during 1967-1982. These magnetic clouds are defined as structures of radial dimension ~ 0.25 AU (at 1AU) in which the magnetic field strength is higher than average and magnetic field direction changes appreciably by means of rotation of one component of the field nearly parallel to the ecliptic plane. The magnetic pressure inside the cloud is higher than the thermal pressure/12/. Magnetic clouds were divided into three classes corresponding to their association with a shock, a stream interface or a cloud magnetic enhancement/10/.

In this paper we have studied the effect of these different categories of magnetic clouds on cosmic ray intensity variations. Plasma and field observations have also been utilized to study the characteristics of large amplitude decreases, generally observed in association with shock associated clouds.

RESULTS AND DISCUSSION

In Fig.1, we show the superposed epoch analysis results of hourly pressure corrected cosmic ray intensity observed at Deep River super neutron neutron monitor. The arrival time (hour) of the cloud observed in 1967-78 at 1AU is taken to be the epoch hour. In this way this analysis is performed, separately, using three classes of clouds i.e.(a) those preceded by shocks,(b) those followed by interaction regions and(c) those associated with cold magnetic enhancements. It is observed that cosmic ray intensity decreases associated with magnetic clouds preceded by shocks is large(Forbush-type) and the decrease starts at the time of arrival of shock front at the earth (Fig.1(a)). Moreover, most of the decrease takes place during the passage of the shocked region between the shock front and the magnetic cloud. The decreases associated with clouds followed by interaction regions (Fig. 1(b)) are smaller in magnitude but larger in duration. The extended depression even after the passage of an interaction region is due to high speed streams presumably coming from coronal holes. Here it may be mentioned that the high speed streams from coronal holes are less effective in transient modulation of cosmic rays than the streams associated with transients on the sun/13.14/. The decrease associated with cold magnetic enhancement clouds is of still smaller amplitude and duration. Moreover, in this case, the intensity recovers to its pre-decrease level only in two days time after the start of decrease (Fig. 1(c)).

In spite of the fact that, for all the three categories of clouds, the maximum field strength is same ($\sim$ 12) and the basic structure is similar/12/, the three types of clouds might be simply different manifestations of a single phenomena e.g. coronal mass ejections/15/, we still found a large difference in the amplitude and time profile of transient decreases associated with these three categories of clouds. One possible reason for this difference is that the shocks are followed by a turbulent sheath of ambient plasma in which, generally, there are large fluctuations in both the strength and the direction of the magnetic field. This shocked region, in turn, is followed by magnetic clouds (of rather ordered field structure) which are driving the shock. These magnetic clouds of ordered structure or the tangential discontinuities (interface region between sheath and the magnetic cloud) do not seem to play an essential role in initiating large Forbush-type decreases.

Fig. 2 shows the amplitude of individual Forbush decreases (due to shock associated clouds) and their relationship to maximum field intensity during these shock associated clouds observed in 1967-78 and 1978-82. The relationship between amplitudes of individual decreases and the speed of magnetic clouds is shown in Fig. 3 (see also, /8/). From these figures we see

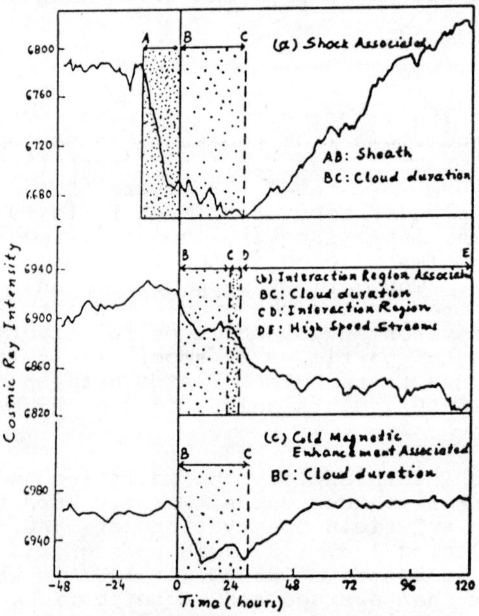

Fig. 1. Superposed epoch analysis results of hourly cosmic ray intensity observed at Deep River neutron monitor. Epoch hour corresponds to the arrival time (hour) of the cloud at 1 AU. Three plots in figure show average behaviour of intensity variation in association with three categories of clouds e.g. (a) shock associated, (b) interaction region associated and (c) cold magnetic enhancement associated clouds.

that the amplitude of decrease is, in general, larger when the field intensity is higher and when the speed of the magnetic clouds is higher.

The field magnitude may not be directly responsible for this modulation of cosmic rays. However, there may be a relationship between the field enhancement and the fluctuations in it, specially when the enhancement is associated with shocks. Moreover the correlation between the field magnitude and the decrease amplitude is poor (it may be inferred from the large scatter in data points in Fig.2). This may be interpreted to indicate that some other factor e.g. field fluctuations may be playing an important role in such intensity modulations. As regards the relationship between the speed of shock associated clouds and the amplitude of decrease, there is a well known correlation between the increase in the solar wind velocity and the magnetic field fluctuations /16/.

From the above arguments and results of Figs. 1,2 and 3, we conclude that it is not the magnetic field strength or topology alone which is responsible for cosmic ray depression. The most likely additonal effect - and perhaps a major effect - is the increased degree of turbulence between the shock front and the magnetic cloud.

In order to understand the process of modulation, the study of recovery characteristic of Forbush decreases is important /17/. In an earlier study Badruddin et al. /13/ studied the relationship of decrease time (from onset to minimum intensity) and reocvery time with the duration for which magnetic cloud engulfs the earth, suing clouds observed during 1967-78. In this paper we have extended this analysis by using data for the period 1978-82 (Fig.4)and found similar results. Thus we confirm our earlier result that the recovery time of the Forbush decrease, increases with the duration of the magnetic cloud engulfing the earth. A simple relationship is obtained i.e.,

$$R_t = 4.36\ C_d - 5.3$$

were R_t is recovery time and C_d is cloud duration (both in hours). This relationship may be used for computing the recovery time of Forbush decreases in the outer helpipshpere as they propagate away from the earth if we know the expansion rate of clouds and the acceleration/ deceleration of these disturbances as they move away.

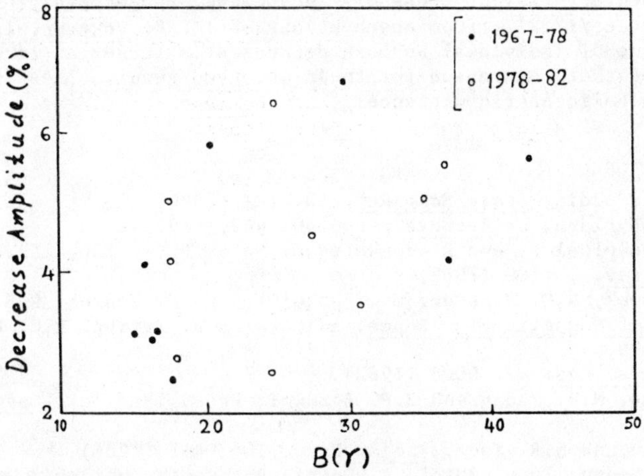

Fig. 2. Amplitude of individual Forbush decreases due to some shock associated clouds and their relationship to maximum field intensity.

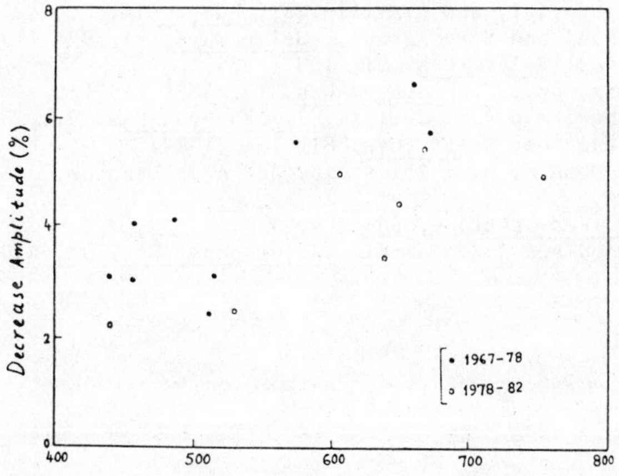

Fig. 3. Amplitude of individual Forbush decreases and their relationship to the speed of clouds.

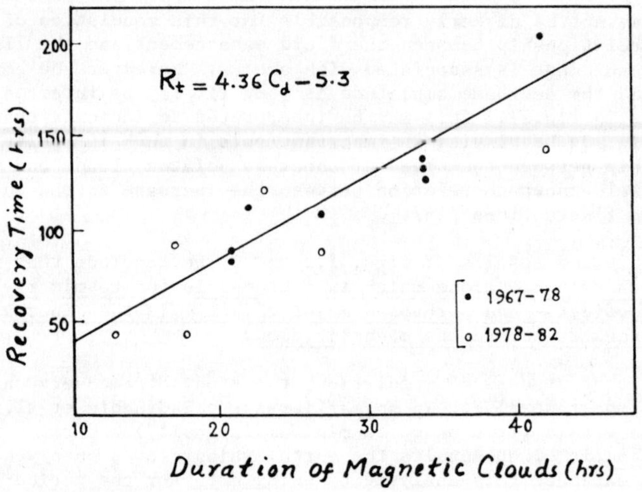

Fig. 4. Relationship between the duration of magnetic clouds and recovery time of Forbush decreases.

The observed duration of the cloud at a point in space will depend upon the size and/or speed of the cloud. Since the clouds are observed to be expanding and retaining their structure to distances of at least ∿ 11 AU /18, 19/, it is expected that the recovery time of these decreases will be longer at larger helio-distances. From observations of cosmic ray intensity at 1, 6.97 and 15.91 AU, Van Allen /20/ found that magnetized plasma cloud moved outwards and that the recovery time at these distances were, respectively, ∿ 6, 22 and 150 days. It has been reported /17,21/, from observations of IMP 8, Voyager, 1,2 and Pioneer 10, that the recovery time of individual Forbush decreases is larger at larger heliocentric distances. They concluded that the reason for their observed results lies in the evolution of the disturbances with heliocentric distance.

REFERENCES

1. D. Venkatesan and Badruddin, Space Sci. Rev. 52, 121 (1990).
2. E. Barouch and L.F. Burlaga, J. Geophys. Res. 80, 449, (1975).
3. E.T. Sarris, C.A. Dadoploulos, and D. Venkatesan, Solar Phys. 120, 153, (1989).
4. T. Gold, Space Sci. Rev. 1, 100 (1962).
5. T.R. Sanderson, J. Beek, R.G. Marsden, C. Tranquille, K.-P. Wenzel, R.B. McKibben, and E.J. Smith, Physics of Magnetic Flux Ropes, editors, E.R. Priest, L.C. Lee, and C.T. Russel, to appear (1990).
6. A. Nishida, J. Geophys. Res. 87, 6003 (1982).
7. Badruddin, R.S. Yadav, N.R. Yadav and S.P. Agrawal, Proc. 19th Int. Cosmic Ray Conf.,La Jolla 5, 258 (1985).
8. Badruddin, R.S. Yadav and N.R. Yadav, Solar Phys. 105, 413 (1986).
9. Badruddin and R.S. Yadav, Proc. 20th Int. Cosmic Ray Conf., Moscow 4, 63 (1987).
10. G. Zhang and L.F. Burlaga, J. Geophys. Res. 93, 2511 (1988).
11. B.T. Thomas and R. Gall, J. Geophys. Res. 89, 2991 (1984).
12. L.W. Klein and L.F. Burlaga, J. Geophys. Res. 87, 613 (1982).
13. N. Iucci, M. Parisi, M. Storini, and G. Villoresi, Nuovo Cim. 2C, 421 (1979).
14. D. Venkatesan, A.K. Shukla, and S.P. Agrawal, Solar Phys. 81, 375 (1982).
15. R.M. Wilson and E. Hildner, Solar Phys. 91, 169 (1984).
16. J.W. Belcher and L. Davis, Jr., J. Geophys. Res. 76, 3534 (1971).
17. J.A. Lockwood, W.R. Webber, and J.R. Jokipii, J. Geophys. Res. 91, 285 (1986).
18. L.F. Burlaga and K.W. Behannon, Solar Phys. 81, 181 (1982).
19. L.F. Burlaga and F.B. McDonald, M.L. Goldstein, and A.J. Lazarus, J.Geophys.Res. 90 12027 (1985).
20. J.A. Van Allen, Geophys. Res. Lett. 6, 566 (1979).
21. W.R. Webber, J.A. Lockwood and J.R. Jokipii, J. Geophys. Res. 91, 4103 (1986).

COSMIC RAY, ENERGETIC ION AND MAGNETIC FIELD CHARACTERISTICS OF MAGNETIC CLOUDS

T.R. Sanderson,* J. Beeck,** R.G. Marsden,* C. Tranquille,* K.-P. Wenzel* and R.B. McKibben***

**Space Science Department of ESA, Estec, Noordwijk, The Netherlands*
***ZARM, Universität Bremen, Bremen, F.R.G.*
****Laboratory for Astrophysics and Space Research, Enrico Fermi Institute, University of Chicago, Chicago, IL 60637, U.S.A.*

ABSTRACT

We present preliminary results of a survey of the relation between magnetic clouds and Forbush decreases during the period August 1978 to May 1982, using energetic ion and magnetic field observations from ISEE-3 to identify the magnetic clouds, and ground-based observations to identify the Forbush decreases. For this analysis we have selected all the events where a magnetic cloud is associated with the passage of an interplanetary shock. We observe a Forbush decrease associated with all except one of these 31 events. We find that many of the Forbush decreases have a two step structure. When the shock arrives at the earth, a slow decrease commences, with the intensity decreasing more or less monotonically whilst the post-shock turbulent region passes over the earth. When the magnetic cloud arrives at the earth, an additional, more rapid decrease is observed, with the minimum of the decrease occuring only a few hours after the arrival of the leading edge of the cloud. We also observe events where the shock and the post-shock turbulent region play only a minor role in the Forbush decrease. In these events, little or no decrease is observed either when the shock arrives at the earth, or when the post-shock turbulent region passes over the spacecraft. A large decrease is observed when the magnetic cloud arrives at the earth, the minimum being observed only a few hours after the arrival of the leading edge of the cloud. We examine in more detail one event of the the second type. In this event, the post-shock turbulent region has a high magnetic field variance, but does not affect the cosmic rays measured by the neutron monitors. A large magnetic cloud, with low magnetic field variance, containing bi-directional energetic ions, and separated from the post-shock region by a discontinuity, follows the cloud. The decrease observed coincides with the arrival at the earth of the discontinuity at the leading edge of this magnetic cloud.

INTRODUCTION

There have been many attempts to identify the origin of Forbush decreases. Both transient events /1/ and long lasting events /2/ have been proposed for the source of this modulation of the cosmic rays. Barouch and Burlaga /1/ and Sarris et al. /3/ have suggested that gradient drift could be responsible for the decreases. Nishida /4/ studied the magnetic structures reported by Bame et al. /5/ and Burlaga et al. /6/ and concluded that the post-shock region is responsible for Forbush decreases. Thomas and Gall /7/ suggested that the post-shock region acted as a barrier to cosmic rays, and that the decrease is caused by adiabatic cooling in the region behind. Badruddin et al. /8/ used a superposed epoch analysis of neutron monitor data, and concluded that the onset of the decreases started before the arrival of the magnetic cloud. Zhang and Burlaga /9/, also using a superposed epoch analysis of shock events, showed that although magnetic clouds could produce Forbush decreases, magnetic clouds preceded by a shock could produce larger decreases than clouds not preceded by a shock, and therefore that the turbulent magnetic fields associated with the sheath between a shock and a magnetic cloud are more effective in modulating the cosmic rays than the strong ordered fields in the magnetic cloud itself.

Detailed studies of large shock events /10, 11, 12, 13/ have shown that many of these events have a turbulent region lasting for several hours immediately after the shock, usually terminated by a discontinuity. This is often followed by a region containing bi-directional fluxes of energetic particles /13, 14/ which is magnetically quiet /15/. This region, where the variance of the magnetic field and the magnetic field power spectral density are lower than average /9, 13/ can last for several days, and is often identified with the driver gas from the flare /16/. Large-scale magnetic loop-shaped regions, identified by the rotation of the magnetic field direction in a plane perpendicular to the ecliptic plane, were called magnetic clouds by Burlaga et al. /6/ and Klein and Burlaga /17/. These regions are often considered to be the signature at 1 AU of a coronal mass ejection /18/. Large-scale structures in the solar wind, some of which are magnetic clouds, have been identified using the bi-directional signature of energetic protons /14,19, 20, 21/, or by the bi-directional signatures of electron heat flux /5, 22, 23, 24/.

In this paper, we present preliminary results of a statistical survey of the relation between Forbush decreases and magnetic clouds, to try to identify whether the post-shock region, or the magnetic cloud is responsible for the Forbush decrease. The survey reveals that some events have a two step decrease, the first step caused by the post-shock region and the second step by the magnetic cloud, whilst other events are caused just by the magnetic cloud. One of the events where just the magnetic cloud appears to be responsible for the decrease is analysed in more detail.

GENERAL CHARACTERISTICS OF FORBUSH DECREASES

For our data set, we used the bi-directional energetic ion events from Marsden et al. /21/ and took those which are associated with interplanetary shocks and a magnetic structure, on the assumption that they represent magnetic clouds. The 31 events so derived, which cover the period from August 1978 to May 1982, coincide reasonably well with the events of Zhang and Burlaga /9/, apart from some differences in start and stop times. This is due mainly to the different identification criteria applied in determining these times.

All but one of the 31 interplanetary shocks which have magnetic clouds associated with them coincide with a Forbush decrease. In addition, we restricted our analysis to those events where the decrease was greater than 2 percent, as measured by the Deep River neutron monitor, giving us a data set of 19 events.

In Figure 1 we show the eight largest Forbush decreases taken from our data set, stacked together with the time of arrival of the magnetic cloud as a reference. The small tick mark on each plot shows the time of arrival of the shock at the spacecraft. In our data set of 19 events, we found that in all but one event the minimum occured after the arrival of the leading edge of the magnetic cloud. The average time between the arrival of the magnetic cloud and the intensity minimum was 5.5 hours. This feature is also clearly visible in the superposed epoch studies /8, 9/. Some features are not apparent from these superposed epoch analyses.

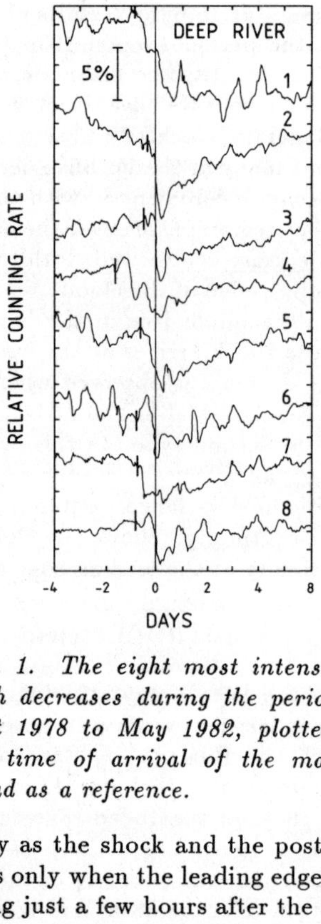

Fig. 1. The eight most intense Forbush decreases during the period August 1978 to May 1982, plotted with the time of arrival of the magnetic cloud as a reference.

From Figure 1 we can see that there are two types of events. In the first type, there is a two step decrease, as has been pointed out by many authors. A slow decrease in intensity begins at the time of arrival of the shock, and continues for several hours whilst the post-shock turbulent region which follows the shock passes over the spacecraft. A second, more rapid decrease begins at the time of arrival of the magnetic cloud. The minimum associated with this decrease is reached only a few hours after the arrival of the leading edge of the cloud. Events 1, 4, 5, and 7 display this behaviour.

In the second type of event, there is little or no change in the intensity as the shock and the post-shock turbulent region pass over the earth. In these events, the decrease begins only when the leading edge of the magnetic cloud arrives at the earth, with the minimum intensity occuring just a few hours after the arrival of the cloud. Events 2, 3, 6 and 8 exhibit this behaviour.

It is clear that both the post-shock turbulent region and the magnetic cloud can play a role in the Forbush decreases, but that in some of the cases, the post-shock turbulent region plays only a minor role. In a simple model of the Forbush decrease proposed by Sanderson et al. /25/, the first decrease is caused by a thick slab-type barrier – the post-shock turbulent region – passing over the observer, causing a slow monotonic decrease in the neutron monitor intensity which lasts throughout the passage of the post-shock turbulent region. The second decrease is caused by a thin barrier passing over the observer, this thin barrier being the discontinuity which marks the leading edge of the magnetic cloud.

THE FEBRUARY 11-12 1982 EVENT

The interplanetary shock event of February 11 1982, event 3 of Figure 1, is one of the events of the second type. Some results of the analysis have already been discussed in Sanderson et al. /26/. In Figure 2 we show the neutron monitor counting rates for the Deep River, Climax, and Huancayo neutron monitors for this event. These neutron monitors have been chosen because their asymptotic arrival directions /27/ at the time of arrival of the magnetic cloud were from an anti-sunward direction, along the interplanetary magnetic field spiral hose direction. The Forbush decrease, which lasted for several days, started at around 0000 UT on February 12, coincident with the arrival of a magnetic cloud at the spacecraft.

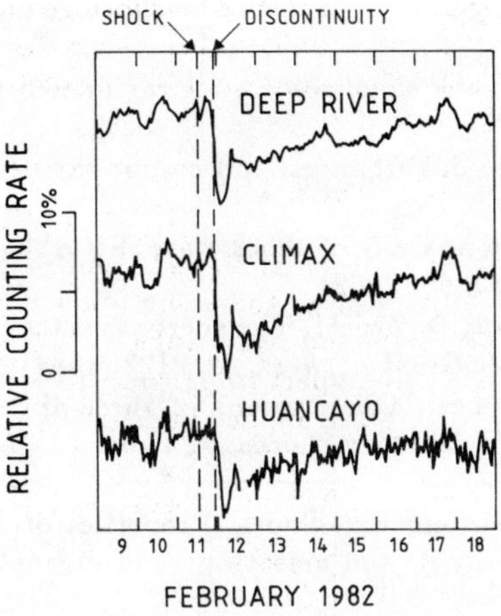

Fig. 2. Relative counting rates of the Deep River, Climax, and Huancayo neutron monitors for the ten-day period commencing on February 9, 1982.

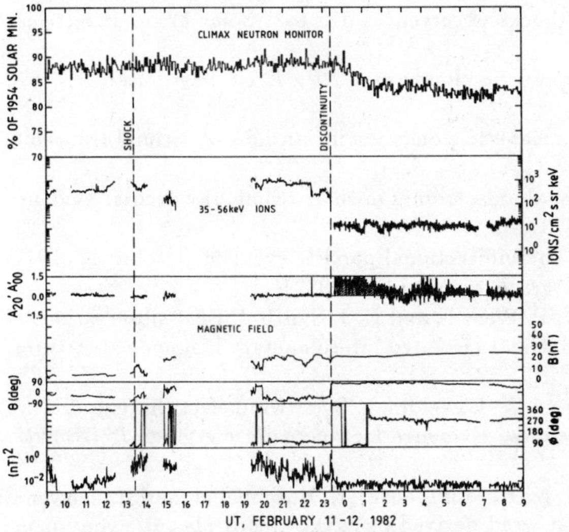

Fig. 3. From top to bottom: Climax neutron monitor counting rate, 35-56 keV proton intensity, 2nd-order field aligned proton anisotropy, A'_{20}/A'_{00}, magnetic field magnitude B, elevation theta, azimuth phi and variance for the 24 hr period beginning at 0900 UT on 9 February 1982.

Figure 3 shows the counting rate of the Climax neutron monitor, plotted together with the 35-56 keV proton intensity, the second harmonic of the proton anisotropy, A'_{20}/A'_{00}, the magnetic field magnitude B, elevation θ, azimuth ϕ, and the variance of the magnetic field. The interplanetary shock event has features which are similar to events attributed to first order Fermi acceleration studied elsewhere /10, 11, 12, 13/. The 35-56 keV intensity increases steadily for several hours prior to the arrival of the interplanetary shock, and stays more or less constant for the duration of the passage of the post-shock turbulent region. A rapid drop in intensity is observed at the edge of the post-shock turbulent region, coincident with the arrival of the leading edge of a magnetic cloud. The arrival of the magnetic cloud can be identified by the sudden appearance of bi-directional fluxes of energetic ions, which can be seen in Figure 3 by the enhancement of the second order anisotropy, A'_{20}/A'_{00}, at 2320 UT on February 11.

The arrival of the cloud can also be identified by changes in the magnetic field characteristics. When the cloud arrives at the spacecraft, the magnetic field starts to rotate slowly, and the variance of the field drops dramatically. The boundary between the post-shock turbulent region and the magnetic cloud is marked by a discontinuity.

From Figure 3, it can be seen that the onset begins at the same time as the discontinuity which marks the leading edge of the magnetic cloud passes the spacecraft. The precise timing must take into account the finite distance of the spacecraft from the earth. The discontinuity was observed when the solar wind velocity was around 600 km/s, so that the transit time from the spacecraft to the earth was around 40 minutes, if the discontinuity was travelling at the solar wind speed. This delay means that the cloud must have arrived at the earth at about 0000 UT on February 12, which is the same time as the onset of the Forbush decrease. Thus this event provides a clear example of a class of Forbush decreases where the dominant cause is the magnetic cloud, and in which the post-shock turbulent region plays a minor role in the event as a whole.

REFERENCES

1. Barouch, E., and L. F. Burlaga, Causes of Forbush decreases and other cosmic ray variations, *J. Geophys. Res., 80,* 449 (1975)
2. Duggal, S. P., M. A. Pomerantz, C. H. Tsao, B. T. Tsurutani, and E. J. Smith, *J. Geophys. Res., 86,* 7473 (1981)
3. Sarris, E. T., C. A. Dodopoulos, and D. Venkatesan, On the E-W asymmetry of Forbush decreases, *Solar Phys. 120,* 153 (1989)
4. Nishida, A, Numerical evaluation of the precursory increase to the Forbush decrease expected from the diffusion convection model, *J. Geophys. Res., 87,* 6003 (1982)

5. Bame, S.J., J.R. Asbridge, W.C. Feldman, J.T. Gosling, and R.D. Zwickl, Bidirectional streaming of solar wind electrons 80eV: ISEE evidence for a closed field structure within the driver gas of an interplanetary shock, *Geophys. Res. Lett., 8,* 173 (1981)
6. Burlaga, L.F., E. Sittler, F. Mariani, and R. Schwenn, Magnetic loop behind an interplanetary shock : Voyager, Helios, and IMP 8 observations, *J. Geophys. Res., 86,* 6673 (1981)
7. Thomas, B. T., and R. Gall, Solar-flare induced Forbush decreases: Dependence on shock geometry, *J. Geophys. Res., 89,* 2991 (1984)
8. Badruddin, R. S. Yadav, and N. R. Yadav, Influence of magnetic clouds on cosmic ray intensity variations, *Solar Phys., 105,* 413 (1986)
9. Zhang, G., and L. F. Burlaga, Magnetic clouds, geomagnetic disturbances, and cosmic ray decreases, *J. Geophys. Res., 93,* 2511 (1988)
10. Borrini, G., J.T.Gosling, S.J. Bame, and W.C. Feldman, An analysis of shock wave disturbances observed at 1 AU from 1971 through 1978, *J. Geophys. Res., 87,* 65 (1982)
11. van Nes, P., R. Reinhard, T.R. Sanderson, K-P. Wenzel, and R.D. Zwickl, The energy spectrum of 35- to 1600 keV protons associated with interplanetary shocks, *J. Geophys. Res., 89,* 2122 (1984)
12. Sanderson, T.R., R. Reinhard, P. van Nes, and K.-P. Wenzel, Observations of three-dimensional anisotropies of 35- to 1000-keV protons associated with interplanetary shocks, *J. Geophys. Res., 90,* 19 (1985)
13. Tranquille, C., T.R. Sanderson, R.G. Marsden, K-P. Wenzel, and E.J. Smith, Properties of a large-scale interplanetary loop structure as deduced from low-energy proton anisotropy and magnetic field measurements, *J. Geophys. Res., 92,* 6 (1987)
14. Sanderson, T.R., R. G. Marsden, R. Reinhard, K.-P. Wenzel, and E.J. Smith, Correlated particle and magnetic field observations of a large scale magnetic loop structure behind an interplanetary shock, *Geophys. Res. Lett., 10,* 916 (1983)
15. Smith, E. J., Observations of interplanetary shocks: Recent progress, *Space Sci. Rev., 34,* 101 (1983)
16. Zwickl, R. D., J. R. Ashbridge, S. J. Bame, W. C. Feldman, J. T. Gosling, and E. J. Smith, Plasma properties of driver gas following interplanetary shocks observed by ISEE-3, Solar Wind Five, *Nasa Conference Publication 2280,* 711 (1983)
17. Klein, L.W., and L. F. Burlaga, Interplanetary magnetic clouds at 1 AU, *J. Geophys. Res., 87,* 613 (1982)
18. Wilson, R.M., and E. Hildner, Are interplanetary magnetic clouds manifestations of coronal transients at 1 AU ?, *Solar Phys., 91,* 169 (1984)
19. Palmer, I.D., F.R. Allum, and S. Singer, Bidirectional anisotropies in solar cosmic ray events: evidence for magnetic bottles, *J. Geophys. Res., 83,* 75 (1978)
20. Kutchko, F.J., P.R. Briggs, and T.P. Armstrong, The bidirectional particle event of October 12, 1977, possibly associated with a magnetic loop, *J. Geophys. Res., 87,* 1419 (1982)
21. Marsden, R. G., T. R. Sanderson, C. Tranquille, K.-P. Wenzel, and E. J. Smith, ISEE-3 observations of low energy proton bidirectional events and their relation to isolated interplanetary magnetic structures, *J. Geophys. Res., 92,* 11009 (1987)
22. Montgomery, M. D., J. R. Ashbridge, S. J. Bame, and W. C. Feldman, Solar wind electron temperature depressions following some interplanetary shock waves: Evidence for magnetic merging, *J. Geophys. Res. 79,* 3103 (1974)
23. Pilipp, W. G., H. Miggenrieder, M. D. Montgomery, K.-H. Muhlhauser, H. Rosenbauer, and R. Schwenn, Unusual electron distribution functions in the solar wind derived from the Helios plasma experiment: Double strahl distributions with an extremely anisotropic core, *J. Geophys Res., 92,* 1093 (1987)
24. Gosling, J. T., D. N. Baker, S. J. Bame, W. C. Feldman, and R. D. Zwickl, Bidirectional solar wind electron heat flux events, *J. Geophys. Res., 92,* 8519 (1987)
25. A study of the relation between magnetic clouds and Forbush decreases, Sanderson, T. R., J. Beeck, R. G. Marsden, C. Tranquille, K.-P. Wenzel, R. B. McKibben, and E. J. Smith, Proceedings 21st International Cosmic Ray Conference, Adelaide, vol 6, p251-254 (1990)
26. Energetic ion and cosmic ray characteristics of a magnetic cloud, Sanderson, T. R., J. Beeck, R. G. Marsden, C. Tranquille, K.-P. Wenzel, R. B. McKibben, and E. J. Smith, Proceedings Chapman Conference on Flux-ropes, Bermuda (1989)
27. Shea, M. A., and D. F. Smart, Asymptotic directions and vertical cut-off rigidities for selected cosmic-ray stations as calculated using the international geomagnetic reference field model appropriate for epoch 1975, *AFCRL-TR-75-0247*, May, 1975.

NONLINEAR EFFECTS OF COSMIC RAY INTERACTION WITH SOLAR WIND IN THE OUTER HELIOSPHERE

V.Kh. Babayan* and L.I. Dorman**

*Erevan Physics Institute, Markarian st. 2, Yerevan, Armenia, 375036, U.S.S.R.

**Institute of Terrestrial Magnetism, Ionosphere and Radio Wave Propagation of Academy of Sciences of the U.S.S.R. (IZMIRAN), Troitsk, Moscow Region, 142092, U.S.S.R.

ABSTRACT

A set of equations is solved which describe the nonlinear interactions of galactic cosmic rays with the moving magnetized solar wind plasma including the inverse effect of cosmic rays on plasma streams. Allowance is also made for solar wind deceleration due to the effect of charge exchange with interstellar neutral atoms. The character of the variations of the Mach number with distance, the conditions of the transition from supersonic to subsonic streaming, and the features of forming the terminal shock wave are studied.

INTRODUCTION

In studying the propagation of galactic cosmic rays in the outer heliosphere, properties of the solar wind are usually considered independent of the cosmic ray intensity. Though it should be noted that the influence of galactic cosmic rays on the supersonic solar wind is comparable with that of other factors (stellar wind, exchange with interstellar neutral atoms, pressure of galactic magnetic field). Then, while investigating the galactic cosmic ray modulation in the outer heliosphere, one must take into consideration that the influence of cosmic rays on the solar wind varies with the intensity of cosmic rays in the time (with solar activity) and in space (with the distance to the Sun), which leads to a change in the velocity of the solar wind and hence, in the cosmic ray intensity. The solar wind velocity in the outer heliosphere will decrease also due to the solar wind proton charge exchange with the neutral interstellar hydrogen. That is, the self-consistent problem should be solved. In the frame of spherically-symmetric model of isotropic diffusion we consider also the character of the transition from supersonic solar wind flow to subsonic flow.

FORMULATION OF PROBLEM AND INVESTIGATION OF EQUATIONS SET

In the spherically symmetrical case the set of stationary equations of hydrodynamics for the solar wind with account of the galactic cosmic rays influence and the exchange of solar wind protons with interstellar hydrogen atoms have the form:

$$(d/dr)(r^2\beta u) = 0 \quad (1)$$

$$\beta u(du/dr) = -(dP_{SW}/dr) - (dP_{CR}/dr) - \beta u \quad (2)$$

$$r^{-2}(d/dr)(r^2 u(\beta u^2/2 + \gamma P_{SW}/(\gamma-1))) = -u(dP_{CR}/dr + Su). \quad (3)$$

Here β is the density of solar wind plasma, P_{SW} is the gaskinetic pressure of solar wind, $S = Q_{ex}\beta u N_H(r)$ is the exchange fasting of neutral interstellar hydrogen in cm^3 per 1 second, Q_{ex} is the cross-section of exchange, $N_H(r)$ is the distribution density of neutral interstellar hydrogen in the heliosphere (which are proportional to its concentration N_{HO} in interstellar metter near solar system), $P_{CR} = (1/3)\int_0^\infty N(T, r)pVdT$ is the pressure of cosmic rays on the distance r from the Sun (here $N(T, r)$ is the differential energy spectrum of cosmic ray concentration; p, V and T are momentum, velocity and kinetic energy of cosmic ray particles), u is the solar wind velocity.

From (1), (2) and (3) by using logarithmic differentiation we obtain for Mach number $M = u/V_s$ (where $V_s = (\gamma P_{SW}/3)^{0.5}$ is the sound velocity in solar wind plasma) the follow differential equation

$$M^{-2}dM^2/dr = (8M^2r^3(u^{-1}dP_{CR}/dr + Q_{ex}\beta u N_H(r)) - 12m(1+M^2/3))(3mr(1-M^2))^{-1}, \quad (4)$$

where $m = \beta_1 u_1 r_1^2$ (index 1 mean that all parameters are for 1 a.u.) and we put $\gamma = 5/3$. Let us consider only long-term cosmic ray variations. Because in this case the time of cosmic ray particles diffusion in heliosphere is very small relative to characteristic time of solar activity change we can use the stationary model. For isotropic spherically symmetric diffusion the equation which determined the change $N(T, r)$ (and therefore $P_{CR}(r)$) is (see Parker, 1965; Dorman, 1967)

$$r^{-2}(\partial/\partial r)(r^2(uN - D\partial N/\partial r)) - (1/3)(\partial/\partial T)(\alpha(T)Tr^{-2}(\partial/\partial r)(r^2 uN)) = 0, \quad (5)$$

where $\alpha(T) = (T + 2m_oc^2)/(T + m_oc^2)$ and m_o is the rest mass of cosmic ray particle. If we introduce the expression for P_{CR} in eq. (2) and take into account eq. (1) we obtain the integrodifferential equation

$$du/dr = -(r^2/3m)\int_0^\infty(\partial N/\partial r)pV dT - Q_{ex}uN_H(r). \quad (6)$$

The simultaneously solution of eqs (5) and (6) gives results about $N(T, r)$ and $u(r)$ in heliosphere for self-consistent problem. The results of numerically calculations $N(T, r)$ and $u(r)$ for some suppositions about N_{HO} and cosmic ray spectrum in interstellar space are obtain in papers Babayan and Dorman (1979a), (1979b). This results very sufficiently depend from value of N_{HO}, cosmic ray spectrum out the heliosphere and $D(T, r)$ in heliosphere. For investigation the character of transition from supersonic solar wind to subsonic solar wind flow let us consider the eq. (4) for Mach number.

TRANSITION FROM SUPERSONIC SOLAR WIND FLOW TO SUBSONIC FLOW

Let us consider the member $u^{-1}dP_{CR}/dr$ in eq. (4). By integrating eq. (5) on r and using the approximations of Jokipii and Parker (1968), Sousk and Lenchek (1969) we obtain at $ur/D \ll 1$ that approximately

$$dP_{CR}/dr \approx uP_{CR}/D \quad . \quad (7)$$

Let us suppose that near critical point $r = r_c$ (where $M^2 = 1$) approximately $P_{CR}/D = kr^j$, where k and j are some constants. Then from (4) we obtain

$$M^{-2}dM^2/dr = ((8/3)M^2r^3(kr^j + mr^{-2}L_{ex}^{-1} - mr^{-3}LL_{ex}^{-1}) - 4m(1+M^2/3))(mr(1-M^2))^{-1}, \quad (8)$$

where $L_{ex} = Q_{ex}N_{HO}$ and L is the distance, on what $N_H(r)$ decrease in e times. The behaviour of integral curves of eq. (8) near $r = r_c$ are determined by characteristic equation

$$s^2 - bs - na = 0, \quad (9)$$

where

$$a = (8/3)(kr_c^{2+j}(3+j) + mL_{ex}^{-1});\ b = (16/3)(kr_c^{3+j} + mr_cL_{ex}^{-1} - mLL_{ex}^{-1}) - 20m/3;\ n = -mr_c. \quad (10)$$

The position of critical point $r = r_c$ are determined by condition which follows from (8):

$$kr_c^{3+j} + mr_cL_{ex}^{-1} - mLL_{ex}^{-1} - 2m = 0\ . \quad (11)$$

According to (9) the eq. (8) have critical point $r = r_c$ (which are determined by eq. (11)) of the type of knot if $b^2 + 4an \geq 0$ and of the type of focus if $b^2 + 4an < 0$. Using (11) and (10) we obtain that if

$$j \leq -2 - (1 + 2LL_{ex}^{-1})/2(2 - r_cL_{ex}^{-1} + LL_{ex}^{-1}), \quad (12)$$

the critical point $r = r_c$ is of type of knot (in this case there are possibility of continuous transition from supersound flow to subsonic flow without formation of shock wave), but if

$$j > -2 - (1 + 2LL_{ex}^{-1})/2(2 - r_cL_{ex}^{-1} + LL_{ex}^{-1}), \quad (13)$$

the critical point $r = r_c$ is of type of focus (in this case the transition from supersonic solar wind flow to subsonic flow must be formated only by shock wave). According to Semar (1970), Mc Donough and Bracc (1971) $L \approx 3\text{-}5$ a.u. $\lesssim 10^{14}$ cm and $L_{ex} \approx 5\cdot 10^{15}$ cm (at $N_{HO} = 0.1$ cm^{-3}) then $2LL_{ex}^{-1} \ll 1$. We expect that $r_c \lesssim 100$ a.u., then $r_cL_{ex}^{-1} < 1$. Then we obtain approximately that at $j \lesssim -2.3$ there are possibility of shockless transition from supersonic flow to subsonic flow. If $j > -2.3$ we have not this possibility and the transition must be formed by shock wave.

DISCUSSION OF THE RESULTS AND CONCLUSION

The behaviour of u(r) and M(r) strongly depend from N_{HO}, from the primary spectrum of cosmic rays in the energy region $\lesssim 10$ GeV/nucleon out of the heliosphere and from D(T, r). If we take into account the results of Dorman, Ptuskin and Zerakashvili (1990) reported on this Colloqium about the self-consistent problem of kinetic stream instability of galactic cosmic rays in the outer heliosphere we expect increasing of cosmic ray concentration at energy ~ 1 GeV about 3 times from r = 50 a.u. to r = 60 a.u., then we expect approximately $P_{CR} \propto r^6$. From this paper we also expect that D(T, r) at T = 1 GeV decreased with increasing r (from $2 \cdot 10^{22}$ cm^2/s at 40 a.u. to $2 \cdot 10^{21}$ cm^2/s at 60 a.u.). Then we certainly expect that $j > 0$ (approximately $j \gtrsim 6$, but it is necessary to made additionally calculations of $P_{CR}(r)$ and D(T, r) near $r = r_c$) and the case (12) can not be realised in the outer heliosphere and we must expect that the transition from supersonic solar wind flow to subsonic flow formated by shock wave.

REFERENCES

Babayan V.Kh., Dorman L.I., 1979a. Proc. 16-th Int. Cosmic Ray Conf., Kyoto, Vol.3, p.117.

Babayan V.Kh., Dorman L.I., 1979b. Izvestiya AN SSSR, ser. phys., Vol. 43, p.793.

Jokipii J.R., Parker E.N., 1968. J. Geophys. Res., Vol.73, p. 3367.

Dorman L.I., 1967. Cosmic Rays, Moscow, Nauka, Vol. 8, p.305.

McDonough T.R., Bracc N.M., 1971. Icarus, Vol.15, p.505.

Parker E.N., 1965. Planet. and Space Sci., Vol. 13, p.9.

Semar C.L., 1970. J. Geophys. Res., Vol.75, p. 6892.

Sousk S.F., Lenchek A.M., 1969. Astrophys. J., Vol. 158, p. 781.

EVOLUTION OF THE IDEAS ABOUT THE HELIOSPHERE AND COSMIC RAY MODULATION IN INTERPLANETARY SPACE

I.V. Dorman* and L.I. Dorman**

*Institute of History of Sciences and Technology of the Academy of Sciences of the U.S.S.R., Staropansky 1/5, Moscow, 113012, U.S.S.R.
**IZMIRAN, Troitsk, Moscow Region, 142092, U.S.S.R.

ABSTRACT

The history of the evolution of ideas, models, and theories concerning modulation of galactic cosmic rays in the interplanetary space is described in brief, including the effects of the Sun's dipolar field, the steady-state diffusion, the corpuscular streams with induced electric fields, the dynamic solar wind, the magnetic inhomogeneity spectrum, the particle energy variations, the isotropic and anisotropic diffusion, the nonlinear and drift effects, the role of shock waves, the hysteresis and estimation of heliospheric dimensions, the neutral current sheet, and the terminal shock wave. The studies of the long-term modulation of galactic cosmic rays are shown to have played an important role in the formation of the present-day concepts concerning the heliosphere, in particular its outer zone.

THE FIRST INVESTIGATIONS OF COSMIC RAY MODULATION IN INTERPLANETARY SPACE

After establishing the extraterrestrial origin of cosmic rays, many researches made attempts to find some temporal variations in cosmic ray intensity (see the review in Dorman, 1981). However, no reliable data on the variations could be obtained in the early times. Only after the first worldwide network of cosmic ray stations were created in the mid-thirties (Compton et al., 1934), did it become possible, within several years, to obtain reliable data on some modulation effects: solar-diurnal and 27-day variations, magnetic storm effects and 11-year variation (Forbush, 1954). These network of stations equipped by shielded ionization chambers and the obtained information about cosmic ray modulation effects covered the primary particle energy range of several tens of GeV. The information in a substantially lower-energy range (3-10 GeV) where the amplitude of all modulation effects is much larger than that inferred from the ionization chamber data, were obtained when the network stations by the IGY (1957-1958) neutron monitors (Simpson et al., 1963) and, later, by the IQSY (1964-1965) neutron supermonitors (Carmichael, 1962) were created. The studies of neutron component data have made possible to find the hysteresis character of the relationships between the variations in solar activity and in cosmic ray intensity (Simpson, 1963; Dorman and Dorman, 1967a). This effect arises from the large size of the heliosphere ($\sim 10^2$ a.u.) and from the delay of the interplanetary processes with respect to the initiating solar processes (Dorman, 1967). Important information about the cosmic ray modulation effects in an even lower-energy range (from $\sim$100 MeV to $\sim$5 GeV) was obtained by regular stratospheric observations on balloons cosmic ray intensity variations from 1957 at several stations in the USSR and in the Antarctics by A.N. Charakhchyan's group. The analysis of these data, together with the cosmic ray neutron monitor data, has revealed the effect of sunspot heliolatitude shift on the long-term modulation variations (Quenby, 1965; Kudo and Wada, 1968; Stozhkov and Charakhchyan, 1969) and the effect of sign reversal of the Sun's general magnetic field on these variations (Charakhchyan et al., 1973).

MODEL FOR THE SUN'S GENERAL MAGNETIC FIELD OF DIPOLAR CHARACTER

The first idea for a mechanism of cosmic ray modulation in interplanetary space is the model of the Sun's dipolar magnetic field (Janossy, 1937). According to Janossy (1937) the east-west effect in such field results in solar-diurnal cosmic ray variation, the dipole axis inclination to the Sun's rotation axis in the 27-day variation, and possibly the change of the dipole magnitude with solar activity cycle - in the 11-year cosmic ray variation. But detailed comparison of observation data with the expected effects in cosmic rays made on the basis of the coupling coefficient method has shown that the discussed model is drastically at variance with experimental results (Dorman, 1957).

MODEL FOR CORPUSCULAR STREAMS IN INTERPLANETARY SPACE

In the early fifties, the model of galactic cosmic ray modulation by corpuscular streams which carry frozen magnetic fields (and induced electric field) was intensively developed: Alfven, 1950, 1954; Dorman and Feinberg, 1955, 1956; Dorman, 1956, 1957; Elliot and Dolbear, 1950; Swann, 1954; Trumpy, 1953. It was suggested two types of corpuscular streams: the first type have frozen magnetic fields $\sim$ some units $\cdot\ 10^{-5}$ Oe and number of them change in solar cycle very weakly; the second type have frozen magnetic fields with strength about 10 times more and number of this type of streams change almost proportionally to sunspot number change in solar cycle (Dorman, 1957). The first type of solar corpuscular streams later was identified with solar wind and the second type streams which caused Forbush-decreases was identified with interplanetary powerful shock waves and magnetic clouds from solar flares.

MODEL FOR COSMIC RAY PARTICLE DIFFUSION WITHOUT CONVECTION

To explain the cosmic ray interplanetary modulation Morrison (1956) proposed the model for particle diffusion due to scattering from magnetic inhomogeneities. The modulation effect arises due to finite time within the particles fill the region occupied by the inhomogeneities. The diffusion model without convection proved to be highly effective for explaining the data of solar cosmic ray propagation but these model proved to be insufficient when explaining the cosmic ray modulation (because of the trend for the volume to be filled rapidly as a result of particle diffusion, which go only in one direction to equality of particle densities into and out of the volume, see Dorman, 1957).

MODEL FOR ISOTROPIC DIFFUSION IN HELIOSPHERE INCLUDING CONVECTION BY SOLAR WIND

Proposed by Parker (1958) convection-diffusion model proved to be an important step in studying the global modulation of galactic cosmic rays in interplanetary space. In the concrete model of Parker (1958), however, a single basic scale of inhomogeneities ($l = 2\cdot10^{11}$ cm) was assumed and, therefore, the modulation spectrum for rigidity $R > R_c = 300\ B_0 l$ (where B_0 is average magnetic field in inhomogeneities) was $\propto R^{-2}$ but this was drastically at variance with experimental data (Dorman, 1959). To eliminate the disagreement, it was assumed by Dorman (1959) that the solar wind contained a broad spectrum of magnetic inhomogeneities which account for the observed dependence of the modulation amplitude $\propto R^{-1}$ (later it was confirm by direct measurements of magnetic fields in interplanetary space).

MODEL FOR ANISOTROPIC DIFFUSION INCLUDING CONVECTION AND ENERGY CHANGE, DRIFT EFFECTS AND INVERSE OF GENERAL SOLAR MAGNETIC FIELD

The generalization of convection-diffusion model on the case of anisotropic propagation in interplanetary spiral magnetic field with inhomogeneities and with take into account particles energy change in expansion solar wind was made in the first phenomenologically by Parker (1965) and Dorman (1965) independently. Later the transport equation of anisotropic diffusion was obtained more strictly from Bolzman kinetic equation by Dolginov and Toptygin (1966). The numerical calculations of expected cosmic ray modulation in the frame of this model was made by Dorman and Milovidova (1973), and then by including the drift effects in regular interplanetary magnetic field and the phenomena of general solar magnetic field reversal by Alaniya and Dorman (1979, 1981), Alaniya et al. (1988). From these results it can be seen that the discussed model proved to be highly effective when explaining the long-term variations of cosmic ray intensity and solar anisotropy, the behaviour of radial and transverse gradients of cosmic ray density.

POSSIBLE BUFFER REGION BETWEEN SUPERSONIC SOLAR WIND AND INTERSTELLAR MEDIUM

By investigations of Dorman and Dorman (1967 b, c) it was shown from an analysis of hysteresis of connection between solar activity change and cosmic ray intensity long-time variations that at small rigidities ($\lesssim 3$ GV) the delay in the recovery of cosmic ray intensity exceeds the theoretical value as the solar activity weakens. This indicates the presence out of the heliosphere with supersonic solar wind (possibly boundared by terminal shock wave, see below) also of a "buffer" region before the interstellar medium. In this region we must to have a subsonic flow with plasma density not depending from heliocentric distance r and with radial velocity $\propto r^{-2}$. From comparison with experimental data Dorman and Dorman (1967b) shows that the thickness of buffer region $\sim$100 a.u. and the diffusion coefficient for small energy particles are small (the transport path for scattering $\lesssim 10^{13}$ cm), that is necessary for explaining supplementary delay about 1 year. It was shown in (Dorman and Dorman, 1967b, c) that in the buffer region particles of solar wind plasma and particles from interstellar medium penetrating in this region can be accelerated to comparatively small energies, which can explain the appearance of an anomalues cosmic ray component.

MODEL FOR RECONNECTION BETWEEN MAGNETIC FORCE OF SOLAR WIND AND INTERSTELLAR MEDIUM NEAR SOLAR SYSTEM

Ahluwalia (1979) supposed that the difference in cosmic ray modulation in solar cycles with opposite direction of general solar magnetic field caused not drift effects but character of reconnection solar wind magnetic field with galactic magnetic field near solar system. In one case we obtain open heliosphere and galactic cosmic ray particles go easy into heliosphere; in another case we obtain closed heliosphere and it is more difficult for cosmic ray particles penetrate into heliosphere. Because the situation in reconnection model do not depend on the sign of particles charge we must expect the same effects for protons and electrons, which is at variance with experimental results. From another side, according to numerical calculations Alaniya et al. (1987) the global modulation of protons and nuclei of galactic cosmic rays in 1947-1957 and in 1971-1981 must be weaker (due to drift fluxes towards the equatorial region) and the anisotropy vector must shift to later hours. In 1958-1970 and in 1982-1993 the modulation must for protons and nuclei enhance and the solar anisotropy vector must shift to earlier hours (relative to model in which the drift effects not take into account). Inverse situation is expected in case of galactic electrons and antiprotons. These results are in good accordance with experimental data (or do not contradict them). From this we concluded that with high probability the reconnection model not realised, because the supersonic solar wind region with spiral magnetic field separated from almost homogeneous local galactic magnetic field by buffer region with turbulent magnetic field.

MODELS FOR NONLINEAR EFFECTS IN THE OUTER HELIOSPHERE

Since the modulation region size is large ($\sim$100 a.u.) the nonlinear interaction of solar wind and galactic cosmic rays must be included. There are three nonlinear effects: 1) solar wind radial deceleration by pressure of cosmic rays caused from positive radial cosmic ray gradient (Axford and Newman, 1965; Dorman and Dorman, 1968; Babayan and Dorman, 1977a, 1979, 1981), 2) transverse compression of solar wind streams by cosmic ray pressure caused from transverse cosmic ray density gradient (Dorman and Dorman, 1969; Babayan and Dorman, 1977b), 3) generation of magnetic inhomogeneities in the outer heliosphere (lead to decrease of cosmic ray diffusion coefficient and to increase of cosmic ray radial gradient) by cosmic ray anisotropy fluxes during stream kinetic instability (Babayan et al., 1987; Dorman et al., 1990). The investigation of these nonlinear effects was made in the frame of solution of self-consistent problems.

CONCLUSION

The evolution of the ideas and models for heliosphere and galactic cosmic ray modulation lead to the formation of much complex conception: self-consistent problem include cosmic ray anisotropic diffusion, convection, particle energy change, drift effects, nonlinear effects, formation of terminal shock wave (see Babayan and Dorman, 1990) and buffer region. It is necessary also to take into account the role in cosmic ray modulation magnetic clouds and shock waves moved with different velocity than solar wind (by generalise of this self-consistent problem on two- or many-component model, see Dorman and Shogenov, 1990).

REFERENCES

Ahluwalia H.S., 1979. Proc. 16-th Int. Cosmic Ray Conf., Kyoto, Vol. 12, p. 216.
Alaniya M.V., Dorman L.I., 1979. Proc. 16-th Int. Cosmic Ray Conf., Kyoto, Vol.3, p.57.
Alaniya M.V., Dorman L.I., 1981. Spatial Distribution of Cosmic Ray Density and Fluxes, Tbilisi.
Alaniya M.V., Dorman L.I., Aslamazashvili R.G., Gushchina R.T., Dzhapiashvili T.V., 1987. Galactic Cosmic Ray Modulation by Solar Wind, Tbilisi.
Alfven H., 1950. Cosmic Electrodynamic, Oxford.
Axford I.W., Newman R.C., 1965. Proc. 9-th Int. Cosmic Ray Conf., London, Vol.1, p.173.
Babayan V.Kh., Dorman L.I., 1977a. Proc. 15-th Int. Cosmic Ray Conf., Plovdiv, Vol. 3, p. 107.
Babayan V.Kh., Dorman L.I., 1977b. Proc. 15-th Int. Cosmic Ray Conf., Plovdiv, Vol. 3, p.113.
Babayan V.Kh., Dorman L.I., 1979. Proc. 16-th Int. Cosmic Ray Conf., Kyoto, Vol. 3, p.117, p.123.
Babayan V.Kh., Dorman L.I., 1981. Proc. 17-th Int. Cosmic Ray Conf., Paris, Vol. 3, p.373.
Babayan V.Kh., Dorman L.I., 1990. Space Res. (in this issue).
Babayan V.Kh., Dorman L.I., Ptuskin V.S., 1987. Proc. Int. Conf. on Plasma Physics, Kiev, Vol. 4, p.228.
Carmichael H., 1962. NM-64 Neutron Monitor Specification, Deep River.
Charakhchyan A.N. et al., 1973. Izvestiya AN SSSR, ser. fiz., Vol. 37, p. 1258.
Compton A. et al., 1934. Rev. Sci. Instr., Vol. 5, p.415.
Dolginov A.Z., Toptygin I.N., 1966. Izvestiya AN SSSR, ser. fiz., Vol.30, p. 1780.

Dorman L.I., 1956. Izvestiya AN SSSR, ser. fiz., Vol. 20, p.24.
Dorman L.I., 1957. Cosmic Ray Variations, Moscow.
Dorman L.I., 1959. Proc. 6-th Int. Cosmic Ray. Conf., Moscow, Vol. 4, p. 328.
Dorman L.I., 1965. Proc. 9-th Int. Cosmic Ray Conf., London, Vol. 1, p. 292.
Dorman L.I., 1967. Cosmic Rays, N 8, p. 305.
Dorman I.V., 1981. Cosmic Rays: Historical Review, Moscow, Nauka
Dorman I.V., Dorman L.I., 1967a. J. Geophys. Res., Vol.72, p. 1513.
Dorman I.V., Dorman L.I., 1967b. Izvestiya AN SSSR, ser. fiz., Vol. 31, p.1239.
Dorman I.V., Dorman L.I., 1967c. Izvestiya AN SSSR, ser. fiz., Vol. 31, p.1273.
Dorman I.V., Dorman L.I., 1968. Geomagnetism and Aeronomy, Vol. 8, p.817, p.1008.
Dorman I.V., Dorman L.I., 1969. Izvestiya AN SSSR, ser. fiz., Vol. 33, p. 1908.
Dorman L.I., Feinberg E.L., 1955. Proc. 4-th Int. Cosmic Ray Conf., Guanajato, p. 393.
Dorman L.I., Feinberg E.L., 1956. Uspekhi Phys. Nauk., Vol.59, p.189.
Dorman L.I., Milovidova N.P., 1973. Proc. 13-th Int. Cosmic Ray Conf., Denver, Vol. 2, p. 713.
Dorman L.I., Ptuskin V.S., Zerakashvili V.N., 1990. Space Res. (in this issue).
Dorman L.I., Shogenov V.Kh., 1990. Space Res. (in this issue).
Elliot H., Dolbear D.W., 1950. Proc. Phys. Soc., Vol. 63, p. 137.
Forbush S.E., 1954. J. Geophys. Res., Vol. 59, p. 525.
Janossy L., 1937. Zs. Phys., Vol. 104, p. 430.
Kudo S., Wada M., 1968. Rep. Ionosph. and Space Res., Japan, Vol. 22, p.137.
Morrison P., 1956. Phys. Rev., Vol. 101, p. 1397.
Parker E.N., 1958. Phys. Rev., Vol. 110, p. 1445.
Parker E.N., 1965. Proc. 9-th Int. Cosmic Ray Conf., London, Vol. 1, p. 26.
Quenby J.J., 1965. Proc. 9-th Int. Cosmic Ray Conf., London, Vol. 1, p. 3.
Simpson J.A., 1963. Proc. 8-th Int. Cosmic Ray Conf., Bombay, Vol. 2, p.152.
Simpson J.A. et al. 1963. Phys. Rev., Vol. 90, p. 934.
Stozhkov Yu.I., Charakhchyan T.N., 1969. Geomagnetism and Aeronomy, Vol.9, p.803.
Swann W., 1954. Phys. Rev., Vol. 93, p. 905.
Trumpy B., 1953. Physica, Vol. 19, p.645.

THE LONG TERM COSMIC RAY VARIATION RELEVANT TO SOLAR WIND STRUCTURE IN THE OUTER HELIOSPHERE

M.V. Alania,* R.G. Aslamazashvili,* T.B. Bochorishvili,*
L.I. Dorman,** R.T. Guschina** and K. Iskra***

**The Institute of Geophysics of the Academy of Sciences of Georgian SSR, Tbilisi, U.S.S.R.*
***IZMIRAN, Troitsk, Moscow Region, U.S.S.R.*
****Siedlce University, Siedlce, Poland*

ABSTRACT

It is shown that the energy spectrum of cosmic ray (CR) variations of isotropic intensity is hard in the minimum periods and is soft in the periods of solar activity maximum. The conclusion is made that the different structures of the magnetic inhomogeneities of solar wind (SW) during the different epochs of solar activity are responsible for such energy spectrum changes. The effective size l of inhomogeneities of solar wind in the periods of near minimum and in the minimum of solar activity is greater than in the periods of solar activity maximum.

INTRODUCTION

The world network of the neutron supermonitors that covers the whole Earth's surface and registers the intensity of cosmic rays in a wide energy range, gives a possibility to define and to study the energy spectrum of the long-term cosmic ray variations. The accuracy of defining the energy spectrum parameters depends on the data continuity and homogeneity. It is high in the periods of solar activity maximum when the amplitude of the relative variation of cosmic rays is significant and becomes low in the minimum epoch when the amplitude of relative variation of cosmic rays is small /1/. The energy spectrum of the long-term variations of cosmic rays of isotropic intensity can be presented in the form

$$\delta D(R)/D(R)=\begin{cases} AR^{-\gamma} & R\le R_{max} \\ 0 & R>R_{max} \end{cases} \qquad (1)$$

where R is the particle rigidity, R_{max} - the upper limiting rigidity beyond which the modulation of cosmic rays vanishes. A - the power of energy spectrum of cosmic ray variation. The intensity change of the secondary component of cosmic rays at the observation point with the geomagnetic cutoff rigidity R_c is defined in the following way:

$$\delta J/J=\int_{R_c}^{\infty} (\delta D(R)/D(R))\ W(R,h)\ dR \qquad (2)$$

where W (R,h) is the coupling coefficient of cosmic rays.

METHOD AND EXPERIMENTAL DATA

For the definition of the energy spectrum parameters of cosmic ray primary variation $\delta D(R)/D(R)$ it is possible to use two methods. The first method is to define the amplitude ratio A_1/A_2 of the variation by the observation data at two stations with different geeomagnetic rigidities R_1 and R_2 of cosmic ray primary radiation /2/. Namely, the amplitudes A_1 and A_2, corrected for the meteorological effects, of the secondary cosmic ray component are expressed in the following way:

$$A_1=\int_{R_1}^{\infty} (\delta D(R)/D(R))\ W_1(R,h)dR \qquad A_2=\int_{R_2}^{\infty}(\delta D(R)/D(R))\ W_2(R,h)dR \qquad (3)$$

where A_1 and A_2 - the amplitudes of cosmic ray variation at different observation stations, $W_1(R_1,h_1)$ and $W_2(R_2,h_2)$ – the corresponding coupling coefficients at different depths h_1 and h_2 of the Earth's atmosphere.

Inserting (1) into (2) and (3) and taking the ratio $A_1/A_2 = M(\gamma)$ we obtain

$$M(\gamma) = \left[\int_{R_1}^{\infty} R^{-\gamma} W_1(R,h)dR\right]\left[\int_{R_2}^{\infty} R^{-\gamma} W_2(R,h)dR\right]^{-1} \quad (4)$$

As it is seen from the expression (4) the ratio M is the function of γ. Hence, the actual magnitude γ corresponds to the definite value of $M(\gamma)$. The data of the two stations Climax (R = 3.03 GV) and Huancayo (R = 13.5 GV) for 1954–1984 period have been used for the definition of the temporal change of the energy spectrum γ of cosmic rays. The calculations have been carried out using the annual means of cosmic ray neutron component intensity and the obtained temporal change of the energy spectrum index is presented in Fig.1a. The error of the energy spectrum index $\Delta\gamma$ is defined by the errors of the ratio of the annual mean values of the amplitudes of cosmic ray isotropic intensity variation . The error $\Delta\gamma$ is smaller in the periods of the solar activity maximum when the amplitudes A_1 and A_2 are relatively large and $\Delta\gamma$ is bigger in the period of the solar activity minimum when A_1 and A_2 are small.

In contrast to the method of two stations the second method is based on the data analysis of the world network of stations with different thresholds of the magnetic cutoff. The data of the neutron monitors for 1964–1987 have been used in this case and stations have been chosen according to a data quality criterium. The number of the stations for the different years was different but not less than 10. The amplitudes of cosmic ray intensity variation with respect to 1965 (the period of solar activity minimum) have been calculated using the annual means at each station. The amplitude of the secondary component intensity variations of cosmic rays corrected for the meteorological effects with the geomagnetic cutoff threshold R_n taking into account (1) is expressed in the form

$$(\delta J/J)_n = A \int_{R_n}^{\infty} R^{-\gamma} W_n(R,h)dR \quad (5)$$

where n – the station number and other notations are as above. The power spectrum parameters A for each station are calculated from the expression (5) for different values of the parameters R_{max} and γ /2/.

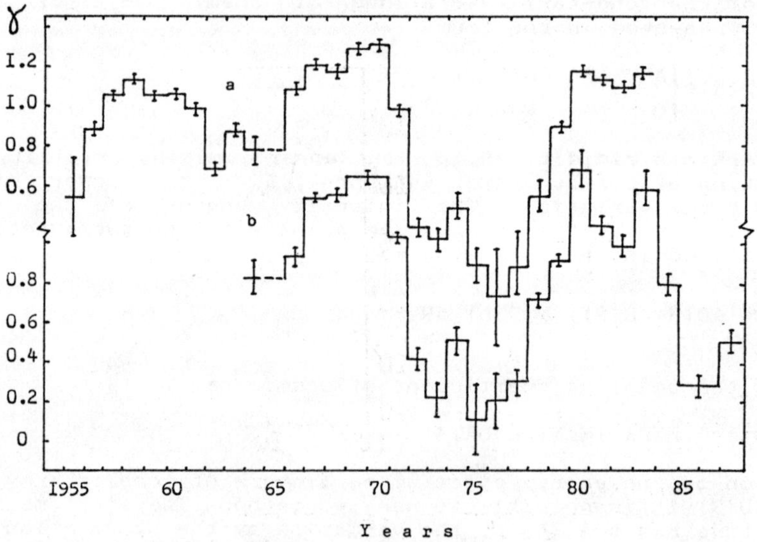

Fig.1 a, b The temporal changes of the energy spectrum index γ
a) For two stations Climax and Huancayo and
b) For the world network of stations

Taking into account that values A_n must be the same for all stations, the least square method $\sum_{n=1}^{m} (A_n - \bar{A})^2 = \min$ (where m – the number of the stations used and $\bar{A} = (1/m)\sum A_n$) was used for determination of γ. The obtained values of

γ for each year in the case R_{max}= 200 GV are presented in Fig. 1b. It is seen from the comparison of the results with Fig. 1a that the energy spectrum indices obtained by two methods within the error limit coincide for 1965 - 1984 period; so it must be expected that for those time intervals 1954-1961 when the data of world network were not used, the calculation results of γ from the data of two stations are quite reliable. The analysis of the temporal changes of the energy spectrum of CR 11-year variations for 1955-1987 period (Fig. 1a,b) shows that the energy spectrum in solar activity maximum is soft and that it is hard in the periods of minimum.

THE INTERPRETATION OF EXPERIMENTAL DATA

The abovementioned character of changes of the energy spectrum is related to the different structures of the magnetic inhomogeneities of SW in different periods of solar activity in the region of the Earth's orbit. So, the distribution of the magnetic inhomogeneities of SW in solar activity maximum is such that the effective size of the magnetic inhomogeneities causing CR particle diffusion in 10-50 GeV energy range is smaller than the Larmor radius ρ of CR particle ($l<\rho$) and the scattering is strongly dependent on the particle rigidity R. In particular, the transport path λ of the particles in the maximum and near solar activity maximum depends on the rigidity in the following way: $\lambda \propto R^{1.5-2.0}$. In the region of minimum and in solar activity minimum, $\lambda \propto R^{0.2-0.5}$ since $l \approx \rho$. Thus, a significant rearrangement of the structure of the magnetic inhomogeneities occurs during the 11-year cycle of solar activity: both the power and the spectral index of SW magnetic fluctuations change. A character of this redistribution can be inferred from the observed changes of cosmic rays 11-year variation energy spectrum and from considerations of the spatial and temporal distribution of solar activity.

Thus, the interaction of the high-speed fluxes with the background SW significantly alters the structure of SW inhomogeneities at a distance from the Sun, especially at 2-10 a.u. The new large-scale inhomogeneities are formed, in particular, as it is shown in /3/, both the power of the spectrum of the magnetic inhomogeneities and the index of this spectrum increase in the low frequency regions. Hence, the intensification of the spectrum power of SW inhomogeneities at low frequencies occurs at a large distance from the Sun in the ecliptic plane, what leads to harder spectrum of CR 11-year variation. A relatively hard energy spectrum is expected for the period of solar activity maximum out of ecliptic plane (in the regions of the medium and high heliolatitudes) since the coronal holes are the sources of the inhomogeneities of large sizes in these regions of the heliosphere.

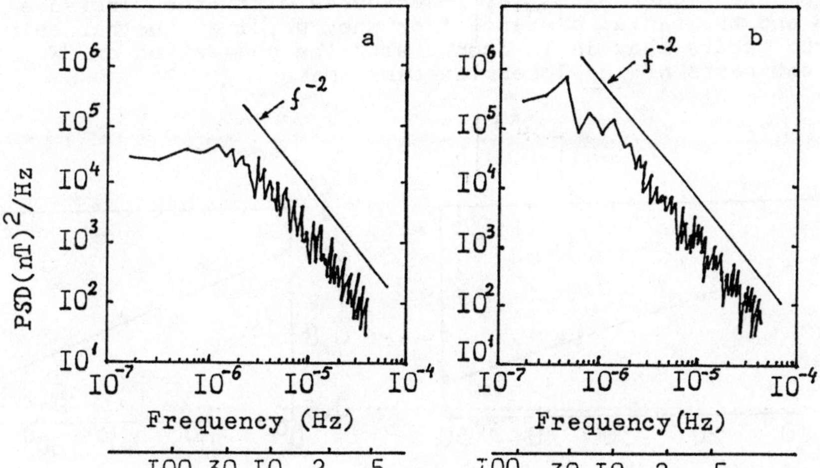

Fig.2 a, b Spectra of the large scale fluctuations in magnetic field strength a) at 1 AU and b) at 6.1 - 8.9 AU /3/. Lower ordinate shows resonant proton energy (GeV)

The spectra of the large-scale fluctuations in the interplanetary magnetic field (IMF) strength at 1 a.u. obtained by ISEE (Fig.2a) and in the heliosphere (6.1–8.9) obtained by Voyager 1 (Fig. 2b) during the period from 1979, day 250, to 1980, day 250 /3/ are presented in Fig. 2a,b. It is seen from the Fig. 2a,b that while PSD of IMF at 1 a.u. for $f<2.10^{-6}$Hz becomes flat (at $P \propto f^{-\alpha}$, $\alpha \simeq 0$), at a distance of 6.1–8.9 a.u. for $f>3.10^{-7}$Hz, $\alpha \simeq 2$.

Since the energy spectrum of cosmic ray isotropic variation is defined, mainly, by the dependence of the diffusion coefficient on the particle rigidity R_i then it is seen from /4/ $\varkappa_{11} = R^{2-\alpha} B_0^{\alpha}/P$ (where B_0 – the average IMF, P is PSD of IMF) that when $\alpha \approx 1.5$–2 the variation spectrum will be hard, i.e. our assumption on cosmic ray variation spectrum hardening is proved at a radial distance from the Sun and out of the ecliptic plane. The significant differences in the energy spectrum of the magnetic inhomogeneities of SW in different parts of the interplanetary space are not expected in solar activity minimum, but, in our opinion, the common picture of redistribution of the spectrum of the magnetic inhomogeneities out of the ecliptic plane and at a large radial distance is preserved.

THE THEORETICAL MODEL

The stationary equation of the anisotropic diffusion /5/ for the definition of the spatial distribution of the expected changes of the energy spectrum index is solved numerically

$$\nabla_i(\varkappa_{ik}.\nabla_k n) - \nabla_i(u_i.n) + \frac{1}{3}\frac{\partial}{\partial R}(R.n).\nabla_i u_i = 0 \qquad (6)$$

where $\varkappa_{ik}$ – the tensor coefficient of the diffusion, u_i – SW velocity, n and R – CR density and particle rigidity, respectively. The equation (6) is solved for the boundary conditions /5/ with constant solar wind velocity U=400 k.m./s and the dependence of the diffusion coefficient on the particle rigidity $\varkappa \propto R^{\alpha(r,\theta)}$ where a) α=0.5–0.7 at the Earth's orbit for the period of minimum and near solar activity minimum, b) α=1.2–1.5 at the Earth's orbit for the period of solar activity maximum (in the cases a) and b) α doesn't depend on the spatial coordinates), c) $\alpha(r,\theta)$ changes with the coordinates in the following way: α= 1.5 at r= 1 a.u. and α = 0.5–0.7 at r = 30 a.u. for any θ; α = 1.5 when θ = 0° and α = 0.5 at θ = 90° for any distance.

The values of γ, obtained on the basis of solution of the anisotropic diffusion equation for the cases a) and b) are in accord with the experimental data both for the maximum period and for the period of solar activity minimum at the Earth's orbit. In our point of view, most interesting is the case c) when the change $\alpha(\tau, \theta)$ is taken into account. The change in γ with the radial distance r and the heliolatitude θ is presented in Fig. 3a,b. It is seen from the Figure 3a,b that γ decreases with the increase of the heliolatitude and the radial distance from the Sun, i.e. the hardening of the energy spectrum occurs that is in accord with the suggestion on SW structure in the different parts of the interplanetary space.

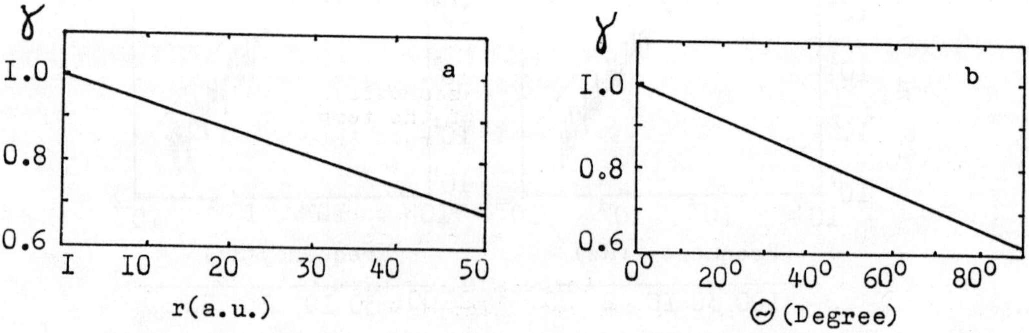

Fig.3 a, b The changes of the energy spectrum index γ as a function: a) of distance r from the Sun and b) of heliolatitude θ.

DISCUSSION

The analysis of the temporal changes of the energy spectrum of 11-year cosmic ray variations gives a possibility to diagnose and predict both the spatial and temporal changes of the structure and the power of the magnetic inhomogeneities of solar wind by the underground directed telescopes. The fact is that cosmic rays of the relatively high energies (150–250 GeV and more) the trajectories of which do not curve in the Earth's magnetic field can be used for the remote sensing of the heliosphere along the ecliptic plane and out of it. The telescope directed along the ecliptic plane will give the information on the structure of the magnetic inhomogeneities of solar wind from the region where the interactions of the transient and corotating fluxes connected with the solar activity centres in the range ± 30° of the heliolatitude play an important role.

SUMMARY

1. The energy spectrum of the isotropic intensity of the long-term variations of cosmic rays in the periods of solar activity maximum is soft (in 1957–1961, γ = 1.1–1.2; in 1967–1970, γ = 1.3–1.4; 1980–1983, γ = 1.2–1.4), and is hard in the periods of minimum and near the minimum of solar activity (in 1954–1955, γ = 0.6–0.7; in 1963–1965; γ = 0.6–0.8; in 1973–1977, γ = 0.2–0.4; in 1985–1987, γ = 0.2–0.4).

2. Both soft and hard energy spectrum of isotropic variations of cosmic rays are conserved on the average, for 5–6 year period.

3. During the increase of solar activity the energy spectrum of isotropic intensity variations of cosmic rays becomes gradually soft (during 4–5 years) while during solar activity decrease the energy spectrum becomes drastically hard (1–2 years).

4. The different structures of the magnetic inhomogeneities of solar wind during the different epochs of solar activity are responsible for such energy spectrum changes. The effective size l of the inhomogeneities of solar wind in the periods of near minimum and in the minimum of solar activity is greater than in the periods of solar activity maximum.

5. The hardening of the energy spectrum of the isotropic intensity variations of cosmic rays must take place with the increasing radial distance from the Sun and out of the ecliptic plane. It must be connected with the changes of the spectra of the large-scale fluctuations in the interplanetary magnetic field in the different parts of the heliosphere.

6. The different structures of the magnetic inhomogeneities of solar wind, which determine the dependence of the diffusion coefficient on the rigidity of cosmic ray particles in different epochs of solar activity, are responsible for 11-year variations of cosmic rays.

ACKNOWLEDGEMENT. The authors are grateful to the personnel of Cosmic Ray Stations from which data were obtained. Committee for Scientific Committee on Solar-Terrestrial Physics Science Council of Japan and National Geophysical Data Center, Boulder, Colorado USA supplied us with cosmic ray data.

REFERENCES

1. K. Iskra, M.V. Alania, R.G. Aslamazashvili, T.B. Bochorishvili, L.I. Dorman and R.T. Guschina, The features of the temporal change of the energy spectrum of 11-year variations of cosmic ray intensity, Proc.20th Int.Cosmic Ray Conf. vol.3, Moscow, 1987.

2. S. Yasue, S. Mori, S. Sakakibara and K. Nagashima, Coupling coefficients of cosmic rays daily variations for neutron stations, Nagoya, 1982.

3. L.F. Burlaga and W.H. Mish, Large-scale fluctuations in the interplanetary medium, J. Geophys. Res. 92 A2, 1261–1266 (1987).

4. J.R. Jokipii, Propagation of cosmic rays in the solar wind. Rev. of Geophys. and Space Phys. 9, 1, 27–87, (1971).

5. M.V. Alania, L.I. Dorman, R.G. Aslamazashvili, R.T. Guschina and T.V. Japiashvili, Modulyatsiya galakticheskich kosmicheskikh luchei solnechnym vetrom., Tbilisi, Metsniereba, 1987.

OUTER HELIOSPHERE: EIGEN PULSATIONS, COSMIC RAYS AND STREAM KINETIC INSTABILITY

L.I. Dorman, V.S. Ptuskin and V.N. Zirakashvili

Institute of Terrestrial Magnetism, Ionosphere and Radio Wave Propagation of Academy of Sciences of the U.S.S.R. (IZMIRAN), 142092 Moscow Region, Troitsk, U.S.S.R

ABSTRACT

The possible large-scale eigen oscillations of the solar wind cavity and the relevant variations of cosmic density are discussed. The expected period of the pulsations is $T = 2R/V_s$ (here R is the size of the cavity, V_s is the sound velocity in the interstellar space near the solar system). The stream kinetic instability of galactic cosmic rays entering the heliosphere is examined. The energy density of excited MHD waves is found. We discuss the modes of cosmic ray propagation in the outer heliosphere.

INTRODUCTION

The penetration of solar wind into interstellar medium is probably accompanied by a shock transition from supersonic to subsonic motion. Solar wind cavity as a whole can have eigen pulsation since the boundary between the wind and the interstellar medium is at the position of a stable equilibrium. Such kind of the natural pulsations of the heliosphere were studied for the first time by Dorman and Ptuskin (1981), Paper I. Here we continue this investigation and consider some search of cosmic ray variations caused by eigen heliosphere pulsations. Another problem is the cosmic ray propagation in the outer heliosphere. Galactic cosmic rays penetrating into the heliosphere interact with the solar wind the plasma and it may be a cause of development of the stream kinetic instability (Babayan et al., 1987).

As a result of this interaction, the MHD-waves are excited. These waves may sufficiently influence the diffusion of relativistic charged particles in outer heliosphere (self-consistent problem).

EIGEN PULSATIONS IN THE OUTER HELIOSPHERE

The problem of heliosphere pulsations was considered in Paper I for the spherically-symmetric solar wind flow. Weak disturbances in the interstellar medium out of the shock transition were treated as sound waves. The equation for the perturbations of the density of the interstellar gas ρ has the form

$$\partial^2\rho/\partial t^2 - V_s^2 \Delta\rho = 0. \qquad (1)$$

Here V_s is the sound velocity in the interstellar medium. Equation (1) is supplemented with the boundary condition at the shock transition

$$\left(\frac{\partial^2\rho}{\partial t^2} - \frac{2P}{R\rho}\frac{\partial\rho}{\partial r}\right)\Big|_{r=R} = 0, \qquad (2)$$

where R is the radius of the cavity, P is the pressure in the interstellar medium. The disturbances do not propagate into the cavity because of the supersonic character of the solar wind flow. Initial perturbation of the boundary leads to the natural pulsations of the cavity with decreasing amplitude. The attenuation is mainly due to the radiation of sound waves in the outer interstellar space. The smallest attenuation was found for one of the second spherical harmonics. The eigenfrequency of this mode is

$$\Omega \approx 1.6\, V_s R^{-1} (1 - 0.08i). \qquad (3)$$

The real structure of the heliopause is more complicated, see Fahr et al. (1986) for review. Nevertheless one can expect the existence of the eigen oscillations of the solar cavity with the period of the order of

$$T = (\mathrm{Re}\,\Omega)^{-1} \approx 5R/V_* . \qquad (4)$$

Here V_* is the velocity of waves determined by the action of the thermal gas pressure (as in usual sound waves), turbulent pressure, magnetic field, tensions and cosmic-ray pressure. The value of R characterizes the dimensions of the cavity. For nonspherical cavity there exists the set of the appropriate values of R and T.

SEARCH OF HELIOSPHERE EIGEN PULSATIONS BY COSMIC RAYS

Depending on the accepted parameters of the interstellar medium one can estimate the value of $T \approx 2 - 30$ yrs for the radius $R = 50$ a.u. ($V_* = 5\cdot10^6 - 5\cdot10^7$ cm/s). Resonant pulsations of the solar wind cavity at the eigenfrequency Ω should be generated by random disturbances in the kinetic energy density of the solar wind. The pulsations, if they exist could lead to the variations of cosmic-ray intensity at the Earth, see Paper I for details. The signature of the considered mechanism is an existence of pronounced peak in the power spectrum of cosmic-ray intensity variations not accompanied by any peculiarities in the power spectrum of solar activity at the same frequency.

We used the data on the area of sunspots (observations made in 1954-1983) and the intensity of green coronal line 5303 A (for 1957-1979) as indicators of solar activity. Neutron monitor data collected at stations Climax (1954-1979), Deep River (1958-1981), Wellington (1957-1983) and Huancayo (1954-1975) were used for analysis of cosmic-ray intensity variations. Our spectral analysis shows no statistically significant peaks in cosmic-ray variation spectrum at $T = 1-10$ yrs not accompanied by the peak in solar activity spectrum (with the confidence level more than 95%).

This result is in agreement with the previous conclusion by Attolini et al. (1987), since they had an indication of the effect at $T \approx 5$ yrs with the confidence level 97% (see also the paper by Otaola and Hurtalo, 1981). The negative result of the search of the natural pulsations of the heliosphere may be due to the global changes of the radius of solar wind cavity during the 11'yrs solar cycle, see Van Allen (1988). We intend to investigate this point in a forthcoming publication.

STREAM KINETIC INSTABILITY OF GALACTIC COSMIC RAYS IN THE HELIOSPHERE (SELF-CONSISTENT PROBLEM)

Let us consider self-consistent problem on galactic cosmic ray propagation in the outer heliosphere and excitement MHD-turbulence. We suppose that the stream kinetic instability of cosmic rays is the main source of turbulence in the outer heliosphere. We suppose also that the spherically-symmetric supersonic flow of solar wind with constant radial velocity $u = 400$ km/s limited by terminal shock wave on the heliocentric distance $R = 60$ a.u. from the Sun. Regular interplanetary magnetic field according to Parker's model have radial H_r and azimuthal H_φ components:

$$H_r = (r_s/r)^2 H_s \;, \quad H_\varphi = (r_s^2 \Omega_s / ur) . \sin\theta . H_s \;, \tag{5}$$

where $r_s = 7\cdot10^{10}$ cm is the radius of Sun's photosphere, $H_s = 2$ Oe is the strength general of magnetic field of the Sun on the level of photosphere, $\Omega_s = 2.7\cdot10^{-6}$ s^{-1} is the angle velocity of the Sun rotation, θ is heliolatitude (from the North pole, $\theta = 90^o$ is helioequator). On the big distances $r \gg 1$ a.u. the interplanetary magnetic field is practically azimuthal and equal

$$H \approx H_\varphi = H_o (r_o/r) . \sin\theta , \tag{6}$$

where $r_o = 1$ a.u., $H_o = 5\cdot10^{-5}$ Oe is the strength of interplanetary magnetic field near Earth's orbit. The angle α between radius-vector and field direction determined by

$$\cos\alpha \approx r_o (r \sin\theta)^{-1} \tag{7}$$

The estimation according to (6) of the value of interplanetary magnetic field near the front of terminal shock wave gives $H(R = 60 \text{ a.u.}) \approx 10^{-6}$ Oe. Therefore after the shock front may be expected $H \sim H_G$, where $H_G \approx 3\cdot10^{-6}$ Oe is the average magentic field in interstellar matter. The structure of the magnetic field in this outer region near the terminal shock wave is not clear. Very strong rotated spiral magnetic field must turn to outer near homogeneous local galactic magnetic field. There are very big uncertainly in description of cosmic ray propagation processes in this region.

In this paper we suppose that the concentration of cosmic ray particles on the front of terminal shock wave is equal to concentration of cosmic rays average in interstellar matter. It means that we neglect some possible effects of cosmic ray modulation in the outer region $r > R$. The diffusion character of cosmic ray particles propagation is correct if $D \gg R.u \approx 10^{22}$ cm^2/s (here D is the radial diffusion coefficient of cosmic ray particles at $r > R$, u is the velocity of matter). Such approximation may be correct for particles with energy $E \gtrsim 100$ MeV/nucleon. The propagation of particles with smallest energy are considered by Berezhko et al. (1988). These consideration included the process of regular particle acceleration near the front of terminal shock wave.

Let us investigate the generation of short (relative to the curve radius of regular interplanetary magnetic field) MHD-waves in the region $r \lesssim R$. We consider waves propagated along the magnetic field $(|k_z| \gg k_\perp)$. The increment of increase of these waves is more and damping is smaller than for inclined waves. Increment $\Gamma(k_z)$ determine the evolution of spectral energy density of MHD-waves with accidentally phases $W(k_z)$:

$$dW(k_z)/dt = 2\Gamma(k_z)W(k_z). \tag{8}$$

The average square of accidentally magnetic field of MHD-waves is $\langle(\delta H)^2\rangle = 4\pi\int dk_z W(k_z)$. For MHD-waves propagated along the magnetic field H the interaction with cosmic ray particles caused by cyclotron resonance on the first harmonic, therefore the particle with charge Ze and with module of momentum p and with cosinus of pitch-angle μ ($\mu = (\vec{p}\vec{H})(pH)^{-1}$) interact with wave only when the condition

$$k_z = \pm ZeH(pc\mu)^{-1} \tag{9}$$

is satisfied (here take into account that the velocity of particles $V \approx c$ is much more than the Alven velocity of wave propagation in motionless matter $V_A = H/\sqrt{4\pi\rho} \approx 5\cdot 10^6$ cm/s). Increment of waves increase are given by equation (see for example Berezinsky et al., 1984)

$$\Gamma(k_z) = \frac{\pi^2 Z^2 e^2 V_A^2}{2c^2}\int d^3p\,(1-\mu^2)\left[\frac{\partial f}{\partial p} + \left(\frac{k_z V}{\omega} - \mu\right)p^{-1}\frac{\partial f}{\partial \mu}\right]\left[\delta\left(k_z\mu - \frac{ZeH}{pc}\right) + \delta\left(k_z\mu + \frac{ZeH}{pc}\right)\right] \tag{10}$$

where f is the momentum distribution function of cosmic ray particles. The concentration of cosmic ray particles is $N = \int d^3pf$. The frequency of wave ω in motionless matter determined by dispersion equation $\omega^2 = k_z^2V_A^2$. The equation (10) shows that for exciting waves (that is satisfying the condition $\Gamma > 0$) a small anisotropy of cosmic rays is sufficient $f^{-1}|\partial f/\partial\mu| > (\gamma + 2)V_A/V \approx 5\cdot 10^{-4}$ (here we suppose that cosmic ray spectrum is power type $f(p) \propto p^{-\gamma-2}$, where $\gamma \approx 2.5$). There are excited waves propagated in motionless matter along magnetic field in direction of cosmic ray flow.

The particles scattering on waves lead to izotropization of cosmic ray distribution function. For convenience let us introduce the effective frequency of particles scattering on waves (see Berezinsky et al., 1984):

$$\nu^\pm = 2\pi^2\omega_H k_{res} W^\pm(k_{res})H^{-2}, \tag{11}$$

where $k_{res} = ZeH(pc|\mu|)^{-1}$, $\omega_H = ZeHV(pc)^{-1}$ is gyrofrequency of relativistic particle in magnetic field H, the signs $\pm$ correspond to different directions of waves propagation. Cosmic ray particles propagation on times $\Delta t \gg \nu^{-1}$ and on distances $\Delta r \gg V\nu^{-1}$ described by diffusion equation for isotropic part of cosmic ray distribution functions $f_o = \int\frac{d\Omega}{4\pi}f$:

$$\frac{\partial f_o}{\partial t} - \nabla_i D_{ij}\nabla_j f_o + U_i\nabla_i f_o - (\nabla_i U_i)\frac{p}{3}\frac{\partial f_o}{\partial p} \tag{12}$$

with tensor of diffusion

$$D_{ij} = D_{\parallel}h_ih_j, \quad D_{\parallel} = \frac{V^2}{2}\int_0^1 d\mu\,\frac{(1-\mu^2)}{\nu^+ + \nu^-}, \quad h_i = H_i/H \tag{13}$$

(we neglect of diffusion across magnetic field because of strong magnetized particles). The eq. (12) described diffusion, convection and regular change of cosmic ray particles energy in solar wind. We neglected effects of possible statistical acceleration of particles and also possible differ convection particles velocity from solar wind velocity u (the order of these effects are V_A^2/u^2 and V_A/u accordingly). The dependence of cosmic ray distribution function from pitch-angle in diffusion approximation determined by equation:

$$\partial f/\partial\mu \approx -V(\nu^+ + \nu^-)^{-1}(h_i\nabla_i)f_o. \tag{14}$$

Galactic cosmic rays penetrating in heliosphere generated waves propagated along H in direction to the Sun, that is $k_z < 0$. Let us consider only this type of waves, and we do not write further the index "-" on value ν^- and W^-. Eqs (12) and (8) in the outer heliosphere in stationary case reduce to (here $\mathcal{E}(k) = kW(k)$):

$$-\frac{1}{r^2}\frac{\partial}{\partial r}\left(\frac{r_0^4 cpVH_0^2}{4\pi^2 Zer^2}\cdot\int_0^1 d\mu \frac{1-\mu^2}{\mathcal{E}\left(\frac{ZeH}{pc|\mu|}\right)}\frac{\partial f_0}{\partial r}\right)+u\frac{\partial f_0}{\partial r}-\frac{2u}{3r}p\frac{\partial f_0}{\partial p}=0, \tag{15}$$

$$u\frac{\partial \mathcal{E}}{\partial r}+\frac{3u}{r}\mathcal{E}-\frac{2\pi V_A r_0^2 ZeH_0}{r^2 ck}\int_{ZeH/ck}^{\infty} dp\, p^2 V\left(1-\left(\frac{ZeH}{pck}\right)^2\right)\frac{\partial f_0}{\partial r}. \tag{16}$$

For obtaining an approximate analytic solution of the system of eqs (15) and (16) we made the supplementary approximations (they are often for solution some problems of particles acceleration on the front of shock wave, see for example Berezhko et al., 1988). Consider the dependence $\mathcal{E}(k)$ as weak, we take out the function $\mathcal{E}$ from the integral in eq. (15) and put $\mathcal{E}(ZeH/pc\mu) \approx \mathcal{E}(ZeH/pc)$. Let us consider the dependence of value $Vp^4 f_0(p)$ from momentum p as weak and take out this function at $p = ZeH/ck$ from the integral in eq. (16). Let us introduce functions $\mathcal{E}(p) = \mathcal{E}(k = ZeH/pc)$ and $\Gamma(p) = \Gamma(k = ZeH/pc)$, then the system (15), (16) reduce to:

$$-\frac{1}{r^2}\frac{\partial}{\partial r}\left(\frac{r_0^3 cpVH_0}{6\pi^2 \sin\theta\cdot Zer\,\mathcal{E}(p)}\cdot\frac{\partial f_0}{\partial r}\right)+u\frac{\partial f_0}{\partial r}-\frac{2u}{3r}p\frac{\partial f_0}{\partial p}=0, \tag{17}$$

$$u\frac{\partial \mathcal{E}(p)}{\partial r}+\frac{3u}{r}\mathcal{E}(p)-\frac{4\pi}{3}\frac{V_A V p^4 r_0}{r\sin\theta}\frac{\partial f_0}{\partial r}=0. \tag{18}$$

Let us consider the solution of eqs (17), (18) in supposition that the space scale of changing values $f_0(r)$ and $\mathcal{E}(r)$ is small, that is

$$\partial f_0/\partial r \gg f_0/r, \quad \partial\mathcal{E}/\partial r \gg \mathcal{E}/r. \tag{19}$$

The conditions, when inequalities (19) are satisfied, will be indicated below. Using (19) we obtain from (18)

$$\mathcal{E}(p) = (4\pi/3)(V_A/u)Vp^4 f_0(r_0/r\sin\theta) \tag{20}$$

and eq. (17) reduce to nonlinear equation

$$\partial f_0/\partial r = ar^2 f_0^2, \tag{21}$$

where

$$a = 8\pi^3 ZeV_A p^3 (cr_0^2 H_0)^{-1}. \tag{22}$$

The solution of eq. (21) with boundary condition $f_0(r = R) = f_e$ (where $f_e(p)$ is cosmic ray distribution function in interstellar space) has a following form:

$$f_0(r,p) = f_e(p)\left(1 + af_e(p)(R^3 - r^3)/3\right)^{-1} \tag{23}$$

At our parameters of model $af_e R^3/3 \approx 3\cdot 10^{11} N_e$, where N_e is the cosmic ray concentration in interstellar space (in cm^{-3}). According to Perko (1987) the cosmic ray intensity differential energy spectrum in interstellar space near the solar system is $I_e(E) = 1.93(V/c)(E + m_p c^2)^{-2.7}$ $(cm^2.s.sr.Gev)^{-1}$, where E is the kinetic energy of cosmic ray particles and $m_p c^2$ is the rest energy of nuclear (this analitical approximation is true for energy $E > 0.03$ GeV/nucleon. It is easy to see that at $E < 10$ GeV/nucleaon the inequality $af_e R^3/3 \gg 1$ is satisfied. At that the solutions (20) and (23) are satisfied to conditions (19) in the outer heliosphere, at $r > 40$ a.u.

The cosmic ray anizotropy calculated according to eq.(14) are directed along the magnetic field and its value is

$$\delta = f_0^{-1}\partial f/\partial\mu = -3(u/V)(r/r_0)\sin\theta, \tag{24}$$

where $r_0 = 1$ a.u. Near the heliosphere boundary at $r = R = 60$ a.u. and for $V \approx c$, $\theta = \pi/2$ we obtain $\delta \approx 20\%$. The density of energy MHD - waves according to (20) provide the radial diffusion coefficient:

$$D_{rr} = D_{\parallel}\cos^2\mathcal{L} = (8\pi^3)^{-1} ucH_0 r_0^2 \left(ZeV_A p^3 f_e(p) R^2\right)^{-1}(R/r)^2\left(1 + af_e(p)(R^3 - r^3)/3\right). \tag{25}$$

From here $D_{rr}(r = R) \approx 3\cdot 10^{10}(p^3 f_e(p))^{-1}$ cm^2/s. For particles with energy $E = 1$ GeV we obtain $D_{rr}(r = R) \sim 10^{21}$ cm^2/s. This value even smaller than typical average diffusion coefficient usually used for heliosphere $D_{rr}(1\ \text{GeV/nucleon}) \sim 10^{22}$ cm^2/s (see for example

Topygin, 1983) that shows the effectivity of considered self-consistent mechanism of cosmic ray modulation. We concluded that kinetic stream instability possibly caused to excited MHD-waves with energy density provide enough effective cosmic ray modulation for particles with energy $E \lesssim 10$ GeV/nucleon in outer heliosphere at $r \gtrsim 40$ a.u. (see Fig.1-3).

CONCLUSION

The problem of heliosphere eigen pulsations now is not clear. We do not know exactly properties of interstellar matter near the solar system, the radius of heliosphere, character of boundary between the solar wind and interstellar matter. It is necessary to continue the search of heliosphere eigen pulsations in cosmic ray variations. We show that kinetic stream instability of galactic cosmic rays leads to essential decreasing of radial diffusion coefficient with increase of heliocentric distance (see Fig. 1), to essential increasing of cosmic ray modulation level and cosmic ray radial gradient in the outer heliosphere (see Fig. 2 and 3).

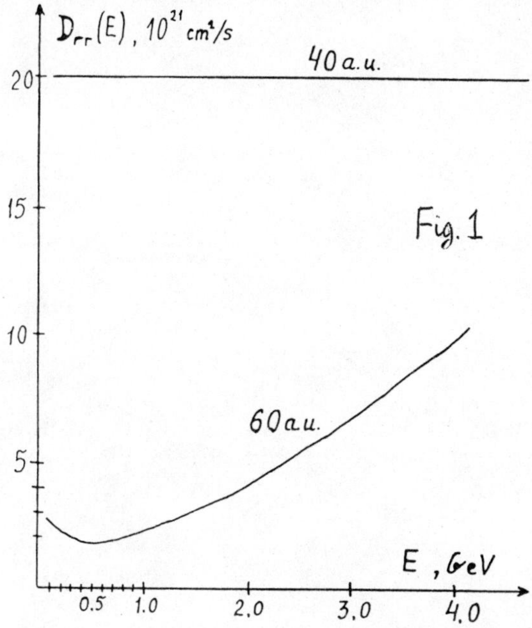

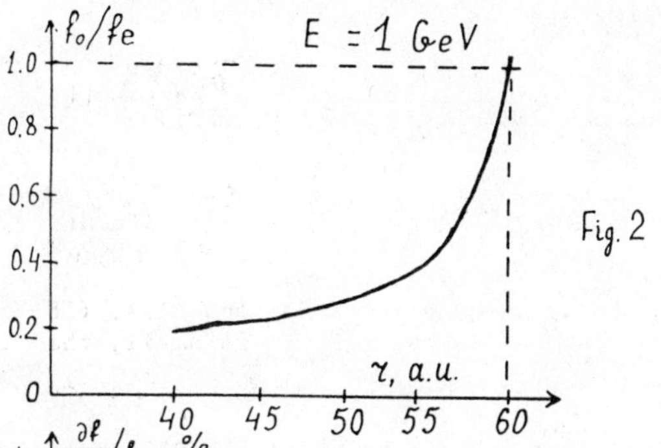

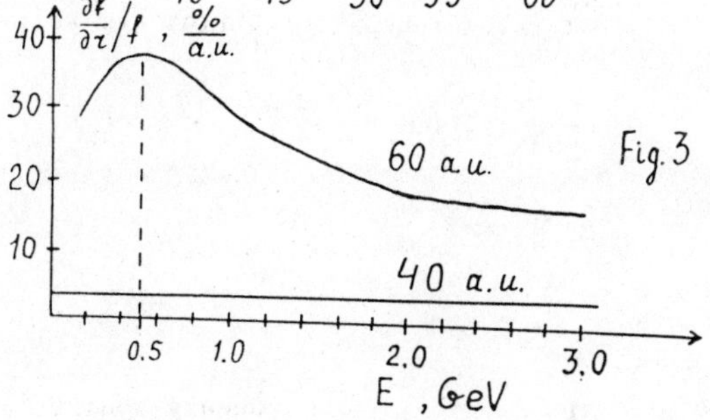

Fig. 1. Cosmic ray radial diffusion coefficient $D_{rr}(E)$ in units 10^{21} cm^2/s at heliocentric distances 40 a.u. (as supposed for inner boundary condition accordingly to space probes measurements) and 60 a.u. (as calculated here from self-consistent model).

Fig. 2. Expected in self-consistent model cosmic ray modulation in outer heliosphere at distances 40-60 a.u. for particles with energy $E = 1$ GeV.

Fig. 3. The same as in Fig. 1, but for cosmic ray radial gradient.

REFERENCES

Attolini M.R., Cecchini S., Galli M., 1987. Astrophys. Space Sci., 134, 103.

Babayan V.Kh., Dorman L.I., Ptuskin V.S., 1987. Proc. Int. Conf. on Plasma Physics, Kiev, USSR, Naukova Dumka, Vol.4, p.228-230.

Berezhko E.G., Elshin V.K., Krymsky G.F., Petukhov S.I., 1988. Cosmic ray generation by shock waves. Novosibirsk, Nauka, p. 182.

Berezinsky V.S., Bulanov S.V., Ginzburg V.L., Dogel V.A., Ptuskin V.S., 1984. Astrophysics of cosmic rays. Moscow, Nauka, p. 360.

Dorman L.I., Ptuskin V.S., 1981. Astrophys. Space Sci., 79, 397. (Paper I).

Fahr H.J., Neutsch W., Grzedzielski S., Macek W., Ratkiewicz-Landovska R., 1986. Space Sci. Rev., 43, 329.

Otaola J.A., Hurtado A., 1981. Proc. 17th Intern. Cosmic Ray Conf., Paris, Vol. 10, p.167.

Perko J.S., 1987, Astrophys. J., 184, p.119-121.

Toptygin I.N., 1983. Cosmic rays in interplanetary magnetic fields. Moscow, Nauka, p.304.

Van Allen J.A., 1988. Geophys. Res. Lett., 15, 1527.

OUTER HELIOSPHERE AS A MANY-COMPONENT MEDIUM FOR COSMIC RAY PROPAGATION

L.I. Dorman* and V.Kh. Shogenov**

**IZMIRAN, Troitsk, Moscow Region, 142092, U.S.S.R.*

***Kabardino-Balkarsk State University, Chernyshevskogo 173, Nalchik, 360004, U.S.S.R.*

ABSTRACT AND INTRODUCTION

The theory developed earlier by the authors for cosmic ray propagation in two-component (Dorman and Shogenov, 1985) and in many-component media (Dorman and Shogenov, 1987) is applied to the outer heliosphere where steady solar wind flow with frozen-in magnetic field occurs together with high velocity fluxes, shock waves and other formations moving at velocities other than the solar wind velocity. A substantial acceleration of particles is shown to occur in the media of this type in addition to Fermi acceleration. The equations have been obtained which describe the cosmic ray propagation and acceleration in a many-component medium and the formation of the anisotropy and the various types of temporal cosmic ray variations. A possibility of generalizing the theory making allowance for nonlinear effects (discussed in Babayan and Dorman (1990) and Dorman et al. (1990)) is noted.

TWO-COMPONENT MODEL FOR COSMIC RAY PROPAGATION IN THE OUTER HELIOSPHERE

After averaging of the kinetic equation (Dolginov and Toptygin, 1966; Toptygin, 1983)

$$\partial f/\partial t + \vec{v}\,\partial f/\partial \vec{r} + (e/c)((\vec{v}-\vec{U})\times\vec{B})(\partial f/\partial \vec{p}) = St\, f \tag{1}$$

on the fluctuations we obtain

$$\partial F/\partial t + \vec{v}(\partial F/\partial \vec{r}) + (E/c^2)((\vec{v}-\vec{U})\times\vec{\beta})(\partial F/\partial \vec{p}) = St\,F + St_1 F\,, \tag{2}$$

where $\vec{U}(\vec{r},t)$ is the velocity of solar wind ($\vec{U}(\vec{r},t) = \vec{U}_o(\vec{r},t) + \vec{U}_1(\vec{r},t)$, $\langle\vec{U}(\vec{r},t)\rangle = \vec{U}_o(\vec{r},t)$), $\vec{B}(\vec{r},t)$ is the magnetic field frozen in the solar wind plasma ($\vec{B}(\vec{r},t) = \vec{B}_o(\vec{r},t) + \vec{B}_1(\vec{r},t)$, $\langle\vec{B}(\vec{r},t)\rangle = \vec{B}_o(\vec{r},t)$), f is the cosmic ray distribution function ($f = F + f_1$, $\langle f\rangle = F(\vec{r},\vec{p},t)$), $\vec{\beta} = ec\vec{B}_o/E$ is the Larmor frequency of the particle with charge e and energy $E = (p^2c^2 + m^2c^4)^{\frac{1}{2}}$. For the distribution function of magnetic clouds $f_{mc}(\vec{r},\vec{V},t) = n(\vec{r},t)\delta(\vec{V}-\vec{V}(\vec{r},t))$ for the collision member in (2):

$$St\,F = n(\vec{r},t)\int|\vec{v}-\vec{V}|\sigma(|\vec{v}-\vec{V}|,\theta)(F(\vec{p}\,')-F(\vec{p}))\,d\Omega'\,, \tag{3}$$

where σ is the effective cross-section for particle scattering by magnetic cloud on angle θ. For collision member St_1F in (2) for particle interactions with solar wind inhomogeneities we obtain according to Toptygin (1983)

$$St_1 F = (\partial/\partial p_i)D^p_{ik}(\partial F/\partial p_k), \tag{4}$$

where D^p_{ik} is the diffusion coefficient in momentum space: $D^p_{ik} = (p^2/2L_B)\,|\vec{v}-\vec{U}|(\delta_{ik} - (\vec{v}-\vec{U})_i(\vec{v}-\vec{U})_k/(\vec{v}-\vec{U})^2)$, L_B - particle transport path in the solar wind plasma. On serie-expanding F in spherical harmonics of momentum space and retaining the first two terms, we obtain

$$F(\vec{r},\vec{p},t) = N(\vec{r},p,t) + (\vec{p}/p)\vec{J}(\vec{r},p,t)\,. \tag{5}$$

After that, the expression (5) is substituted in (2), taking into account (3) and (4), and we obtain the diffusion approximation for cosmic ray propagation in two-component model:

$$\partial N/\partial t = (\partial/\partial x_k)(D_{ki}(\partial N/\partial x_i) - (pq_k/3(\alpha^2+\beta^2))(\partial N/\partial p)) + p^{-2}(\partial/\partial p)(p^3\vec{q}_A\cdot$$
$$\cdot\vec{\nabla}N(3(\alpha^2+\beta^2)^{-1} + p^4/3v^2)(\alpha_A\langle V^2\rangle + \alpha_B\langle U^2\rangle + d)(\partial N/\partial p), \tag{6}$$

where $\alpha_A = v/L_A$ is the frequency of particle collisions with magnetic clouds (L_A is the particle transport path relative to magnetic clouds), $\alpha_B = v/L_B$ the same for collisions in solar wind plasma, $\alpha = \alpha_A + \alpha_B$, the coefficient of diffusion in space $D_{kl} = \alpha v^2(3(\alpha^2 + \beta^2))^{-1}(\delta_{kl} + \beta_k\beta_l\alpha^{-2} - e_{kil}\beta_i\alpha^{-1})$ (here e_{kil} is antisymmetric unity tensor); $\vec{q} = \alpha((\vec{U}_o + \vec{W}_o)x\vec{\beta}) - \alpha^2\vec{W}_o - \beta^2\vec{U}_o + \vec{\beta}(\vec{U}_o\vec{\beta} - \vec{W}_o\vec{\beta})$; $\vec{q}_A = \vec{q} - 2\alpha((\vec{U}_o - \vec{W}_o)x\vec{\beta})$; $d = (\alpha^2 + \beta^2)^{-1} \cdot (\alpha\beta^2U_o^2 - \alpha(\vec{U}_o\vec{\beta})^2 - 2\alpha\beta^2\vec{U}_o\vec{W}_o + 2\alpha(\vec{U}_o\vec{\beta})(\vec{W}_o\vec{\beta}) - \alpha^3W_o^2 - \alpha(\vec{W}_o\vec{\beta})^2)$; $\vec{W}_o = \alpha^{-1}(\alpha_A\vec{V}_o + \alpha_B\vec{U}_o) = (L_A + L_B)^{-1}(L_A\vec{U}_o + L_B\vec{V}_o)$ is the effective convective velocity in two-component model. It is easy to see that at $\vec{U} = \vec{V}$ eq.(6) turn to equation of anisotropy diffusion in usually used one-component model (Dorman and Shogenov, 1985).

COSMIS RAY PARTICLES CHANGE ENERGY IN HELIOSPHERE

Let us consider for simplicity the case $\vec{B}_o = 0$. Then from (6) we obtain

$$\partial N/\partial t = (\partial/\partial x_k)(D(\partial N/\partial x_k) + (p/3)W_{ok}(\partial N/\partial p)) + p^{-2}(\partial/\partial p)(-(p^3/3)\vec{W}_o\vec{\nabla}N + p^2D^p(\partial N/\partial p)), \quad (7)$$

where D is the coefficient of isotropic diffusion in space, D^p is the diffuse coefficient in momentum space. Here $D^p = D^p_A + D^p_B + D^p_{AB}$, where $D^p_A = p^2\langle\Delta V^2\rangle/3vL_A$ determines the Fermi acceleration by magnetic clouds, $D^p_B = p^2\langle\Delta U^2\rangle/3vL_B$ determines the Fermi acceleration by inhomogeneities of solar wind and $D^p_{AB} = p^2(\vec{V}_o - \vec{U}_o)^2/3v(L_A + L_B)$ determines the new type of regular cosmic ray particles acceleration in the heliosphere, the effectivity of which is determined by characteristic time of acceleration $T^p_{AB} = p^2/4D^p_{AB} = 3v(L_A + L_B)/4(\vec{V}_o - \vec{U}_o)^2$. The time T^p_{AB} decrease (the effectivity increase) with increasing $(V_o - U_o)^2$ and with decreasing $v(L_A + L_B)$. Then we conclude that this regular acceleration mechanism may be effective in the heliosphere for small energy particles. If we take into account the effect of kinetic stream instability (Dorman et al., 1990) in the outer heliosphere, we come to the conclusion that in the outer heliosphere may be effective the Fermi acceleration of low energy particles by moving magnetic inhomogeneities in the solar wind and the regular acceleration caused by difference $\vec{U}_o$ and $\vec{V}_o$ (the diffusion coefficients in momentum space D^p_B and D^p_{AB} respectively).

THE POSSIBILITY OF SHOCK TYPE ACCELERATION WITHOUT SHOCK WAVE

Let us suppose that at some distance $r = r_c$ the values $\vec{U}_o$ and $\vec{V}_o$ are continue. If at $r = r_c$ L_A or L_B (or L_A and L_B) have some jump then at $r = r_c$ the jump must have also the effective convective velocity $\vec{W}_o = (\vec{U}_oL_A + \vec{V}_oL_B)/(L_A + L_B)$. It is easy to see from eq. (6), that in this case shock type acceleration mechanism works and generates a spectrum of accelerated particles $\propto p^{-\gamma}$, where $\gamma = 3W_{o1}/(W_{o1} - W_{o2})$ (here W_{o1} and W_{o2} are the convective velocities from two sides near r_c; point r_c moved in space). For details see Dorman and Shogenov (1987a).

MANY-COMPONENT MODEL FOR COSMIC RAY PROPAGATION IN THE OUTER HELIOSPHERE AND NONLINEAR EFFECTS

Let us consider many-component model for the outer heliosphere: solar wind with the velocity U and n component of magnetic clouds with velocities V_j ($j = 1, 2, \ldots, n$). The diffusion equation for the general case was obtained in Dorman and Shogenov (1987). It is complicated and we here consider only the case of isotropic diffusion. As for (7) we obtain

$$\partial N/\partial t = (\partial/\partial x_k)(D(\partial N/\partial x_k) + (p/3)W_{ok}(\partial N/\partial p) + p^{-2}(\partial/\partial p)(-(p^3/3)\vec{W}_o\vec{\nabla}N + (p^4/3v^2)(\alpha_B U_o^2 + \sum_{j=1}^{n}\alpha_j V_j^2 - \alpha W_o^2)(\partial N/\partial p)), \quad (8)$$

where $D = v^2/3$, $\vec{W}_o = \alpha_B\vec{U}_o/\alpha + \alpha^{-1}\sum_{j=1}^{n}\alpha_j\vec{V}_j$, $\alpha = \alpha_B + \sum_{j=1}^{n}\alpha_j$. In Dorman and Shogenov (1987b) it was supposed that the properties of solar wind and magnetic clouds moving are not dependent on the cosmic ray energy spectrum and the anisotropy changes. But the kinetic stream instability lead to a significant increase of the cosmic ray gradient (Dorman et al., 1990) and respectively to an increase of cosmic ray pressure on the solar wind and magnetic clouds moving in the outer heliosphere (Babayan and Dorman, 1990). Then it is necessary to consider the self-consistent problem: eq.(8) together with equations of papers Dorman et al. (1990) and Babayan and Dorman (1990). In this case $\vec{U}_o$ and $\vec{V}_j$ in (8) will be determined by solution of self-consistent problem.

REFERENCES

Babayan V.Kh., Dorman L.I., 1990. Space Res. (in this volume).
Dolginov A.Z., Toptygin I.N., 1966. ZhETF, Vol.51, p.1771.
Dorman L.I., Ptuskin V.S., Zirakashvili V.N., 1990. Space Res. (in this volume).
Dorman L.I., Shogenov V.Kh., 1985. ZhETF, Vol.89, p.1624.
Dorman L.I., Shogenov V.Kh., 1987a. Geomagnetism i Aeronomiya, Vol.27, p.132.
Dorman L.I., Shogenov V.Kh., 1987b. Proc. 20-th Int. Cosmic Ray Conf., Moscow, Vol. 3, p.182.
Toptygin I.N., 1983. Cosmic rays in interplanetary magnetic fields, Nauka, Moscow.

[illegible]

[illegible] (1980) and Rasayan and Dzhaman (1980). In [illegible] ... and [illegible] (5) will be [illegible] by [illegible] "[illegible]" program.

REFERENCES

[illegible]
[illegible]
[illegible] (In [illegible])
Dzhaman, [illegible], Shakerov, [illegible] (19[illegible]). [illegible], Vol. [illegible]
Dzhaman, L., Shakerov, A. M. (1979). [illegible], Vol. 27, p. 352.
[illegible] (1982). [illegible]
[illegible] (19[illegible]). [illegible] in [illegible] Fields, Nauka, [illegible]

ON THE PROPAGATION VELOCITY OF COSMIC RAY MODULATION WAVE IN THE OUTER HELIOSPHERE

L.I. Dorman,* A.G. Zusmanovich** and O.A. Kryakunova**

**IZMIRAN, 142092 Moscow Region, Troitsk, U.S.S.R.*
***Institute of Ionosphere of the Academy of Sciences of Kazah SSR, Alma-Ata, U.S.S.R.*

ABSTRACT

The expected propagation velocity of cosmic ray modulation effects in the outer heliosphere is calculated analytically and numerically. In case of a harmonic or symmetric wave of cosmic ray modulation, the propagation velocity of the modulation has been found to be $U_M = 2u$, where u is solar wind velocity (with good agreement with Dorman's earlier work of 1967). In case of an asymmetric wave of cosmic ray modulation, its phase velocity of propagation is $U_M \neq 2u$. The U_M value has been found for different shapes of the cosmic ray modulation wave. The calculation results are compared with Voyager-1,2 and Pioneer-10,11 measurement data. A good agreement has been obtained.

INTRODUCTION

Observed long-term interplanetary modulation of galactic cosmic rays is an integral effect determined by modulation properties of solar wind in all volume of heliosphere. It is usually supposed that cosmic ray modulation effects are transported in heliosphere with velocity of solar wind, but it is not correct. Dorman (1967) considered spherically-symmetrical model of galactic cosmic ray isotropic diffusion with radial solar wind convection, when diffusion coefficient D and solar wind velocity u change in time t harmonically with angle frequency β and when this disturbance propagates in heliosphere with solar wind velocity u, that is:

$$u(r,t)/D(r,t) = (u/D)_{max} - ((u/D)_{max} - (u/D)_{min})(1 + \cos\beta(t - r/u))/2 . \quad (1)$$

After integration on whole region of modulation from r to heliosphere boundary r_o Dorman (1967) obtained

$$N(r,R,t)/N_e(R) = \exp(-K\cos\beta(t - (r_o + r)/2u)), \quad (2)$$

where $N(r, R, t)$ is the cosmic ray concentration of particles with rigidity R on the distance r from the Sun at the moment of time t (counted from the moment $(u/D)_{min}$) and $N_e(R)$ is the cosmic ray concentration in interstellar space near solar system. In expression (2) K is the constant part of modulation parameter determined by values of r_o, β, $(u/D)_{max}$ and $(u/D)_{min}$ (see for detail Dorman, 1967 or Dorman, 1975). From (2) follows that the extremum of cosmic ray modulation obtained at moment $t = (r_o + r)/2u$; from here the velocity of modulation extremum propagation is

$$dr/dt = 2u . \quad (3)$$

Here we consider the case when $D \propto r$ and the case of propagation not harmonic modulation waves.

PROPAGATION OF HARMONIC MODULATION WAVE IN THE CASE $D \propto r$

In the case $D \propto r$ we obtain for spherically-symmetric isotropic diffusion with convection the follow expression for relative cosmic ray modulation :

$$N(r,R,t)/N_e(R) = \exp\left(-K_1 - K_2\int_r^{r_o} r^{-1}\cos\beta(t - r/u)\,dr\right), \quad (4)$$

where K_1 and K_2 are modulation parameters not depended from time t. Let us consider the integral in (4):

$$\int_r^{r_0} r^{-1}\cos\beta(t-r/u)dr = \cos\beta t\int_r^{r_0} r^{-1}\cos(\beta r/u)dr + \sin\beta t\int_r^{r_0} r^{-1}\sin(\beta r/u)dr =$$
$$= \cos\beta t.(\ln(r_0/r) - \beta^2(r_0^2-r^2)(u^2.2.2!)^{-1} + \beta^4(r_0^4-r^4)(u^4.4.4!)^{-1} - \ldots) + \tag{5}$$
$$+ \sin\beta t.(\beta u^{-1}(r_0-r) - \beta^3(r_0^3-r^3)(u^3.3.3!)^{-1} + \beta^5(r_0^5-r)(u^5.5.5!)^{-1} - \ldots).$$

For long-term modulation $\beta r_o/u < 1$ and therefore for the first approximation we can take into account only first members of the series (5). We obtain

$$\int_r^{r_0} r^{-1}\cos\beta(t-r/u)dr \approx \cos\beta t.\ln(r_0/r) + \sin\beta t.\beta(r_0-r)/u =$$
$$= (\ln^2(r_0/r) + \beta^2(r_0-r)^2u^{-2})^{1/2}.\cos(\beta t - \operatorname{arctg}(\beta(r_0-r)(u\ln(r_0/r))^{-1})). \tag{6}$$

The argument of the function arctg in (6) is small relative to 1, therefore we obtain from (4) and (6) approximately:

$$N(r,R,t)/N_e(R) \approx \exp(-K_1 - K_3\cos\beta(t-(r_0-r)(u.\ln(r_0/r))^{-1})). \tag{7}$$

From (7) we see that the difference in time between extremums of cosmic ray intensity and solar activity is equal

$$t = (r_0-r)/(u.\ln(r_0/r)) \tag{8}$$

and the velocity of cosmic ray modulation wave

$$dr/dt = u.\ln^2(r_0/r)(r_0/r - 1 - \ln(r_0/r))^{-1}. \tag{9}$$

This velocity increase with increasing r from $\sim$0.3u at $r_o/r = e^4$ (near the Earth's orbit) to $\sim$ u at $r \rightarrow r_o$ (near the boundary of heliosphere). Therefore the investigation of the velocity propagation of cosmic ray harmonically modulation wave can give some information about the dependence of diffusion coefficient from the distance to the Sun.

VELOCITY PROPAGATION OF NOT HARMONICALLY MODULATION WAVE

Dorman et al. (1989) analysed the propagation in interplanetary space of different kind of periodical modulation waves (not harmonic type) by Fourier expansion. The main result is: if the change in time of cosmic ray modulation parameters described by functions for which Fourier sinus-imagine are equal to zero then we obtain result (3) for D not depend from r, and result (9) for $D \propto r$; if Fourier sinus-imagine is not equal to zero (in the case imaginary part of Fourier-intergal is different from zero), we obtain supplementary member caused by phase change and the velocity of modulation wave propagation depend on the shape of the wave. Let us consider for example the case when u/D linear increased in time from t = 0 to t = T and then suddenly decrease to zero (that is the change of solar activity as straight triangle). The coefficients of Fourie serie of this function are $a_k = 0$, $b_k = -A/k$, where A is the maximum of function change. In this case for k-th harmonic we obtain instead (3) and (2) for the lag time

$$t = (r_0+r)/2u + kT/4. \tag{10}$$

This result lead us to conclusion that in not harmonic and not symmetrical case the velocity propagation of modulation wave differ from (3) and the shape of wave changed during the propagation process because the time lag of cosmic ray intensity extremum depend not only from r_o, r and u but also from the frequency of the modulation parameters changes.

NUMERICALLY CALCULATIONS OF COSMIC RAY MODULATION WAVE PROPAGATION

For verification of above analytical results and consideration of more complicated cases with modulation parameters changing in time (according to real changing of solar activity) let us obtain a numerical solution for cosmic ray transport equation for spherically-symmetric model of isotropic diffusion with convection and energy particles change:

$$u^{-1}\partial N/\partial t = r^{-2}(\partial/\partial r)(r^2(D/u)(\partial N/\partial r) - r^2N) + (2R/3r)(\partial N/\partial R). \tag{11}$$

Let us introduce dimensionless parameter $y_{ik} = (u_{ik}/D_{ik})\Delta r$ which characterised the influence of solar activity on galactic cosmic rays in the k-th moment of time in the spherical layer Δr on the distance r_i from the Sun (the detail methodic of solution (11) see in Zusmanovich et al., 1987). Let us consider two cases. In the first case y_{ik} changed monotonous with symmetrical increase and decrease. Calculations made for radius of heliosphere r_o = 76 a.u. and average solar wind velocity u = 400 km/s. Expected cosmic ray modulation waves on the distances from the Sun 1 a.u., 41.5 a.u., 69.1 a.u. are shown on Fig. 1 (here

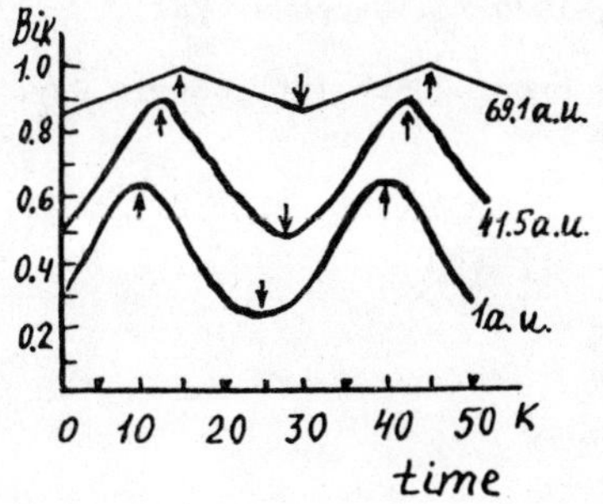

Fig. 1. Results of numerically calculations of cosmic ray symmetric modulation wave propagation in heliosphere. Tringles shows the time of solar activity extrema, arrows - of cosmic ray intensity extrema at distances 1 a.u., 41.5 a.u. and 69.1 a.u. (r_o = 76 a.u., value k in monthes).

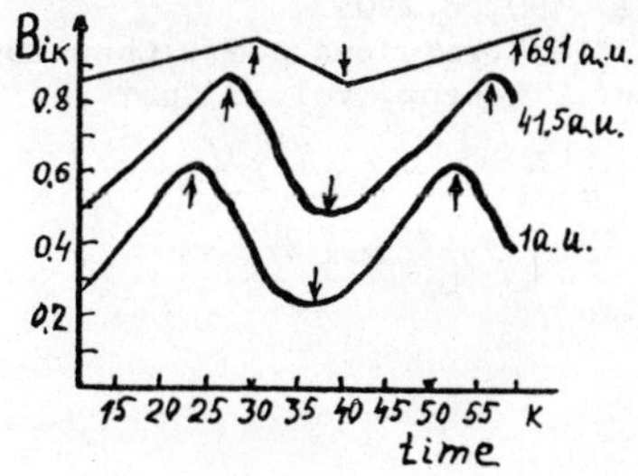

Fig. 2. The same as on Fig. 1, but for not symmetric wave.

$B_{ik} = N_{ik}/N_e$ and the step of k is 1 month). From Fig.1 follows that the propagation velocity of maximum and minimum B_{ik} is the same and equal 800 km/s = 2u in agreement with (3). In the second case y_{ik} changed monotonous but not symmetrical: we suppose that the time of decrease y_{ik} is two times longer than increase time (near as in real solar activity cycle). Results are shown on Fig. 2 for the same distances from the Sun. In this case the maximum of B_{ik} propagated in heliosphere slower than minimum of B_{ik}: the velocity of maximum is 571 km/s and the velocity of minimum is 1333 km/s.

COMPARISON WITH OBSERVATIONS

The obtained above result for not symmetrical case give natural explanation for some observations of time lag long-term cosmic ray intensity variations relative to changes of solar activity from the phase of 11-year solar cycle (Simpson, 1963; Dorman and Dorman, 1967; Kane and Winckler, 1969; Dorman et al., 1979). There are some space-probe experimental data about the velocity of modulation wave propagation: on the base of measurements on 1 and 10 a.u. Venkatesan et al. (1984) obtained for this velocity 700 km/s; by cosmic ray measurements on distances 1, 10 and 25 a.u. Venkatesan et al. (1985) observed two minimums in long-term cosmic ray modulation wave, the first moved with velocity 500 km/s, and the second - with velocity 700-800 km/s (the possibility of the change with distance from the Sun of the modulation effects phase velocity was shown by Dorman et al., 1989).

CONCLUSION

The propagation velocity of cosmic ray modulation effects in heliosphere is not equal to solar wind velocity. For symmetrical modulation wave its velocity propagation is two times the solar wind velocity. For not symmetrical modulation wave the propagation velocity of minimum is more than of maximum. In this case the shape of cosmic ray modulation wave changed with removing from the Sun. The time lag of cosmic ray minimum relative to solar activity maximum is large than the time lag for cosmic ray intensity maximum relative to solar activity minimum (see Fig. 2).

REFERENCES

Dorman L.I., 1967. Cosmic Rays N 8, p.305, Moscow, Nauka.
Dorman L.I., 1975. Variations of galactic cosmic rays, Moscow, University Press.
Dorman I.V., Dorman L.I., 1967. Cosmic Rays N 8, p.100, Moscow, Nauka.
Dorman L.I., Zusmanovich A.G., Churunova L.F., 1979. Proc. 16-th Int. Cosmic Ray Conf., Kyoto, Vol. 3, p.380.
Dorman L.I., Zusmanovich A.G., Kryakunova O.N., Nikolaevsky N.F., 1989. Geomagnetism and Aeronomy, Vol. 29, N 2, p.192.
Kane S.P., Winckler J.P., 1969. J. Geophys. Res., Vol.74, p.6247.
Simpson J.A., 1963. Proc. 8-th Int. Cosmic Ray Conf., Denver, Vol.2, p.155.
Venkatesan D., Decker R.B., Krimigis S.M., 1984. J. Geophys. Res., Vol.A89, p.3735.

Venkatesan D., Decker R.B., Krimigis S.M., Van Allen J., 1985. J. Geophys. Res., Vol. A90, p.2905.
Zusmanovich A.G., Kryakunova O.N., Churunova L.F., 1987. Proc. 20-th Int. Cosmic Ray Conf., Moscow, Vol. 3, p.94.

COSMIC RAYS AND MAGNETOSONIC INSTABILITIES OF SOLAR WIND FLOW NEAR THE HELIOSPHERIC SHOCK WAVE

S.V. Chalov

Institute for Problems in Mechanics, U.S.S.R. Academy of Science, Pr. Vernadskogo 101, 117526 Moscow, U.S.S.R.

ABSTRACT

It is known that the flow of a thermal plasma interacting with energetic particles (cosmic rays) in front of a sufficiently strong shock wave is unstable with respect to short-wavelength magnetosonic oscillations. The results of the general theory of this instability are applied to the solar wind flow in front of a heliospheric shock wave close to the equator. It is shown that unstable magnetosonic waves with the greatest growth rate propagate at right angles to the equatorial plane. The direction of propagation depends on the solar magnetic field polarity. When the polarity changes sense, the direction of propagation reverses. This process may result in a shift in the peak intensity of anomalous component.

INTRODUCTION

The estimates of the cosmic-ray energy density in the interstellar medium lead us to conclusions about the necessity to consider the influence of the cosmic-ray pressure on cosmic plasma dynamics. The most detailed analysis of such influence has been carried out for the case of a shock flow. This question is connected with the cosmic-ray acceleration problem.

The dynamic interaction of cosmic rays with a thermal plasma results in the formation of a smooth transition region describing a deceleration of the flow in front of the shock wave. This region is usually called the cosmic-ray shock precursor. In the case when the shock wave is sufficiently strong the precursor is unstable with respect to fast magnetosonic oscillations /1 - 5/. The wavelength of these unstable oscillations is small in comparison with the precursor length scale.

In the present report the general results of such an instability /5/ are applied to the solar wind flow in front of the heliospheric shock wave close to the equatorial plane.

THE PROPERTIES OF UNSTABLE WAVES

It is supposed that the current sheet is plane and it coincides with the equatorial plane. At a great distance from the Sun the interplanetary magnetic field is almost longitudional at low latitudes. Let us suppose that the magnetic field vector is directed along the shock front tangent. That is, we assume that the radial diffusion of the cosmic rays near the shock is determined mainly by the transverse component of the diffusion tensor. Because we consider only short-wavelength oscillations the shock may be regarded as a plane in a zeroth-order approximation. In this case in a moving frame of reference the most unstable fast magnetosonic waves (as it follows from the general results /5/) propagate at right angles to the magnetic field and plasma velocity vectors, that is, at right angles to the equatorial plane. In Figure 1 the direction of propagation of the most unstable waves coincides with y-axis. *

* The unstable waves propagate along the positive direction of y-axis. The damping waves are oppositely directed.

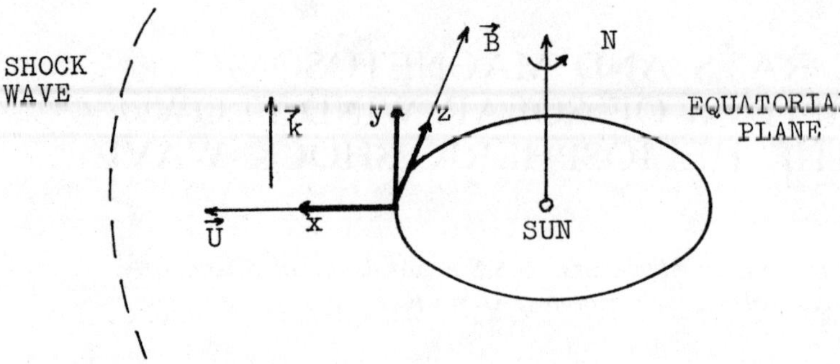

Fig. 1. The schematic picture of the heliosphere.

The last conclusion is true if the ratio of the parallel diffusion coefficient to the perpendicular one is large compared with unity. There are reasons to think that this condition is valid in the outer heliosphere /6/.

The velocity and magnetic field amplitudes of the most unstable waves may be written in the following form:

$$\tilde{U}_x = 0, \quad \tilde{U}_y \propto \tilde{\rho}, \quad \tilde{U}_z = 0,$$
$$\tilde{B}_x = 0, \quad \tilde{B}_y = 0, \quad \tilde{B}_z \propto \tilde{\rho}, \tag{1}$$

where $\tilde{\rho}$ is the wave amplitude of the solar wind plasma density. To determine the wavelength of these waves we must take into account the effects of plasma viscosity, heat and electroconductivity, and Landau damping. The estimates of these dissipative effects show that unstable magnetosonic waves with the greatest growth rate in the distant heliosphere have a wavelength of about 1 A.U. and a period of about two months /7/. If such oscillations are detected by spacecraft in the distant heliosphere then this fact may be considered as experimental corroboration of the theory.

THE CONNECTION WITH THE ANOMALOUS COMPONENT

After the middle of 1980 (when the solar magnetic field polarity reversed) a shift in the peak of the intensity of the anomalous component towards higher energies has been detected /8/. There are several possible explanations of this shift. It is possible to show that the shift may be connected with the instability under consideration.

The unstable waves with the greatest growth rate propagate at right angles to the equatorial plane. The direction of propagation depends on the solar magnetic field polarity /7/. Figure 2A corresponds to the case when the solar magnetic field in the nothern hemisphere is out from the Sun. In this case the unstable waves propagate away from the equatorial plane. When the magnetic polarity changes, the direction of propagation reverses. In the case B the unstable waves propagate towards the equatorial plane and consequently the disturbances of a solar wind plasma close to the equator will be stronger than in the case A. It follows from this that in the near-equatorial region the cosmic-ray diffusion coefficient in case B is smaller than in case A.

Let us suppose now that acceleration of the anomalous component takes place mainly at the near-equatorial part of the termination shock. Then the shift in the peak of the intensity towards higher energies after the middle of 1980 (when polarity changed) may be accounted for by the decrease of the cosmic-ray diffusion coefficient in front of the shock, because the decreasing of this coefficient results in increasing of the efficiency of cosmic-ray acceleration. As regards the injection of particles into the shock to be accelerated, this is not a problem /9/.

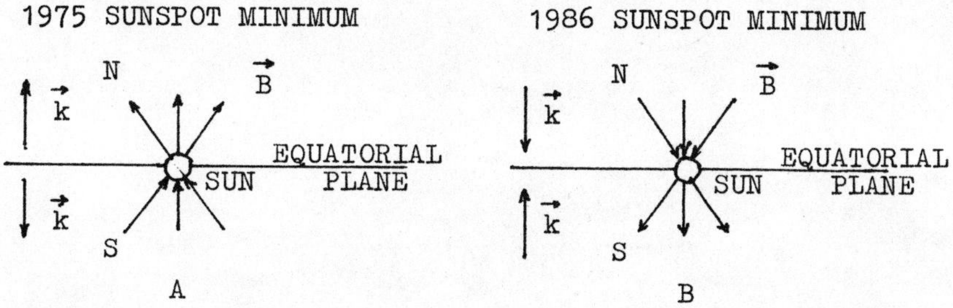

Fig. 2. The meridional section of the heliosphere.

SUMMARY

Because the solar wind Mach number is sufficiently large the solar wind flow in front of the heliospheric shock wave close to the equator most probably is unstable with respect to fast magnetosonic waves. In the case when the radial diffusion of the cosmic rays near the shock is defined by the transverse component of the diffusion tensor and when the ratio of the parallel diffusion coefficient to the perpendicular one is large compare with unity, the waves with the greatest growth rate propagate at right angles to the equatorial plane. These waves have wavelength of about 1 A.U. and period of about two months. The direction of propagation depends on the solar magnetic field polarity.

REFERENCES

1. E.A.Dorfi and L.O'C.Drury, A cosmic ray driven instability, in: Proc. 19th Int. Cosmic Ray Conf., La Jolla, USA, 3, 121 (1985)

2. G.P.Zank and J.F.McKenzie, Instabilities in decelerating supersonic flows with applications to cosmic ray shocks, in: Proc. 19th Int. Cosmic Ray Conf., La Jolla, USA, 3, 111 (1985)

3. L.O'C.Drury and S.A.E.G.Falle, On the stability of shocks modified by particle acceleration, Mon. Not. Roy. Astron. Soc. 223, 353 (1986)

4. E.G.Berezhko, Instability in a shock wave propagating in gas with cosmic rays, Pis'ma v Astron. Zh. 12, 842 (1986)

5. S.V.Chalov, Instability of the structure of strong oblique MHD cosmic-ray shocks, Astrophys. Sp. Sci. 148, 175 (1988)

6. J.R.Jokipii and J.M.Davila, Effects of particle drift on the transport of cosmic rays. 4.More realistic diffusion coefficients, Astrophys. J. 248, 1156 (1981)

7. S.V.Chalov, Cosmic rays and magnetosonic instability of solar wind flow near the heliospheric shock, Geomagnetizm i Aeronomia 30, 27 (1990)

8. A.C.Cummings, E.C.Stone and W.R.Webber, Changes in the energy spectrum of anomalous oxygen during 1977 - 1985, J.Geophys.Res. 91, 2896 (1986)

9. L.A.Fisk, The anomalous component, its variation with latitude and related aspects of modulation, in: The Sun and the Heliosphere in Three Dimensions, ed. R.G.Marsden, D.Reidel Publ. Comp., Dordrecht, Holland 1986, p. 401.

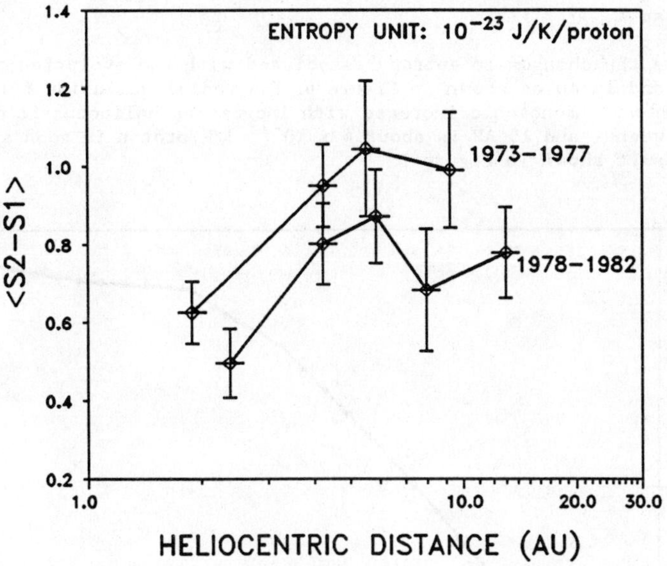

Figure 3

We study shocks observed from Voyager 1 and 2 and Pioneer 10 and 11 in 1973-1982. A total of 413 shocks identified from the hourly average plasma and magnetic field data with the jump in flow speed greater than or equal to 10 km/s are used in this study. All the shocks satisfy the condition that the number density, temperature, pressure and entropy increase across the shock. The selection criteria include the increase of the magnetic field across the shock if the magnetic field data are available. Thus all shocks under study are fast shocks.

We divide the observed shocks into two groups. The first group consists of 210 shocks observed during a 5-year period 1973-1977 near a sunspot minimum. In this period the yearly mean sunspot numbers were no greater than 38.0. Almost all the shocks (92%) in this group were observed from Pioneers spacecraft. A small fraction of the shocks (8%) were observed from Voyager spacecraft inside of 2 AU. The second group consists of 203 shocks observed during 1978-1982 near a sunspot maximum. During this period the yearly mean sunspot numbers were greater than 92.0. All the shocks inside 6 AU in this group were observed from the two Voyager spacecraft. On the other hand all the shocks outside 15 AU in this group were observed from Pioneer 10, no magnetic field data are available for these shocks. The shock strength, measured by p_2/p_1 or n_2/n_1, increases with the heliocentric distance outside 1 AU, reaches a maximum near 5 AU, and then decreases with the distance. Our study also shows that shocks are stronger during the year of low solar activity.

We also carry out a statistical study of the entropy increases, $s_2 - s_1$, for the shocks observed during the two 5-year periods as shown in Figure 3. The average entropy increases across shocks during the years of low solar activity are slightly greater than those during the year of high solar activity. Their maximum values which occur at near 5 AU are in the order of 10^{-23} J/K/proton. When a shock propagates through the solar wind, the shock heating increases the entropy of the solar wind plasma by approximately 0.8×10^{-23} J/K/proton.

MHD SIMULATION FOR SHOCK HEATING OF THE SOLAR WIND

An MHD model /2/ is used to study the increase in solar rotational average of entropy of the solar wind between 1 AU and 15 AU shown in Figure 1. During the three-month period starting November 1977 the same solar wind was observed at different heliocentric distances by five radially aligned spacecraft: IMP 7 and 8 at 1 AU, Voyager 1 and 2 at near 1.5 AU, and Pioneer 10 at 15.1 AU /3/. Three large streams (A, B and C) were observed by IMP and by Voyager. IMP observed two weak shocks at the leading edge of stream B: a forward shock and a reverse shock. Pioneer 10 observed the same solar wind at 15 AU having a distinctly different solar wind structure. The principal structures observed at 15 AU are four shock waves. We may identify them as FB, RA, FC, and RB. Pioneer 10 observed no traces of the streams.

Plasma and magnetic field data from IMP are used to generate the input function for the simulation. The simulation result shows in details how the solar wind structure evolves from a series of three large streams at 1 AU to a new structure consisting of four shocks at 15

AU /3/. The results identify that FB and RB are the forward-reverse shock pair associated with stream B. RA is a reverse shock associated with stream A and FC is a forward shock associated with stream C.

We can also calculate the changes in entropy associated with the evolution of the solar wind structure between 1 and 15 AU as shown in Figure 4. The radial variation for the rotational average of entropy shows a monotonic increase with increasing heliocentric distances. The entropy increase between 1 and 15 AU is about 4 x 10^{-23} J/K/proton in good agreement with the observational result shown in Figure 1.

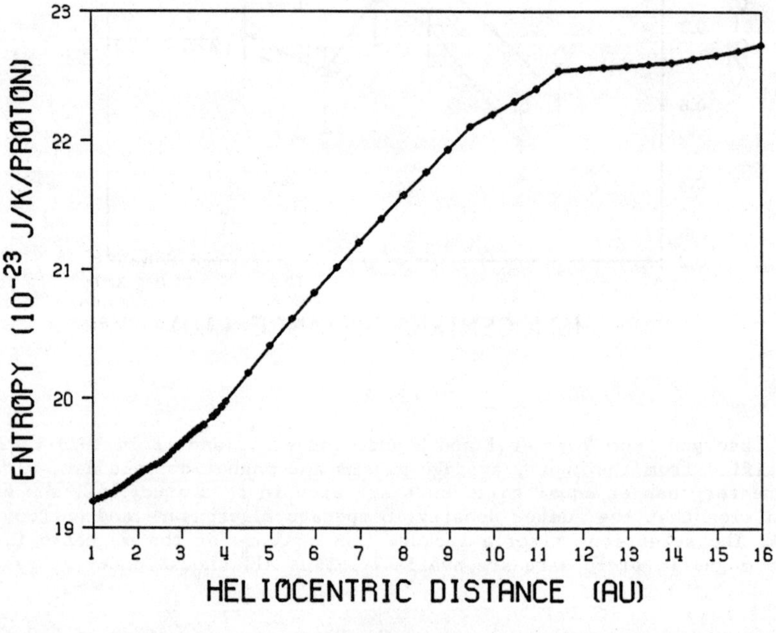

Figure 4

Several mechanisms have been suggested for heating of the solar wind. Heating by the interaction between solar wind ions and interstellar neutral ions, turbulent heating, heating by the damping of compressive waves, and shock heating. This simulation model uses the Rankine-Hugoniot solution to calculate the jumps in flow properties at all shock crossings. Only shock heating is included in this simulation model. We interpret the good agreement between the simulation result and the observation as to indicate that shocks are chiefly responsible for the heating of the solar wind plasma between 1 and 15 AU.

ACKNOWLEDGMENTS

The Voyager 1 and 2 plasma data are from the experiment of H. Bridge and J. Belcher. The Pioneer 10 and 11 plasma data are from the experiment of J. Wolfe. The Voyager 1 and 2 magnetic field data are from the experiment of N. Ness. The Pioneer 11 magnetic field data are from the experiment of E. Smith. The work at the Catholic University of America was supported by the National Aeronautics and Space Administration under grant NAGW-1595 and by the Atmospheric Sciences Section of the National Science Foundation under grant ATM-8614990.

REFERENCES

1. Y. C. Whang, K. W. Behannon, L. F. Burlaga, and S. Zhang, Thermodynamic properties of the heliospheric plasma, J. Geophys. Res. 94, #A3, 2345-2364 (1989)

2. Y. C. Whang, The forward-reverse shock pair at large heliocentric distances, J. Geophys. Res. 89, #A9, 7367-7379 (1984)

3. Y. C. Whang and L. F. Burlaga, Evolution and interaction of interplanetary shocks, J. Geophys. Res. 90, #A11, 10,765-10,778 (1985)

RADIAL EVOLUTION OF INTERACTION REGIONS

Y.C. Whang* and L.F. Burlaga**

**Department of Mechanical Engineering, Catholic University of America, Washington, DC 20064, U.S.A.*
***Laboratory for Extraterrestrial Physics, NASA–Goddard Space Flight Center, Greenbelt, MD 20771, U.S.A.*

ABSTRACT

Near and outside 1 AU, corotating interaction regions bounded by shocks form at the leading edges of high-speed streams. This paper describes the radial evolution of interaction regions, the merging of interaction regions belonging to neighboring streams, and the coalescence of interaction regions belonging to successive solar rotations.

COROTATING INTERACTION REGIONS

Near and outside 1 AU, corotating shocks form at the leading edges of high-speed streams. The large differences in flow speeds compresses the plasma and the magnetic field at the leading edge. As the stream moves outwards, the pressure fronts steepen to form a forward-reverse shock pair. The region bounded by the shock pair is a corotating interaction region (CIR). This paper describes the radial evolution of interaction regions based on plasma and magnetic field data and MHD simulation model.

Outside several AU high-speed streams give way to large interaction regions. A CIR is bounded by a pair of well developed shocks, large pressure jumps exist at the boundaries of the CIR. The pressure of the shocked plasma inside CIR is substantially greater than that outside. From a study of 413 shocks observed from Voyagers and Pioneers in 1973-1982, we found that the shock strength measured by the pressure ratio p_2/p_1 increases with the heliocentric distance outside 1 AU, reaches a maximum near 5 AU, and then decreases with the distance /1/. The forward and the reverse shocks propagate in opposite directions in the solar wind frame of reference. This causes the spatial dimension of CIR to expand with increasing heliocentric distance r.

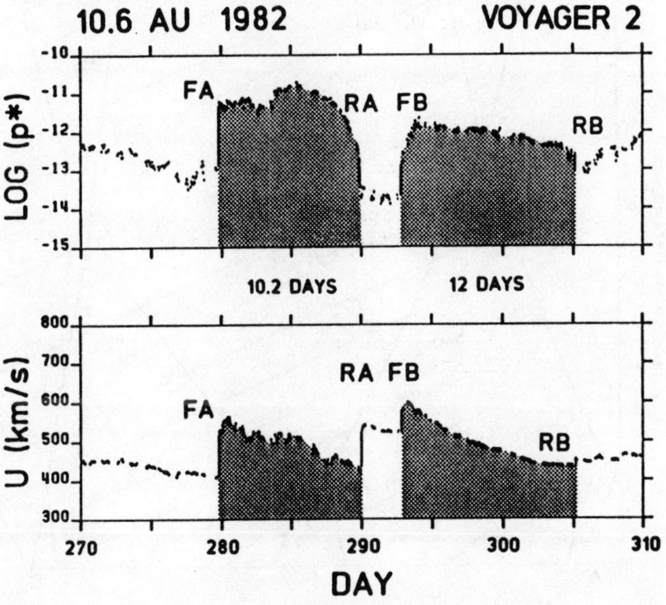

Figure 1

MERGING OF INTERACTION REGIONS BELONGING TO NEIGHBORING STREAMS

Figure 1 shows the observation at 10.6 AU of two large interaction regions, A and B, belonging to two neighboring streams. With reference to the unshocked plasma in the region between them, the two shocks RA and FB propagate in opposite directions facing each other. The two shocks are on a colliding course. After the collision, the two shocks RA and FB are weakened, and the two interaction regions coalesce to form a merged interaction region bounded by FA and RB. Inside a merged interaction region all plasmas have been shocked at least once.

Next, we will use a simulation result to demonstrate the radial evolution of streams and interaction regions between 1 and 15 AU. In a three-month period starting November 1977, the same solar wind was observed at different heliocentric distances by five radially aligned spacecraft: IMP 7 and 8, Voyager 1 and 2, and Pioneer 10. These spacecraft observed the solar wind with a radial separation of 14 AU. Their observations provide a rare opportunity to study the radial evolution of the interaction regions. Three large streams (A, B and C) were observed near 1 AU. Pioneer 10 observed the same solar wind at 15 AU having a distinctly different structure. Pioneer 10 observed no traces of the streams, The only significant solar wind structures that can be identified at 15 AU are four shock waves FB, RA, FC, RB. The results of a simulation study /2/ show that FB and RB form the forward-reverse shock pair associated with stream B, RA is a reverse shock associated with stream A and FC is a forward shock associated with stream C.

Figure 2 shows the trajectories of shocks associated with the three streams A, B and C. The formation of an interaction region associated with stream B between 1 AU and 4 AU is shown near the center, lower part. At 1 AU, IMP observed a compression region bounded by a pair of weak shocks. At 1.6 AU, the interaction region has a complicated structure containing one forward shock, two reverse shocks, and another new shock pair. Near 2 AU, a forward and a reverse shock collide. At 2.2 AU, the interaction region contains a sequence of five shocks (F-F-R-R-R). Near 2.5 AU the three reverse shocks merged to form a single reverse shock RB. At 2.8 AU, the solar wind structure contains two forward shocks and a reverse shock RB. At 3.6 AU the two forward shocks merge and the leading edge region of stream B finally evolved into a simple interaction region bounded by a shock pair FB and RB. The width of the CIR continues to expand at increasing r. At 9.5 AU, RB collides with FC, the forward shock of stream C. This causes the two CIRs associated with stream B and C to coalesce. At 11.4 AU FB collides with RA, the reverse shock of stream A. The CIR associated with stream B now coalesce with the CIR associated with stream A. This simulation result identifies the sequence of shocks that pass a radially aligned spacecraft Pioneer 10 at 15 AU as FB, RA, FC and RB. It also provides the detailed evolution process for large-scale solar wind structures in the vast unobserved region between 1 AU and 16 AU.

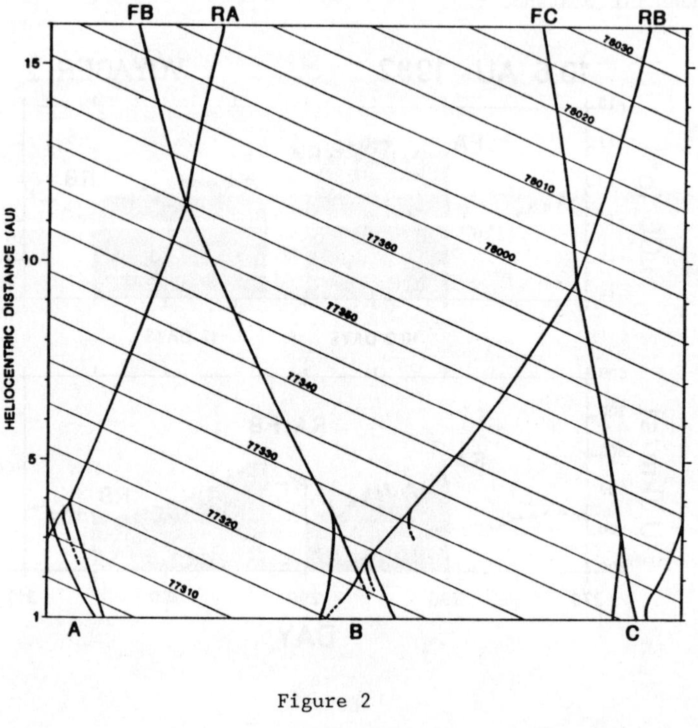

Figure 2

Many dominating solar wind structures are quasi-stationary in a frame of reference corotating with the Sun. During the decline and minimum of solar activity, the pattern of corotating features may last for a few solar rotations. Figure 3 displays the structure of a same solar wind over five solar rotations observed from three spacecraft at 1 AU, near 4.5 AU, and near 6 AU. The data set indicates that (a) during the first four solar rotations, the quasi-recurrent corotating streams at 1 AU have two streams per solar rotation; (b) about a score of new shocks form in the solar wind between 1 AU and 6 AU; (c) the amplitudes of the speed variation significantly change from 500 km/s at 1 AU to less than 300 km/s at 6 AU; (d) perhaps, the solar wind structure has evolved from two streams per rotation at 1 AU to a new single structure per rotation at 6 AU. We may take the solar rotation C as an example to examine the radial evolution of the solar wind structure. No well-defined shocks were observed at 1 AU. However, two shock pairs and two CIRs can be clearly identified at 4.5 AU and at 6 AU. The amplitude of the flow speed variation decreases significantly between 1 AU and 6 AU. The amplitude of pressure variation increases by a factor of 10 or more. The width of each CIR increases at increasing r. The CIRs associated with two neighboring streams will eventually coalesce to form a merged interaction region outside 10 AU. As a result the recurrent solar wind structure will evolve from two large streams per solar rotation at 1 AU to a new recurrent structure containing a single interaction region per solar rotation outside 10 AU.

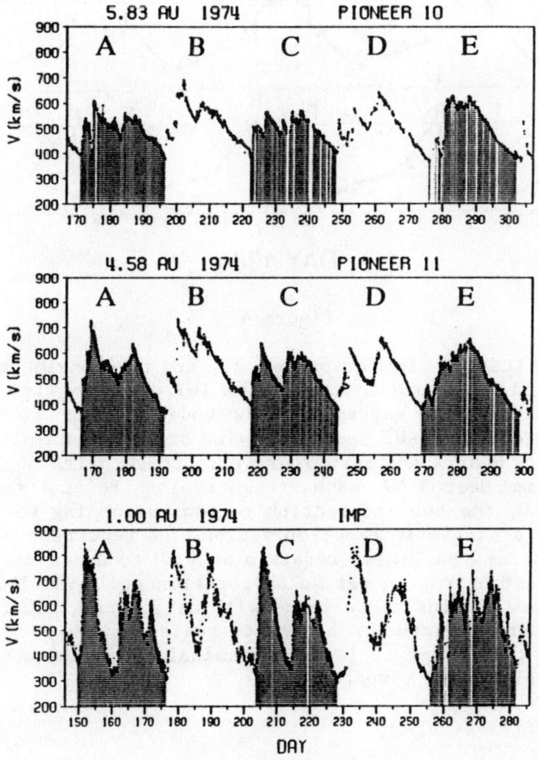

Figure 3

COALESCENCE OF INTERACTION REGIONS BELONGING TO SUCCESSIVE SOLAR ROTATIONS

During the decline and minimum of solar activity the corotating solar wind structure near 15 AU is more likely to consist of a single interaction region per solar rotation than two /3/. In March and November of 1984, Voyager 2 observed two sets of nearly recurrent interaction regions in the outer heliosphere, each with a period of 26 days. Large variations in pressure are present associated with the interaction regions. In order to understand the radial evolution of recurrent interaction regions outside 15 AU in the outer heliosphere, we carried out a simulation study /4/ based on an idealized recurrent interaction regions generated from the Voyager 2 data of March 1984 as shown in Figure 4. This idealized recurrent solar wind structure consists of one interaction region per solar rotation. Each interaction region is bounded by a shock pair. The two shocks propagate in opposite directions relative to the solar wind. The width of the interaction region increases with increasing r. Eventually, the forward shocks collide with the reverse shocks belonging to previous solar rotation and the interaction regions belonging to successive solar rotations coalesce.

The forward and the reverse shocks propagate in opposite directions relative to the solar wind. If we follow a forward shock at 14 AU. This forward shock collides at 34 AU with the reverse shock belonging to the previous solar rotation. The two shocks are weakened by the collision and they continue to propagate in opposite directions away from each other. At 79 AU, the forward shock encounters another reverse shocks belonging to a previous solar rotation. A third collision takes place at 124 AU. A forward shock collides with a reverse shock at about every 45 AU. Shocks are weakened after each collision.

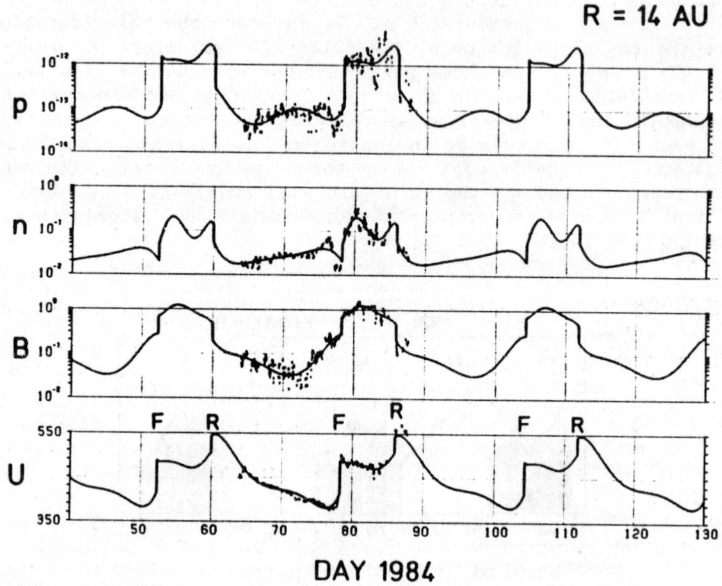

Figure 4

Based on a series of quantitative simulation, we may try to describe the global scale evolution of the solar wind structure observed from IMP during a five solar rotation period of 1974 (shown in Figure 3). We can extrapolate our understanding for the radial evolution of the solar wind structures. At 1 AU, the solar wind structure mainly consists of two large streams per solar rotation. Near 2 AU, forward-reverse shock pairs begin to form at the leading edge of each stream. Near 5 AU, each stream evolves to an interaction region bounded a shock pair. Outside 10 AU, the two interaction regions belonging to the same solar rotation coalesce to form a single interaction region. The recurrent solar wind structure evolves from two large streams per solar rotation at 1 AU to a new recurrent structure containing a single interaction region per solar rotation at 15 AU. Near 40 AU, the solar wind structure of rotation C begins to coalesce with neighboring solar wind structure B and D. Near 80 AU, the solar wind structure C begins to coalesce with 4 neighboring structures A, B, D and E. This study shows that it is the dynamical interaction that dominates the radial evolution of the solar wind structure.

ACKNOWLEDGMENTS

The Voyager 1 and 2 plasma data are from the experiment of H. Bridge and J. Belcher. The Pioneer 10 and 11 plasma data are from the experiment of J. Wolfe. The work at the Catholic University of America was supported by the National Aeronautics and Space Administration under grant NAGW-1595 and by the Atmospheric Sciences Section of the National Science Foundation under grant ATM-8614990.

REFERENCES

1. Y. C. Whang, S. Liu and L. F. Burlaga, Shock heating of the solar wind plasma, This volume, (1990).

2. Y. C. Whang and L. F. Burlaga, Evolution and interaction of interplanetary shocks, J. Geophys. Res. 90, #A11, 10,765-10,778 (1985)

3. L. F. Burlaga, Period doubling in the outer heliosphere, J. Geophys. Res. 93, #A5, 4103-4106, (1988).

4. Y. C. Whang and L. F. Burlaga, Evolution of recurrent solar wind structures between 14 AU and the termination shock, J. Geophys. Res. 93, #A6, 5446-5460 (1988).

NONLINEAR SURFACE WAVE PROPAGATION ON HELIOSPHERIC CURRENT SHEET

M.S. Ruderman

Institute for Problems in Mechanics, U.S.S.R. Academy of Sciences, 101 Vernadsky Ave., 117526 Moscow, U.S.S.R.

ABSTRACT

We consider surface wave propagation on heliospheric current sheet. The current sheet is modelled by tangential magnetohydrodynamic discontinuity. At this discontinuity the density and the direction of magnetic field can change abruptly. The plasma is taken to be cold. Therefore the condition of full pressure balance leads to the continuity of magnetic induction value. The anisotropic ion viscosity is taken into account. With the use of reductive perturbation method we derive nonlinear integro-differential equation governing the small-amplitude surface wave propagation. The numerical solutions of this equation show that nonlinearity causes the steepening of initially sinusoidal waves. If the viscosity is small the steepening is very strong and there are large gradients of plasma parameters inside the wave. As a result the large increase of wave energy dissipation takes place. The dissipation rate can be several orders of magnitude higher than that calculated in the linear approximation. The dependence of the wave energy flux on the distance from wave source is calculated at various values of nonlinearity.

INTRODUCTION

According to recent concepts the large-scale structure of heliospheric magnetic field contains the current sheet placed near the ecliptic. At present it is not established whether the angle of magnetic field rotation inside the current sheet is equal to 180° or somewhat less. But in accordance with observations /1/ the change of magnetic field inside the current sheet is mainly due to the rotation of magnetic induction vector under small variation of its value. In addition there are evidences /1/ that for the ratio of the plasma pressure to the magnetic one the estimation $\beta \lesssim 1$ is valid. Hence it seems suitable to describe the current sheet dynamics in the approximation of cold plasma.

In the linear approximation the wave propagation on the current sheet and the interaction of waves and current sheet have been studied in detail in /2/. In so doing the structure of current sheet have been taken into account by means of explicit definition of magnetic field variation inside the current sheet.

In the present paper we study the nonlinearity effect on surface wave propagation on the current sheet. The consideration of nonlinearity complicates the problem significantly. Hence we simplify the model in comparison with /2/. Namely we model the current sheet with the tangential magnetohydrodynamic discontinuity at which the density and the magnetic field direction are discontinuous. Such discontinuity is called magnetic interface. The value of magnetic induction is continuous due to the condition of full pressure balance.

The wave propagation on magnetic interface in the linear approximation has been studied in a lot of works. The most comprehensive study has been carried out in /3/. Nonlinear wave propagation on a magnetic interface has been considered in /4,5/ in the approximation of incompressible fluid and in /6/ taking into account the compressibility, but under the restriction that waves propagate along the undisturbed magnetic field. In /5/ the viscosity has been taken into account.

The surface waves propagate on a magnetic interface without the dispersion. Therefore it could be expected that the nonlinearity leads to the wave overturning. This hypothesis is confirmed by numerical calculations /4,6/. Taking into account the viscosity stops wave steepening and makes it possible to calculate wave evolution up to almost full its damping.

One of the main purposes of this paper is to calculate the dependence of wave energy flux on the distance from the source. The dissipation leads to the decrease of wave energy flux. That is why we take into account the dissipative processes.

Beyond the Earth orbit the collision interval obtained with the use of classical formula for Coulomb's collisions is more than or of the order of 10^7 s. But taking into account the collective plasma processes can substantially decrease the collision interval. As a result we can expect that waves with the petiods of the order of 10^5 s can be described by magnetohydrodynamic equations for collisional plasma. Hence we use classical formulas for coefficients of anisotropic viscosity and electrical conductivity /7/. As the plasma is supposed to be cold the thermoconductivity does not effect on wave propagation.

Formula for tensor of anisotropic ion viscosity contains five coefficients. But the ratios of the last four coefficients to the first one are no more than $(\omega_i \tau_i)$, where ω_i is the gyrofrequency of ions, τ_i is the collision interval of ions. As we have $\omega_i \gtrsim 10^{-2} s^{-1}$ for heliocentric distances less than 50 A.U., then $\omega_i \tau_i \gg 1$ under the condition $\tau_i \gg 100$ s. More than likely the last estimation is valid. Therefore we can take the last four coefficients in the formula for tensor of anisotropic viscosity are equal to zero. As a result we have the following expression for tensor of ion viscosity

$$\pi_{jl} = \mu(b_j b_l - \frac{1}{3}\delta_{jl})(3\vec{b}\cdot(\vec{b}\cdot\nabla)\vec{v} - \nabla\cdot\vec{v}) \tag{1}$$

where $\vec{v}$ is the velocity, $\vec{b}$ is the unit vector directed along the magnetic field, $\mu \approx kn_i T_i \tau_i$, k is the Boltzmann's constant, n_i and T_i are the concentration and the temperature of ions.

Finally, estimations show that the effects of electron viscosity and Joule dissipation can be neglected as well as the Hall effect provided $\tau_e/\tau_i \sim (m_e/m_i)^{1/2}$ where τ_e is the collisional interval of electron, m_e amd m_i are the masses of electron and ion.

THE BASIC EQUATIONS

In accordance with said above the motion of medium is governed by the magnetohydrodynamic equations for cold plasma. The anisotropic ion viscosity is taken into account, the viscosity tensor being written in the form (1). These equations are as follows

$$\begin{gathered} \frac{\partial \rho}{\partial t} + \nabla\cdot(\rho\vec{v}) = 0; \quad \frac{\partial \vec{B}}{\partial t} = \nabla\times(\vec{v}\times\vec{B}) \\ \rho\left(\frac{\partial \vec{v}}{\partial t} + (\vec{v}\cdot\nabla)\vec{v}\right) = \frac{1}{4\pi}(\nabla\times\vec{B})\times\vec{B} + \\ + \mu(\vec{b}(\vec{b}\cdot\nabla) - \frac{1}{3}\nabla)(3(\vec{b}\cdot\nabla)(\vec{b}\cdot\vec{v}) - \nabla\cdot\vec{v}) \end{gathered} \tag{2}$$

where ρ is the density; $\vec{v}$ the velocity; $\vec{B}$ the magnetic induction.

In an undisturbed state the magnetic interface coinsides with the xy-plane of Cartesian coordinate system x, y, z (see Figure 1). The coordinate system moves with the velocity of solar wind, so in the undisturbed state $\vec{v} = 0$. In addition $\rho = \rho_{0-}$, $\vec{B} = \vec{B}_{0-}$ at $z < 0$, $\rho = \rho_{0+}$, $\vec{B} = \vec{B}_{0+}$ at $z > 0$, $\vec{B}_{0-}$ and $\vec{B}_{0+}$ being parallel to the xy-plane and $|\vec{B}_{0+}| = |\vec{B}_{0-}|$. In the disturbed state the interface is defined by the equation $z = \eta(t,x)$. All variables are supposed to be independent of y.

At $z = \eta(t,x)$ the kinematic boundary conditions and the condition of full pressure balance must be satisfied

$$v_z = \frac{\partial \eta}{\partial t} + v_x \frac{\partial \eta}{\partial x} \ ; \quad \left[\frac{B^2}{8\pi} + \mu(\vec{b}(\vec{b}\cdot\nabla)\vec{v} - \frac{1}{3}\nabla\cdot\vec{v}) \right] = 0 \tag{3}$$

where v_x and v_z are x and z-components of velocity and square brackets denote the jump of value through the interface.

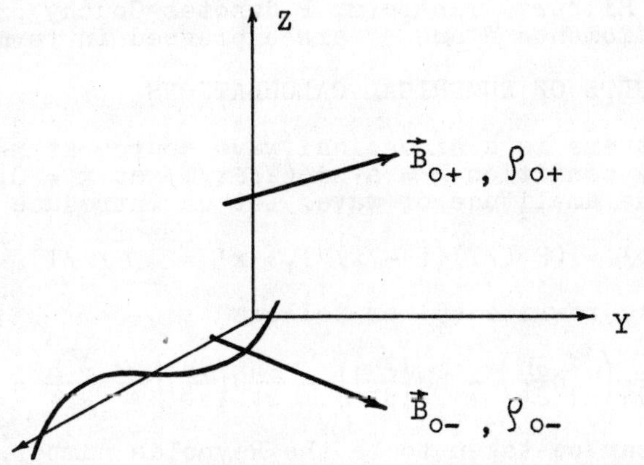

Fig.1. The undisturbed state. The wavy line shows the intersection of the disturbed magnetic interface with xz-plane.

Below we fit with the subscripts + and - variables at $z > \eta$ and $z < \eta$ respectively. Let us note that in fact the first equation (3) represents two boundary conditions because it is true for $\vec{v}_+$ and $\vec{v}_-$ simultaneously.

In addition to the boundary conditions (3) we require all dependent variables vanish at $|z| \longrightarrow \infty$. The system of equations (2) and boundary conditions (3) is basic for study of nonlinear surface wave propagation.

DERIVATION OF THE EQUATION FOR NONLINEAR WAVES

We consider disturbances with non-dimensional amplitude $\varepsilon \ll 1$. It means that disturbances of ρ , $\vec{v}$ and $\vec{B}$ are of the order of $\varepsilon\rho_o$, εV and εB_o respectively, where V is the phase velocity when the viscosity is absent. It is our intention to derive the equation describing cooperative effect of nonlinearity and viscosity. So we suppose the ratio of the scale length of disturbances to the viscous damping length calculated in the linear approximation is of the order of ε . Now we follow the reductive perturbation method /8/. We introduce the stretching transformation $\tau = \varepsilon(t - x/V)$, $\xi = \varepsilon^2 x$, $\zeta = \varepsilon z$ and the perturbation expansion of all dependent variables in the form $f = f_o + \varepsilon f^{(1)} + \varepsilon^2 f^{(2)} + \dots$, where f_o denotes the undisturbed value.

In the first approximation we get from (2) and (3) the system of equations and boundary conditions coinsiding with the linear approximation without viscosity. The condition of consistency of this system is the dispersion equation defining V. To solve this equation in common case is not so easy, but in the important case $\rho_{o-} = \rho_{o+}$ the solution is straightforward

$$V^2 = \frac{1}{2} V_a^2 (\cos^2\alpha_+ + \cos^2\alpha_-) \tag{4}$$

where $V_{a\pm}^2 = B_{o\pm}^2 / 4\pi\rho_{o\pm}$, $\alpha_\pm$ is the angle between $\vec{B}_{o\pm}$ and the direction of wave propagation.

Proceeding to the next order, we have the equations and the boundary conditions that are linear with respect to the second approximation variables. Their left sides coinside with the equations and the boundary conditions of the first approximation. The right sides are expressed in terms of the first approximation variables. The system of second order approximation is consistent provided the right sides satisfy the compatibility condition. This condition is the equation governing the evolution of $\eta^{(1)}(\tau, \xi)$. Transforming this equation to the initial independent variables and introducing $h = \varepsilon\eta^{(1)}$ we write it as follows

$$\frac{\partial h}{\partial x} + \frac{1}{V}\frac{\partial h}{\partial t} - W\left\{\frac{\partial}{\partial t}\left(H\left(h\frac{\partial h}{\partial t}\right) - hH\left(\frac{\partial h}{\partial t}\right)\right) - \frac{\partial h}{\partial t}H\left(\frac{\partial h}{\partial t}\right)\right\} - \gamma\frac{\partial^2 h}{\partial t^2} = 0$$

$$H(h) = \frac{P}{\pi}\int_{-\infty}^{\infty}\frac{h(t')dt'}{t' - t} \qquad (5)$$

Here H(h) is the Hilbert transform, P denotes Couchy principal value of integral. The coefficients W and γ are expressed in terms of ρ_0, B_0, α and μ.

THE RESULTS OF NUMERICAL CALCULATIONS

We suppose that there is a sinusoidal wave source at x = 0. This means that there is boundary condition $h = a\cdot\sin(2\pi t/T)$ at x = 0, where T and a are the period and the amplitude of wave. Let us introduce non-dimensional variables

$$t' = (2\pi/T)(t - x/V), \quad x' = 4\pi^2\gamma x/T^2, \quad h' = h/a$$

Omitting primes we rewrite (5) as follows

$$\frac{\partial h}{\partial x} - R\left\{\frac{\partial}{\partial t}\left(H\left(h\frac{\partial h}{\partial t}\right) - hH\left(\frac{\partial h}{\partial t}\right)\right) - \frac{\partial h}{\partial t}H\left(\frac{\partial h}{\partial t}\right)\right\} - \frac{\partial^2 h}{\partial t^2} = 0 \qquad (6)$$

where $R = aW/\gamma$ can be taken to be the Reynolds number. The value R determines the ratio of nonlinearity effect to the viscosity one. At large values of R the nonlinearity leads to the wave steepening. As a result large gradients of plasma parameters appears and the rate of wave damping increases. Let us Q stands for the mean energy flux through the band of plane perpendicular to x-axis having unit width on y. This flux is equal to x-component of Poynting's vector averaged on time and integrated on z. In Figure 2 the dependences of $\log(Q/Q_0)$ on the distances from wave source at different values of R is shown. Here Q_0 is the value of Q at x = 0.

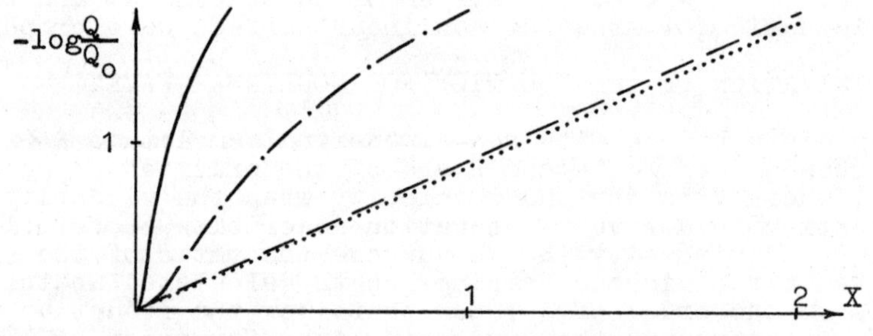

Fig.2. The dependences of Q on the distances from the wave source. The dotted, dashed, chain-dotted and solid lines refer to R = 0, 1, 10 and 50.

As we see the larger R the bigger the rate of wave damping. The nonlinear dissipation of surface waves can lead to warming up of the current sheet.

REFERENCES

1. K.W.Behannon, F.M.Neubauer, and H.Barnstorf, J.Geophys.Res. 88, 3273 (1981)

2. Z.E.Musielak, and S.T.Suess, Astrophys.J. 330, 456 (1988)

3. C.Uberoi, and A.S.Narayanan, Plasma Phys.Contr.Fusion 28, 1635 (1986)

4. M.S.Ruderman, Fluid Dynamics 20, 85 (1985)

5. M.S.Ruderman, Fluid Dynamics 21, 925 (1986)

6. M.S.Ruderman, Plasma Phys.Contr.Fusion 30, 1117 (1988)

7. V.B.Baranov and K.V.Krasnobaev, Gidrodinamicheskaya teoriya kosmicheskoi plasmy, Nauka, Moscow, 1977.

8. T.Taniuti, Suppl.Progr.Theor.Phys. #55, 1 (1974)

MAGNETIC FIELDS IN THE HELIOSPHERE: PIONEER OBSERVATIONS

E.J. Smith

Jet Propulsion Laboratory, California Institute of Technology, Pasadena, CA 91109, U.S.A.

ABSTRACT

Major physical processes are reviewed that involve the magnetic field as measured by Pioneers 10 and 11. Topics discussed are spatial gradients out to 20 AU, the evolution of individual solar wind streams from 1 to 5 AU, the merging of solar wind structures (corotating streams and coronal mass-ejections) from 5 to 20 AU, the search for waves near 10 AU associated with interstellar ions, the acceleration of energetic particles to MeV energies and both short duration decreases and solar cycle modulation of galactic cosmic rays. Attention is given to departures from expectation and phenomena that were not anticipated.

INTRODUCTION

In keeping with the theme of this symposium, the following review emphasizes major physical processes in the outer heliosphere involving the magnetic field. The discussion is based on Vector Helium Magnetometer observations made by Pioneers 10 and 11. The contributions of the Voyagers to our understanding of this important scientific frontier is the subject of a separate review.

The various topics and processes are approached from the standpoint of increasing heliocentric distance and/or phases of the Pioneer mission which now extend from 1 to beyond 40 AU and cover almost two sunspot cycles. The phase of solar activity when the observations were made is basic because of a correlation between radial distance and time as represented principally by the solar cycle. The initial observational phase occurred during declining solar activity; average properties and the evolution of recurrent solar wind streams received the most attention along with the interplanetary acceleration of particles to MeV energies and the study of recurrent cosmic ray decreases. With increasing distance and the transition to solar maximum, the merging of solar wind structures, as distinct from the evolution of individual streams, became evident. The causes of cosmic ray modulation also became of great interest because of the uniqueness of the observations. Throughout the mission, evidence of an interaction of the solar wind with neutral interstellar gas has been sought. Limitations of space preclude even a brief discussion of other important topics such as waves, discontinuities, and/or turbulence, the structure of the heliospheric current sheet, etc. Knowledge of these items as revealed by Pioneer must be sought elsewhere /1, 2/.

RESULTS

Expansion of the Solar Wind

The hydrodynamic expansion of the corona to create the solar wind, as originally formulated by Parker /3/, assumes radial, spherically symmetric flow that is time-stationary or without structure. Stresses associated with the frozen-in solar magnetic fields are ignored everywhere, including those in the source region where, in fact, the magnetic field may be strong enough to inhibit the flow. It is now clear that the solar wind does not originate from a spherically symmetric corona but comes from specific open field regions such as polar coronal holes. Furthermore, the solar wind can exhibit temporal variability on a variety of scales which can be both shorter and longer than the solar rotation period.

In spite of these shortcomings, the Heliospheric Magnetic Field (HMF) properties as observed far from the sun and averaged over many solar rotations agree remarkably well with the predictions of this simple

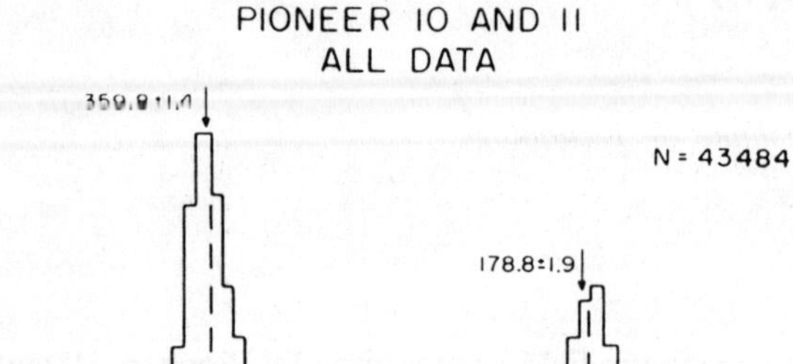

Fig. 1. Histogram or probability distribution of the azimuth angle between 1 and 8.5 AU in coordinates with two axes along and across the Parker spiral direction.

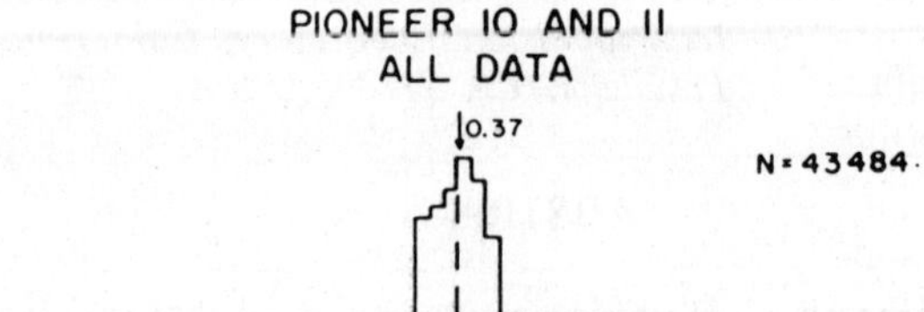

Fig. 2. Histogram of the elevation or latitude angle between 1 and 8.5 AU.

model. To compare the model with observations, it is necessary to suppress the ever-present structure by averaging the measurements or by forming probability distributions over long time intervals. One must also take account of the slow changes associated with the 11 year or 22 year solar cycle by comparing the distant observations with those obtained simultaneously at 1 AU. It is then possible to investigate the spatial gradients predicted by the model.

A basic feature of the model is the field spiral angle, given by $\tan\psi = B\phi / Br = \Omega r / v$. The model does not predict or depend on Br, but given the speed of the supersonic solar wind, requires that $B\phi$ and Br be related as above. Figure 1 is a histogram of the observed azimuth angle over an extended interval obtained by Pioneer 10, 11 at heliocentric distances of 1 to 8.5 AU /4/. In the coordinates used, the x axis was continually rotated to lie along the spiral direction. The tendency of the field to lie at $\psi = 0°$ or $180°$ is evident. The widths of the two distributions centered on the spiral angle is about $\pm 45°$, approximately the same as at 1 AU. This disordering of the field is attributable to a combination of effects, including waves and discontinuities (alternatively, turbulence) and perhaps stochastic motions of the field lines in the photosphere-chromosphere. The dependence of the spiral angle on changes in the solar wind velocity is weak ($\delta(\tan\psi) \propto \delta v/v^2$). However, Alfvén waves can produce large changes in the direction of B at the expense of only modest changes in V, since $\delta B/B \cong \delta v/C_A \cong 1$. The similarity of the distributions to those obtained at 1 AU is noteworthy since it might be supposed that some form of reconnection or interactions between contiguous solar wind structures would lead to increasing disorder.

The two peaks correspond to the two magnetic sectors or, equivalently, to the field polarities above and below the heliospheric current sheet (HCS). With the solar field outward in the north and with Pioneer at north heliographic latitudes, more positive (0°) than negative (180°) fields should be observed as is evident from the figure.

Another important feature of the Parker theory is the absence of an average north-south or polar-latitudinal field component. This requirement follows directly from $\nabla x E = 0$. Figure 2 is a histogram of the field co-latitude angle, θB, for the same interval as Figure 1. The most probable value and the average are both consistent with zero as required. The disorder is real, is comparable to that associated with the spiral angle and is attributable to the same causes.

Histograms of the field magnitude, when compared with expectations based on observations at 1 AU, show a modest but significant departure from the model /5/. Figure 3 contains a histogram of the field

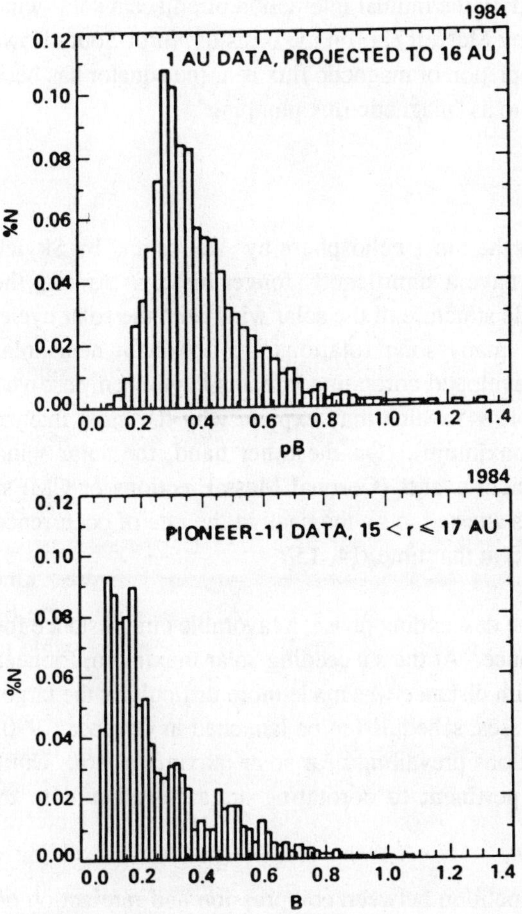

Fig. 3. Histograms of the field magnitude at 16 AU as measured by P11 and extrapolated outward using ICE measurements at 1 AU.

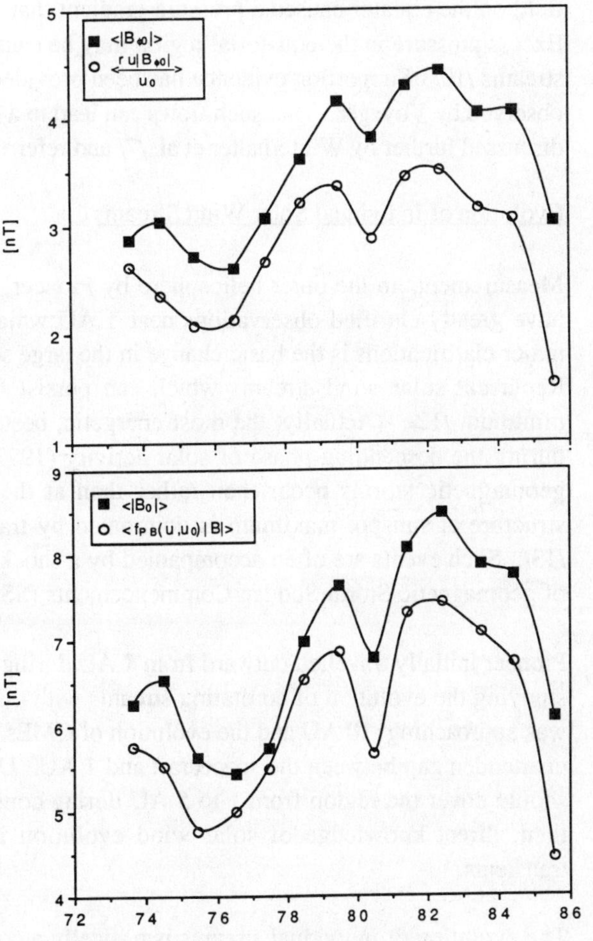

Fig. 4. Comparison of annual averages of the total fields and azimuthal components at P11 and 1 AU.

magnitude observed at 1 AU by ICE and extrapolated out to P11 and a histogram of the P11 observations. There is an overall shift of the P11 magnitudes to lower values than predicted.

This discrepancy is characteristic of the P11 measurements throughout the mission as in Figure 4 /6,7/. Solar cycle changes in B and $B\phi$ at 1 AU are correlated with simultaneous changes at P11. However, the P11 annual averages extrapolated back to 1 AU are everywhere less than the 1 AU values, indicating a deficit in magnetic flux in the outer heliosphere amounting to ≅20%. For reasons which are not understood, this deficit has not been seen or reported in simultaneous Voyager observations (from 1977 onward) /8/. It had been proposed that the deficit in the P11 data was caused by variations in the solar wind velocity with time or latitude (since $B \propto B\phi \propto v^{-1}$). However, the P11 data presented here have been corrected using the velocity measured at P11 and at 1 AU without significantly reducing, let alone eliminating, the deficits. We are now collaborating with the Voyager investigators in an effort to account for the apparent discrepancy in the two sets of measurements.

Several physical explanations have been advanced to account for such a departure from the Parker model. The original suggestion was by Suess and Nerney /9/ that the maximum in the strength of the spiral field on the equator caused a pressure gradient that would cause the field to diverge to higher latitudes. Excess pressure in the equatorial region may be caused by the mutual interaction of adjacent solar wind streams /10/. Supporting evidence has been provided by McNutt /11/ on the basis of North-South flows observed by Voyager. That such flows can lead to a depletion of magnetic flux near the equator has been discussed further by Winterhalter et al. /7/ and referred to as "magnetic flux pumping".

Evolution of Individual Solar Wind Streams

Measurements in the outer heliosphere by Pioneer, in the inner heliosphere by Helios and by Skylab have greatly clarified observations near 1 AU which have a significantly longer history. Among the major clarifications is the basic change in the large scale structure of the solar wind with the solar cycle. Recurrent solar wind streams which can persist for many solar rotations are dominant near solar minimum /12/. (Actually, the most energetic, best developed corotating streams have been observed during the descending phase of solar activity (1973, 1984) which may explain why the most intense geomagnetic storms occur then rather than at the maximum). On the other hand, the solar wind structure at sunspot maximum is dominated by transient events (Coronal Mass Ejections or CMEs) /13/. Such events are often accompanied by a shock as attested to by the peak in the rate of occurrence of geomagnetic Storm Sudden Commencements (SSCs) at that time /14, 15/.

Pioneer initially travelled outward from 1 AU during the descending phase, a favorable circumstance for studying the evolution of corotating streams with distance. At the succeeding solar maximum, Pioneer was approaching 10 AU and the evolution of CMEs with distance was made more difficult by the large, unattended gap between the spacecraft and 1 AU. Ulysses, scheduled to be launched in October, 1990, should cover the region from 1 to 5 AU during conditions prevailing near solar maximum /16/. Until then, direct knowledge of solar wind evolution is pertinent to corotating streams rather than to transients.

The evolution of individual streams is basically a competition between compression and rarefaction or expansion. In regions of decreasing solar wind speed, i.e., the trailing part of streams, a steady expansion of the field and plasma is superposed on the overall radial expansion discussed above. The average field strength and plasma density and temperature are below the values predicted by the Parker model and the level of fluctuations is low.

In regions in which the solar wind speed is increasing near 1 AU (the stream front), compression occurs, the situation is much more dynamic and the physical processes which dominate the stream evolution are operating /17,18/. The overtaking of slow solar wind by faster wind, in the absence of interpenetration or intermingling (because the magnetic field in an adjacent region cannot penetrate into a perfect conductor), causes a pressure "pulse" to develop at the interface. Stresses propagate away from the interface as large amplitude waves which steepen into a pair of shocks, a forward shock preceeding, and a reverse (sunward-propagating) shock trailing, the interface. The resulting well-defined Corotating Interaction Regions (CIRs) are a prominent feature of the solar wind beyond the Earth's orbit. B, n, T and p are greatly enhanced relative to their values based on a simple Parker-type expansion as is the level of fluctuation in all parameters /19/.

The stream now consists of these two distinct regions but, with increasing distance, the expansion region is attacked and systematically eroded by the shocks. By about 10 AU, the velocity structure of individual streams has been almost completely obliterated although significant structure remains in the field and plasma. There is very little, if any, radial gradient in the average solar wind speed.

The magnetic field plays an important role in this evolution. In addition, the streams have a characteristic magnetic polarity as at 1 AU, with streams of opposite polarities coming from opposite solar hemispheres. The HCS would be expected to lie within CIRs, as at 1 AU. Figure 5 shows an extended time interval when two streams were present with opposite polarities /20/. As is evident, the current sheet (sector boundary) did lie within the CIR except for a few cases (when the stream structure shifted abruptly).

The effect of the CIRs and the stream-stream dynamics on the spiral angle is surprising. The spiral is more obvious inside CIRs than in the adjacent expansion region, a curious result considering how large the fluctuations are in the interaction region.

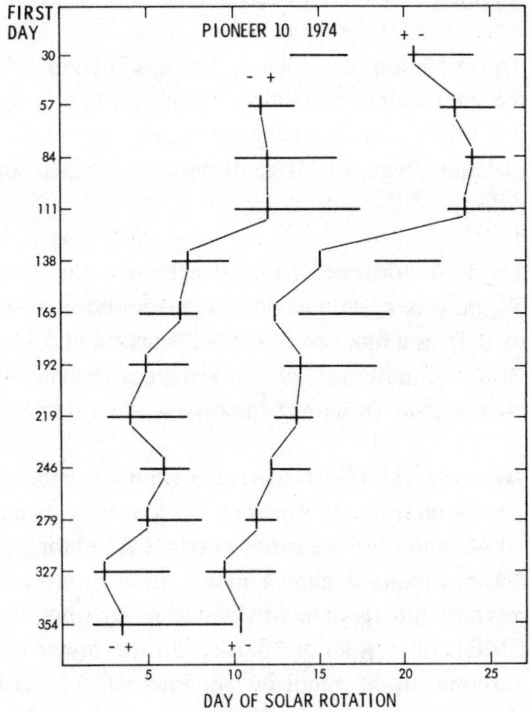

Fig. 5. Comparison of the locations of the HCS or the sector/boundary (vertical lines) and a sequence of CIRs (horizontal lines).

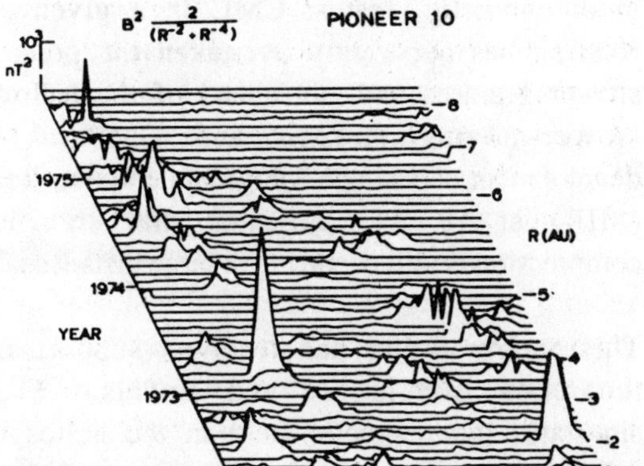

Fig. 6. A plot of B as measured by P10 over successive solar rotations and increasing radial distance.

Merging of Solar Wind Structures

The evolution of corotating streams beyond 1 AU implied that the CIRs would grow at the expense of the intervening rarefaction regions or, in other words, that adjacent streams would merge. One simple possibility was that merging would result in a single more-or-less homogeneous, i.e., structure-less but turbulent, region at large distances. It was a surprise to find instead that merging led to a single large pulse or CIR.

The first evidence of this effect was seen in the Pioneer 10 data between 1 and 8 AU (1972-1976). Figure 6 is a 3D plot of field magnitude (corrected to its equivalent value at 1 AU using the Parker model) as a function of radial distance and solar longitude. A clear trend is discernible with an apparent "drift" of multiple CIRs toward a convergence into a single CIR near 8 AU. At that distance, most of the magnetic flux was compressed into a single region.

This general effect was also obvious near solar maximum from comparisons of Pioneer 11 data near 10 AU and ISEE measurements at 1 AU during 1979 /21/. Figure 7 shows many more compression regions, attributable to corotating streams and/or CMEs, at 1 than at 10 AU. In the lower panel, maxima in B recurred at Pioneer 11 with a characteristic period of ≅27.5 days which departs significantly from the Pioneer synodic period of 26 days. At 1 AU, any periodicity is much less obvious.

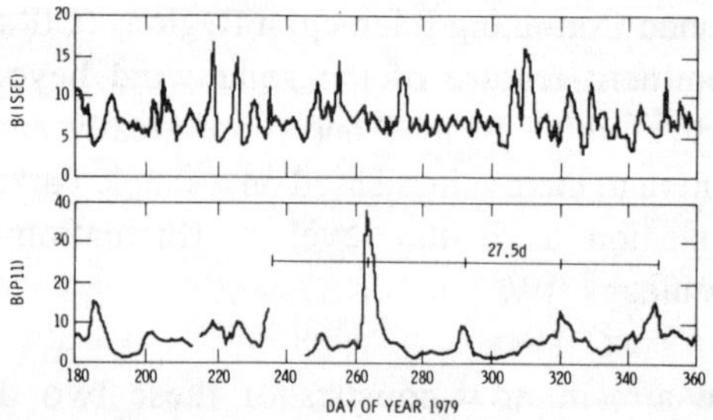

Fig. 7. Comparison of the number of interaction regions evident in B at 1AU and at P11.

This apparent periodicity does not imply that the features being observed were corotating. Instead, the most energetic, fastest CME in a given solar rotation has apparently overtaken the preceding slower features and run ahead of the following slower-moving streams and CMEs. The development of a single Merged Interaction Region (MIR) at 10 AU indicates that significant compression is still occurring at large distances.

This interpretation has received support from theoretical simulations and models. Fig. 8 simulates the magnetic field in the heliosphere following several very large solar events /22/. The dark "concentric" regions are compressions of the spiral field which alternate with lighter rarefaction regions. Superposed on this pre-existing structure are a half-dozen CMEs emitted at various solar longitudes. The dark "bubbles" represent MIRs while the trailing light regions are the large rarefactions, characterized by weaker and more radial fields, as observed in the Pioneer plasma and field data. The same qualitative explanation, that the pile-up of draped and distended magnetic fields results in

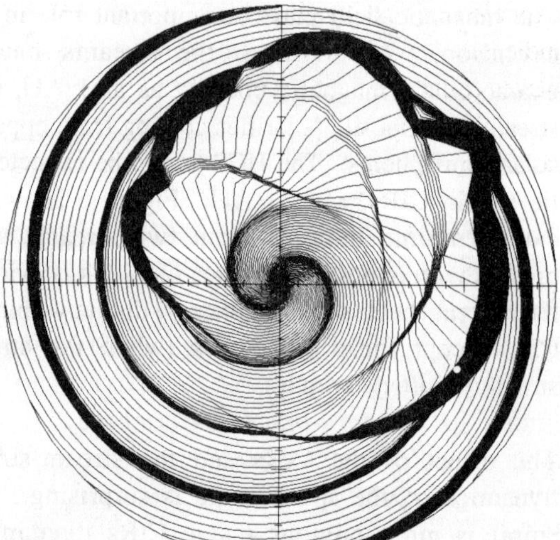

Fig. 8. A simulation of magnetic field structure in the outer heliosphere resulting from a series of coronal mass ejections.

isolated large "pulses", has also been proposed by McComas et. al /23/. The models and observations are qualitatively consistent with continued attachment of field lines to the sun.

Interaction of the Solar Wind with Interstellar Neutrals

At very large distances, the expanding solar wind is restrained by the pressure of the local interstellar medium, specifically the interstellar magnetic field and plasma. Structures resulting from this interaction, the heliopause and terminal shock, are still beyond the present radial distance of the Pioneers and Voyagers. However, interstellar neutrals penetrate into the heliosphere and approach the sun until ionized by solar radiation or charge exchange /24/.

The newly-created ions respond to the convected heliospheric magnetic field and begin to execute a cycloidal motion. They constitute a hot halo co-existing with the cooler solar wind core /25/. The physics involved is very similar to that associated with the pick up of cometary ions which leads to the generation of intense ion cyclotron waves. It might be expected that waves would be generated by the interstellar ions and could couple the ion thermal energy to the solar wind, heating it in the process.

Theoretical calculations indicate that the flux of hydrogen ions exhibits a broad maximum near 10 AU /26/. Quasi-linear theory applied to the resulting wave-particle interaction indicates that rapid amplification of the ever-present interplanetary waves is likely /27/. Estimates of the wave spectrum have been derived and predict outward propagating waves with power approximately an order-of-magnitude above the background and centered about the proton gyrofrequency, fgp.

The presence of these waves is likely to be the most sensitive indicator of the solar wind-interstellar neutral interaction present in the magnetic field data. Accordingly, a search for the waves has been carried out when Pioneer 11 was in the vicinity of 10 AU (1980) /1/. Pioneer spectra show the typical power law dependence throughout the frequency interval without any apparent enhancement near fgp. Waves have been sought at other times and other distances without success.

This negative finding indicates that either the flux of interstellar neutrals and/or ions is much less than expected or that the instability is much less effective in generating waves than the specific model indicates. Inward-propagating waves are heavily damped and coupling between them and outward-going waves could drastically reduce the estimated wave amplitude (M. Lee, private communication).

Acceleration of Energetic Particles

Particle acceleration is a universal phenomenon both inside and outside the heliosphere and acceleration by shocks is frequently resorted to as an explanation. Although much can be learned from the study of planetary bow shocks, conditions in interplanetary space are more favorable to acceleration studies. A bow shock has significant curvature and the magnetic fields are attached to it for a short time.

Prior to the Pioneer Mission, recurrent streams of energetic particles had been observed for many years at 1AU with a typical period of 27 days /28/. The particles were thought to be more or less continuously accelerated at the sun and emitted into space. The energetic particle investigators on Pioneer were studying these streams when shocks and CIRs were becoming evident in the magnetic field and plasma data. Correlation of the two sets of observations proved extremely fruitful.

The energetic particle streams were highly correlated with the CIRs (Fig. 9) /29,30/. Furthermore, their intensities showed an increase with radial distance. Observations at 1 AU revealed that the particles were streaming back into the inner solar system /31/. The corotating shocks were probably involved and it was possible to study large numbers of cases. Fortunately, theoretical developments had kept pace with observation and superior particle observations were also being obtained at 1 AU by IMP and ISEE.

Theorists had identified four possible acceleration processes /32/. Shock drift acceleration depended on the motion of particles parallel to the electric field at the shock. Fermi acceleration resulted from the convergence of two magnetic "walls", the shock and waves approaching the shock from upstream. Stochastic (second order Fermi) acceleration was thought to occur in the presence of waves propagating simultaneously in opposite directions. Transit-time damping depended on strong fluctuations in field magnitude over an extended region such as downstream of a shock. All four mechanisms involved magnetic field fluctuations either directly or indirectly (gradient drift acceleration benefited from multiple shock crossings made possible by wave scattering).

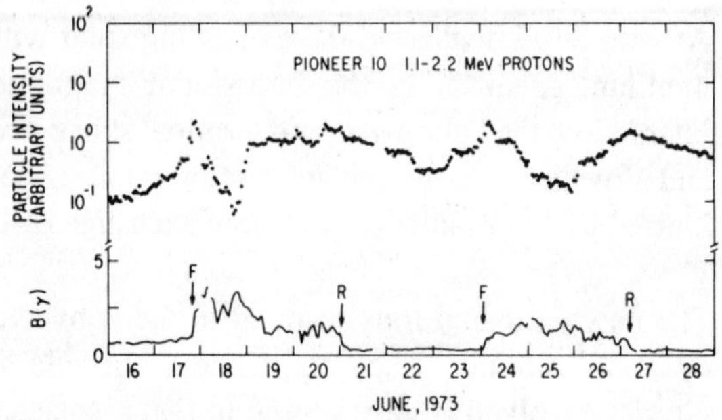

Fig. 9. Energetic particle streams and CIRs with Forward-Reverse shock pairs.

The ensuing studies led to several important conclusions /33,34/. Transit time damping was largely ruled out by the observed wave levels which were less intense than needed. Peaks in the particle intensities were observed in the vicinity of both the forward and reverse shocks with a distinct minimum inside the CIR. Examples of the particle anisotropies were shown to be consistent with shock drift acceleration. The observed spectra of the particles and waves were often consistent with Fermi acceleration. It was generally concluded that shock drift and Fermi acceleration were both operating.

A major unresolved issue was the origin of the particles being accelerated. A large gap existed in the Pioneer measurements between the upper range of the solar wind measurements and the low energy range of the particle measurements. Two widely-discussed possibilities were that the "seed" particles originated as the superthermal tail of the solar wind distribution or were solar particles accelerated to intermediate energies as a first stage /30/. Observations at 1 AU over an extended energy range have provided an example consistent with diffusive/stochastic acceleration of solar wind particles /35/.

Recurrent Cosmic Ray Decreases

The early phase of the Pioneer Mission favored the study of phenomena typical of solar minimum. An obvious feature in the energetic particle data was the recurrence of transient decreases in cosmic rays at intervals of 27 days. A correlation was found between the decreases and shocks/CIRs (Figure 10) which was further evidence of a relation noted at 1AU between cosmic ray decreases and enhanced magnetic fields ("blobs") /36/.

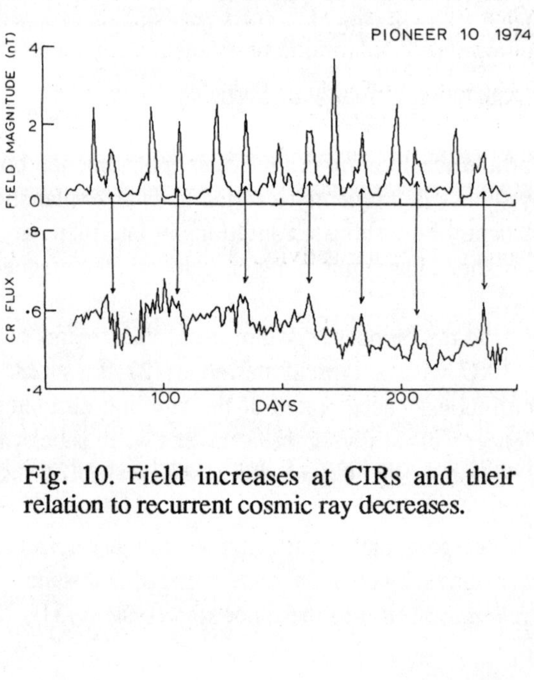

Fig. 10. Field increases at CIRs and their relation to recurrent cosmic ray decreases.

Cosmic ray decreases measured on Earth were compared with CIRs observed by Pioneer and corotated back to 1 AU /37/. An overall correspondence was seen with an intriguing twist. CIRs in which the HCS was embedded were much more effective in producing decreases than CIRs which did not include a change in magnetic polarity (Fig. 11).

Several features of CIRs could affect cosmic rays: the shocks, the turbulent fields up-and-down-stream of the shocks, the interface (tangential discontinuity?) between the slow and fast solar wind, or some combination of these features. The relative importance of these processes may be different than for Forbush decreases which are typically associated with CMEs and for which the global geometry is very different. The possible influence of the HCS on cosmic rays has not been discussed in this context. Is the current sheet directly involved or is the effect global and associated, for example, with the presence of solar wind streams from opposite hemispheres?

The correlation with CIRs suggests that the spatial regions in which the decreases occur have the same spiral geometry as the CIRs. Perhaps, the cosmic ray particles are sensitive to both the enhanced field strengths and increased turbulence of the CIR and respond to them as to a barrier. If so, the physics may be analogous to the confinement of relativistic jovian electrons by CIRs to the rarefaction regions between them.

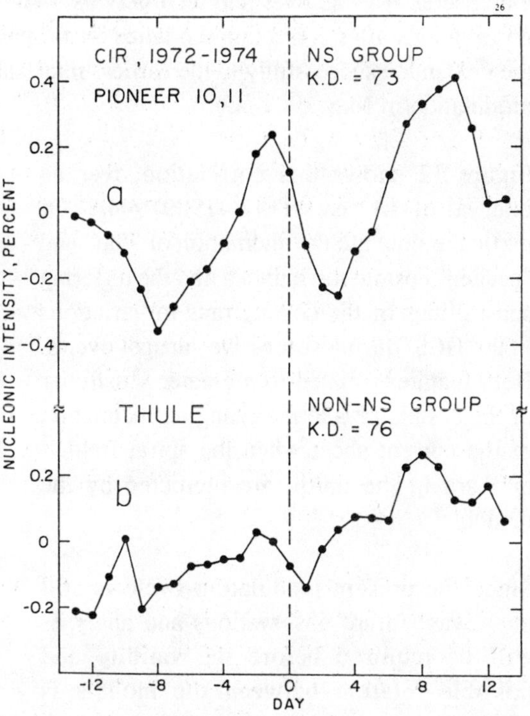

Fig. 11. Superposed epoch analysis of cosmic ray measurements at Earth (Thule) with the key date being the passage of CIRs with (NS for neutral sheet) and without (non-NS) a magnetic polarity reversal.

Modulation of Galactic Cosmic Rays

When Pioneer 10, 11 were launched, GCR intensities at different energies were near maximum as appropriate to solar minimum. Interesting exceptions were decreases reminiscent of modulation referred to as "mini-solar-cycle variations". These decreases ceased after 1 to 2 years. Modulation was then essentially absent until 1978 when it resumed as an abrupt onset attributed to a large solar event observed at the sun, 1 AU, and at Pioneer /38/. Comparisons showed that the onset had propagated outward at solar wind speeds. Modulation continued for the next several years through solar maximum with an apparent recovery in 1981. However, a large decrease occurred in mid-1982 accompanied by a resurgence in solar activity. The intensity then recovered steadily, reaching a maximum in early 1987.

The availability of Pioneer and Voyager observations in the outer heliosphere motivated a Cosmic Ray Workshop. Two hypotheses were actively pursued, in addition to an early suggestion that modulation is caused by the superposition of Forbush decreases, which are much more frequent at solar maximum. One hypothesis envisioned a series of interaction regions caused by CMEs emitted at various longitudes which merge in the outer heliosphere to form barriers to the entry of cosmic rays /39/. The other hypothesis developed from the gradient drift model of cosmic ray transport by the inclusion of a tilted, wavy current sheet /40,41/.

Support for the gradient drift-current sheet model has come from various observations, including the test of a predicted correlation between GCR intensity and the tilt or maximum latitude of the HCS /42/. The latter was successful, with an excellent correlation being found between the solar cycle decrease and increase as well as over shorter intervals. The correlation included both ground-based and Pioneer observations after correcting the latter for propagation of the current sheet at solar wind speed. It was then extended successfully to the earlier "mini-solar-cycle" variations and, more recently, to the onset of modulation in May 1987 /43/.

Figure 12 shows this correlation over an interval of 13 years (1976-1989) /44/. Of particular note are the alternation of "flat" and "peaked" cosmic ray maxima and the differing sensitivities of the cosmic rays to variations in the HCS during successive sunspot cycles. Both features, related to a greater sensitivity of the cosmic rays to the changing inclination of the current sheet when the spiral field is outward in the north, are predicted by the model.

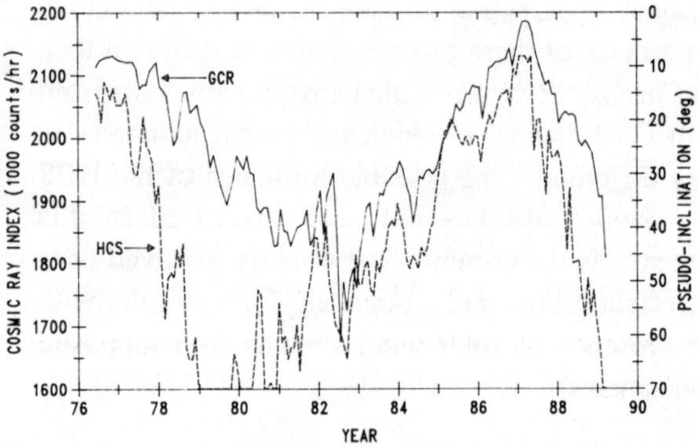

Fig. 12. Cosmic ray measurements at Earth (Deep River) compared with the maximum latitudinal extent or psuedo-inclination of the HCS.

Since the present modulation cycle is still underway, future observations and analyses will be required before the validity and possible relation between the models is resolved.

SUMMARY

Many of the major questions foreseen at the beginning of the Pioneer mission have now been answered. However, significant departures from expectation, and phenomena that were not predicted, have also been revealed and have given rise to new questions.

Although the solar wind expansion conforms to the Parker model in most respects, the average field strength has been found to be systematically lower than expected. The postulated shocks associated with corotating solar wind streams have been seen, however, the continued existence of significant structure at large distances as regions of enhanced field strength was unexpected. Merging of streams and CMEs was foreseen to some extent, but not the formation of giant structures consisting of a large field compression followed by a very long rarefaction. The interaction of the solar wind with interstellar neutrals was expected to become evident at 5 to 10 AU, however, the interaction is weaker and harder to observe than previously supposed.

Acceleration of particles at shocks was anticipated, but the origin of the seed particles which undergo acceleration is uncertain. Recurrent cosmic ray decreases were known to be correlated with the presence of strong magnetic fields in the solar wind, however, the influence of the Heliospheric Current Sheet was unexpected. Cosmic ray modulation was thought to be the result of solar cycle changes in convection and diffusion, i.e., changes in the solar wind speed and in the level of magnetic fluctuations. Instead, large scale structures have turned out to be the major influence, either in the form of CMEs or the HCS and its changing inclination.

Obviously, the above list is representative rather than comprehensive. No doubt other mysteries and surprises are still waiting to be discovered by the four spacecraft now travelling through the outer heliosphere.

ACKNOWLEDGEMENT

Two previously unpublished figures (6,10) were prepared by B.T. Thomas. The cosmic ray flux in Figure 10 is by courtesy of J.A. Simpson. Research described in this publication was carried out by the Jet Propulsion Laboratory, California Institute of Technology, under a contract with the National Aeronautics and Space Administration.

REFERENCES

1. E.J. Smith, Interplanetary magnetic field over two solar cycles and out to 20 AU, Adv. Space Res., 9, 159 (1989)

2. E.J. Smith, J.A. Slavin, and B.T. Thomas, The heliospheric current sheet: 3 dimensional structure and solar cycle changes, in: ESLAB Symposium on the Three Dimensional Structure of the Heliosphere, ed. R.G. Marsden, D. Reidel, Hingham, Mass. 1986, p. 267.

3. E.N. Parker, Interplanetary Dynamical Processes, Wiley-Interscience, New York, 1963.

4. B.T. Thomas, and E.J. Smith, The Parker spiral configuration of the interplanetary magnetic field between 1 and 8.5 AU, J. Geophys. Res. 85, 6861 (1980)

5. D. Winterhalter, E.J. Smith, and J.A. Slavin, The radial gradient in the interplanetary magnetic field between 1 and 23 AU, in: Solar Wind 6, eds. V.J. Pizzo, T.E. Holzer, and D.G Sime, NCAR/TN-306, HAO/NCAR, Boulder, Colorado, U.S.A. 1988, p. 587.

6. B.T. Thomas, J.A. Slavin, and E.J. Smith, Radial and latitudinal gradients in the interplanetary magnetic field: Evidence for meridional flux transport, J. Geophys. Res. 91, 6760 (1986)

7. D. Winterhalter, E.J. Smith, J.H. Wolfe, and J.A. Slavin, Spatial gradients in the heliospheric magnetic field: Pioneer 11 observations between 1 AU and 24 AU, and over solar cycle 21, J. Geophys. Res., 95, 1 (1990)

8. L.W. Klein, L.F. Burlaga, and N.F. Ness, Radial and latitudinal variations of the interplanetary magnetic field, J. Geophys. Res. 92, 9885 (1987)

9. S.T. Suess, and S.F. Nerney, The global solar wind and predictions for Pioneers 10 and 11, Geophys. Res. Lett., 2, 75 (1975)

10. V.J. Pizzo, and B.E. Goldstein, Meridional transport of magnetic flux in the solar wind between 1 and 10 AU: A theoretical analysis, J. Geophys. Res. 92, 7241 (1987)

11. R.L. McNutt, Jr., Possible explanation of north-south plasma flow in the outer heliosphere and meridional transport of magnetic flux, Geophys. Res. Lett., 15, 1523 (1988)

12. J.T. Gosling, J.R. Asbridge, S.J. Bame, and W.C. Feldman, Solar wind speed variations: 1962-1974, J. Geophys. Res., 81, 5061 (1976)

13. R.A. Howard, N.R. Sheeley, Jr., D.J. Michels, M.J. Koomen, The solar cycle dependence of coronal mass ejections, in: The Sun and the Heliosphere in Three Dimensions, ed. R.G. Marsden, D. Reidel, Dordrecht, 1986, p. 107.

14. J.K. Chao, and R.P. Lepping, A correlative study of ssc's, interplanetary shocks, and solar activity, J. Geophys. Res., 79, 1799 (1974)

15. E.J. Smith, J.A. Slavin, R.D. Zwickl, and S.J. Bame, Shocks and storm sudden commencements, Solar Wind-Magnetosphere Coupling, eds. Y. Kamide, and J.A. Slavin, Terra Scientific Publishing Company, Tokyo, 1986, p. 345.

16. K.-P. Wenzel, R.G. Marsden, D.E. Page, and E.J. Smith, Ulysses: The first high-latitude heliospheric mission, Adv. Space Res., 9, 25 (1989)

17. G.L. Siscoe, Structure and orientation of solar wind interaction fronts: Pioneer 6, J. Geophys. Res. 77, 27 (1972)

18. A.J. Hundhausen, Some macroscopic properties of shock waves in the heliosphere, in: Collisionless Shocks in the Heliosphere: A Tutorial Review, eds. R.G. Stone, and B.T. Tsurutani, Geophys. Monogr. Ser. 34, AGU, Washington, D.C., 1985, p. 37.

19. E.J. Smith, and J.H. Wolfe, Fields and plasmas in the outer solar system, Space Sci. Rev. 23, 217 (1979)

20. B.T. Thomas, and E.J. Smith, The structure and dynamics of the heliospheric current sheet, J. Geophys. Res., 86, 105 (1981)

21. E.J. Smith, Interplanetary shock phenomena beyond 1 AU, in: Collisionless Shocks in the Heliosphere: Reviews of Current Research, eds. B.T. Tsurutani and R.G. Stone, Geophys. Monogr. Ser. 35, AGU, Washington, D.C., 1985, p. 69.

22. S.-I. Akasofu, and K. Hakamada, Solar wind disturbances in the outer heliosphere, caused by six successive solar flares from the same active region, Geophys. Res. Lett., 10, 577 (1983)

23. D.J. McComas, J.T. Gosling, D. Winterhalter, and E.J. Smith, Interplanetary magnetic field draping about fast coronal mass ejecta in the outer heliosphere, J. Geophys. Res., 93, 2519 (1988)

24. T.E. Holzer, Interaction of the solar wind with the neutral component of the interstellar gas, J. Geophys. Res. 77, 5407 (1972)

25. P.A. Isenberg, Evolution of interstellar pickup ions in the solar wind, J. Geophys. Res. 92, 1067 (1987)

26. V.M. Vasyliunas, and G.L. Siscoe, On the flux and the energy spectrum of interstellar ions in the solar system, J. Geophys. Res. 81, 1247 (1976)

27. M.A. Lee, and W.-H. Ip, Hydromagnetic wave excitation by ionized interstellar hydrogen and helium in the solar wind, J. Geophys. Res. 92, 11041 (1987)

28. D.A. Bryant, T.L. Cline, U.D. Desai, and F.B. McDonald, Continual acceleration of solar protons in the MeV range, Phys. Rev. Lett., 14, 481 (1965)

29. C.W. Barnes, and J.A. Simpson, Evidence for interplanetary acceleration of nucleons in corotating interaction regions, Astrophys. J., 210, L91 (1976)

30. M. Scholer, G. Morfill, and M.A.I. Van Hollebeke, On the origin of corotating energetic particle events, J. Geophys. Res., 85, 1743 (1980)

31. F.E. Marshall, and E.C. Stone, Persistent sunward flow of ~1.6 MeV protons at 1 AU, Geophys. Res. Lett., 4, 57 (1977)

32. M.A. Forman, and G.M. Webb, Acceleration of energetic particles, in: Collisionless Shocks in the Heliosphere: Reviews of Current Research, eds. B.T. Tsurutani, and R.G. Stone, AGU, Washington, DC, 1985, p. 91.

33. B.T. Tsurutani, E.J. Smith, K.R. Pyle, and J.A. Simpson, Energetic protons accelerated at corotating shocks: Pioneer 10 and 11 observations from 1 to 6 AU, J. Geophys. Res., 87, 7389 (1982)

34. M.E. Pesses, B.T. Tsurutani, J.A. Van Allen, and E.J. Smith, Acceleration of energetic protons by interplanetary shocks, J. Geophys. Res., 84, 7297 (1979)

35. J.T. Gosling, J.R. Asbridge, S.J. Bame, W.C. Feldman, R.D. Zwickl, G. Paschmann, N. Sckopke, and R.J. Hynds, Interplanetary ions during an energetic storm particle event: The distribution function from solar wind thermal energies to 1.6 MeV, J. Geophys. Res., 86, 547 (1981)

36. E. Barouch, and L.F. Burlaga, Causes of Forbush decreases and other cosmic ray variations, J. Geophys. Res., 80, 449 (1975)

37. S.P. Duggal, B.T. Tsurutani, M.A. Pomerantz, C.H. Tsao, and E.J. Smith, Relativistic cosmic rays and co-rotating interactions regions, J. Geophys. Res., 86, 7473 (1981)

38. R.B. McKibben, K.R. Pyle, and J.A. Simpson, The galactic cosmic-ray radial intensity gradient and large-scale modulation in the heliosphere, Astrophys. J., 254, L23 (1982)

39. L.F. Burlaga, F.B. McDonald, N.F. Ness, R. Schwenn, A.J. Lazarus, and F. Mariani, Interplanetary flow systems associated with cosmic ray modulation in 1977-1980, J. Geophys. Res., 89, 6579 (1984)

40. J.R. Jokipii, and B.T. Thomas, Effects of drifts on the transport of cosmic rays, IV, Modulation by a wavy interplanetary current sheet, Astrophys. J. 243, 1115 (1981)

41. J. Kotá, and J.R. Jokipii, Effects of drift on the transport of cosmic rays, VI, A three-dimensional model including diffusion, Astrophys. J. 265, 573 (1983)

42. E.J. Smith, and B.T. Thomas, Latitudinal extent of the heliospheric current sheet and modulation of galactic cosmic rays, J. Geophys. Res. 91, 2933 (1986)

43. T. Saito, and D.B. Swinson, The inclination of the heliospheric neutral sheet and cosmic ray intensity at the earth, J. Geophys. Res., 91, 4536 (1986)

44. E.J. Smith, The heliospheric current sheet and modulation of galactic cosmic rays, J. Geophys. Res., in press (1990)

RADIO NOISE IN THE HELIOSPHERIC CAVITY

W.S. Kurth

Department of Physics and Astronomy, University of Iowa, Iowa City, IA 52242, U.S.A.

ABSTRACT

The Pioneer 10 and 11 and Voyager missions are providing us with the capability of viewing the heliosphere from a more global perspective than previously possible. One aspect of the heliosphere which is especially intriguing in an astrophysical sense is the fact that the heliosphere is the source of a wide variety of radio emissions. These emissions, in order of relative intensity, are generated at the sun, in planetary magnetospheres, and in the heliospheric cavity, itself. The focus of this paper will be the latter category, specifically, low frequency interplanetary emissions since these might possibly be generated as a result of the interaction of the solar wind with the interstellar medium. We will summarize the observations by the Voyager spacecraft of these low frequency emissions and discuss the various possible sources of the emission. The most exciting possibility is that these radio emissions are generated at either the terminal shock or the heliopause providing the first remote evidence for the termination of the interplanetary medium.

INTRODUCTION

In this paper we regard the solar wind as the outer layer of the sun's atmosphere and consider the interaction of that atmosphere with the interstellar medium. In effect, we wish to view the sun's interaction with its environment from a more global perspective than usual. From our vantage point at the Earth, it is too easy to think of the corona as the upper atmosphere of the sun since we are very distant from that visible part of the sun's atmosphere. In fact, all the planets are immersed in the most distant portion of the sun's atmosphere. From the Earth, the lowest frequency radio emissions associated directly with the sun are type III solar bursts, excited in the inner heliospheric plasma as solar flare electrons stream through the solar wind. The Voyager spacecraft, carrying the first plasma and radio wave receivers significantly beyond 1 AU, however, provide us with an even more remote view of 'solar' radio emissions which propagate at frequencies of only a few kiloHertz, well below the plasma frequency at 1 AU. Indeed, in order to detect such low frequency emissions, it is necessary to travel well beyond 10 AU where the local plasma frequency falls to below the propagation frequency of the waves.

In this review, we discuss the observations of the low frequency interplanetary radio emissions /1,2,3,4,5/ including recent observations which suggest the Voyagers have ceased (at least temporarily) detecting them. We present the model proposed for explaining the emissions. We recount the other possible sources for the emissions, including recent suggestions /6,7/ that the emissions could be a phenomenon local to the spacecraft. We discuss the suitability of the various possible sources, spending a bit more effort on the new suggestion of a local phenomenon, and finally suggest by a process of elimination that a most likely interpretation of the emissions is a signature of the solar wind interaction with the interstellar medium--that is, a 'solar' emission in the outermost region of the sun's influence.

OBSERVATIONS OF THE LOW FREQUENCY INTERPLANETARY RADIO EMISSIONS

Figures 1 and 2 provide a synopsis of the observations of the low frequency interplanetary radio emissions from the Voyager 1 and 2 plasma wave receivers, respectively. This synoptic view is an extension of that provided by Kurth et al. /4/. These figures are in the form of frequency-time spectrograms wherein the amplitude of radio emissions are plotted using a gray scale in which black represents the most intense wave activity and white the least intense. The intensities are plotted as a function of frequency (ordinate) and time (abscissa). The time axis covers about 6 years from the beginning of 1983. The tick marks at the top of each figure indicate times when data are actually available. Each tick mark

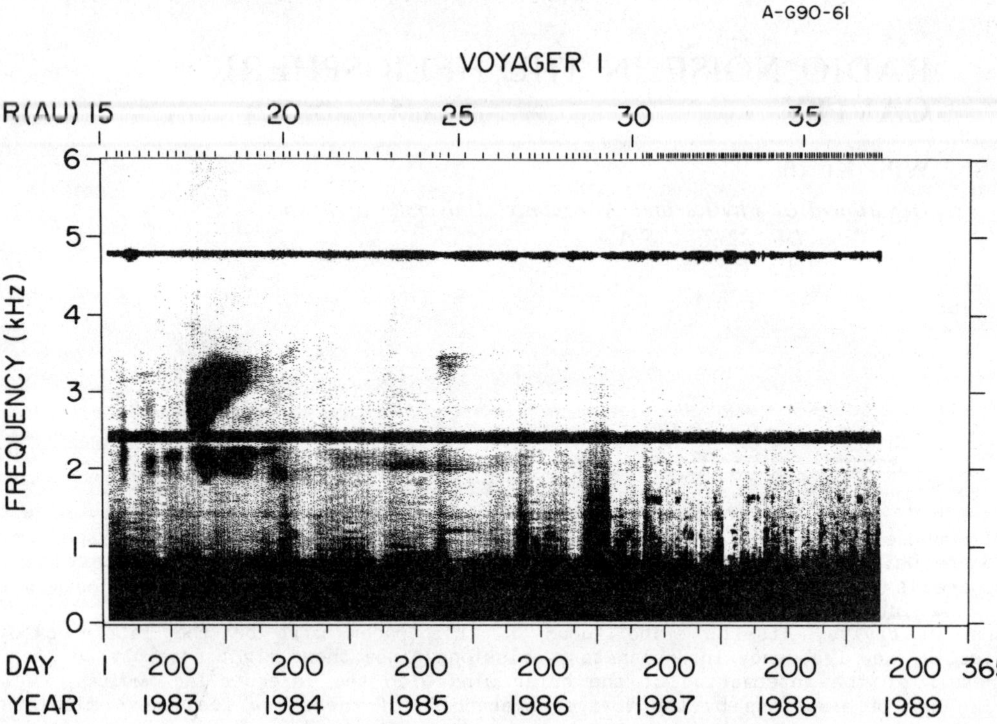

Fig. 1. Dynamic spectrogram of Voyager 1 observations from 1983 through 1989 showing the temporal evolution of low frequency interplanetary radio emissions.

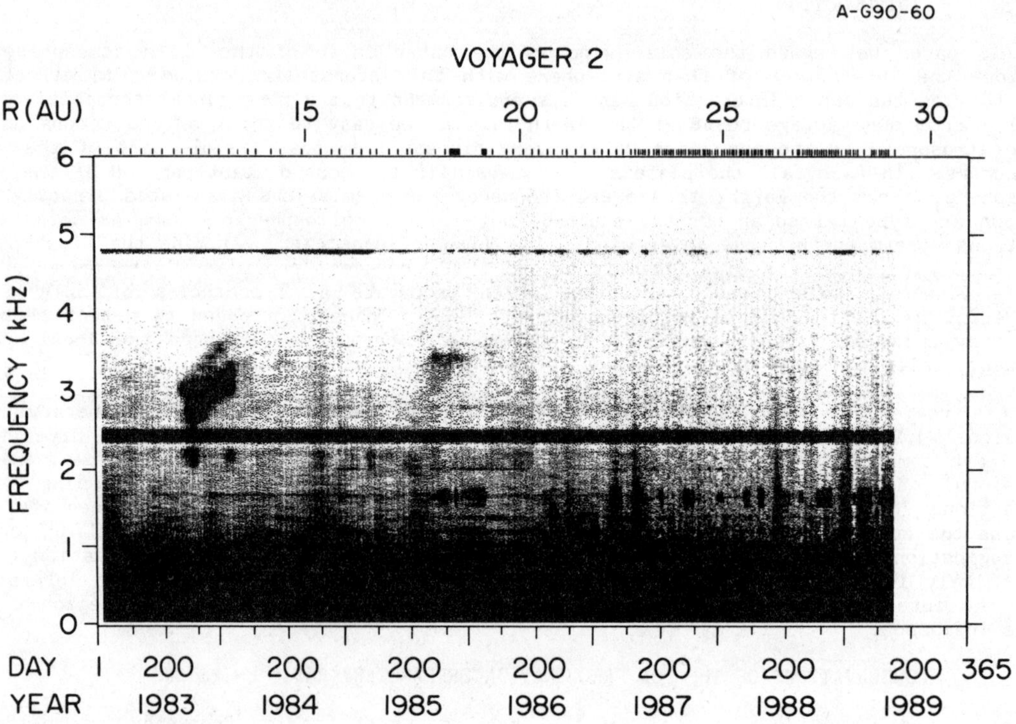

Fig. 2. A dynamic spectrogram similar to that in Figure 1 except that these data are from Voyager 2 for approximately the same time interval. The similarity between the two spectrograms while the spacecraft are separated by large distances implies the emissions are freely propagating radio waves.

indicates times when one or two 48-second wideband frames were obtained. The data from the cleanest (from an interference point of view) 15 seconds of each frame are Fourier transformed and averaged in order to form a single spectral scan. In the figures, these scans are stacked side by side in order to give the impression of the extended time history of variations in the frequency and intensity of the emissions.

In each of Figures 1 and 2, there are narrowband emissions at 2.4 and 4.8 kHz which are interference from the spacecraft power supply. Occasionally, burst-like emissions are seen at about 1.7 kHz; these are interference from the gyros on board the spacecraft. Finally, the intense noise below about 1 kHz is due to other interference on the spacecraft, most notably the tape recorder used to record the high rate waveform measurements. The radio emissions of interest in Figures 1 and 2 commence just prior to day 300 of 1983 although brief periods of weaker activity were observed as early as the summer of 1982. In late 1983 Voyager 1 was beyond 17 AU from the sun and well above the ecliptic plane. At the same time, Voyager 2 was in the ecliptic at a distance of just under 13 AU from the sun. The geometry of the observations is summarized in Figure 3 (taken from /5/). The two spacecraft began observing the intense event at about 3 kHz at the same time (within the two-week uncertainty imposed by the infrequent samples of wideband data on Voyager 2). On Voyager 1 nearly continuous observations are available with the 3.11-kHz spectrum analyzer channel, however this part of the Voyager 2 receiver is not usable for observations of such weak emissions due to a failure in the Flight Data System on board the spacecraft.

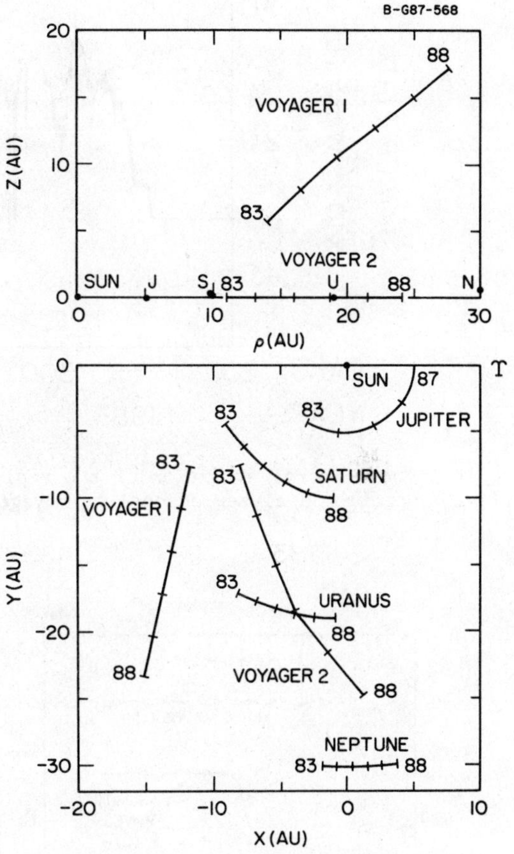

Fig. 3. Two projections of the relative positions of the Voyager spacecraft and the outer planets for the interval 1983 through 1988 (from /5/).

The 3-kHz emission has a duration of 6 months or more and reappears for a few months in late 1985 and early 1986. A second component can be seen at 2 kHz which appears to be somewhat steadier in both amplitude and frequency and can be seen as late as the second half of 1987 from both spacecraft. The 2-kHz component is somewhat correlated with the higher frequency component in that it has peak intensity during the peak at 3 kHz; however, the 2 kHz emission is almost continuously observable for three and one-half years. The frequency of the higher frequency component drifts upward at a rate of about 1 kHz/yr during the 1983 occurrence; a frequency drift in the late 1985 event is not so apparent. There is little or no evidence of radio emission activity since the latter half of 1987.

The overall temporal behavior of the 3-kHz component can be studied using the 3.11-kHz channel on Voyager 1. Figure 4 (from /5/) shows the Voyager 1 data from late 1983 through mid-1984. The radio event has a very clear 'sudden' onset (taking only a few weeks to rise from background to peak) and a gradual decay lasting about 6 months. Superimposed on this overall envelope are unmistakable periodic decreases in amplitude with a period of about 26 days. This periodicity is strongly suggestive of a solar rotational effect and has been speculated as being due to increases in the solar wind density in the vicinity of the spacecraft sufficient to temporarily attenuate or block the propagation of the waves to the spacecraft. Studies of variations in the magnetic field strength thought to be proportional to the plasma density have supported this hypothesis /8/.

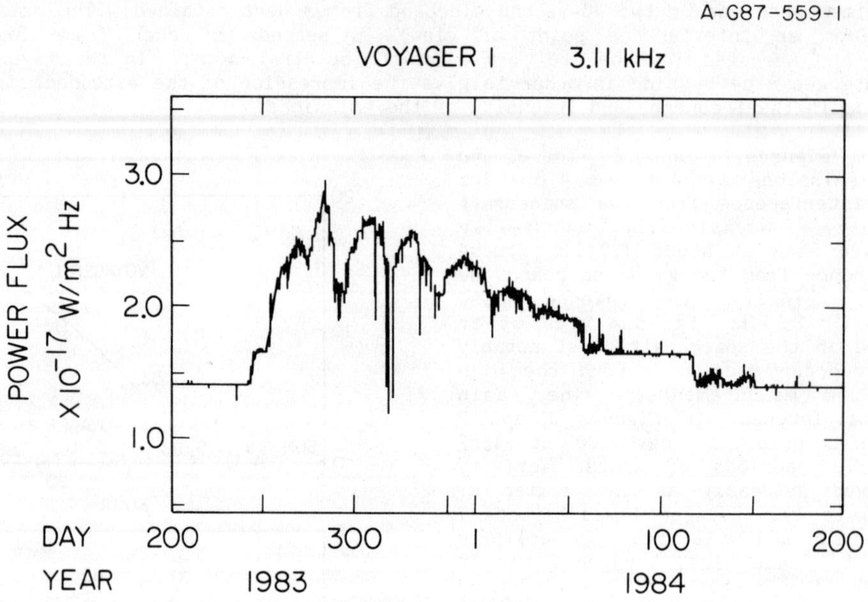

Fig. 4. The temporal evolution of the amplitude of the 3-kHz component as observed by the 3.11-kHz spectrum analyzer channel on Voyager 1. (from /5/).

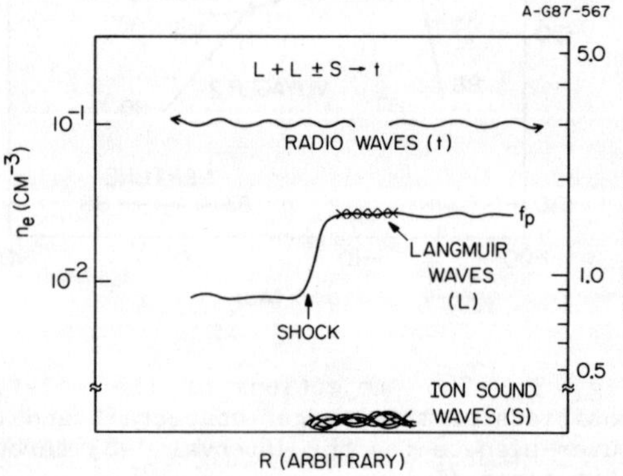

Fig. 5. A model for the generation of low frequency interplanetary radio emissions at the terminal shock (from /5/).

THE MODEL

Figure 5 taken from Kurth et al. /5/ depicts the model originally proposed to explain the low frequency interplanetary emissions. The basis for the model is that the bow shock at the Earth and some interplanetary shocks are the sources of relatively weak, narrowband radio emissions at twice the plasma frequency in the vicinity of the shock. In the present application, however, the shock under consideration is the terminal shock where the solar wind becomes subsonic due to the external pressure of the interstellar medium. This model produces an emission over a limited frequency range (subject to the variations in conditions at the shock) and provides the possibility of an emission region distributed over a large area, hence, solid angle. The large source places a much less stringent constraint on the minimum emissivity for the source; since a radio emission from a source extended over an infinite plane suffers no decrease in amplitude as a function of distance from the source, the use of an extended source distributed over a large solid angle at the terminal shock allows the use of a relatively weak emission mechanism such as the $2f_p$ emission from shocks (f_p is the electron plasma frequency).

The model presented above leads to a prediction for the distance to the shock based on the assumptions of a strong shock and density jump of a factor of four plus a model for the density of the solar wind as a function of distance ($n \propto 1/R^2$). The predictions from the observations as presented in Kurth et al. /1/ and Suess and Dessler /9/ range from 30 to 46 AU, depending only on the density of the solar wind at some fixed distance from the sun. The value of 30 AU is derived by Suess and Dessler for a more reasonable density model since it is based on an average value at 1 AU. Both Voyager 1 and Pioneer 10 are well beyond 30 AU and Pioneer 10 is further than 46 AU and no evidence for a terminal shock is in hand, so there is an obvious problem with the predictions of distance to the shock. The first obvious source of the problem is perhaps related to variations in the distance with time. The shock may be further out today than it was in 1983. Or, the concept that the density falls off as $1/R^2$ is in error. The loss of solar wind momentum due to interactions with the interstellar neutrals has been suggested as a mechanism for slowing down the solar wind, hence decreasing the density gradient. Though no evidence for the slowing of the solar wind is seen in the observations in the distant heliosphere in hand, some such effect may yet be detected. Finally, there could be a problem with the model, itself. If the source is not at the shock but at the heliopause, as suggested by Fahr et al. /10/, the density model does not take into account the terminal shock or subsonic solar wind and the distance predictions are not valid.

Other complications introduced by the observations include the fact that there appear to be two somewhat poorly correlated frequency components at about 2 and 3 kHz. The frequency spacing is not conducive to models involving emissions at successive harmonics of a characteristic frequency, specifically the fundamental and second harmonic of the plasma frequency. Kurth et al. /4/ suggested that asymmetries in the shape of the heliosphere (e.g. a latitudinal asymmetry) could lead to emission at two different frequencies associated with two different source locations. A second suggestion was made by Fahr et al. /10/ that one of the components could be generated at the heliopause. The more distant source would have to be generated at the higher frequency to propagate through the subsonic solar wind, hence, the lower frequency emission at 2 kHz could then be generated at the shock. Consideration of the lower frequency component as being generated at the shock results in moving the shock further from the sun than had been previously calculated. For example, Suess and Dessler /9/ using the 3 kHz emission deduced a shock distance 30 AU. The use of the same model for the solar wind, but using an emission frequency of 2 kHz gives a shock location of about 40 AU. This eases the prediction problem somewhat, but does not eliminate it. Macek /11/ has incorporated the heliopause source into his model of the solar wind-interstellar medium interaction. The interstellar density used by Macek is about 0.04 cm^{-3}, corresponding to a plasma frequency of about 1.8 kHz. Hence, second harmonic emission from such a plasma would yield a radio emission at a frequency very close to the higher frequency component observed. In Macek's model, however, the terminal shock is much further out than 40 AU, hence, the 2 kHz component would not correspond to emission from the shock; another explanation for the lower frequency component is required.

There has been no detailed modeling of the proposed radio generation mechanism and virtually no speculation of how the waves might be excited. A recent paper by McNutt /12/ suggests a very high speed solar wind stream might in some way be responsible for exciting or at least intensifying the radio emission generation mechanism. McNutt reports one high speed stream with a speed of nearly 800 km/sec which passed Voyager 2 some 265 to 292 days prior to the appearance of the primary radio event in late 1983. (An additional very high speed stream was seen in 1979, but the spacecraft were immersed in a plasma which was too dense to allow the detection of 3 kHz radio waves). McNutt suggests the high speed stream triggers the radio emission when it arrives at the shock. Based on a constant stream speed, the distance traversed by the plasma in some 270 days would be about 130 AU; the distance could be as little as 72 AU if the stream speed slowed to about 400 km/s. McNutt's suggestion, then, is that the source is at a distance of 70 - 130 AU. High speed streams of this magnitude are only seen once or twice in a 10-year period, hence, the mechanism might explain the long time between intensifications. The model proposed by McNutt predicts a reappearance of the emission some time after another very high speed stream is observed and, therefore, should be testable over the next several years.

A feature of the observations which has escaped adequate explanation until recently is the frequency drift noted primarily in the late 1983 3-kHz emission. The only previous attempts to explain this drift focussed on ad hoc changes in the source electron density in order to change the frequency of the emission. For a terminal shock source, this necessitates changing density on a time scale much longer than usual solar wind density variations (less than 25 days). Czechowski and Grzedzielski /13,14/ have now offered an interesting explanation for the frequency drift which makes no special requirements on the source and is a natural expectation for radio waves trapped in the heliospheric cavity. The suggestion is that Fermi scattering occurs with the walls of the cavity; the velocity of the wall changes the frequency of the wave upon reflection. In the heliospheric cavity this has the unusual result that a net increase in frequency is expected because the waves will usually reflect off of density structures moving outward from the sun at the solar wind speed. Reflections off the outer boundary (heliopause) presumably have no net effect in increasing or decreasing the frequency assuming the outer boundary moves inward and outward with equal probability.

The Czechowski and Grzedzielski model for the increasing frequency of the low frequency interplanetary radio emission results in an independent measure of the size of the heliosphere. Since the frequency gain at any one reflection is well know simply from the speed of the expanding solar wind, the only remaining first order quantity needed to account for the 1 kHz/yr drift rate is the number of reflections against the expanding solar wind per unit time. This results in an estimate of the distance to the heliopause of less than 150 AU and most likely less than 100 AU. This result depends only on the assumption that the radio waves are trapped within the heliopause; the local interstellar medium plasma frequency must be greater than about 3 kHz in order for this to hold. Otherwise, there is no need to assume a particular source for the radio emissions. Given that there is still a great deal of speculation on the true source, the frequency drift result is particularly useful.

OTHER POSSIBLE SOURCES FOR THE RADIO EMISSIONS

The original suggestion of the terminal shock as a source for the low frequency interplanetary radio emissions /1/ came as a result of a process of elimination of other possible sources. It is illustrative to review the list of other possible sources in view of the lack of definitive proof of the terminal shock source and in view of the recent suggestion of an additional source (or perhaps a slight modification of one of the original possibilities). Kurth et al. /1/ discussed plasma oscillations in the solar wind, quasi-thermal electrostatic plasma emissions (a local interaction between the plasma and the antenna), Jupiter, Saturn, Uranus, and a pulsar as possible sources in addition to the terminal shock. Kurth et al. /2/ considered an additional source which is basically an extended source in the heliosphere at a distance of somewhat greater than 10 AU where the plasma frequency passes through about 3 kHz.

Compact sources such as the outer planets' magnetospheres are unlikely candidates because the flux of radio waves detected at the two well-separated spacecraft is very similar. One would expect a $1/R^2$ law to produce significantly different fluxes at the two observation points. The pulsar is ruled out primarily because of the increase in frequency of the 3-kHz component would imply an increase in the rotation rate of the pulsar, which would seem implausible. The extended source near 10 AU (or where the solar wind plasma frequency passes through 3 kHz) seems unlikely because (1) there is little or no evidence of electron plasma oscillations in the Voyager data set at the appropriate distances (or frequencies) to indicate that radio waves might be generated in the distant solar wind and (2) one would expect significant variations in plasma densities as a function of heliocentric distance, latitude, and time (i.e. due to the sector structure) which would not be consistent with a very steady emission frequency (or one which evolves on time scales large with respect to the solar rotation period) and a relatively narrow band of emissions.

Aside from the terminal shock/heliopause source, the above process of elimination leaves only quasi-thermal electrostatic plasma emissions /15/ to consider. This possibility was discounted originally by Kurth et al. /1/ by arguing that the Voyager antennas are short with respect to the Debye length in the distant heliosphere and the theory /15/ does not predict a peaked spectrum for the thermal plasma noise. We will take additional care here to revisit this possibility in view of a recent comment by Meyer-Vernet and Perche /6/ that questions whether an "emission...attributed to a radiation coming from the outer boundary of the heliosphere is due instead to a negative antenna resistance." A more complete discussion of this idea is given by Meyer-Vernet /7/. The theory for the negative antenna resistance grows out of the consideration of the thermal plasma noise discussed by Hoang et al. /15/. In the recent comments, Meyer-Vernet and Perche point out that for a wire antenna which is short with respect to the Debye length, a negative resistance can arise in the antenna if the drift velocity of the plasma exceeds the thermal velocity of the plasma. The effect for short wire antennas is greatest for an antenna alignment parallel to the drift velocity. For the case of Voyager in the outer heliosphere, the antenna is short compared to the Debye length and there are regions, particularly at greater distances where the solar wind flow at about 400 km/s exceeds the thermal velocity. However, the Voyager antenna electrical axis is <u>perpendicular</u> to the flow velocity, hence, there is little reason to expect the Voyager observations to be affected by the negative resistance effect. To illustrate this point, Figure 6 (taken from /5/) shows the variation in amplitude of the 3-kHz component as a function of spacecraft yaw angle. The electric antenna axis is approximately perpendicular to the sun-spacecraft line when the yaw phase angle is 0°. These observations were made during a maneuver in which the spacecraft was rotated several times about its yaw axis. It is clear that the maximum signal occurs when the electric antenna is oriented perpendicular to the solar wind velocity, hence, opposite to the prediction of the negative resistance model.

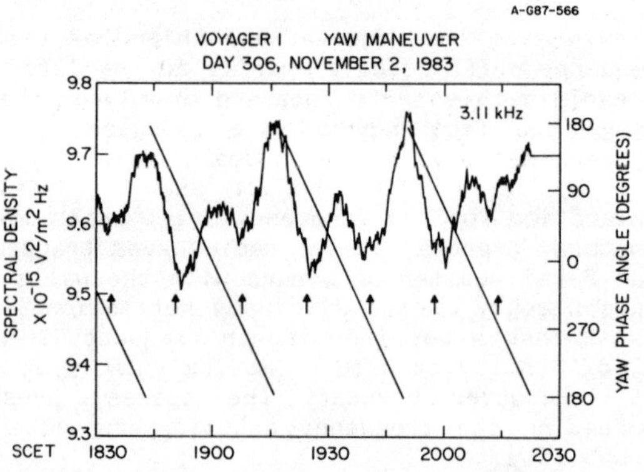

Fig. 6. The variation of radio emission amplitude at 3.11 kHz as a function of the yaw angle of Voyager 1 during a maneuver showing that the minimum in amplitude occurs when the angle is near 90° and 270° (see arrows) (from /5/). This result is the opposite of that predicted for the negative resistance effect /6/, hence, demonstrates that the emissions cannot be explained by a negative resistance in the antenna.

It is worth emphasizing other evidence which exists to rule out the negative resistance effect or <u>any</u> mechanism which is local to the spacecraft. First, the onset of the 1983 event occurred simultaneously (to within the two week timing on Voyager 2 afforded by the separation of wideband observations) at the two spacecraft separated by several astronomical units. Such timing is either a dramatic coincidence for a purely local phenomenon or else is evidence of a temporal variation (increase in amplitude) of the same, remotely-sensed freely propagating radio emission. Second, nearly simultaneous observations obtained by both spacecraft, again separated by several AU and several degrees in solar longitude and latitude (see Figure 3) show strikingly similar, complex spectra as shown in Figure 7 (taken from /5/). How can a local phenomenon produce such similarities at the two widely separated locations in the solar wind? Third, the local effects suggested by Meyer-Vernet and Perche should occur at the local plasma frequency of the solar wind surrounding the spacecraft. Figure 8 shows the results of a recent study based on the observations of the low frequency emission observed by Voyager 2 and on-board local measurements of the plasma frequency by the Voyager plasma investigation /16/. In Figure 8 we have plotted the low frequency cutoff of the radio emission as a function of the local plasma frequency as determined by the plasma instrument. The observed plasma density would have to be 75% low in order to explain the discrepancy in the figure if the cutoff frequency is the local plasma frequency as would be predicted by Meyer-Vernet and Perche. The accuracy of the plasma density determination is 30% or better. Unfortunately, for the times when wideband observations are present the plasma density variations are not large, hence, it is difficult to determine whether there is any correlation between the local conditions and the frequency of the cutoff.

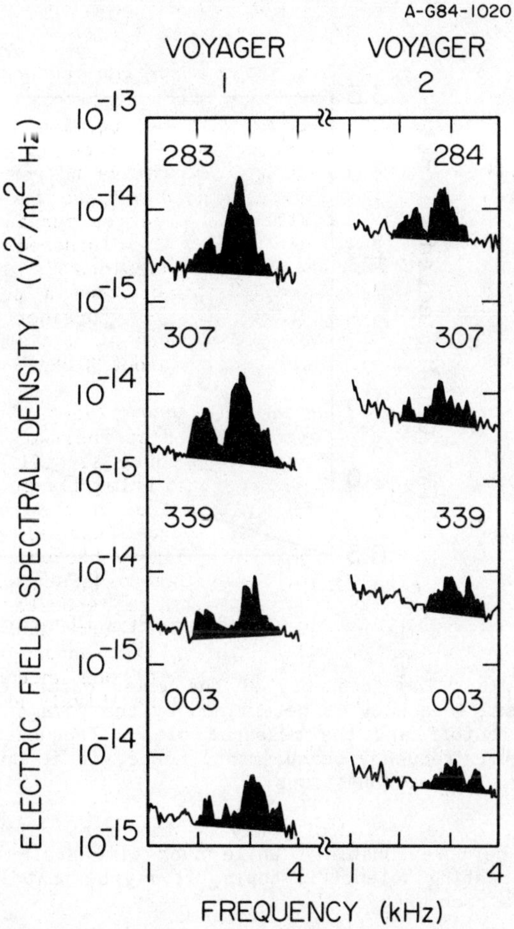

Fig. 7. A series of spectra taken by Voyagers 1 and 2 at nearly the same times showing the similarities in amplitude and spectral character of the emissions observed at the two well-separated observation points (from /5/). These similarities strongly argue against a local phenomenon being responsible for the radio emissions.

DISCUSSION AND CONCLUSIONS

The conclusion we come to now is similar to that of several years ago /1/. The terminal shock or heliopause source for low frequency interplanetary radio emissions still appears to be the best interpretation of the observations. However, the case has still not been proven. What is clear is that the emission is certainly a propagating radio wave and not a local effect. It is disconcerting that the emission peaked in 1983 and the spacecraft have been in a region where the emission could have been detected for more than six years, now, with little or no evidence of reoccurrence. We remind the reader that the sudden, nearly simultaneous onset of the emission at both spacecraft in 1983 is evidence of a temporal variation in amplitude, not a spatial variation. Further, we remind the reader that the amplitude at peak is only about a factor of two above the sensitivity threshold of the instruments. In view of the fact that most solar system radio emissions we are aware of often exhibit several order of magnitude variations in their amplitudes, it is quite likely that the 1983 event was simply an abnormally large amplitude for the emission. Further, if McNutt /12/ is correct and the emission is triggered by extremely fast solar wind streams which only occur one to a few times per decade (or any other triggering mechanism with a long time constant), then it is not unreasonable to expect to wait several years between events. In fact, noting that the 1983 event itself showed amplitude variations on a half-year time scale implies that the source does not operate on time scales as short as others that we are accustomed to observing in the heliosphere. One might even conclude that the

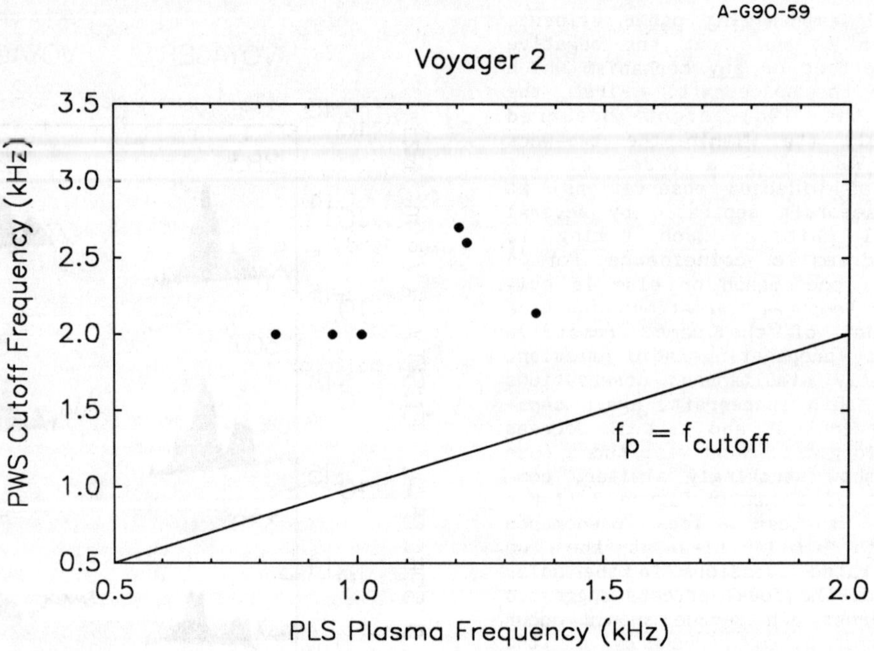

Fig. 8. The frequency of the lower cutoff of the radio emission as a function of local plasma frequency as determined by the Voyager 2 plasma instrument /16/. The difference in the cutoff and the measured plasma frequency is much greater than the accuracy of the plasma frequency measurement, hence, it is not likely that the emission is related to the local plasma conditions.

solar cycle periodicity may be important which suggests that a decade between events would not be unreasonable. While such time scales are difficult to deal with when studying an interesting scientific topic, it may be that little progress will be made except by waiting for a reoccurrence.

What is important to consider here is the fact that the radio emissions have been observed and that they have been undetectable for a few years does nothing to diminish the importance of understanding the source of the emissions as they likely represent the lowest frequency radio emission of the solar system and probably are a result of the interaction of the solar wind with the interstellar medium. The emissions would, therefore, provide valuable remotely detectable information on the solar wind-interstellar medium interface. An example of this are the several independent distance estimates of the terminal shock or heliopause. The original estimates by Kurth et al. /1/ and Suess and Dessler /9/, the suggestion of McNutt /12/ based on the propagation time for high speed streams, and the new explanation of the frequency drift by Czechowski and Grzedzielski /13,14/ all give heliosphere sizes that are similar, especially if certain limits are considered. The maximum heliospheric radius consistent with all of these measures is 150 AU, but 100 AU is a more likely figure.

The Voyager Interstellar Mission is just underway and we anticipate receiving data from both spacecraft for another 25 years out to distances of 130 to 140 AU. This should provide ample time to re-detect the low frequency interplanetary radio emissions, but more importantly, should assure passage through the terminal shock and in situ samples of the plasma and fields in the subsonic solar wind. Perhaps, the Voyagers will even make the first in situ observations of the interstellar medium and settle several ongoing debates on the nature of the plasmas just outside of the solar wind.

ACKNOWLEDGEMENTS

We would like to thank J. Belcher for the direct observations of the plasma densities used in creating Figure 8. This research was supported by the National Aeronautics and Space Administration through Contract 957723 with the Jet Propulsion Laboratory.

REFERENCES

1. W.S. Kurth, D.A. Gurnett, F.L. Scarf, and R.L. Poynter, Detection of a radio emission at 3 kHz in the outer heliosphere, Nature, 312, 27 (1984)

2. W.S. Kurth, D.A. Gurnett, F.L. Scarf, and R.L. Poynter, A new radio emission at 3 kHz in the outer heliosphere, in: Proceedings of a Course and Workshop on Plasma Astrophysics (ESA SP-207), Varenna, Italy, August 28-September 7, 1984, p. 285.

3. W.S. Kurth, D.A. Gurnett, and F.L. Scarf, Recent observations of the very low frequency interplanetary radio emission, Adv. Space Res., 6, #1, 379 (1986).

4. W.S. Kurth, D.A. Gurnett, F.L. Scarf, and R.L. Poynter, Long-period dynamic spectrograms of low-frequency interplanetary radio emissions, Geophys. Res. Lett., 14, 49 (1987).

5. W.S. Kurth, The low frequency interplanetary radio emission: Evidence of the solar wind-interstellar wind interaction? in: Proceedings of the Sixth International Solar Wind Conference, Vol. II, (NCAR/TN-306+Proc) ed. V.J. Pizzo, T.E. Holzer, and D.G. Sime, National Center for Atmospheric Research, Boulder, Colorado 1988, p. 667.

6. N. Meyer-Vernet and C. Perche, Tool kit for antennae and thermal noise near the plasma frequency, J. Geophys. Res., 94, 2405 (1989).

7. N. Meyer-Vernet, Electric antennae in the outer heliosphere: The importance of being stable, Astron. Astrophys., 224, L5 (1989).

8. L. Burlaga, private communication (1987).

9. S.T. Suess and A.J. Dessler, Probing the local interstellar medium, Nature, 317, 702 (1985).

10. H.J. Fahr, W. Neutsch, S. Grzedzielski, W. Macek, and R. Ratkiewicz-Landowska, Plasma transport across the heliopause, Space Sci. Rev., 43, 329 (1986).

11. W.M. Macek, Reconnection pattern at the heliopause, Adv. Space Res., this issue.

12. R.L. McNutt, A solar-wind "trigger" for the outer heliosphere radio emissions and the distance to the terminal shock, Geophys. Res. Lett., 15, 1307 (1988).

13. A. Czechowski and S. Grzedzielski, Frequency drift of 3 kHz interplanetary radio emissions: Evidence of Fermi accelerated trapped radiation in small heliosphere? Adv. Space Res., this issue.

14. A. Czechowski and S. Grzedzielski, Frequency drift of 3 kHz interplanetary radio emissions: Evidence of Fermi accelerated trapped radiation in small heliosphere? Nature, in press (1990).

15. S. Hoang, J.L. Steinberg, G. Epstein, P. Tilloles, J. Fainberg, and R.G. Stone, The low-frequency continuum as observed in the solar wind from ISEE 3: Thermal electrostatic noise, J. Geophys. Res., 85, 3419 (1980).

16. J. Belcher, private communication (1989).

SOLAR WIND VORTEX FLOW IN THE OUTER HELIOSPHERE

I.S. Veselovsky

Nuclear Physics Institute, Moscow State University, Moscow, U.S.S.R.

ABSTRACT

The structure of the T. von Karman vorticity wake type near the heliomagnetic equator is considered. Linearized equations are used for polytropic gas on the background of axially symmetric supersonic flow with the velocity minimum at the equatorial plane win the heliosphere.

1. According to the contemporary ideas, the solar wind velocity increases with distance from the surface of the warped heliospheric current sheet. The existence of higher velocities of the flow at high heliolatitudes in years of solar activity minimum was established by radioastronomic data. The typical velocities near the heliomagnetic equator are those of 400 km/s and far from it, 600-700 km/s in years of low solar activity. The free energy in this differential super-Alfvén and supersonic motion is rather large and may cause the development of a wide spectrum of waves in the interplanetary plasma. More over nonstationary and inhomogeneous boundary conditions near the rotating Sun are also a source of perturbations propagating in the solar wind. Besides potential disturbances there may exist vortex motions. The below presented consideration was undertaken to investigate a regular structure of vortex streams in the solar wind with the smallest radial velocity. The result reminds a well known in hydrodynamics vortex wake behind a body flown around and qualitatively corresponds to the stream structure observed on board spacecrafts near the equatorial plane in the outer heliosphere /1-3/.

2. Let us consider the equations of motion of polytropic gas

$$\begin{aligned} &\dot{\rho} + \nabla(\rho\vec{v}) = 0, \\ &\rho\,[\dot{\vec{v}} + (\vec{v}\nabla)\,\vec{v}] = -\nabla p, \\ &\dot{S} + (\vec{v}\nabla)\,S = 0, \\ &S = p\,\rho^{-\gamma}. \end{aligned} \tag{1}$$

Let an axially symmetric radial expansion be in the zero approximation: $v_{r0}(\theta)$ is the given function of polar angle θ independent of r, φ,

$$v_\theta = v_\varphi = 0\,, \qquad r^2\rho\, v_{r0} = \text{const}, \qquad S = \text{const}.$$

For perturbation ρ_1, $\vec{v}_1$, p_1 one always has near $\theta = \Pi/2$ in the local linear approximation:

$$\rho_1,\ \vec{v}_1,\ p_1 \propto \exp\, i(kr-\omega t),$$

$$(\omega-kv_{r0})\rho_1 + ik\rho_0 v_{r1} + \frac{2}{r}\,v_{r0}\rho_1 + \left[(\Pi/2-\theta)\rho_0 v_{\theta 1} + \rho_0{}' v_{\theta 1} + \rho_0 v_{\theta 1}'\right] r^{-1} = 0$$

$$\rho_0\left\{-i(\omega-kv_{r0})v_{r1} + \frac{v_{r0}{}'\,v_{r1}}{r}\right\} = ikp_1, \tag{2}$$

$$\rho_o\left\{-i(\omega-k\upsilon_{ro})\upsilon_{\theta 1}+\frac{\upsilon_{ro}'\,\upsilon_{\theta 1}}{r}\right\}=-\frac{p_1'}{r}$$

$$-i\left[\omega-k\upsilon_{ro}\right]\rho_o^{-\gamma}\left[p_1\ c_s^2\rho_1\right]-\gamma\upsilon_{ro}\rho_o^{-\gamma-1}\frac{\partial\rho_o}{\partial r}\left[p_1-c_o^2\rho_1\right]-2\upsilon_{ro}\rho_o^{-\gamma}c_s\frac{\partial c_s}{\partial r}\rho_1-\frac{S_o'\upsilon_{\theta 1}}{r}=0.$$

Here the trait denotes differentiation over θ,

$$\rho_o\approx\rho_o(\Pi/2)+\frac{1}{2}\rho_o''(\theta-\Pi/2)^2,$$

$$\upsilon_{ro}\approx\upsilon_{ro}(\Pi/2)+\frac{1}{2}\upsilon_{ro}''(\theta-\Pi/2)^2,$$

$c=(\gamma p/\rho)^{1/2}$ is the adiabatic sound velocity.

Let us examine now a fairly short-wave perturbation of the acoustic type $p_1=c_s^2\rho_1$, $\upsilon_{\theta 1}\sim$ const $+\ o(\theta-\Pi/2)$, $kr \gg 1$ near $\theta=\Pi/2$. The set of equations (2) can be rewritten as

$$-(\omega-k\upsilon_o)\rho_1+ik\rho_o\upsilon_{r1}+\left[(\theta-\Pi/2)\rho_o\upsilon_{\theta 1}+\rho_o''(\theta-\Pi/2)\upsilon_{\theta 1}\right]=0,$$

$$\rho_o\left\{i(\omega-k\upsilon_{ro})\upsilon_{r1}+\upsilon_{ro}''(\theta-\Pi/2)\upsilon_{\theta 1}r^{-1}\right\}=ikp_1, \tag{3}$$

$$\rho_o\left\{(\omega-k\upsilon_{ro})\upsilon_{\theta 1}\right\}=-\frac{p_1'}{r},$$

$$p_1=c_s^2\rho_1.$$

Perturbations v_{r1}, p_1, $\rho_1\propto(\theta-\Pi/2)$ can be expressed through $\upsilon_{\theta 1}$:

$$p_1=i(\omega-k\upsilon_{ro})(\theta-\Pi/2)\rho_o\upsilon_{\theta 1},$$

$$\rho_1=c_s^{-2}p_1. \tag{4}$$

In the case when the angular width of a sheet occupied by a slow wind satisfies the condition $\alpha \ll (M\,k\,r)^{-1}$, $M=\upsilon_{ro}/\upsilon_{TP}$ – the Mach's number for protons, υ_{TP} being their thermal velocity, Eq. (3) yields

$$\omega=k\left(\upsilon_{ro}-\upsilon_{ro}''\frac{\rho_o}{\rho_o''}\right),$$

$$\upsilon_{r1}=\frac{i\rho_o''}{kr\rho_o}(\theta-\Pi/2)\upsilon_{\theta 1}. \tag{5}$$

If $k\,r\,\alpha > 1$, the dispersion relation reads

$$\omega\approx k(\upsilon_{ro}\pm c_s) \tag{6}$$

and perturbations υ_{r1} and p_1 are expressed through $\upsilon_{\theta 1}$

$$\upsilon_{r1}=ikr(\theta-\Pi/2)\,\upsilon_{\theta 1},$$

$$p_1=\pm ikr\,\rho_o c_s(\theta-\Pi/2)\,\upsilon_{\theta 1}. \tag{7}$$

If $M^{-1} \ll k\,r\,\alpha \ll 1$, then

$$\omega\approx k\upsilon_{ro}\left[1-(krM\alpha)^{-2/3}\right] \tag{8}$$

The available imaginary parts prove to be of a higher order of smallness so strong instabilities in the local approximation are absent. Similar investigations of large-scale perturbations beyond the local approximation are still lacking.

3. Formulae (7) describe two sets of vortex perturbations of the acoustic type driven by the stream with the velocity and propagating respective it with the velocity upstream and downstream. Fig.1 shows the qualitative form of the summary field of velocities in the coordinate system moving with velocity . This velocity field has the shape of vortex with zero points of two types.

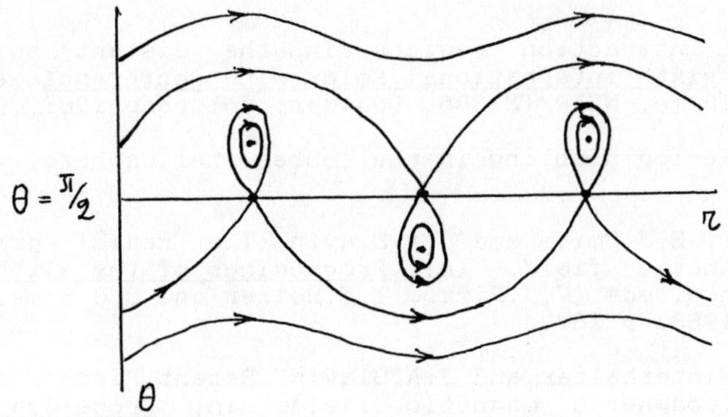

Fig.1. The qualitative pattern of a velocity field in the propagating vortex structures near the solar equatorial plane in the solar wind.

At $\theta = \Pi/2$ The x-type zero points are observed, at some value of $\theta_\pm$ there are o-type zero points. The o-type zero points represent the vortex centres located at the distance $z = \pm\, 2kr^2 v_{\theta 1} / v''_{r0} = (\theta_\pm - \Pi/2) r$ from the plane $\theta = \Pi/2$. This distance can be seen to increase with the perturbation amplitude, with the distance from the Sun and with decreasing slope of the profile $v_{r0}(\theta)$. At the centres of vortices propagating downstream (the sign "+" in (6), (7)) there is a pressure minimum as in "cyclones". At the centres of vortices propagating upstream (the sign "−" in (6),(7)) there is a pressure maximum as in "anticyclones".

4. It is of interest to compare qualitatively the results obtained with observations of oscillations in the solar wind stream near the helioequator plane at distances up to 25 a.u. /1,2/. Firstly, the correlated character of changes (see Fig.2 in /2/ and Fig.3 in /3/), having the same phase according to (7) and the period approximately equal to the Sun's rotation period attract attention. Secondly, in the "Voyager-2" spacecraft data under consideration one can also observe a correlated character of variation of v_θ and v_r, however changed strongly compared with a simple linear theory, due to perturbations being nonlinear. A rapid steepening of the fronts of the radial velocity and a large amplitude of density and pressure variations can be seen. The frozen-in magnetic field calculated in kinematic approximation is equal to $B_{\varphi 1} = -B_{\varphi 0} v_{r1}/v_{r0}$, where $B_{\varphi 0}$ is the unperturbed spiral field.

Our linear gas-dynamic consideration indicates only the possibility of vortex motions in the solar wind. A more complete theory should regard nonlinearity and the role of the magnetic field as well as a real geometry of the flow. The role of these factors is discussed in /3-8/ however at present a real pattern remains insufficiently clear both in theory and observations. In particular, it is still partially unclear why the stream structure observed at large distances in the heliosphere on board the "Voyagers" and "Pioneers" spacecrafts seems to be simpler than at the Earth's orbit and comprises the Sun's rotation period as the main time scale, while near the Earth's orbit a smaller-scale structures are prevailing.

REFERENCES

1. A.J.Lazarus and J.Belcher, Large-scale structure of the distant solar wind and heliosphere, Proceedings of the sixth international solar wind conference, eds. V.J.Pizzo, T.E.Holzer and D.G.Sime, NCAR/TN-306 Boulder, Colorado 1988, v.2, p.533.

2. A.J.Lazarus, B.Yedidia, L.Villanueva, R.L.McNutt, Jr., J.W.Belcher, U.Villante and L.F.Burlaga, Meridional plasma flow in the outer heliosphere, Preprint CSR-P-88-10, Cambridge, 1988.

3. R.L.McNutt, Jr., Possible explanations of North-South plasma flow in the outer heliosphere and meridional transport of magnetic flux, Preprint CSR-P-88-7, Cambridge, 1988.

4. V.J.Pizzo and B.E.Goldstein, Meridional transport of magnetic flux in the solar wind between 1 and 10 AU: a theoretical analysis, J.Geophys.Res. 92, 7241 (1987).

5. L.F.Burlaga, Interaction regions in the distant solar wind, in: Proceedings of the sixth international solar wind conference, eds. V.J.Pizzo, T.E.Holzer and D.G.Sime, NCAR/TN-306, Boulder, Colorado 1988, p.547.

6. L.F.Burlaga, Period doubling in the outer heliosphere, J.Geophys.Res. 93, 4103 (1988).

7. D.Winterhalter, E.J.Smith and J.A.Slavin, The radial gradient in the interplanetary magnetic field, in: Proceedings of the sixth international solar wind conference, eds. V.J.Pizzo, T.E.Holzer and D.G.Sime, NCAR/TN-306, Boulder, Colorado 1988, p.587.

8. E.J.Smith, D.Winterhalter and J.A.Slavin, Recent Pioneer 11 observations of the distant heliospheric magnetic field, in: Proceedings of the sixth international solar wind conference, eds. V.J.Pizzo, T.E.Holzer and D.G.Sime, NCAR/TN-306, Boulder, Colorado 1988, p.581.

TRAPPED RADIATION IN THE OUTER HELIOSPHERE

A. Czechowski and S. Grzedzielski

Space Research Centre, Polish Academy of Sciences, Ordona 21, 01–237 Warsaw, Poland

ABSTRACT

Time evolution of the electromagnetic radiation trapped in the outer heliosphere and interacting with the solar wind of fluctuating intensity is studied within linear theory in the geometrical optics approximation. A net energy transfer from moving solar wind fluctuations to trapped radiation is found, resulting in the increase of the radiation frequency at the rate of ~1 kHz/year. We discuss the implications of this result for understanding the origin of the 2–3 kHz signals observed by the Voyager missions.

INTRODUCTION

The motivation for this study was the specific time dependence of the 3 kHz interplanetary radio emissions detected by the Voyagers /1,2,3,4/. The signal exhibits ~1 kHz/yr upward frequency drift and decays after ~6 months /3,4/. We have found /5/ that such behavior arises quite naturally if, as suggested by /6,7/, the signal is trapped between the heliopause and the inner solar wind (SW). Comparing the observed drift rate with the result of model calculations we estimated the average distance to the heliosphere to be 60÷100 AU /5/.

Here we concentrate on those aspects of our investigation which were not covered in /5/. Apart from remarks concerning the interpretation of the drift mechanism, we discuss the effect of absorption and the problem of confining the 3 kHz signal.

METHOD

We restrict attention to the radiation in kHz range. The wavelength (~10^2km) is then « than the typical scale of variations of plasma density in the outer heliosphere and we can use the geometrical optics approximation. This regards the signal as a classical particle with the Hamiltonian = $(c^2k^2+\omega_P^2)^{1/2}$ which is equal to the frequency ω of the signal. The wave vector $\mathbf{k}$ and the local plasma frequency $\omega_P(\mathbf{r},t)$ correspond respectively to the momentum and the effective mass of the particle. The equations of motion are the usual Hamilton's equations :

$$\dot{\mathbf{r}} = \frac{\partial\omega}{\partial\mathbf{k}} \qquad \dot{\mathbf{k}} = \frac{\partial\omega}{\partial\mathbf{r}} \tag{1}$$

THE FREQUENCY DRIFT

Consider radiation trapped in a box with reflecting walls. If a wall moves inwards at the speed V, the frequency increases at the rate O(V/c) due to Doppler effect; however, since the cavity shrinks, this cannot be sustained. If instead the walls oscillate with average $\langle V\rangle=0$ and $\langle V^2\rangle\neq0$, there is a sustained upward frequency drift due to Fermi mechanism /8/, but the rate is $O(V^2/c^2)$. We shall see that in the case of the heliospheric cavity the drift is both O(V/c) and sustained.

The drift is caused by interaction with plasma density fluctuations moving with the SW. Assume for simplicity that the SW is spherically symmetric with the density profile $r^{-2}\alpha(r-Vt)$ where α consists of step functions (Fig.1). The frequency of the signal changes whenever a moving step in SW density is encountered. Let $\tilde{\omega}_{+,-}=(\omega_{+,-}^2+c^2k_T^2)^{1/2}$ where $\omega_+(\omega_-)$ denotes the SW plasma frequency just inside (outside) the step and $\underset{\sim}{k}_T$ the tangent part of the wave vector at the moment of collision. If $\omega > \tilde{\omega}_+, \tilde{\omega}_-$ the signal crosses the step and the frequency changes by $(V/c)(\omega^2-\tilde{\omega}_-^2)^{1/2}-(V/c)(\omega^2-\tilde{\omega}_+^2)^{1/2}$. If $\omega < \tilde{\omega}_+$, it is reflected with frequency gain $2(V/c)(\omega^2-\tilde{\omega}_-^2)^{1/2}$.

The net effect, as follows from computer simulations, is the average frequency gain /5/. To understand how this happens it is instructive to compare the change in frequency of a signal reflecting from two (moving) sample density profiles (Fig.1) chosen so that one of them complements the other: that is each density step is replaced by a step in the opposite direction. Although $\Delta\omega$ is >0 (<0) when a step with $\omega_+-\omega_->0$ (<0) is passed, the O(V/c) shifts in frequency corresponding to these two profiles will not generally sum to zero because the limits of penetration of the signal will be different (Fig.1). At the limit the signal is reflected back with $\Delta\omega$ either >0 (reflection from a density step) or =0.

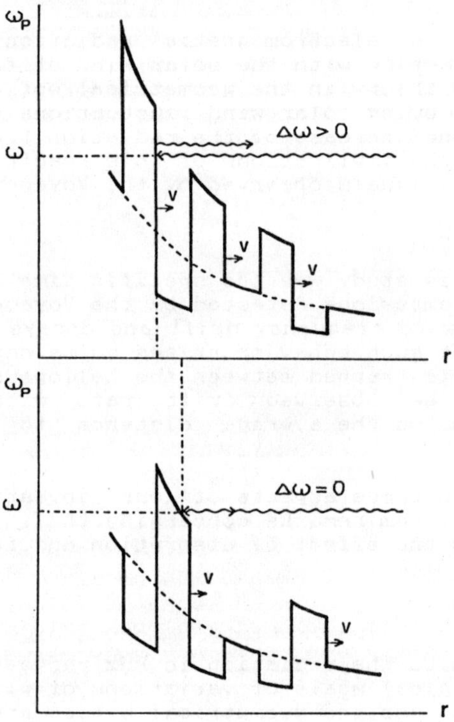

Fig.1. Reflection from two sample mutually complementary density profiles. The local plasma frequency ω_P is plotted versus the radial distance r. The difference of the signal penetration limits is shown.

DRIFT RATE IN THE MODEL HELIOSPHERE

A simple model of a heliosphere with trapped radiation was introduced in /5/. Both the nonspherical shape of the cavity and the presence of the termination shock are taken into account. To determine the drift rate the average frequency shift $\langle\Delta\omega\rangle$ per reflection cycle and the average time length $\langle\Delta t\rangle$ of the cycle are calculated. The cavity is characterized by the average distance to the shock (R) and to the heliopause (L). The SW profile is parametrized by P: the probability of reflection from a moving density step and a: the relative amplitude of fluctuations of local plasma frequency in the SW. As noted in /5/, the values $P\approx 1$, $a\approx 2\div 4$ are indicated by Voyagers data /9,10/.

Fig.2 shows the lines of constant drift rate obtained using the approximate formula from /5/. The values of R, L and a consistent with the observed drift rate can be read out.

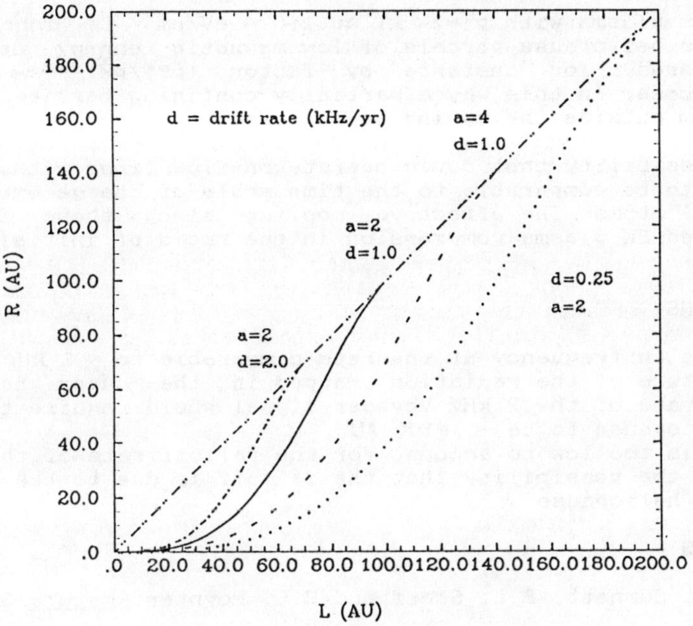

Fig.2. Lines of constant drift.

ABSORPTION

The absorption coefficient due to electron-ion free-free collisions is given by

$$\mu(\text{cm}^{-1}) = 10^{-2}\frac{N^2}{T^{3/2}\omega^2}\left[\,17.7 + \ln\frac{10^3 T}{\omega}\,\right] \qquad (2)$$

N (cm^{-3}) is the local electron density, T (K) the electron temperature and ω (Hz) the frequency of the radiation. The optical thickness after one reflection cycle is equal to the integral $2\int dr(\mu/n)$ with the integration limits appropriate to the trajectory of the photon; $n=(1-\omega_p^2/\omega^2)^{1/2}$ is the refraction coefficient. The average optical thickness after time t in the model heliosphere: $\langle\Lambda(t)\rangle$ can be calculated using the method described in the previous Section. The results for t=4 months and ω=3 kHz are presented in Table 1 ($\langle\Lambda(4\text{months})\rangle=1$ would account for the falloff of the 3 kHz Voyager signal). The temperature is assumed to be $T=10^4$K inside the shock and $T'=3\cdot10^4$K outside. It is evident that the absorption is rather insignificant even for the low assumed value of T': the typical lifetime of a trapped photon is > 1 yr. The main reason is that the angular momentum of the photon keeps it away from the dense inner SW region. The values shown in the Table 1 are too low to be the cause of the falloff of the 3 kHz signal.

TABLE 1 Average Optical Thickness after 4 Months.

L (AU)	R (AU) 30	40	50	60	70
40	0.77				
50	0.79	0.26			
60	0.79	0.26	0.13		
70	0.79	0.26	0.12	0.07	
80	0.79	0.26	0.11	0.06	0.05

CONFINEMENT

Trapping of radiation with frequency $\lesssim$ 3.5 kHz requires external electron density $\gtrsim$ 0.15 cm^{-3}. This exceeds the expected LISM values $N^{\infty}\approx(1\div5)\cdot10^{-2}\text{cm}^{-3}$. Note, however, that for R≈50 AU as recently suggested by the Ly-α scattering data /11,12/, the stagnation pressure has to build up to $p^{st}\approx$ 4 eV/cm^3. This holds true irrespective of nature of the dominant stresses, whether thermal, magnetic, turbulent etc. Suppose for instance that the distant interstellar plasma with $p^{\infty}=2N^{\infty}kT^{\infty}\approx 0.1\div 0.4$ eV/cm^3 is inhomogeneous. Upon approaching the heliopause parcels of low magnetic energy density may become much compressed, for instance by factor $(p^{st}/p^{\infty})^{3/5}\approx 4\div 9$ in the simplest adiabatic case. In this way a partially confining barrier of dense cloudlets could form outside the upwind heliopause.

There is another possibility that could operate on flow lines with flow time scales long enough to be comparable to the time scale of charge exchange with interstellar neutral atoms. The effective cooling along these flow lines would lead to shocked SW plasma compression in the ratio of initial and final temperatures.

CONCLUSIONS

(1) The upward drift in frequency at the rate comparable to ~ 1 kHz/yr can be regarded as a signature of the radiation trapped in the outer heliosphere. The observed drift rate of the 3 kHz Voyager signal would require the average distance to the heliopause to be ~60÷100 AU.
(2) The absorption is too low to account for the falloff rate of the Voyager signal. This leaves the possibility that the falloff is due to the escape of photons through the heliopause.

REFERENCES

1. W.S. Kurth, D.A. Gurnett, F.L. Scarf and R.L. Poynter, Nature 312, 27-31 (1984)
2. W.S. Kurth, D.A. Gurnett and F.L. Scarf, Adv.Space Res. 6, 379-383 (1986)
3. W.S. Kurth, D.A. Gurnett, F.L. Scarf and R.L. Poynter, Geophys. Res. Lett.14, 449-452 (1987)
4. W.S. Kurth, this issue.
5. A. Czechowski and S. Grzedzielski, Nature 344, 640-641 (1990)
6. W.I. Axford, Solar Physics 100, 575-586 (1985)
7. H.J. Fahr, W. Neutsch, S. Grzedzielski, W. Macek and R. Ratkiewicz - Landowska, Space Sci. Rev. 43, 329-381 (1986)
8. D.D. Barbosa, Astrophys. J. 243, 1076 (1981)
9. R.L. McNutt, Geophys. Res. Lett. 15, 1307-1310 (1988)
10. L.F. Burlaga, Space Sci. Rev. 39, 255-316 (1984)
11. P. Gangopadhyay, H.S. Ogawa and D. Judge, Astrophys. J. 326, 2012-2021 (1989)
12. D.L. Judge, P. Gangopadhyay and S. Grzedzielski, this issue.

Session V

Interaction of Solar Wind with the Interstellar Medium

(Curator — H. J. Fahr)

INTERACTION OF THE SOLAR WIND WITH THE EXTERNAL PLASMA

V.B. Baranov

Institute for Problems in Mechanics, U.S.S.R. Academy of Sciences, prosp. Vernadskogo 101, 117526 Moscow, U.S.S.R.

ABSTRACT

A critical survey of the present-day situation in gasdynamical models of solar wind interaction with the local interstellar medium /LISM/ is presented . Our attention is concentrated on the main limitations of the gasdynamical models,in particular,two-shocks model taking into account the influence of the H-atoms,moving from the LISM,on the interface plasma structure due to effects of the resonance charge exchange /1,2/.The results of the first step of an iterative method using Monte Carlo simulation for describing the motion of the H-atoms from the LISM are given /3/.These results show us that the effect of the LISM plasma component on the H-atoms penetration into the heliosphere can give rise to reinterpretation of data on the scattered solar Ly_α-radiation,if the degree of the LISM ionization is sufficiently large.
The effects of solar cycle /non-stationary problem/ and neutral component of the solar wind /4/ on the gasdynamical models are estimated.

INTRODUCTION

The problem of the interaction of the solar wind with the interstellar medium has long been a topic of interest /5-9/.Nevertheless it has become especially actual after at the beginning of the seventies,when experiments on scattered solar radiation at wave lengths λ 1216Å and λ 584Å proved us /10--15/ that the H and He-atoms of local interstellar medium /LISM/ are moving with a supersonic velocity $V_\infty \sim 20$ kms^{-1} relative to the Sun and that the direction of this motion is almost in the ecliptic plane and does not coincide with the direction of the solar motion with respect to nearest stars /apex direction/.
At present the scientists vanity is excited by,first,the possibility of direct investigations of the solar system periphery by "Voyager-1/2" and "Pioneer-10/11" spacecraft and,second,by the great achievements of space astronomy.
The construction of a theoretical model for the solar wind interaction with the interstellar gas which should be able to explain the physical phenomena is a significant problem for the following three reasons.First,from a purely theoretical viewpoint,the solution for the solar wind should be consistent with those for the interstellar medium.Secondly,the correct interpretation of observable phenomena,such as the solar radiation scattering at wave lengths λ 1216Å /H-atoms/ and λ 584Å /for He-atoms/,can only be made on the basis of an adequate theoretical model.And,thirdly,an internally consistent model makes it possible to determine,through reliable measured quantities, those parameters which are measured very poorly,e.g.,the degree of the LISM ionization.
E.Parker /6/ was the first to develop a quantitative gasdynamic model for the interaction of the moving interstellar medium /interstellar wind/ with stellar winds,including solar wind.He assumed that dynamic pressure of the interstellar wind is small as compared with the static one,i.e.

$$\rho_\infty V_\infty^2 \ll p_\infty , \qquad (1)$$

where ρ_∞ , V_∞ , p_∞ -the mass density,bulk velocity and static pressure of the interstellar wind respectively.In this case the interstellar wind is subsonic / $M_\infty \ll 1$, where M_∞ -Mach number of the interstellar wind/, and the shape of the solar wind termination shock /heliospheric shock/ is a sphere with a radius r_{HS}. It is determined by formula

$$r_{HS} = r_E \left[\frac{\gamma + 3}{2(\gamma+1)} \; \frac{\rho_E V_E^2}{p_\infty} \right]^{\frac{1}{2}}.$$

Here ρ_E, $V_{SW} = V_E$ -the mass density and solar wind velocity at the Earth's orbit / $r = r_E$ /, γ -specific ratio.

We think that Parker's model of the solar wind's interaction with subsonic interstellar wind is not real, because the inequality (1) is realized for $V_\infty \sim 20 kms^{-1}$ if the temperature of the interstellar medium $T_\infty \gg 10^4 K$. Scattering experiments of the solar radiation showed us that $T_\infty \sim 10^4 K$. Therefore we shall consider the gasdynamic model of the solar wind interaction with the supersonic interstellar wind.

THE SOLAR WIND INTERACTION WITH THE SUPERSONIC INTERSTELLAR WIND

The gasdynamic model of the solar wind interaction with supersonic interstellar wind was first proposed and calculated by Baranov, Krasnobaev and Kulikovsky /9/ in the thin layer approximation under the assumption of cylindrical symmetry. This model was based on the assumption that the hypersonic solar wind is damped mainly on the charged component /called below the plasma component/ of the interstellar medium /electrons and protons/. It is this assumption that makes it possible to describe the model using Euler's equations.

However solar Ly_α -scattering experiments show that there is a penetration of hydrogen atoms into the solar wind from the interstellar medium. The resonance charge exchange and photoionization processes give rise to a situation in which some of these atoms turn into protons. In this case the "new" protons change the total plasma component's momentum and energy by means of their "pick-up" process. In such a way these processes can act on the solar wind's deceleration, distorting its spherical symmetry /16/, as well as on the structure of the interface separating the solar wind and the plasma component of the interstellar medium /17/. A qualitative picture of the flow considered is shown in Figure 1. Here BS is the bow shock, which is the shock through which the plasma component of the interstellar gas passes, HS is the heliospheric shock /the solar wind termination shock/, CD is the contact discontinuity /the boundary of heliosphere/ separating the compressed plasma component of the interstellar wind and compressed solar wind.

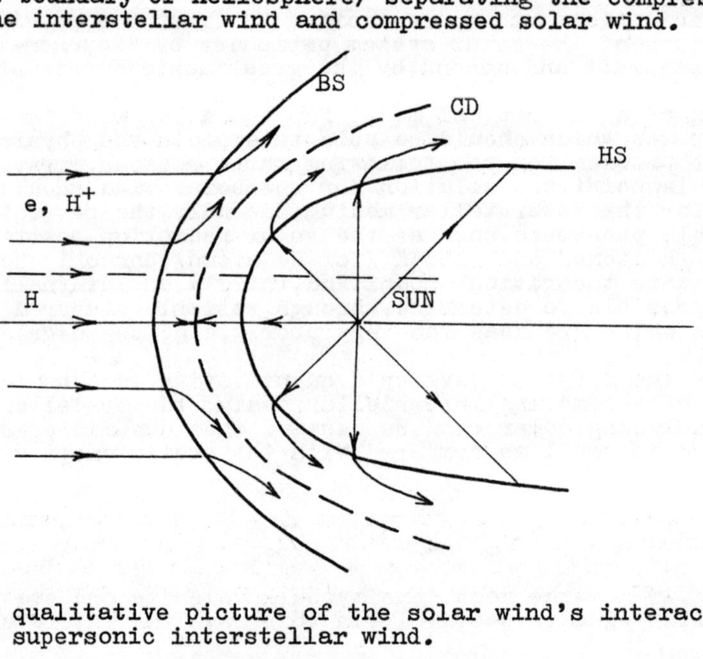

Fig.1. A qualitative picture of the solar wind's interaction with a supersonic interstellar wind.

The effect of the resonance charge exchange in the region between BS and HS /the interface/ on the H-atoms penetration from the LISM into the solar wind was calculated in papers /17-21/ without considering the effect of this process on the motion of the plasma component. Theoretical models were not self-consistent. For developing a self-consistent model it is necessary to change the Euler's equations for plasma component taking into account the effects of the resonance charge exchange. For the stationary problem the conservation equations will have a form

$$(\vec{V}\cdot\nabla)\vec{V}+\frac{1}{\rho}\nabla p=\vec{\mathcal{F}}_1(n_H,\vec{V}_H,T_H),$$

$$\nabla\cdot\rho\vec{V}=0, \tag{2}$$

$$\frac{p}{\gamma-1}\vec{V}\cdot\nabla\left(\ln\frac{p}{\rho^{\gamma}}\right)=\mathcal{F}_2(n_H,\vec{V}_H,T_H)$$

where p, ρ, $\vec{V}$ are the pressure, mass density and bulk velocity of the plasma component, n_H, T_H, $\vec{V}_H$ -the number density, temperature and bulk velocity of the H-atoms, $\vec{\mathcal{F}}_1$, $\mathcal{F}_2$ are the functions describing the change of momentum and energy of the plasma component due to collisions between H-atoms and protons accompanied by resonance charge exchange.

The self-consistent, two dimensional /cylindrical symmetry/ problem was solved in /1,2/ by using the following formulas for $\vec{\mathcal{F}}_1$ and $\mathcal{F}_2$ /16,22/

$$\vec{\mathcal{F}}_1=\nu_c\vec{U}=\nu_c(\vec{V}-\vec{V}_H),\quad \nu_c=n_H\sigma U_*,$$

$$U_*=\left(U^2+\frac{128k(T+T_H)}{9\pi m_H}\right)^{\frac{1}{2}}, \tag{3}$$

$$\mathcal{F}_2=\rho\nu_c\left(\vec{U}\cdot\vec{V}+\frac{V^2}{2}+\frac{3kT_H}{m_H}-\frac{3kT}{m_p}\right).$$

Here σ is the effective charge exchange cross-section, ν_c is the frequency of collisions for resonance charge exchange processes. It was also assumed that parameters of H-atoms are described by the equations

$$\vec{V}_H=\vec{V}_\infty=const,\quad T_H=T_\infty=const,$$

$$\nabla\cdot n_H\vec{V}_H=-\rho\nu_c. \tag{4}$$

Some numerical solutions of the equations (2)-(4) for the problem of the solar wind interactions with the supersonic interstellar wind are plotted in Figure 2-Figure 5. Figure 2 and Figure 3 demonstrate the shapes and the locations of the bow shock BS, contact discontinuity CD and the heliospheric shock HS as a function of the Mach number and degree of the LISM ionization.

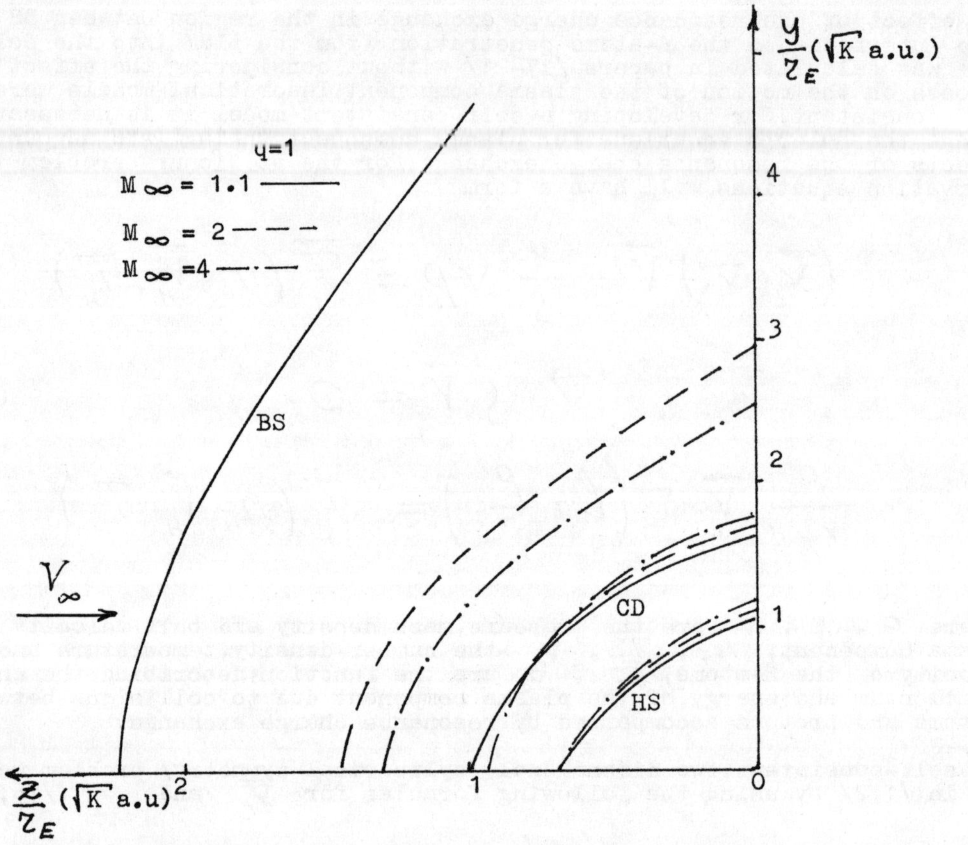

Fig.2. Effect of the Mach number on the position and shape of discontinuity surfaces.

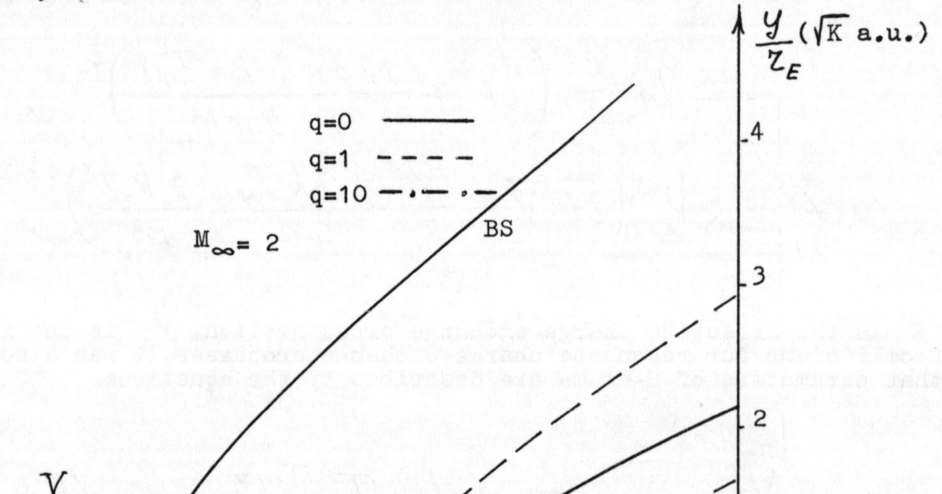

Fig.3. Effect of the LISM ionization on the position and shape of the discontinuity surfaces.

Four dimensionless parameters determine the problem formulated above /the specific heat ratio $\gamma = 5/3$ for the plasma component/

$$M_\infty = \frac{V_\infty}{a_\infty} \left(a_\infty = \sqrt{\gamma \frac{2k}{m_p} T_\infty} \right), \varphi = 6 z_E n_{e\infty} K^{\frac{1}{2}},$$

$$K = \frac{\rho_E V_E^2}{\rho_\infty V_\infty^2} = \frac{n_E V_E^2}{n_{e\infty} V_\infty^2}, \quad q = \frac{\rho_{H\infty}}{\rho_\infty} = \frac{n_{H\infty}}{n_{e\infty}}$$

$$(n_{e\infty} \approx n_{p\infty}),$$

where φ characterizes the resonance charge exchange process, M_∞ is the interstellar wind Mach number, K is the ratio of the solar wind kinetic energy at $z = z_E$ and the kinetic energy of the interstellar wind. The parameter q characterizes the degree of the LISM ionization.

It can be seen from the data in Figure 2 that a change in the Mach number of the interstellar wind affects only the size of the region between the bow shock BS and the contact discontinuity CD and has hardly any influence on the position and size of the region between CD and the heliospheric shock HS in the solar wind. As can be seen from Figure 3, the region between BS and HS moves strongly toward the Sun with increasing parameter q /parameter $\varphi = 0.8$ for Figure 2 and Figure 3/. For $q = 0$ these results coincide with these of paper /18/.

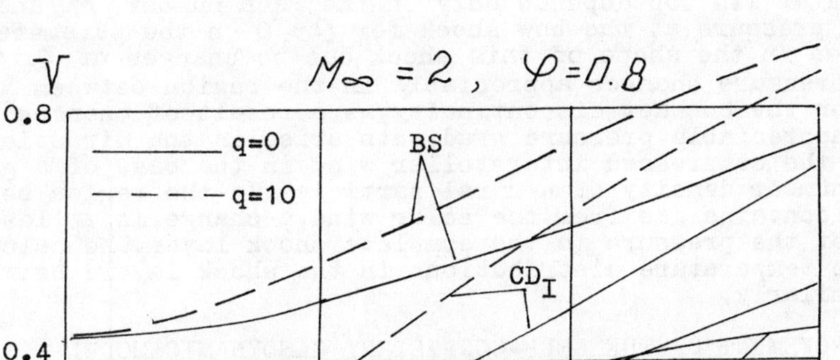

Fig.4. The absolute value of the velocity along the shocks BS and HS and contact discontinuity CD.

In Figure 4 and Figure 5 the distributions of the velocity /absolute value/ and pressure along the shock waves /BS and HS/ and contact discontinuity CD as a function polar angle θ are plotted. In Figure 4 the velocity is divided by V_∞ in the region between CD and BS and by V_E in the region between CD and HS, symbols CD_I and CD_{II} refer to outer and inner sides of CD.

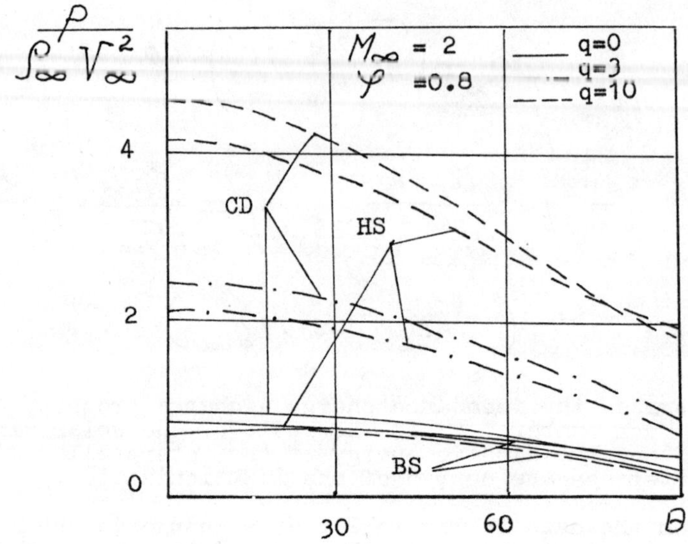

Fig.5. The distribution of the pressure along the shocks BS and HS and the contact discontinuity CD.

As can be seen from Figure 5 the pressure at the bow shock depends weakly on the number density $n_{H\infty}$ of the H-atoms in the LISM /the pressure behind the bow shock at its top depends only on the Mach number M_∞ and the dependence of the pressure at the bow shock for $\theta > 0$ on the parameter q is due to changes in the shape of this shock due to changes of q /.At the same time the pressure changes appreciably in the region between BS and CD,in particular,on the contact discontinuity,as a result of changes in q.Because of this,appreciable pressure gradients arise in the direction across the layer of the compressed interstellar wind in the case of a gas flow with a high number density of neutral particles.In the region between CD and HS,which contains gas from the solar wind,a change in q leads to uniform change of the pressure in the complete shock layer.The calculations show that the temperature distributions in the shock layers between BS and HS behave similarly.

IMPROVEMENT OF THE SELF-CONSISTENT GASDYNAMIC MODEL

The most important limitations of the self-consistent interface model considered above /1,2/ are connected with the assumptions introduced by equations (3) and (4).In particular,it is assumed here that H-atoms,moving from the LISM with velocity V_∞ ,disappear through their charge exchange with protons.However,a hydrogen atom,"born" in this process,moves with the velocity of the "killed" proton.These "fresh" H-atoms are not taken into account in the equations (4).The "born" H-atoms can again be effected by charge exchange and so on /see,for example,/23//.The effects of multiple resonance charge exchange can give rise to violation of the assumptions introduced in deriving the expressions (3),for example,the assumption on the Maxwell distribution of the hydrogen atoms /22,24/.

It is impossible to solve these problems on the basis of hydrodynamical equations for H-atoms,because their mean free path is comparable with the characteristic length of the problem /L $\sim$ 100a.u./.

In Figure 6 we have plotted the right side of the continuity equation for H-atoms /last equation of (4)/ calculated by Monte Carlo method /25/ as a function of the distance from the Sun for the case of the spherically-symmetric solar wind.For comparison solid lines are calculated on the basis of formulas (3).We see from Figure 6 that the processes of multiple charge exchange can be especially significant for large polar angles and at heliocentric distances $r \lesssim$ 10a.u.,i.e.in the region where the solar Ly_α - scattered radiation is formed.

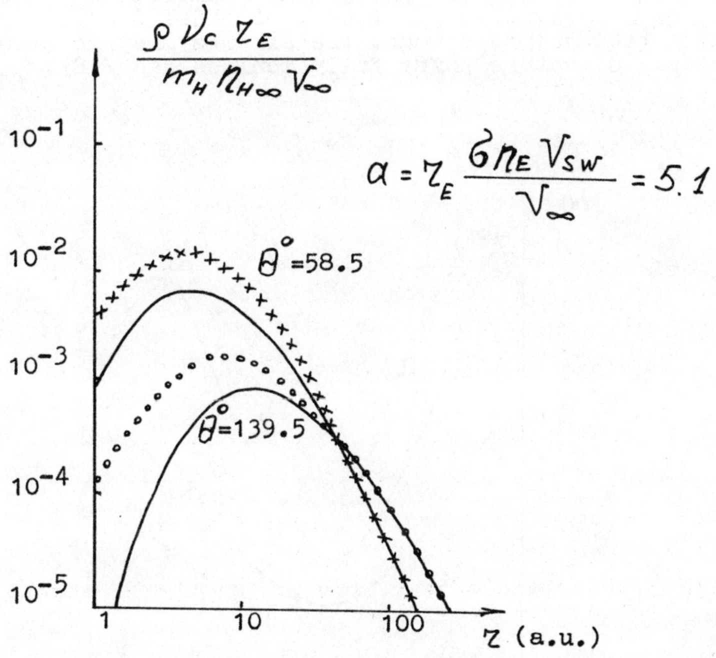

Fig.6 The right-hand side of the continuity equation for H-atoms from the LISM.

To correct the defects introduced by equations (3) and (4) we decided to use the Monte Carlo method for the description of H-atoms penetration through the plasma interface /3/.It was assumed that the distribution function of H-atoms in the LISM is maxwellian that shifted around the bulk velocity $\vec{V}_\infty$ with temperature T_∞ and a number density $n_{H\infty}$.We also assumed that the H-atoms motion is determined by solar gravitational force F_g ,radiation pressure F_r ,photoionization and resonance charge exchange processes. The last process esentially depends on the parameters distribution in the interface of the plasma component.

To solve the self-consistent problem we are going to use an iterative method.On the first step of this method we calculate H-atoms trajectories bythe Monte Carlo method by using the distribution of the plasma parameters obtained in the model for fully ionized hydrogen /18/.Then we can determine mass,momentum and energy sources caused by the resonance charge exchange and photoionization processes in the hydrodynamical equations (2) /if we take into account the last process,right-hand side of the continuum equation for plasma component is also non-zero/.In the next iteration the hydrodynamical equations must be solved with these sources.A new distribution of plasma parameters can be used for next Monte Carlo iteration for H-atoms and so on.It is necessary to continue this process of iterations until the results of two subsequent iterations will practically be coincided.The H-atoms parameters distribution obtained under such iterative procedure can be used for calculating the scattered solar Ly_α -radiation.

To illustrate the effect of the LISM plasma component on the H-atoms penetration into the solar system we illustrate below some results of the first step of the iteration /3/ /these results were also communicated by Baranov at the Simposium on the Outer Heliosphere,May 21-24,1989,Durham,USA/.

It can be noted that first step of the iteration has already been obtained in /26/.Nevertheless the "imitation" Monte Carlo technique which was used in this paper is not effective from numerical point of view and for this reason H-atoms parameters distribution had not been obtained.

Figure 7 and Figure 8 /see /3/ / show us the intensity (I) of the solar Ly_α -scattered radiation and reduction factor /RF/ as a function of the scanner angle /27/. The dots denote here experimental data on "Prognoz-6" for observation S1 /27/, the curve 1 is the result obtained by Monte Carlo method taking into account the spherically-symmetric solar wind only /this result coincides ideally with the result of model in /27//. curves 2 and 3 are our results taking into account the axial-symmetric structure of the interface /plasma structure / for fully ionized gas /18/.

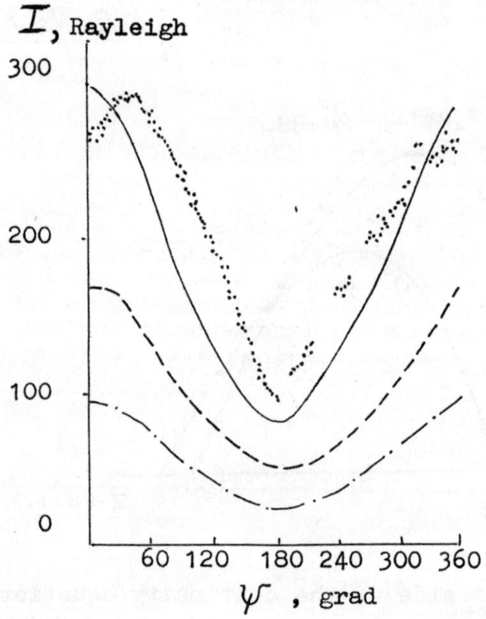

Fig.7. The scattered solar Ly_α -radiation intensity as a function of the scanner angle /27/.

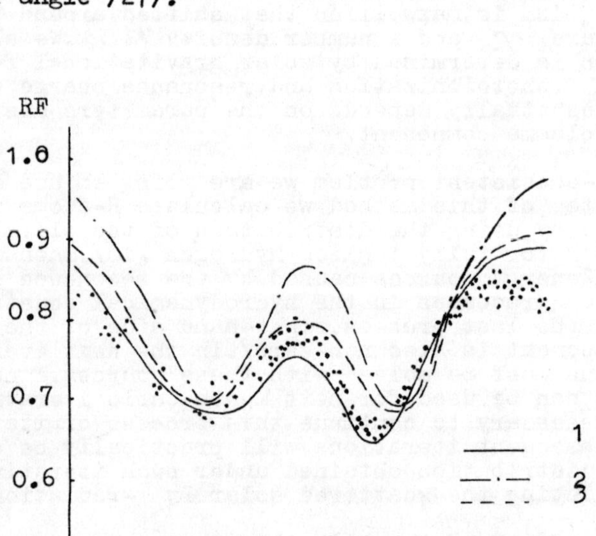

Fig.8. The reduction factor as a function of the scanner angle /27/.

We used the following values of parameters

$$M = \frac{F_r}{F_g} = 0.75, \; n_E = 7\,cm^{-3}, \; V_E = 450\,kms^{-1};$$

$$T_E = 10^5 K, \; n_{H\infty} = 3 \cdot 10^{-2}\,cm^{-3}, \; n_{e\infty} = 7 \cdot 10^{-2}\,cm^{-3}.$$

Besides, for the curves 1 and 2

$$T_\infty = 8 \cdot 10^3 K, \; V_\infty = 20\,kms^{-1} \quad (M_\infty = 1.34)$$

for curve 3

$$T_\infty = 7.4 \cdot 10^3 K, \; V_\infty = 23\,kms^{-1} \quad (M_\infty = 1.6).$$

From the results showed in Figure 7 and Figure 8 it seems, at first sight, that the model, which does not take into account the plasma structure of the interface, provides the better coincidence with observable data than the model taking into account this structure. However, a "good" coincidence of the Bertaux's at al theoretical results /27/ with the observable data arises from selection of a number of model's parameters.

The distributions of the hydrogen number density and bulk velocity in this iteration are showed in Figure 9 and Figure 10 /28/. It is seen from Figure 7-Figure 10 that the interface of the plasma component must influence on

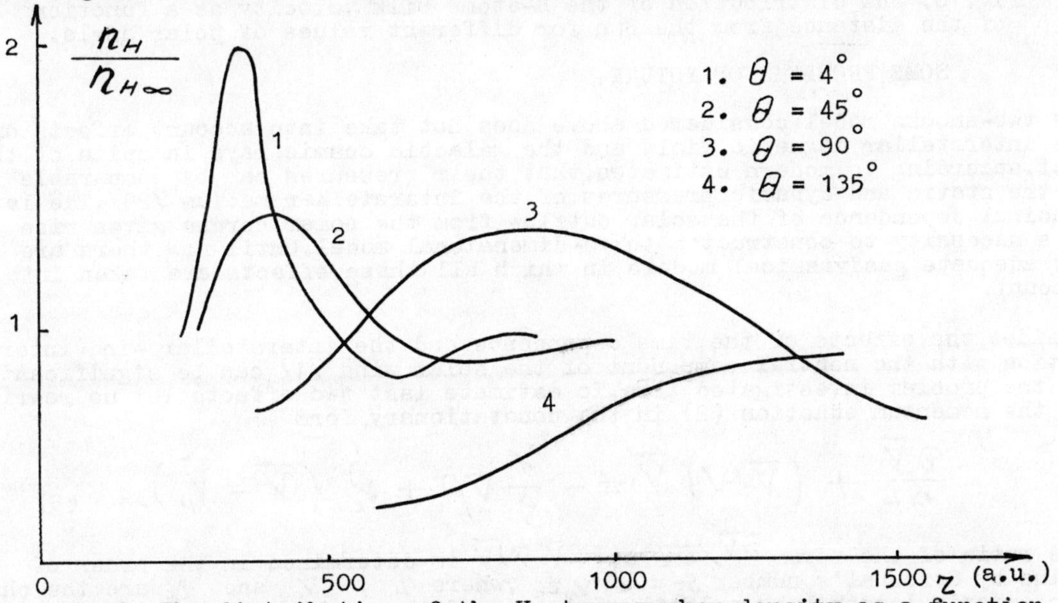

Fig.9. The distribution of the H-atoms number density as a function of the distance from the Sun for different values of polar angle.

the penetration of H-atoms into the solar wind and on the interpretation of the LISM parameters on the basis of the scattered solar radiation /curves in Figure 9 and Figure 10 are calculated for the same values of parameters which are used for calculating curve 3 in Figure 7 and Figure 8/. This influence can be significant in the presence of sufficiently large degree of ionization. Peaks of the H-atoms number density, which we see in Figure 9 as a result are connected, obviously, with processes of the resonance charge exchange near the critical point of the heliopause /contact discontinuity/.

But these results do not take into account reverse influence of H-atoms on the plasma structure of the interface. We hope to solve the self-consistent problem in the nearest future.

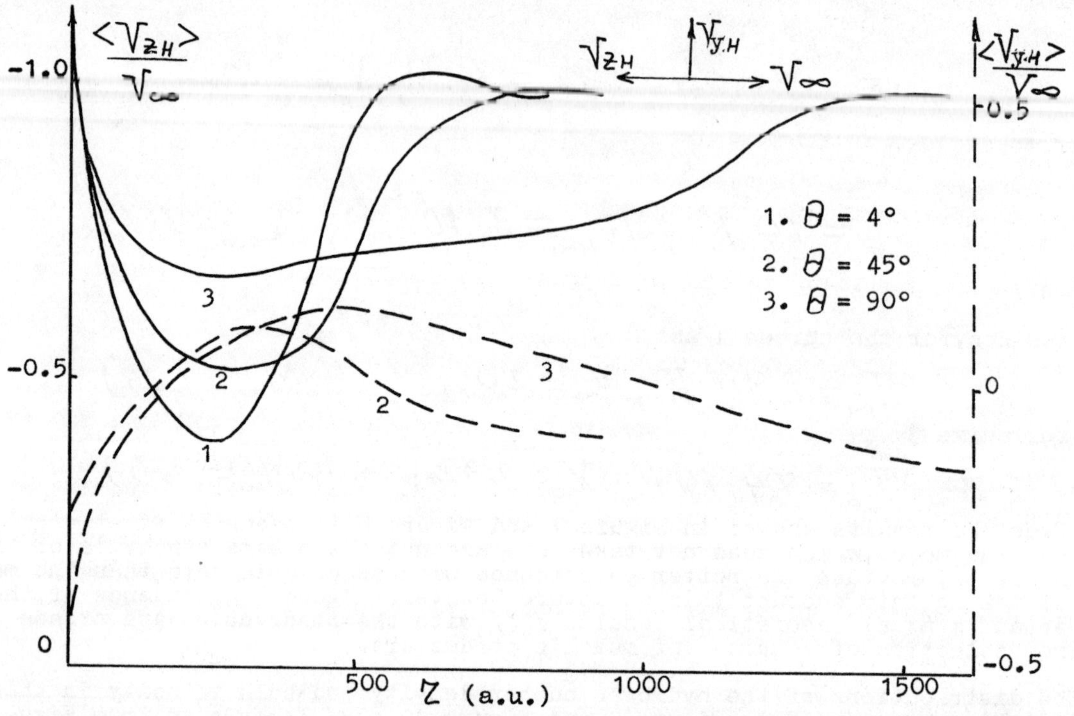

Fig.10. The distribution of the H-atoms bilk velocity as a function of the distance from the Sun for different values of polar angle.

SOME PROBLEMS OF FUTURE

The two-shocks model considered above does not take into account effects of the interstellar magnetic field and the galactic cosmic rays in spite of the fact, according to modern estimates, that their pressures can be comparable to the static and dynamic pressures of the interstellar medium /29/. The latitudinal dependence of the solar outflow from the solar corona gives rise to a necessity to construct a three-dimensional model. Until now there are not adequate gasdynamical models in which all these effects are taken into account.

Besides the effects of the time dependence and the interstellar wind interaction with the neutral component of the solar wind /4/ can be significant in the problem investigated here. To estimate last two effects let us rewrite the momentum equation (2) in the nonstationary form

$$\frac{\partial \vec{V}}{\partial t} + (\vec{V}\cdot\nabla)\vec{V} = -\frac{1}{\rho}\nabla p + \nu_c\,(\vec{V} - \vec{V}_H). \qquad (5)$$

The ratio of the term $\partial\vec{V}/\partial t$ to $(\vec{V}\cdot\nabla)\vec{V}$ is determined in the order of value by Strouhal's number $St = L/V_\infty t_o$, where L, V_∞ and t_o are the characteristic length, bulk velocity and time of the problem. The nonstationary problem we can consider as stationary one if $St \ll 1$. But for $V_\infty \sim 2\cdot 10^6$ cms^{-1}, $t_o \sim 11$ years we have $St \ll 1$ only for $L \ll 10^{15}$ cm. Characteristic size of the heliopause $L \sim 10^{15}$ cm and therefore we can not, strictly speaking, consider the problem of the solar wind interaction with the LISM as stationary problem.

The second effect is determined by the ratio of the last term of the equation (5) to the term $(\vec{V}\cdot\nabla)\vec{V}$. The solar wind H-atoms, created by the resonance charge exchange of the solar wind protons with the H-atoms from the LISM, penetrate into the LISM plasma component with the velocity of the solar wind / $V_H \sim V_{sw} \sim V_E$ /. After their charge exchange with the LISM protons the "new" energetic protons are generated. They can decrease then the momentum and increase the inner energy of the interstellar wind due to "pick-up" processes. This effect is not significant if the change of the velocity ΔV

is small as compared with the velocity V_∞ due to "pick-up" processes. From equation (5) we have,using the equations (3),

$$\frac{\Delta V}{V_\infty} \sim \nu_c \frac{V_E L}{V_\infty^2} \sim n_H V_E \frac{\sigma V_E L}{V_\infty^2} \tag{6}$$

If to use the estimation of the H-atoms flux from the solar wind /4/

$$n_H V_E \approx n_E V_E \left(\frac{r_E}{r} \right)^2$$

and to put it in (6) we can obtain the estimation

$$\frac{\Delta V}{V_\infty} \sim n_E V_E \left(\frac{r_E}{L} \right)^2 \frac{\sigma V_E L}{V_\infty^2} \quad (r \sim L).$$

For $n_E \sim 5\,cm^{-3}$, $V_E \sim 4 \cdot 10^7\,cms^{-1}$, $V_\infty \sim 2 \cdot 10^6\,cms^{-1}$, $\sigma \sim 2 \cdot 10^{-15}\,cm^2$ we have $\Delta V / V_\infty \ll 1$,if $L \gg 2 \cdot 10^{15}\,cm$,i.e.,in this case the effect considered /4/ is not significant.If this inequality is not fulfiled,the interstellar wind before the bow shock BS will be disturbed.

REFERENCES

1.V.B.Baranov,M.K.Ermakov and M.G.Lebedev,Astron.Zh./Letters/,v.7,p.372, /1981/.
2.V.B.Baranov,M.K.Ermakov and M.G.Lebedev,Fluid Dynamics /Engl.Transl./, No.5,p.754 /1982/.
3.V.B.Baranov,M.G.Lebedev and Yu.G.Malama,Astrophys.J.,in press /1989/.
4.M.A.Gruntman,Astron.Zh./Letters/,v.8,p.48 /1982/.
5.L.Davis,Phys.Rev.,v.100,p.1440 /1955/.
6.E.Parker,Astrophys.J.,v.134,p.20 /1961/
7.W.I.Axford,A.Dessler and B.Gottlieb,Astrophys.J.,v.137,p.1268 /1963/.
8.J.Brandt,Planetary Space Sci.,v.12,p.650 /1964/.
9.V.B.Baranov,K.V.Krasnobaev and A.G.Kulikovsky,Dokl.Akad.Nauk SSSR,v.194, p.41 /1970/.
10.V.G.Kurt,Issledovanija Kosmicheskogo Prostranstva,v.3,p.576 /1956/.
11.J.Bertaux and J.Blamont,Astron.Astrophys.,v.11,p.200 /1971/.
12.G.Thomas and R.Krassa,Astron.Astrophys.,v.11,p.218 /1971/.
13.P.Blum and H.Fahr,Astron.Astrophys.,v.4,p.280 /1970/.
14.H.Fahr,Space Sci.Rev.,v.15,p.483 /1974/.
15.C.Weller and R.Meier,Astrophys.J.,v.193,p.471 /1974/.
16.S.Grzedzielski and R.Ratkiewicz,Acta Astron.,v.25,p.177 /1975/.
17.M.Wallis,Nature,v.254,p.207 /1975/.
18.V.B.Baranov,M.G.Lebedev and M.S.Ruderman,Astrophys.Space Sci.,v.66, p.441 /1971/.
19.V.B.Baranov and M.S.Ruderman,Astron.Zh. /Letters/,v.5,p.615 /1979/.
20.H.Ripken and H.Fahr,Astron.Astrophys.,v.122,p.181 /1983/.
21.H.Fahr and H.Ripken,Astron.Astrophys.,v.139,p.551 /1984/.
22.T.Holzer,J.Geophys.Res.,v.77,p.5407 /1972/.
23.M.Burgin,Comments Astrophys.,v.9,p.157 /1981/.
24.T.Holzer and P.Banks,Planetary Space Sci.,v.17,p.1074 /1969/.
25.Yu.G.Malama,IVth Intern.Workshop on the plasma in Space,Poland,Warsaw, September /1987/.
26.S.Bleszynski,Astron.Astrophys.,v.180,p.201 /1987/.
27.J.Bertaux,R.Lallement,V.Kurt and E.Mironova,Astron.Astrophys.,v.150, p.1 /1985/.
28.Yu.G.Malama,private communication /1989/.
29.W.I.Axford and W.Ip,Adv.Space Res.,v.6,p.27 /1986/.

SIGNATURE OF A VISCOUS INTERACTION AT THE HELIOPAUSE

H. Pérez-de-Tejada

Instituto de Geofísica, Universidad Nacional Autónoma de México, Ensenada, Baja California, México

ABSTRACT

Observations of the solar wind interaction with weakly magnetized bodies in the solar system (Venus, Mars, comets) are used to define guidelines applicable to the region of interaction between the local interstellar plasma and the solar wind. In particular, a configuration based on a supersonic flow past a plasma obstacle is considered. It is argued that anomalous viscous processes may produce a downstream deformation of the heliospheric cavity and the stratification of the interstellar plasma that flows around it. In addition to the interface (heliopause) that bounds the solar wind from the interstellar plasma, and the external bow shock that develops upstream from that boundary, there should be a third transition within the interstellar flow located in the region between them. This additional (intermediate) boundary marks the outermost extent of the region affected by friction processes at the heliopause and divides the postshock interstellar flow into layers of different properties. Downstream from the intermediate transition the plasma should be hotter, more rarifed and move slower than that upstream from it.

INTRODUCTION

The in-situ examination of the region of interaction of the solar wind with Venus, Mars and comets Giacobini-Zinner and Halley has provided a large body of experimental evidence of the behavior of a magnetized supersonic flow past a plasma obstacle /1-9/. This evidence has shown that despite the collisionless character of the solar wind there is a highly complex set of processes which produce an intricate response of both, the streaming flow and the obstacle's plasma, to their interaction. In all cases there is ample information that the obstacle forces the solar wind to stream around it, and in so doing becomes compressed showing a well-defined termination of its plasma extent. Measurements have further revealed that the solar wind becomes contaminated with added mass far from the obstacle /8,10/ and, as a result, may substantially change its composition. All these changes have the effect of slowing down the solar wind in both a gradual and in an abrupt manner including a bow wave transition that develops far upstream from the obstacle's boundary.

The continued examination of the data has revealed that in addition to these two plasma transitions there is evidence of a third boundary located in the region between them. At the present time there is sufficient information to clearly identify that transition as a permanent steady state feature in the Venus plasma environment /1-3,11/, and it should also be noted that a transition with a similar signature is also present in the plasma data of the fly-by passes near comets Giacobini Zinner and Halley /6-9/. On Mars, however, there is only a suggestion of a similar feature from the early Mars 5 data /12/ and it is still necessary to await for the analysis of the recent Phobos orbiter plasma data /4,5/ to confirm its presence. The plasma changes seen across the intermediate transition are peculiar in that they reflect variations that affect the dynamics of the solar wind as well as that of the contaminating plasma particles that the solar wind has assimilated. Most notable is a sudden change in the decrease rate of the solar wind speed with decreasing distance from the obstacle. At the same time the slowed-down plasma exhibits higher temperatures but, unlike at a bow shock crossing, its density is lower downstream /1-3,13-15/. While the details of the available data differ from case to case they provide, as a whole, a body of information which supports the view that similar phenomena are common to all of them. The contention here is that those phenomena are characteristic of the interaction of two large-scale independent bodies of plasma in space. If this is the case they should be also operative in other astrophysical situations and, in particular, within the context of a plasma-plasma interaction configuration at the heliopause.

This issue is raised in the present study with the purpose of examining the structure of the interstellar plasma that streams around the heliospheric cavity. If the relative motion of the heliosphere with respect to the local interstellar medium results in a supersonic and superalfvenic flow regime then we could presumably expect a flow configuration similar to that documented from solar wind interaction geometries in the solar system. The discussion centers, primarily, on a review of the existing experimental evidence for the flow conditions seen at Venus, Mars, and comets Giacobini Zinner and Halley, and follows with their direct application to the conditions expected at and in the vicinity of the heliopause.

VENUS

Because of the extensive probing of the Venus plasma environment over the last 20 years there is at the present time an adequate, if not complete, understanding of the main physical properties of the solar wind–Venus interaction region. From the early Mariner 5 fly–by /1,2/ and the Venera 9 and 10 orbiters /3/ experiments it was learned that, in the absence of a significant intrinsic planetary magnetization, the planet's ionosphere receives the full impact of the dynamic pressure of the solar wind. Under such conditions the Venus ionosphere is compressed below a sharp boundary (ionopause) and deformed toward the nightside. The comprehensive examination of the Venus plasma environment that was conducted with the Pioneer Venus Orbiter (PVO) further revealed that the ionopause is the site of a complex electrodynamic process which results in the accumulation of the solar wind magnetic fluxes around the dayside ionosphere /16/. This process is less severe at and near the magnetic polar regions of the ionopause (defined with respect to the time–dependent IMF orientation) where the magnetic field lines skim undeflected past the planet /17/. Consequently, it is in these regions where a direct plasma–plasma interaction can be expected to occur, and where its effects on the Venus ionosphere should be more evident (thoughout most of the dayside ionopause the accumulated solar wind magnetic field lines reduce the direct contact between both bodies of plasma). The latter include the formation of ionospheric clouds and filamentary tail–like structures which extend far downstream from the planet /18/. Equally important is the observation of the general nightward displacement of the topside Venus ionosphere across the terminator producing a strong plasma flow in that direction /19/. All these features are important because they reflect the effects of friction processes between the solar wind and the ionospheric plasma /20/, a contention that is further supported by the observation of a thick velocity shear in the solar wind that streams around the ionosphere. The latter was initially revealed by the Mariner 5 and the Venera 9 and 10 plasma data /1–3/, and has been repeatedly inferred from the PVO measurements /21/.

Figure 1 shows the magnetic field intensity, thermal speed, ion density and bulk speed of the solar wind measured during the Mariner 5 fly–by pass near Venus. In addition to the bow shock crossings (at t = E – 158 min (inbound) and t = E + 20 min (outbound), E being the time of closest approach) which are characterized by lower bulk speeds and higher densities and thermal speeds downstream, there is evidence of an even stronger deceleration and more intense thermalization of the plasma deep within the post–shock flow region (at t = E – 100 min (inbound) and at t = E + 8 min (outbound)). The strong decrease of the density and flow speed seen downstream from this latter transition reflects the effects of processes not operating further outside, and implies an important deficit of the momentum flux of the streaming plasma. This effect is seen more dramatically outbound where the sharp speed jump that identifies that transition marks the most important feature of the speed profile after the spacecraft exited the wake region (at t = E – 15 min). Similar signatures have also been reported from the Venera 9 and 10 and PVO data /3,11,21/, thus confirming the steady state stratification of the post–shock flow around Venus. A circumstance that emphasizes the existence of a strong coupling between the shocked solar wind and the cross–terminator flow of the Venus upper ionosphere is the remarkable balance that exists between the height–integrated momentum flux of the latter and the height–integrated deficiency of momentum flux of the solar wind within the velocity shear adjacent to the ionosphere /20/. What this means is that the total kinetic energy of the ionospheric cross–terminator flow can be accounted for adequately in terms of the total deficiency of kinetic energy exhibited by the shocked solar wind that streams outside.

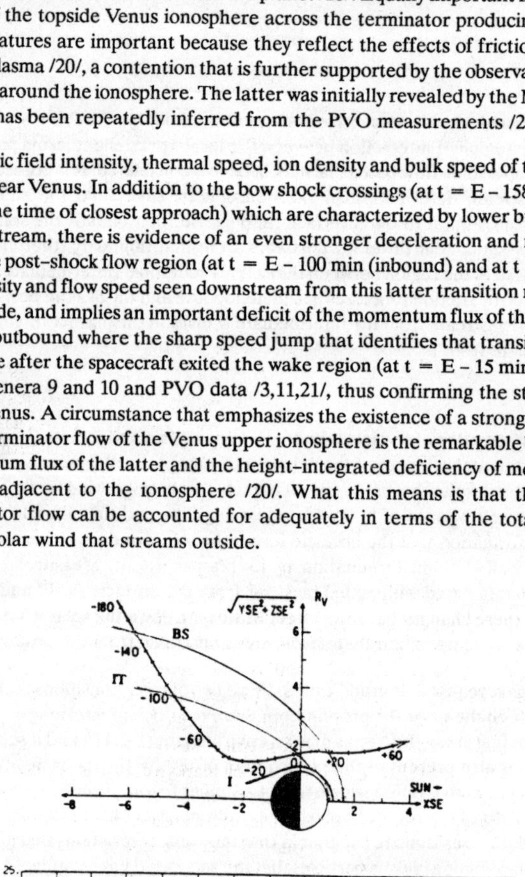

Fig. 1. (upper panel) Trajectory of the Mariner 5 spacecraft near Venus. (lower panel) Magnetic field intensity, thermal speed, ion density and bulk speed measured during the fly–by pass of the Mariner 5 near Venus. The curves labeled BS, IT, and I indicate, respectively, the bow shock, the intermediate transition and the ionopause (from /2/).

An important implication of the bulk speed profile shown in Figure 1 is that the upstream deceleration of the post-shock flow does not necessarily occur in a gradual accumulated manner as would be expected from the sole action of mass-loading processes. As noted above, exospheric neutral particles located upstream from the ionopause can be assimilated into the solar wind following their photoionization by the solar radiation. The effect of this process is the gradual deceleration of the oncoming solar wind, and should be more intense the closer the solar wind comes to the planet. While there is clear evidence of the contamination of the solar wind that streams around the Venus ionosphere /10/ the data shown in Figure 1 suggests that its impact on the solar wind speed does not seem to be the dominant feature of the observed profile. In particular, it is not evident that this process should produce the sudden deceleration exhibited by the post-shock flow at t = E + 8 min, and the maintenance a nearly uniform speed between that transition and the wake boundary downstream. The observed behavior is more in accord with the onset of friction phenomena at the ionopause and with the distribution of their effects along the sides of the ionosheath. In this view the region influenced by the viscous action is restricted within a well-defined layer by the supersonic speed of the re-expanded shocked flow. No disturbances produced at the ionopause can reach beyond that layer (upstream from the transition detected at t = E – 100 min inbound, and at t = E + 8 min outbound) because they are washed out by the local supersonic flow. The overall shape of that transition around and behind Venus has been determined from the extended analyses of the PVO data. These have revealed that near the terminator plane it is usually located closer to the ionopause than to the bow shock and that, as suggested from the Mariner 5 measurements /1/, it may not originate from the subsolar region but from a position at the ionopause located not far upstream from the terminator.

Although much of the physical process responsible for the collisionless transfer of momentum of the solar wind to the upper ionospheric plasma remains to be identified there are a number of studies from which it is possible to derive information as to the manner in which this may come about. The numerical modeling of the motion of pick-up ions in the mass-loaded solar wind indicates, in particular, that in the presence of strong magnetic turbulent conditions the diffusion of energy of those particles proceeds in time scales comparable to their own gyromotion /22,23/. This result implies that when a pick-up ion moves across a region in which there is a velocity shear (such as that present outside the Venus ionopause) the particle will exchange its energy excess with the local flow thus completing, in effect, a net non-collisional energy transfer. Since the pick-up ions derive their energy from the solar wind momentum it is clear that as they execute gyrations which partially enter the ionosphere, they will deliver in the process some of that momentum to the ionospheric plasma, thus driving the latter along their own direction of motion. The studies on which these concepts are based examine, specifically, the conditions observed in cometary plasma environments where the solar wind becomes highly turbulent as it becomes heavily mass-loaded with particles of cometary origin. On Venus the mass-loading is not as extensive but it also leads to highly turbulent plasma conditions /24/. While this circumstance validates the application of the comet models to the region in the immediate vicinity of the Venus ionopause it is significant that, as shown below, the flow geometry around cometary plasma obstacles includes a feature which strongly resembles the intermediate transition of the Venus ionosheath.

COMETS

In the absence of any significant gravitational force the atmospheres and ionospheres of comets attain extremely large dimensions and, as a result, produce severe modifications on the solar wind at far away distances from them. From the fly-by passes of the Giotto and Vega spacecraft near comet Halley it is now known that the solar wind suffers substantial contamination with cometary material at distances as far as a few million kilometers from the nucleus, and that a bow shock develops upstream from it at a subsolar distance of about 5 10^5 km (bow shock crossings were reported at nearly 10^6 km from the nucleus along the nearly east-west trajectories of the Giotto and Vega spacecraft /8, 25/). Downstream fron the bow shock there are a number of additional plasma transitions whose origin has not yet been clearly established. Most important is a boundary, detected at about 1.5 10^5 km from the nucleus, which marks a dramatic change in the composition of the plasma, from a solar wind proton-dominated population upstream, to a mostly cometary ion population downstream /26/. In the vicinity of this boundary, named the cometopause, there are also notable changes in the dynamics of the flow, including an outward deflection of the solar wind around the comet, and a severe decrease of its speed accompanied by the increase of the local temperature /8,26/. These variations are analogous to those seen outside the Venus ionosphere and thus suggest that, to a certain extent, the cometopause may be considered as the outer boundary of the comet's effective plasma obstacle /27/.

In addition to these observations there is in the cometary plasma data strong evidence of a third boundary, located nearly half way between the bow shock and the cometopause, with properties identical to those of the intermediate transition of the Venus ionosheath. This boundary, named the mystery transition /8,9/, was detected at about 5 10^5 km from the nucleus during the inbound and outbound passes of the Giotto spacecraft, and projects to a subsolar position located ~ 3 10^5 km upstream from the comet. A schematic diagram of the geometry of the bow shock, the cometopause, and the mystery transition, implied by the Giotto measurements, is reproduced in Figure 2 (right panel), together with plots (left panel) of the bulk speed, density, and thermal speed, of alfa particles measured during the inbound pass. Since these particles are mostly of solar wind origin their behavior is more representative of the changes that affect the solar wind flow as a whole, rather than the variations produced by the contaminating cometary material. The transitions at 19.5 ST, 21.8 ST, and 23.5 ST, mark, respectively, the bow shock, the intermediate transition and the cometopause. As in Venus, the intermediate (mystery) transition is characterized by a sudden drop of the bulk speed and density of the flow, and a net increase of the local temperature. These changes are again consistent with those reflecting the effects of friction processes at an inner boundary, and suggest that similar phenomena occur in both cases. In particular, a viscous interaction at the cometopause is expected to slow down the streaming flow, to increase its temperature through the dissipation of energy, and to lower the local density so that the plasma pressure matches that of the external (inviscid) flow.

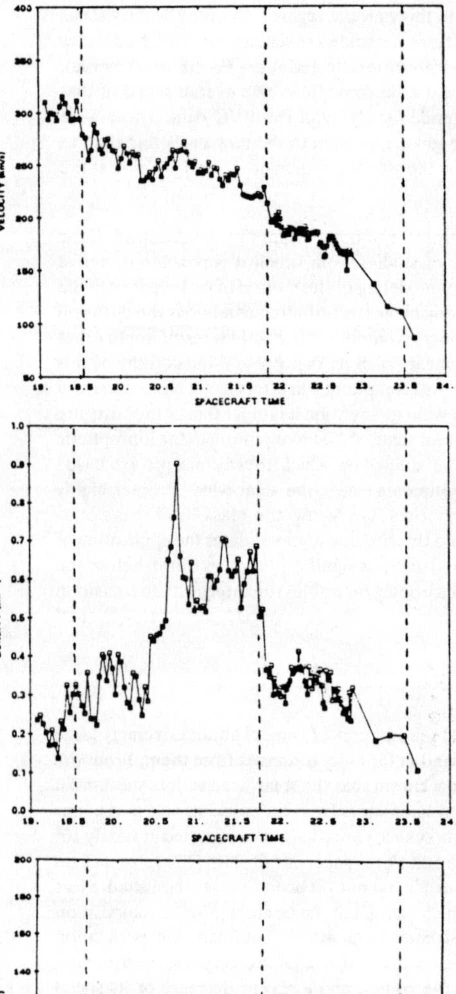

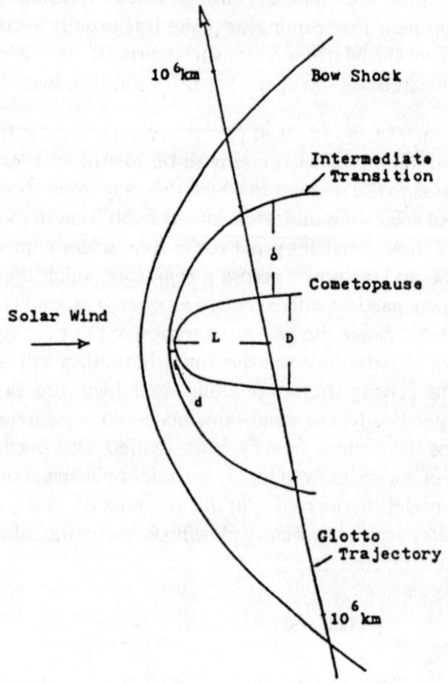

Fig. 2. (left panel) Bulk speed, density, and thermal speed of alfa particles measured near comet Halley along the inbound pass of the Giotto spacecraft. The bow shock, the intermediate transition and the cometopause are identified by the plasma changes seen near 19.5 ST, 21.8 ST, and 23.5 ST, respectively (from /13/). (right panel) Schematic diagram of the geometry of comet Halley's bow shock, intermediate transition and cometopause inferred from the Giotto measurements (from /27/).

In this interpretation the intermediate transition marks the outer extent of the post–shock flow region influenced by the friction produced at the cometopause. The supersonic character of the re-expanded post-shock flow prevents perturbations from reaching upstream from that transition. It should be noted, however, that unlike Venus, the cometary viscous boundary layer seems to develop from the subsolar region and thus to extend across the entire frontside of the comet's effective ionospheric obstacle. This condition may arise from the ability of the solar wind to directly interact with the cometary plasma at the cometopause despite the presence of the piled-up interplanetary magnetic field that extends downstream from this boundary /28/. In the outset, the different geometry of the intermediate transitions of the Venus and comet Halley's ionosheaths may be due to the different contamination levels of the solar wind, a circumstance that is evidenced by the strong gradual erosion of the flow speed shown in Figure 2.

Further and independent support of the stratification of cometary ionosheaths is also available from measurements conducted at comet Giacobini–Zinner with the International Cometary Explorer (ICE) spacecraft. In this case a cross tail pass revealed the existence of two different regimes within the shocked flow region. An irregular but gradual decrease of the flow speed with decreasing distance from the comet is seen in an outer layer, and a much steeper speed decrease characterizes the inner region that surrounds the comet's plasma tail /6,7/. This variation is analogous to that revealed from the Giotto measurements in comet Halley's outer plasma environment, and is again reminescent of the flow speed signature across the intermediate transition of the Venus ionosheath.

MARS

Despite a weak intrinsic planetary magnetic field the Martian plasma environment exhibits features similar to those seen at Venus. Most notable is a well-defined velocity shear within the shocked solar wind that streams by the sides of the magnetosphere, and the severe erosion that this flow exerts on the planet's ionospheric plasma. The early plasma measurements conducted with the Mars 5 spacecraft /29/ showed that, like Venus, the speed and the density of the shocked solar wind in the region adjacent to the flanks of the Martian magnetosphere is significantly lower than that measured further outside and, at the same time, its temperature is increased over the values measured upstream. These variations are different from those seen outside the flanks of the earth's magnetosphere where nearly uniform speed, density and temperature profiles seem to characterize the shocked solar wind in that region /30/. Not surprisingly, the earth's magnetic field represents an entirely different obstacle to the solar wind than the small-scale highly contaminated ionospheric environments of Venus and Mars, and thus prevents the direct plasma–plasma interaction that takes place in the latter cases. In the absence of a strong Martian magnetic field the interaction of the solar wind with the upper ionosphere should also result in the onset of frictional processes and, as in Venus, propel the ionospheric plasma along its own direction of motion. These concepts were explored in an earlier study /12/ where it was further suggested that a flow of magnetospheric dimensions is necessary to carry-off the momentum flux lost by the solar wind as it interacts with the planet's ionospheric plasma. The measurements recently conducted with the Phobos spacecraft have confirmed the presence of a plasma flow of planetary origin within the entire cross-section of the Martian magnetosphere /4,5/. The speed and density of this flow are comparable to those inferred from considerations of a viscous transfer of momentum across the boundary of the magnetosphere, and reveal that a large fraction of the momentum flux of the solar wind is indeed used to drive that motion. This result further emphasizes the relevance of viscosity in the solar wind interaction with planetary plasmas as a phenomenon which can adequately account for the geometry and dynamics of ionospheric flows. Further studies are still necessary, however, to determine the position of the intermediate transition around the flanks of the Mars magnetosheath as well as its geometry in front of the dayside magnetosphere.

VISCOUS FLOW AT THE HELIOPAUSE

The experimental evidence reviewed so far illustrates what seems to be a general property of the solar wind interaction with plasma obstacles, namely, the stratification of the post-shock flow region. The intermediate transition seen in the ionosheaths of Venus and comets Giacobini–Zinner and Halley marks the outer extent of the region influenced by perturbations produced at the boundary of their corresponding plasma obstacles. Since this feature occurs around obstacles of very different dimensions it is possible that it may also develop in other astrophysical situations where similar physical conditions are present. Such may be the case for the interaction between the heliospheric (solar wind) and the interstellar plasmas. Here the relative motion of the heliosphere with respect to the local interstellar medium may produce effects similar to those observed around plasma obstacles in the solar wind.

Current models of the flow conditions at the heliopause /31/ suggest that this boundary should be located at $\gtrsim$ 100 AU, and that the density and magnetic field intensity of the shocked solar wind in its vicinity should be of the order of 10^{-3} cm^{-3} and 10^{-6} G, respectively. Somewhat lower (10^{-7} G) values are expected for the magnetic field intensity in the local interstellar medium, but with a higher 10^{-1} cm^{-3} plasma density. With an expected temperature of the order of 10^4°K the sound velocity in the latter should be comparable to the (~20 km/s) bulk speed with respect to the heliosphere, so that it is possible that supersonic flow conditions may characterize their interaction /32–34/. If this is the case a geometry similar to those described before can also be predicted here since the magnetic field intensity, on either side of the heliopause, is not strong enough to make the flow subalfvenic. In this respect the interaction at that boundary should resemble more the conditions seen at weakly magnetized obstacles in the solar wind (Venus, Mars, comets) than at planetary magnetospheres where the interaction is essentially with a magnetic obstacle. It should be noted, however, that this condition does not necessarily imply that a region of strong magnetic fluxes may not be present in the vicinity of the heliopause. In fact, like the (superalfvenic) solar wind flow around Venus and comets, a region of piled-up magnetic field lines is likely to protect almost entirely the heliospheric obstacle from the interstellar plasma and curtail, to a certain extent, the direct interaction between them. An exception to this situation should occur over the magnetic polar regions of the heliosphere (defined with respect to the orientation of the interstellar magnetic field) where the magnetic field lines graze over the obstacle without accumulating around it.

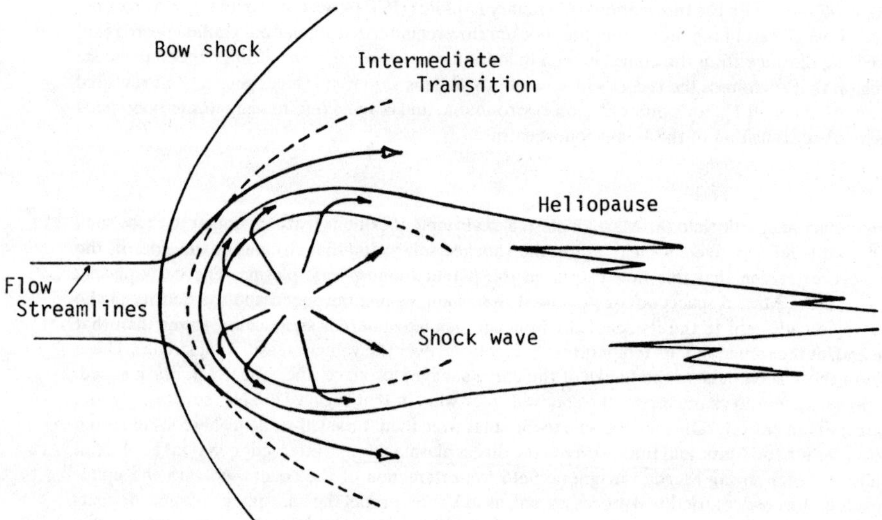

Fig. 3. Schematic diagram of the interstellar flow around the heliosphere. The intermediate transition marks the outer extent of the interstellar post-shock flow region influenced by friction processes at the heliopause. The plasma within that region experiences a local deceleration, thermalization and expansion.

Further complications to the dynamics of the local flow should also arise from the contamination produced by the photo and collisional ionization of the interstellar neutral gas. These processes, as well as charge exchange collisions with the plasma particles, will ultimately be responsible for the onset of turbulent conditions which should increase the efficiency of local transport processes. Thus, we can advance the suggestion that an effective transport of momentum between the interstellar and the heliospheric plasmas may occur across the heliopause, at least in the region around the heliospheric magnetic poles. A viscous flow region similar to that present around the Venus and cometary plasma obstacles can be predicted for the flow external to the heliopause and, depending on the efficiency of the friction process, it is possible that the viscous layer extends forward all the way to the nose region. A schematic view of the proposed flow configuration, projected on the plane perpendicular to that formed by the interstellar magnetic field vector and the (relative) flow velocity vector, is shown in Figure 3 (the draping of the interstellar magnetic field lines occurs along directions perpendicular to the figure). Like in cometary plasma obstacles it is possible that the magnetic pile-up region be strongly asymmetric with respect to magnetic latitude and that near the heliospheric magnetic poles, where the component of the interstellar flow velocity directed against the heliosphere becomes small, the accumulation of magnetic field lines be neglegible /35/. The (dashed) curve drawn between the external bow wave and the heliopause is the postulated intermediate transition which marks the outer extent of the post-shock flow region influenced by friction processes at the heliopause. The effects of such processes should be more noticeable along the heliospheric tail which may even be fragmented into filamentary structures similar to those seen behind the Venus ionosphere /18/.

A viscous interaction at the heliopause has been suggested in the past in studies related to the frontside confinement of the heliosphere /36/. While the process suggested should work in the desired direction, it is worth pointing out a restriction on the value of the effective kinematic viscosity coefficient υ that is implied by the external parameters of the problem. In turbulent flows that coefficient is related to the local velocity fluctuations (δu) through /37,38/:

$$\upsilon \sim (\delta u)^2 / w$$

where w denotes the characteristic frequency of the turbulent motion. By taking δu = 20 km/s as upper limit for the largest possible velocity fluctuations that can be considered within the context of an organized flow, and by adopting $w \sim 10^{-3}$ Hz as suitable for waves in resonance with the gyration motion of interstellar protons in a local 10^{-7} G magnetic field we can conclude that υ is at the most of the order of $10^{15} - 10^{16}$ cm^2/s. These values are somewhat larger than those derived for oxygen ions in the viscous layer around Venus and comets /27/ (a result that reflects the different scale-size of the mean free path of the efective collisions for each case) but are also substantially smaller than that derived in /36/. It is possible, therefore, that the effects of viscosity on the shape of the heliosphere on the frontside region may not be as important as along its flanks.

As it is the case for the interaction of the solar wind with plasma obstacles, further studies are necessary to incorporate viscous processes in the formulation of flow models suitable to the heliopause. This will allow a better examination of the implications that they may have on the in-situ and earth-based observation of the plasma behavior.

Acknowledgements. I wish to thank H. Sauer, M. Ness and V. Hill of the Space Environment Laboratory at NOAA in Boulder, Colorado for technical assistance provided.

REFERENCES

1 – Bridge, H. S, et al., Science, 158, 1669, (1967)
2 – Sheffer, R., et al., J. Geophys. Res. 84, 2109, (1979)
3 – Romanov, S., et al., Cosmic Res., 16, 603, (1979)
4 – Lundin, R., et al., Nature, 341, 609, (1989)
5 – Rosenbauer, H., et al., Nature, 341, 612, (1989)
6 – Bame, S., et al., Science, 232, 356, 1986.
7 – Baker, D., et al., Geophys. Res. Lett., 3, 271, (1986)
8 – Johnstone, A., et al., Nature, 321, 344, (1986)
9 – Reme, H., et al., Astron. Astrophys. 187, 33, (1987)
10 – Intriligator, D. S., Geophys. Res. lett., 9, 727, (1982)
11 – Pérez-de-Tejada, H., et al., Geophys. Res. Lett, 11, 31, (1984)
12 – Pérez-de-Tejada, H., J. Geophys. Res., 92, 4713, (1987)
13 – Goldstein, B., et al., Heidelberg Conf., ESA SP-250, 229, (1986)
14 – Thomsen, M. et al., Astron. Astrophys. 187, 141, (1987)
15 – Perez-de-Tejada H., et al., J. Geophys. Res., 90, 1759, (1985)
16 – Russell, C, in "Venus", Ed. by D. Hunten, U, Arizona Press p. 873 (1983)
17 – Pérez-de-Tejada, H., Science, 207, 981, (1980)
18 – Brace, L. H., et al., J. Geophys. Res., 92, 15, (1987)
19 – Knudsen, W. C., et al., J. Geophys. Res., 85, 7803, (1980)
20 – Pérez-de-Tejada, H., J. Geophys. Res., 91, 6765, (1986)
21 – Mihalov, J., et al., J. Geophys. Res., 85, 7613, (1980)
22 – Wu, C. S., et al., Geophys. Res. Lett., 13, 865, (1986)
23 – Gaffey, J. D., et al., J. Geophys. Res., 93, 5470, (1988)
24 – Scarf, F., et al., J. Geophys. Res., 85, 7599, (1980)
25 – Gringauz, K., et al., Nature, 321, 282, (1986)
26 – Gringauz, K., et al., Geophys. Res. Lett., 13, 613, (1986)
27 – Perez-de-Tejada, H., J. Geophys. Res., 94, 10131, (1989)
28 – Neubauer, F., Heidelberg Conf., ESA SP-250, 35, (1986)
29 – Vaisberg, O., et al., NASA, SP-397, Ed. by F. Ness, p. 21, (1976)
30 – Rosenbauer, H., et al., J. Geophys. Res, 80, 2723, (1975)
31 – Fahr, H., et al., Space Sci. Rev., 43, 329, (1986)
32 – Baranov, V., et al., Astrophys. Space Sci., 41, 491, (1976)
33 – Baranov, V., et al., Astrophys. Space Sci., 66, 441, (1979)
34 – Wallis, M. and M. Dryer, Astrophys. J., 205, 895, (1976)
35 – Pérez-de-Tejada, H., Adv. Space Res., 9, 3(319), (1989)
36 – Grzedzielski, G., et al., Adv. Space Res., 9, 4(247), (1989)
37 – Khorzov, N. et al., et al., Planet. Space Sci., 32, 1169, (1984)
38 – Saffman, P., Proc. Royal Soc., London A317, 417, (1970)

ON USING THE ACR TO PROBE THE LISM/HELIOSPHERE INTERFACE

W.-H. Ip

Max-Planck-Institut für Aeronomie, D-3411 Katlenburg-Lindau, F.R.G.

ABSTRACT

A spherically symmetric model of diffusive shock acceleration is used to derive the energy spectra of several species in the anomalous cosmic rays. It is found that, if the efficiency factor for converting the kinetic energy of the supersonic solar wind to the ACR is on the order of 0.2, the solar wind termination shock would probably locate at $r_s \approx 120$ AU. This estimate is consistent with the location of the heliopause obtained by the argument of pressure balance, but is much larger than the value (55 AU) derived from a consideration of the radial gradients of galactic cosmic rays in interplanetary space.

1. INTRODUCTION

The question of the location of the heliopause perhaps will not be settled until the first crossing of this plasma boundary of our solar system either by the Pioneer 11 or by one of the Voyager spacecraft in the next century. Nevertheless, it is of interest to find ways to approximate the radial distances of the termination shock of the solar wind (r_s) and the heliopause (r_p) between the shocked solar wind flow and the interstellar medium. One way is to compare the total pressure of the galactic cosmic rays (GCR), and the interstellar gas and magnetic field with the solar wind ram pressure such that the point where the two values balance each other be r_s. As discussed by Axford and Ip /1/, the distance of the solar wind termination shock is expected to be $\approx$ 120 -150 AU if the nominal values of the interstellar electron density ($N_e = 0.03 cm^{-3}$) and interstellar magnetic field ($B \approx 3$ microgauss) are used. For $r_s \approx 50$ AU, a much enhanced interstellar electron density ($N_e \approx 0.5$ cm^{-3}) or a larger value of $B(\approx 8$ microgauss) would be necessary.

Webber and Lockwood /2/ used interplanetary radial gradient measurements of galactic cosmic rays to estimate the cosmic ray modulation boundary and found that the r_s could be as small as 55 AU.

In this short note, a simple model for the diffusive shock acceleration of the anomalous cosmic rays (ACR) is used to probe the location of the solar wind termination shock. The basic idea is that, within the context of the shock acceleration theory /3-5/, the energy density of the accelerated particles is a fraction (ξ) of the kinetic energy of the supersonic plasma flow just upstream of the shock transition. Therefore, if the initial energy spectra of the anomalous cosmic rays are known, the efficiency factor, ξ, and the site of the diffusive shock acceleration may be inferred.

2. THE 1-D MODEL CALCULATION

The ACR is most likely the result of ionization of the interstellar neutrals in the heliosphere /6/. After creation, the new- borne ions (i.e., H^+, He^+, N^+, O^+, etc.) will be picked up by the solar wind and convected to the outer heliosphere /7-9/. At the solar wind termination shock, these interstellar ions will supply a suprathermal population as the seed particles for diffusive shock acceleration /10-11/. To simulate this process, we follow the steady-state spherically symmetric model developed by Portgieter and Moraal /11/ to compute the energy spectra of different ACR ions. The basic steps are to solve the cosmic-ray diffusion-convection equation,

$$\frac{\partial}{\partial r}\left[r^2(K\frac{\partial f}{\partial r} - Vf)\right] + \frac{1}{3r^2}\frac{\partial}{\partial r}(r^2V)\frac{\partial f}{\partial lnp} = 0 \quad (1)$$

subject to the boundary condition:

$$\left.\frac{\partial f}{\partial r}\right|_{r=r_s} = \frac{1}{3}\left(\frac{V}{K}\right)_1 \left(1 - \frac{V_2}{V_1}\right)\frac{\partial f}{\partial lnp} \tag{2}$$

where f is the differential number density of the ACR, K the radial diffusion coefficient, V the solar wind expansion velocity, p the particle momentum. The index "1" refers to the quantitites upstream of the termination shock and "2" for quantities downstream. In the computation, $V_1/V_2 = 4$, $K_2 = 0$ and

$$K_1 = K_o \beta K(P) \tag{3}$$

where K(P) = P/(1GV) if P the rigidity > 0.4 GV and K(P) = 0.4 if P < 0.4 GV.

Application of the Crank-Nicholson scheme to Eq.(1) with Eq.(2) as the boundary condition at r_s allows the automatic generation of the energy spectra at the termination shock /11/. Two numerical examples, one for the oxygen ions and the other for the helium ions are illustrated in Fig.1. The energy spectra at smaller radial distances are also shown depicting the effect of solar wind modulation. Observations from Pioneer 10 and Voyager 1 and 2 /12/ are used to constrain the possible combinations of r_s and K_o. The position of the solar wind termination shock adopted in this calculation is r_s = 70 AU.

Note that while the Pioneer 11 data at r = 40 AU and the Voyager 2 data at r = 20 AU can be fitted reasonably well by the theoretical curves, the experimental points from Voyager 1 at 28 AU, however, show large discrepencies. This effect may reflect the importance of latitudinal dependence of the ACR flux - as Voyager 1 was at a much higher latitude ($\theta = 28^o$) than those of Pioneer 10 ($\theta = 3.7^o$) and Voyager 2 ($\theta = 1.3^o$). A two (or three) -dimensional model, with r- and θ- dependences taken into consideration, would give a better approximation /10/. As we shall discuss in the following, the present 1-D spherically symmetric model can nevertheless provide some very interesting insights.

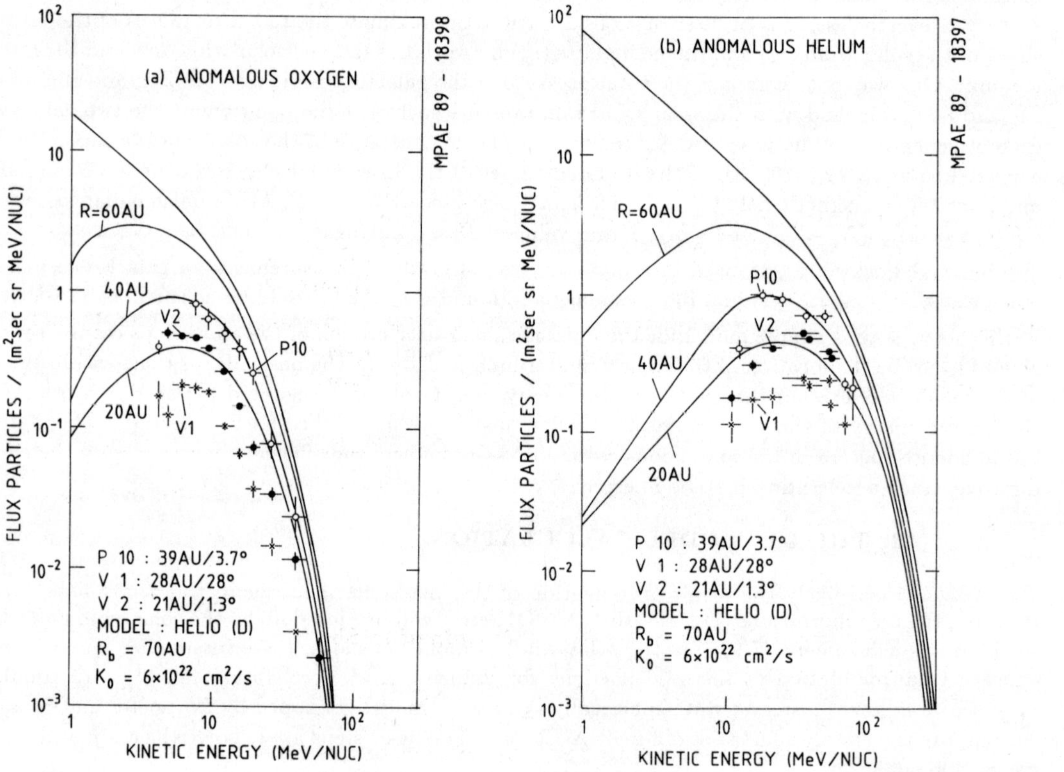

Fig. 1. The theoretical energy spectra of the ACR He^+ and O^+ ions according to the present 1-D spherically symmetric model of diffusive shock acceleration. The observational data from the Pioneer 11 and Voyager spacecraft are used to constrain the values of r_s and K_o. In this particular example, r_s = 70 AU and $K_o = 6 \times 10^{22}$ cm^2 s^{-1}.

3. THE ACR ENERGY SPECTRA

One consequence of the diffusive shock acceleration using the simple expression of the radial diffusion coefficient given above is that the energy spectra of different ions have cutoffs of the total energy (E_{cutoff}) at the same value. This is simply because in the present formulation, the particle energy spectra can be described by a power law for the rigidity P $< P_{cutoff}$. For $P >> P_{cutoff}$,

$$V_1 r_s / K_1 << 1 \tag{4}$$

and the shock transition can no longer be approximated as a planar structure. This condition, in turn, will determine the upper limit of the rigidity and total energy for efficient shock acceleration. For $P > P_{cutoff}$, all ACR energy spectra would follow an exponential fall-off.

This means, except for the normalization factors, the energy spectra for all species as a function of energy per nucleon could be obtained by appropriately displacing the ACR He^+ (or O^+) energy spectrum in Fig. 1. For instance, if the number density of the proton component in the ACR is a factor of 20 larger than that of the He^+ ions; see /13/, the energy spectrum of the ACR protons can be derived by shifting the He^+ energy spectrum as shown in Fig. 2.

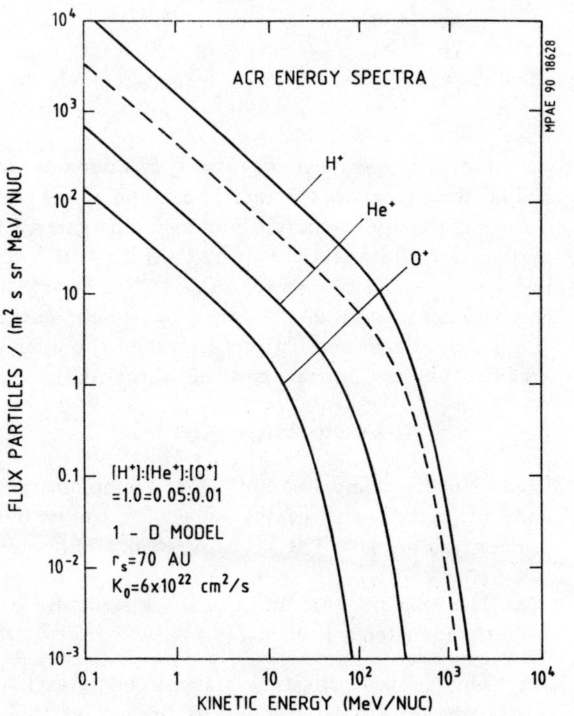

Fig. 2. A comparison of the energy spectra of the ACR H^+, He^+ and O^+ ions at the solar wind termination shock. The He^+ and O^+ abundances relative to H^+ are assumed to follow the relation: $[H^+]:[He^+]:[O^+] = 1.0:0.05:0.01$. The dashed curve is for the ACR H^+ in case $[H^+]:[He^+] = 1.0:0.17$.

4. Location of the Solar Wind Termination Shock

With the energy spectra of the most abundant ACR ions (i.e., H^+ and He^+) in hand, the total energy density of the anomalous cosmic rays (ϵ_{ACR}) at $r \approx r_s$ can be estimated. Note that our procedure of computing the energy densities from individual species is defined as:

$$\epsilon_i = A_i \int_{1keV}^{\infty} \frac{f(E_n) E_n dE_n}{V_n} \tag{5}$$

where A_i is the atomic mass of the ith species. E_n the energy per nucleon and V_n the particle velocity. In such a scheme, we have $\epsilon(H^+) : \epsilon(He^+) : \epsilon(O^+) = 1.0:0.12:0.06$, with $\epsilon(H^+) \approx 0.07$ eV cm^{-3}.

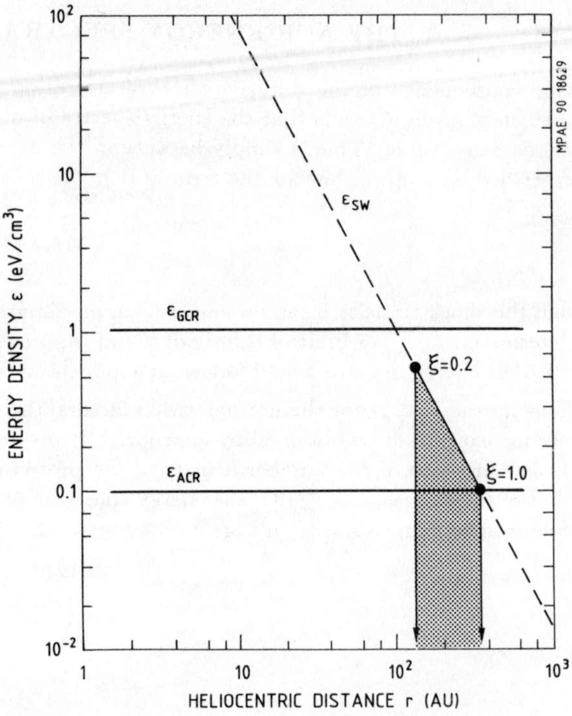

Fig. 3. A comparison of the total energy density of the ACR with that of the solar wind flow at different radial distances. The energy density of the galactic cosmic rays is also shown as ϵ_{GCR}.

We find that the value of $\epsilon_{ACR} (= \sum \epsilon_i)$ remains nearly the same if we increase r_s from 70 AU to 150 AU. In Fig. 3 we compare the radial variation of the solar wind kinetic energy density , ϵ_{sw}, and the energy density of the interstellar medium including galactic cosmic rays and interstellar magnetic field. For $\xi = 1$, we find $r_s \approx 400$ AU; and $r_s \approx 120$ AU if $\xi \approx 0.2$. The first value, requiring 100% efficiency for extracting the solar wind kinetic energy to the ACR is clearly too large. The latter value of $r_s \approx 120$ AU implies that only a small fraction of the solar wind kinetic energy is available for ACR acceleration. A modest value of ξ is indeed expected if the structure of the solar wind termination shock is significantly modified (e.g., smoothed) by the pressure gradient of the ACR /5/.

5. Discussion

Because of the many assumptions and simplifications adopted in the present calculation, it is likely that different (and more accurate) values of r_s will be obtained by other model calculations. In any event, the present result of $r_s \approx 120$ AU with $\xi \approx 0.2$ points to the following interesting aspects.

(a) The solar wind termination shock structure could be modified by the diffusive shock acceleration of the interstellar neutrals/14,15/.

(b) The characteristic time scale of charge exchange between the interstellar ions and the interstellar hydrogen atoms is about 10^8 sec. This time scale is comparable to the solar wind transit time of 3.2×10^7 sec. Charge exchange loss of the interstellar pickup ions during their journey to the termination shock therefore could be significant. The possible loss of the accelerated ions via charge exchange effect - during the diffusive shock acceleration with a time scale of about one year /11/ - is also an interesting issue to be investigated.

(c) The abundance ratio of hydrogen to oxygen in the ACR is $H/O < 16$ and the relative abundances of ACR N, O, and Ne are in agreeement with the solar system values /13/. The relatively large amount of ACR oxygen component came as a surprise because a fraction of the interstellar oxygen atoms are expected to be screened from the heliosphere as a result of charge exchange process in the subsonic interstellar plasma flow /16,17/.

Acknowledgment:

I thank Josef Kota, Bruce McKibbin and Romana Ratkiewicz for useful information and discussions on this problem.

References

1. W.I. Axford and W.-H. Ip, Adv. Space Res., 6, No.2, 27 (1986).
2. W.R. Webber and J.A. Lockwood, Astrophys. J., 317, 534 (1987).
3. L. O'C. Drury and H.J. Völk, Astrophys. J., 248, 344 (1981).
4. D. Eichler, Astrophys. J., 229, 419 (1979).
5. D.C. Ellison, Monte Carlo Simulation of Collisionless Shock Acceleration, Ph.D. Thesis, Univ. of Maryland, 1981.
6. L.A. Fisk, B. Kozlovsky and R. Ramaty, Astrophys. J. Lett., 190, L35 (1974).
7. W.I. Axford, in Solar Wind, Eds. C.P. Sonett, P.J. Coleman, and J.M. Wilcox, NASA SP-308, 609 (1972).
8. H.J. Fahr, Space Science Rev., 15, 483 (1974).
9. T.E. Holzer, Ann. Rev. Astron. Astrophys.,27, 194 (1989).
10. J.R. Jokipii, J. Geophys. Res., 91, 2929 (1986).
11. M.S. Potgieter and H. Moraal, Proc. 20th Intern. Cosmic Ray Conf., Moscow, Vol. 3 (SH), 299 (1987).
12. A.C. Cummings and E.C. Stone, Proc. 20th Intern. Cosmic Ray Conf., Moscow, Vol. 3 (SH), 413 (1987).
13. A.C. Cummings, E.C. Stone, F.B. McDonald, W.R. Webber, and N. Lal, Proc. 20th Intern. Cosmic Ray Conf., Moscow, Vol.3 (SH), 417 (1987).
14. M.A. Lee, L.A. Fisk and W.I. Axford, Proc. 20th Intern. Cosmic Ray Conf., Moscow, Vol.2, 187 (1987).
15. C. M. Ko and G. M. Webb, Astrophys. J., 323, 657 (1987).
16. Fahr, H.J. and H.W. Ripken, Astron. Astrophys., 139, 551 (1984).
17. S. Bleszynski, Astron. Astrophys., 180, 201 (1987).

MHD SOLAR WIND–INTERSTELLAR PLASMA INTERACTION: 3D FORMULATION BY THE PROJECTED CHARACTERISTICS METHOD AND THE STABILITY ANALYSIS

R. Ratkiewicz

Space Research Centre, Polish Academy of Sciences, Ordona 21, 01–237 Warsaw, Poland

ABSTRACT

The method of projected characteristics is developed for solving the nonlinear MHD solar wind – interstellar gas interaction problem in three-dimensional Cartesian geometry. For practical application numerical solutions are to be obtained on the basis of that formulation. The von Neumann stability analysis for a linearized and simplified version of the numerical procedure is performed. In the case discussed here the time step required to reach stability is the same as required by the Courant-Friedrichs-Levy criterion.

INTRODUCTION

One of the essential problems of heliospheric physics is the determination of the shape of the interface caused by the interaction between the solar wind and the local interstellar medium. This problem has been widely discussed in the recent years by many authors /1-5,8-10/.

Since the heliosphere is created by the interaction between two plasmas with frozen-in magnetic fields, for the description of such a flow field it is necessary to solve the fully nonlinear magnetohydrodynamic (MHD) equations. The system of MHD equations (basing on the conservation laws of mass, momentum and energy and supplemented by the induction equation for the magnetic field) is hyperbolic /12/ hence, the solution can be found by the method of characteristics /7/. This method, however, when directly applied to the problem, leads to fairly cumbersome numerical procedures due to the nonisotropic nature of the propagation of the MHD waves /13/.

This difficulty can be overcome by the use of a new method of analysis of the MHD initial-boundary value problem presented by Nakagawa in a series of papers /12-14/.The method of projected characteristics is proposed to simplify the procedure by projecting the real characteristics on planes containing one spatial coordinate and parallel to the time axis.

BASIC EQUATIONS

For simplicity (without losing any essence of this method) a set of ideal MHD equations is used (i.e., all dissipation terms are ignored),defined in a system of Cartesian coordinates (x,y,z) in the following matrix form:

$$\frac{\partial \mathbf{W}}{\partial t} = -\mathbb{A}\,\frac{\partial \mathbf{W}}{\partial x} - \mathbb{B}\,\frac{\partial \mathbf{W}}{\partial y} - \mathbb{C}\,\frac{\partial \mathbf{W}}{\partial z} \qquad (1)$$

where $\mathbf{W}$ is the column vector of the MHD variables, $\mathbb{A}$, $\mathbb{B}$, $\mathbb{C}$ are 8×8 matrices given below, and in their elements ρ is the density, p is the pressure, a is the adiabatic speed of sound, u,v,w are the components of the velocity $\mathbf{V}$ and B_x, B_y, B_z are the components of the magnetic field $\mathbf{B}$ in the x,y,z directions, respectively.

$$\mathbf{W} = \begin{vmatrix} \rho \\ u \\ v \\ w \\ p \\ B_x \\ B_y \\ B_z \end{vmatrix} \qquad \mathbb{A} = \begin{vmatrix} u & \rho & 0 & 0 & 0 & 0 & 0 & 0 \\ 0 & u & 0 & 0 & 1/\rho & 0 & \frac{B_y}{\rho} & \frac{B_z}{\rho} \\ 0 & 0 & u & 0 & 0 & 0 & \frac{-B_x}{\rho} & 0 \\ 0 & 0 & 0 & u & 0 & 0 & 0 & \frac{-B_x}{\rho} \\ 0 & a^2\rho & 0 & 0 & u & 0 & 0 & 0 \\ 0 & 0 & 0 & 0 & 0 & u & 0 & 0 \\ 0 & B_y & -B_x & 0 & 0 & 0 & u & 0 \\ 0 & B_z & 0 & -B_x & 0 & 0 & 0 & u \end{vmatrix}$$

$$\mathbb{B} = \begin{vmatrix} v & 0 & \rho & 0 & 0 & 0 & 0 & 0 \\ 0 & v & 0 & 0 & 0 & \frac{-B_y}{\rho} & 0 & 0 \\ 0 & 0 & v & 0 & 1\backslash\rho & \frac{B_x}{\rho} & 0 & \frac{B_z}{\rho} \\ 0 & 0 & 0 & v & 0 & 0 & 0 & \frac{-B_y}{\rho} \\ 0 & 0 & a^2\rho & 0 & v & 0 & 0 & 0 \\ 0 & -B_y & B_x & 0 & 0 & v & 0 & 0 \\ 0 & 0 & 0 & 0 & 0 & 0 & v & 0 \\ 0 & 0 & B_z & -B_y & 0 & 0 & 0 & v \end{vmatrix} \qquad \mathbb{C} = \begin{vmatrix} w & 0 & 0 & \rho & 0 & 0 & 0 & 0 \\ 0 & w & 0 & 0 & 0 & \frac{-B_z}{\rho} & 0 & 0 \\ 0 & 0 & w & 0 & 0 & 0 & \frac{-B_z}{\rho} & 0 \\ 0 & 0 & 0 & w & 1\backslash\rho & \frac{B_x}{\rho} & \frac{B_y}{\rho} & 0 \\ 0 & 0 & 0 & a^2\rho & w & 0 & 0 & 0 \\ 0 & -B_z & 0 & B_x & 0 & w & 0 & 0 \\ 0 & 0 & -B_z & B_y & 0 & 0 & w & 0 \\ 0 & 0 & 0 & 0 & 0 & 0 & 0 & w \end{vmatrix}$$

METHOD OF SOLUTION

It is known /7/ that in the method of characteristics, the compatibility equation is derived along a selected characteristic. Following Nakagawa /13/ in the present study, this equation is replaced by the compatibility equation along each projected characteristic in the (x,t), (y,t), (z,t) planes.

For eight physical variables the eight directions of projected characteristics are obtained as the eigenvalues of matrices $\mathbb{A}$, $\mathbb{B}$, $\mathbb{C}$ in each of the (x,t), (y,t), (z,t) planes. Then the eight independent left-eigenvectors associated with these eigenvalues are found. Next, the compatibility equations are obtained in each plane by simple multiplications of the left-eigenvector to equation (1) i.e., eight equations for the determination of the variations of the eight physical variables in each of the (x,t), (y,t), (z,t) planes. In order to achieve the final determination of physical variables the average of variations obtained in each projected plane is taken.

On the basis of the present formulation the final equation in the following difference form /13/ is found:

$$\Delta_t \mathbf{W} = -\mathbb{A}(\Delta_x \mathbf{W}) - \mathbb{B}(\Delta_y \mathbf{W}) - \mathbb{C}(\Delta_z \mathbf{W}) + \mathbb{D}(\delta_x \mathbf{W}) + \mathbb{E}(\delta_y \mathbf{W}) + \mathbb{F}(\delta_z \mathbf{W}) \tag{2}$$

where at the time step $p\Delta t$ and the spatial grid point (l,m,n) (where l refers to x, m to y, n to z coordinate, respectively)

$$\Delta_t W^p_{l,m,n} = \frac{W^{p+1}_{l,m,n} - W^{*p}_{l,m,n}}{\Delta t},$$

$$\Delta_x W^p_{l,m,n} = \frac{W^p_{l+1,m,n} - W^p_{l-1,m,n}}{2\Delta x}, \quad \text{etc.}$$

$$\delta_x W^p_{l,m,n} = \frac{W^p_{l+1,m,n} - 2W^p_{l,m,n} + W^p_{l-1,m,n}}{2\Delta x}, \quad \text{etc.}$$

$$W^{*p}_{l,m,n} = \frac{1}{6}(W^p_{l+1,m,n} + W^p_{l-1,m,n} + W^p_{l,m+1,n} + W^p_{l,m-1,n} + W^p_{l,m,n+1} + W^p_{l,m,n-1})$$

The terms associated with $(\Delta_x W)$, $(\Delta_y W)$ and $(\Delta_z W)$ reproduce the original equation. The elements of the matrices $\mathbb{D}$, $\mathbb{E}$, $\mathbb{F}$ are obtained according to the rules described by Nakagawa /12/.

Here only the matrix $\mathbb{D}$ is presented as the example; matrices $\mathbb{E}$, $\mathbb{F}$ are of the similar form.

The non-zero elements (which should be multiplied by $\frac{1}{3}$) of the matrix $\mathbb{D}$ are:

$D_{12}=(\rho(a+U_a)u)/(aU_{fs})$
$D_{13}=(B_xB_yu)/(aU_aU_{fs})$
$D_{14}=(B_xB_zu)/(aU_aU_{fs})$
$D_{15}=(a+U_a)/(aU_{fs})$
$D_{17}=B_y/U_{fs}$
$D_{18}=B_z/U_{fs}$

$D_{22}=(\rho a(a+U_a)+B_{yz})/(\rho U_{fs})$
$D_{23}=(-B_xB_y)/(\rho U_{fs})$
$D_{24}=(-B_xB_z)/(\rho U_{fs})$
$D_{25}=(u(a+U_a))/(\rho aU_{fs})$
$D_{27}=(B_yu)/(\rho U_{fs})$
$D_{28}=(B_zu)/(\rho U_{fs})$

$D_{32}=D_{23}$
$D_{33}=U_a(B_y^2(a+U_a)+B_z^2U_{fs})/(B_{yz}U_{fs})$
$D_{34}=(-B_yB_zU_a(U_{fs}-U_a-a))/(B_{yz}U_{fs})$
$D_{35}=B_xB_yu/\rho^2aU_aU_{fs}$
$D_{37}=(-B_xu/(\rho U_a^2))D_{33}$
$D_{38}=(-B_xu/(\rho U_a^2))D_{34}$

$D_{42}=D_{24}$
$D_{43}=D_{34}$
$D_{44}=U_a(B_y^2U_{fs}+B_z^2(a+U_a))/(B_{yz}U_{fs})$
$D_{45}=B_xB_zu/\rho^2aU_aU_{fs}$
$D_{47}=D_{38}$
$D_{48}=(-B_xu/(\rho U_a^2))D_{44}$

$D_{52}=\rho^2a^2D_{25}$
$D_{53}=\rho^2a^2D_{35}$
$D_{54}=\rho^2a^2D_{45}$
$D_{55}=a(a+Ua)/U_{fs}$
$D_{57}=a^2B_y/U_{fs}$
$D_{58}=a^2B_z/U_{fs}$

$D_{72}=\rho D_{27}$
$D_{73}=\rho D_{37}$
$D_{74}=\rho D_{47}$
$D_{75}=D_{57}/\rho a^2$
$D_{77}=(B_y^2(aU_a+b^2)+B_z^2U_{fs}U_a)/(B_{yz}U_{fs})$
$D_{78}=B_yB_z(aU_a+b^2-U_aU_{fs})/(B_{yz}U_{fs})$

$D_{82}=\rho D_{28}$
$D_{83}=\rho D_{38}$
$D_{84}=\rho D_{48}$

$D_{85}=D_{58}/\rho a^2$
$D_{87}=D_{78}$
$D_{88}=(B_y^2U_aU_{fs}+B_z^2(aU_a+b^2))/(B_{yz}U_{fs})$

where

$b^2 = B_x^2/\rho+B_y^2/\rho+B_z^2/\rho,\qquad U_a=|b_x|,\qquad b_x=B_x/\sqrt{\rho},$

$U_s^2 = \frac{1}{2}\{a^2+b^2-[(a^2+b^2)^2-4a^2b_x^2]^{1/2}\},$

$U_f^2 = \frac{1}{2}\{a^2+b^2+[(a^2+b^2)^2-4a^2b_x^2]^{1/2}\},$

$U_{fs} = U_f+U_s,\qquad B_{yz}=B_y^2+B_z^2$

STABILITY ANALYSIS

For the stability analysis the equation (1) is linearized i.e., the matrices $\mathbb{A}$, $\mathbb{B}$, $\mathbb{C}$ are regarded as locally constant. To proceed with a von Neumann analysis /6,11,15/ a Fourier component of any dependent variable is written as:

$$\mathbb{W}(t)\ \exp[i(k_1x+k_2y+k_3z)]$$

The substitution of this expression into the linearized numerical scheme (2) yields the amplification matrix of the form:

$$\mathbb{G} = \frac{1}{3}\,\mathbb{I}\,(\cos\xi + \cos\eta + \cos\zeta) - \frac{\Delta t}{\Delta x}\, i\,(\mathbb{A}\sin\xi + \mathbb{B}\sin\eta + \mathbb{C}\sin\zeta) + \frac{\Delta t}{\Delta x}\,[\mathbb{D}\,(\cos\xi-1) + \mathbb{E}\,(\cos\eta-1) + \mathbb{F}\,(\cos\zeta-1)]$$

where $\mathbb{I}$ is the unit matrix, $\xi=k_1\Delta x$, $\eta=k_2\Delta y$, $\zeta=k_3\Delta z$ and equal spacing is assumed ($\Delta x=\Delta y=\Delta z$), $i=\sqrt{-1}$.

Stability is assured if it can be shown that all of the eigenvalues λ_j, of $\mathbb{G}$ satisfy the von Neumann condition $|\lambda_j|\le 1 + O(\Delta t)$.

It is very difficult to obtain the eigenvalues of matrix $\mathbb{G}$ in its general form. Therefore the consideration is limited to the matrix

$$\mathbb{G}' = \frac{1}{3}\,\mathbb{I}\,(\cos\xi + \cos\eta + \cos\zeta) - \frac{\Delta t}{\Delta x}\, i\,(\mathbb{A}\sin\xi + \mathbb{B}\sin\eta + \mathbb{C}\sin\zeta)$$

under the assumption that the rejected terms can only improve the stability. Such an assumption is justified by the fact that according to Nakagawa /13/ the contribution of terms associated with the matrices $\mathbb{D}$, $\mathbb{E}$, $\mathbb{F}$ is similar to a contribution from artificial viscosity. Thus, it can be expected that the inclusion of these terms in the analysis would only enhance stability /6,11/.

After some algebraic manipulations the following eigenvalues of the matrix $\mathbb{G}'$ are found:

$$g' = \frac{1}{3}(\cos\xi + \cos\eta + \cos\zeta) - \frac{\Delta t}{\Delta x}\, i\sqrt{\sin^2\xi+\sin^2\eta+\sin^2\zeta}\begin{Bmatrix} u' \\ u' \\ u'-U_a \\ u'+U_a \\ u'-U_s \\ u'+U_s \\ u'-U_f \\ u'+U_f \end{Bmatrix}$$

where $u' = u\cos\alpha + v\cos\beta + w\cos\gamma$ is the component of the velocity $\mathbf{V}$ in the direction defined by the direction cosines:

$$\cos\alpha=\frac{\sin\xi}{M},\quad \cos\beta=\frac{\sin\eta}{M},\quad \cos\gamma=\frac{\sin\zeta}{M},\quad M=\sqrt{\sin^2\xi+\sin^2\eta+\sin^2\zeta}$$

analogously $B'=-B_x/\rho\cos\alpha-B_y/\rho\cos\beta-B_z/\rho\cos\gamma$,

and now $U_a=|B'|\sqrt{\rho}$,

$$U_s^2=\frac{1}{2}\{a^2+b^2-[(a^2+b^2)^2-4a^2\rho B'^2]^{1/2}\},\qquad U_f^2=\frac{1}{2}\{a^2+b^2+[(a^2+b^2)^2-4a^2\rho B'^2]^{1/2}\}$$

In order to meet the von Neumann condition, the inequality

$$|g'|^2= \frac{1}{9}(\cos\xi + \cos\eta + \cos\zeta)^2 + \left(\frac{\Delta t}{\Delta x}\right)^2(\sin^2\xi+\sin^2\eta+\sin^2\zeta)\begin{Bmatrix} u' \\ u' \\ u'-U_a \\ u'+U_a \\ u'-U_s \\ u'+U_s \\ u'-U_f \\ u'+U_f \end{Bmatrix}^2 \le 1$$

must be true for all values of ξ, η, and ζ.
The largest value of the quantities in the curly bracket is $|\mathbf{V}|+|U_f|$. Hence $|g'|^2 \le 1$ if

$$\frac{\Delta t}{\Delta x} \le \sqrt{\frac{1-1/9\,(\cos\xi+\cos\eta+\cos\zeta)^2}{\sin^2\xi+\sin^2\eta+\sin^2\zeta}}\;\frac{1}{|\mathbf{V}|+|U_f|} \tag{3}$$

It is easy to verify that the minimal value of the right-hand side of the inequality (3) is $1/[\sqrt{3}\,(|\mathbf{V}|+|U_f|)]$. Finally, the condition (3) can be written as:

$$\Delta t \le \frac{\Delta x}{\sqrt{3}\,(|\mathbf{V}|+|U_f|)}$$

expressing the Courant-Friedrichs-Levy requirement for the time step in the three-dimensional case.

Acknowledgements. *R.R. is grateful for the hospitality extended to her by Professor W.I. Axford, Dr.W.-H. Ip and his colleagues during her stay at the Max Planck Institut für Aeronomie, Lindau, during which this work was initiated. This work has been largely supported by an MPAE grant.*

REFERENCES

1. W.I. Axford, Solar Wind, NASA SP-308, 609-658 (1972)
2. W.I. Axford, Solar Phys. 100, 575-586 (1985)
3. W.I. Axford and W.-H. Ip, Adv.Space Res. 6, No. 2, 27-32 (1986)
4. M. Banaszkiewicz and R. Ratkiewicz, Adv.Space Res. 9,No.4, 251-254 (1989)
5. V.B. Baranov, M.G. Lebedev and M.S. Ruderman, Astrophys.Space Sci. 66, 441-452 (1979)
6. S.Z. Burstein, J.Comp.Phys. 2, 198-222 (1967)
7. R. Courant and D. Hilbert, Method of Mathematical Physics Vol.2, Interscience Publishers, 1962
8. H.J. Fahr and W. Neutsch, Astron.Astrophys. 118, 57-65 (1982)
9. H.J. Fahr, W. Neutsch, S. Grzedzielski, W. Macek and R. Ratkiewicz-Landowska, Space Sci.Rev. 43, 329-381 (1986)
10. H.J. Fahr, S. Grzedzielski and R. Ratkiewicz, Annales Geophys. 6,(4), 337-354 (1988)
11. R.W. MacCormack, Proceedings of the Second International Conference on Numerical Methods in Fluid Dynamics, Springer-Verlag 1971, pp. 151-163
12. Y. Nakagawa, Ap. J. 240, 275-299 (1980)
13. Y. Nakagawa, Ap. J. 247, 707-718 (1981)
14. Y. Nakagawa, Ap. J. 247, 719-733 (1981)
15. Y.W. Shin and C.A. Kot, J.Comp.Phys. 28, 211-231 (1978)

THE INTERACTIONS OF THE SOLAR WIND DISCONTINUITIES IN THE VICINITY OF THE HELIOPAUSE

S.A. Grib,* E.A. Pushkar** and A.A. Barmin***

**LOIZMIRAN, 23, Line 2, V.O. Leningrad 199053, U.S.S.R.*
***MASI (VTUZ-ZIL), Avtozavodskaya 16, Moscow 109068, U.S.S.R.*
****Institute of Mechanics, Moscow University, Michurinsky Pr. 1, Moscow 117192, U.S.S.R.*

ABSTRACT

The normal and oblique interactions of different solar wind discontinuities such as double shocks and tangential discontinuities with the heliospheric shock, heliopause and hypothetical interstellar shock are considered. The theory recently developed for the terrestrial magnetosphere is generalized for this case. The probable generation of secondary fast rarefaction waves and slow shock waves is indicated. It is shown that for the anisotropic gas pressure the proton concentration increases may differ substantially from the magnetic field intensity rise across the oblique terminal shock wave front. The significance of the secondary waves in energy transfer and their effect on the gradient regions of the heliosphere is emphasized. A series of exact solutions is obtained. It makes possible to evaluate the effect of the magnetic field orientation on the discontinuities parameters and on the qualitative picture of the emerging flow in the vicinity of the heliopause.

INTRODUCTION

Though the MHD-approach to the solar wind heliospheric events is limited the continiuum magnetohydrodynamics has been successfully applied to the contemporary field of heliospheric space plasma research in evaluating the macroscopic picture of the solar wind phenomena /1/. A lot of pertubations of the solar wind flow may be considered as the MHD- or quasi-MHD (including pressure anisotropy) discontinuities. The solar wind discontinuities accompanying coronal mass ejections and solar flares may carry a special kind of information from the vicinity of the Sun to the boundary of heliopause. The problems of the solar wind discontinuities propagating and interacting with the gradient boundaries was studies in /2 - 5/. The interaction of solar wind shock with the bow shock - magnetopause system was considered /3,4/. It was shown that the secondary waves appear in the magnetosheath before the terrestrial magnetosphere and this result (especially the rarefaction waves generation) was confirmed by the spacecraft data /4/. The effect of oblique interplanetary magnetic field (IMF) on the solar wind shock wave interaction with the Earth's magnetosphere was considered in /6/. The problem of the shock interactions inside the free solar wind flow was also studied in ideal MHD-approximation /7/ and there was emphasized that we may have the combination $\underleftarrow{R}_+\underleftarrow{S}_-C\underrightarrow{R}_-\underrightarrow{S}_+$ (R_+, R_- - fast and slow rarefaction waves, S_+, S_- - fast and slow shocks, C - contact discontinuity) in the result of $\underrightarrow{S}_+\ \underrightarrow{S}_+$ interaction. In this case the profile of the IMF change will have the rise-type form. In /8/ more general problem (cf. with /3/) was considered for the regular interactions of oblique shock waves on the basis of Barmin-Pushkar method /9/. The significant feature of the solar wind plasma - the interplanetary current sheet - nowadays is dealt by many authors and it is considered either as the tangential /5/ or the rotational /10/ discontinuity. Here we consider the problem of normal and oblique solar wind discontinuities interactions in the free flow of the solar wind and in the vicinity of the heliopause on the basis of procedures described in /3, 6, 8, 9/.

THE ANALYSIS

Usually the interplanetary shock waves have Mach number M_S = 2-3 and Alfvenic-Mach numbers M_A = 1.3-6 with the density ratio 1.5-3.9 /11/. At first we try to study the problem of the interaction between the solar wind rotational A , and contact C discontinuities (which might become unstable).

Then we shall deal with the problem of solar wind shock wave interaction (S_+) with the heliopause and 'bow' shocks (S_{+1}, S_{+2}) around it (Figure 1).

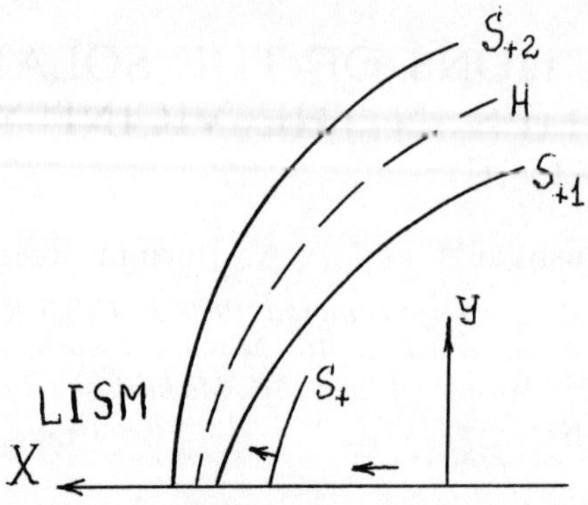

Fig. 1. A schematic view of the vicinity of the heliopause H with two 'bow' shock waves S_{+1}, S_{+2} and a nonstationary solar wind shock S_+.

For the vicinity of the heliopause it's possible to use the parameters of the solar wind and the interstellar flow (LISM) approximately characterizing this region /12/. At the beginning we use the plasma parameter $\beta = 2p/B^2 \gg 1$ for the free solar wind coming to the terminal shock, $\beta \sim 1$ in the heliosheath, and $\beta \lesssim 1$ – for the interstellar flow before the heliopause. It's known that 1/3 – 1/4 flare shock waves may come to the distant heliosphere /13/ and the recurrent solar wind structures and double shock waves may appear also there /14/. For the local interstellar medium we take (just for an example) the parameters /15/: concentration $n_H = 0.07\ cm^{-3}$, $n_{He} = 0.01\ cm^{-3}$, $n_p \approx n_e = 0.03\ cm^{-3}$, the velocity $V^\infty = 22.5\ kms^{-1}$ and for the magnetic field intensity $H = 2 \cdot 10^{-10}$ G. For the solar wind we take the typical values as in /16/.

Let's deal with the problem of an oblique interaction of the strong magnetohydrodynamic discontinuity (fast, slow shocks $S_{+,-}$ or rotational discontinuity A) with the contact surface (here contact discontinuity) C or a shock. A lot of inhomogeneities in the free flow of the solar wind may contain contact surfaces (unstable contact discontinuities or the tangential discontinuities), rotational and shock discontinuities. For the discontinuity C we have the invariance of all values except the density. Its magnitude may be characterized by the ratio of densities. Suppose also that there is no flow before the coming discontinuity and the velocity of the one is given.

For making the problem stationary we'll use the reference system XOY connected with the point of intersection. At XOY the falling discontinuity is at rest, the problem is selfsimilar. So we may find the solution as a combination of the selfsimilar stationary solutions diverging from the point O. The combination contains the oblique shock waves S_+ and S_- (or the simple rarefaction waves R_+ and R_-) and the Alfvenic (rotational) discontinuities A. There is also the contact discontinuity C between the two ensembles of the fast, Alfvenic and slow waves moving one after another. The shock waves S_+, S_- are characterized by the angle $\varphi_\pm$ to the velocity, and the simple rarefaction waves R_+, R_- by the angle $\varphi_\pm$ for the last characteristic in rarefaction sector. There is the continuous dependence of each wave (shock or rarefaction) on the parameter $\varphi_\pm$. The angles, determining the intensities of fast and slow waves may be varied. The oblique shock wave may be characterized by the 'generalized shock's adiabate'/17/.The rarefaction waves are defined by the system of the ordinary differential equations /18/. At the final contact discontinuity C the pressure, velocity and magnetic field must be continuous so the boundary conditions for C have the form of four equations for four unknown angles $\varphi_\pm^{1,2}$. The upper indexes (1), (2) refer to the flow above and under the C. We must add that for each system of reflected or refracted waves the primal flow must be hyperbolic for the existence of the selfsimilar solution.

To solve this problem in the frame of the selfsimilar consideration means to define the types of the appearing waves satisfying the boundary conditions on the contact surface and at the same time to find their intensities. The procedure of the calculations is organized in such a way that the qualitative aspect of the solution is defined in the process of resolution. We use the method of generalized polars /18/ and the iterative method.

Dealing with the problem of the interaction between one fast wave S^*_{+1} running through the solar wind flow with another S^*_{+2} running before it (Figure 2), we must have the initial value of the plasma parameter β, the orientation of the shock wave front in reference to the magnetic field, the angle between S^*_{+1} and S^*_{+2}, the Mach numbers M_1 and M_2 for the shocks.

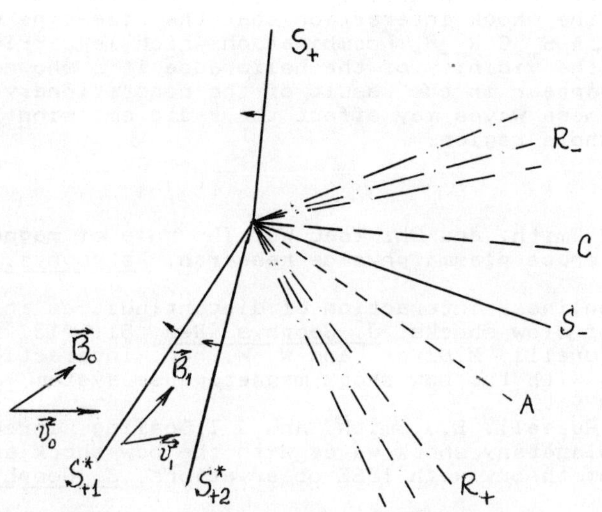

Fig. 2. The schematic view of the nonstationary shock waves S^*_{+1} S^*_{+2} interaction.

It's useful to note that there is some anisotropy of pressure in reference to the interplanetary magnetic field in the solar wind flow ($p_\perp = p_\parallel$) and so it's possible to take for the plasma parameter β the value /19/: $\beta = 2\mu(2p_\perp + p_\parallel)/3H^2$ which is close to unit at 1 AU. In that case the shock wave may change significantly the intensity of the magnetic field without significant change of the proton concentration /20/.

THE MAIN RESULTS

1. Dealing with the rotational (A) and contact (C) discontinuities interaction in the solar wind flow we obtained the result /9, 21/: $\underrightarrow{A}C \rightarrow \underleftarrow{R}_+AS_-C'S_-A\underrightarrow{R}_+$ for $\psi = 30^\circ$, $\varphi = 175^\circ$, $\beta \sim 1$, with the concentration change or the primal contact discontinuity $n/n_0 < 1$. For $\psi = 45^\circ$ we have two final shock waves: $\underrightarrow{A}C \rightarrow \underleftarrow{S}_+R_-C'R_-A\underrightarrow{S}_+$ and it's found here the existence of the peculiar interval for the catastrophic nonlinear reconctruction for $\varphi = 165^\circ - 175^\circ$: $AC \rightarrow \underleftarrow{R}_+\underleftarrow{S}_-C'\underrightarrow{S}_-\underrightarrow{A}R_+ \rightarrow \underleftarrow{R}_+\underleftarrow{A}S_-C'\underrightarrow{S}_-\underrightarrow{A}R_+ \rightarrow \underleftarrow{R}_+\underleftarrow{A}R_-C'\underrightarrow{S}_-\underrightarrow{A}R_+ \rightarrow \underleftarrow{S}_+\underleftarrow{R}_-C'\underrightarrow{R}_-\underrightarrow{A}S_+$, $\psi = 60^\circ$, $n/n_0 \approx 10$.

2. For shock interaction in the free solar wind we have /8/(Figure 2): $\underrightarrow{S}_+\underrightarrow{S}_+ \rightarrow \rightarrow \underleftarrow{R}_+\underleftarrow{A}S_-CR_-\underrightarrow{S}_+$. Thus one fast shock wave has disappeared. For perpendicular case of normal collision of the shock waves we have /3/: $S_+S_+ \rightarrow \underleftarrow{R}_+\underleftarrow{A}S_-\underrightarrow{C}R_-\underrightarrow{S}_+$.

3. The result of the interaction between solar wind shock waves and the terminal shock wave S_{+1} for $\varphi = 175^\circ$, $\psi = 15^\circ$ and $\beta \sim 1$ may be illustrated thus: $\underrightarrow{S}_+ \underleftarrow{S}_{+1} \rightarrow \underleftarrow{S}'_{+1}R_-CR_-\underrightarrow{A}S'_+$, where S'_{+1} is a modified terminal shock wave, S'_+ – shock wave refracted in the heliosheath. If the magnetic field is stronger (e.g. inside the magnetic bottle) for $\beta = 0.6$ there is no refracted shock wave at all: $\underrightarrow{S}_+\underleftarrow{S}_{+1} \rightarrow \underleftarrow{S}'_{+1}R_-CR_-A\underrightarrow{R}_+$. In the normal perpendicular case we'll have $\underrightarrow{S}_+ \underleftarrow{S}_{+1} \rightarrow \underleftarrow{S}'_{+1}C\underrightarrow{S}'_+$. The terminal shock wave is weakened and moves outward after the collision (cf. with /14/).

4. Considering the interaction of the refracted waves with the heliopause H treated as the tangential discontinuity we have (in normal perpendicular case): $\underrightarrow{S}'_+H \rightarrow \underleftarrow{S}''_+H'\underrightarrow{S}'''_+$, where S''_+ is the reflected from the heliopause shock wave and S'''_+ – fast shock wave refracted into the interstellar medium. In oblique case we have slow shock waves S_-, slow rarefaction waves R_- and rotational discontinuities A.

5. The shock wave S''_+ will interact with the modified terminal shock wave $\underleftarrow{S}'_{+1}\underleftarrow{S}''_+ \rightarrow \underleftarrow{S}''_{+1}TT\underrightarrow{R}'_+$ and the secondary rarefaction wave R'_+ appears in the heliosheath as in /3/. In the result of the similar interaction with the hypothetical interstellar shock wave S_{+2} the rarefaction wave appears between the heliopause H and the interstellar shock wave front S_{+2}.

CONCLUSIONS

The slow shock waves and the double shock wave ensemble may appear in the result of the collision between the rotational and the contact discontinuities. It's found that there is a special interval for the parameters where the nonlinear catastrophic reconctruction takes place for the $\underrightarrow{A}$ C interaction. It's shown for the shock interaction that the rise-type waves configuration may appear ($\underleftarrow{R}_+$ A S_- C $\underrightarrow{R}_-S_+$ combination which is carried outward by the solar wind). For the vicinity of the heliopause it's shown that secondary rarefaction waves appear in the result of the nonstationary solar wind shock wave arrival. These waves may effect the radio emission /22, 23/ and the dynamics of the whole region.

REFERENCES

1. M.Dryer, Z.Kopal Smith, and Shi Tsau Wu, The role of magnetohydrodynamics in heliospheric space plasma physics research, Astrophys. Space Sci., 144, 407 (1988).
2. F.M.Neubauer, Nonlinear interaction of discontinuities in the solar wind and the origin of slow shocks, J. Geophys. Res. 81, #13, 2248 (1976).
3. S.A.Grib, B.E.Brunelli, M.Dryer, and W.-W.Shen, Interaction of interplanetary shock waves with the bow shock-magnetopause system, J. Geophys. Res. 84 #10, 5907 (1979).
4. N.C.Zhuang, C.T.Russell, E.J.Smith, and J.T.Gosling, Three-dimensional interaction of interplanetary shock waves with the bow shock and magnetopause: a comparison of theory with ISEE observations, J. Geophys. Res. 86, #A7, 5590 (1981).
5. S.J.Schwartz, R.L.Kessel, C.C.Brown, J.C.Woolliscroft, M.W.Dunlop, C.J.Farrugia, and D.S.Hall, Active current sheets near the Earth's bow shock, J. Geophys. Res. 93, #A10, 11295 (1988).
6. S.A.Grib, Interaction of non-perpendicular/parallel solar wind shock waves with the Earth's magnetosphere, Space Sci. Rev. 32, 43 (1983).
7. S.A.Grib, On the interactions and the evolution of the strong discontinuities in the solar wind plasma, Solar Maximum Year, Proc. Intern. Workshop, ed. V.N.Obridko. IZMIRAN, M., 1981, p.34.
8. A.A.Barmin and E.A.Pushkar, The interaction of the oblique shocks in MHD: Application to solar wind around the Earth's magnetosphere, 5th Intern. Workshop 'Cosmic Gas Dynamics', M. (1988).
9. A.A.Barmin and E.A.Pushkar, The oblique interaction of strong discontinuities in magnetodynamics, in: Mechanics. Actual Problems. M., 1987, p. 100.
10. G.G.Paschmann, G.Haerendel, N.Sckopke, E.Möbius, N.Lühr, and C.W.Carlson, Three-dimensional plasma structures with anomalous flow directions near the Earth's bow shock, J. Geophys. Res. 93, # A10, 11.279 (1988).
11. M.R.Sheely, Jr., R.A.Howard, M.J.Koomen, D.H.Michels, R.Schwenn, K.H.Muhlhauser, and H.Rosenbauer, Associations between coronal mass ejections and interplanetary shocks. Publ. 41-83-3 (April 1983).
12. H.J.Fahr, S.Grzedzielski, and R.Ratkiewicz, Magnetohydrodynamics modelling of the 3-dimensional heliopause using the Newtonian approximation, Annales Geophys., 6, 4, 337 (1988).
13. J.D.Mihalov, Distant heliospheric results on interplanetary shock propagation, J. Geophys Res. #A90, 1, 210 (1985).
14. Y.C.Whang and L.F.Burlaga, Evolution of recurrent solar wind structures between 14 AU and the termination shock, J. Geophys.Res. 93, #A6, 5446 (1988).
15. R.Chassefiere, J.L.Bertraux, R.Lallement, and V.G.Kurt, Astron. Astrophys. 160, 229 (1986).
16. W.Macek, Reconnection at the Heliopause, Adv. Space Res. 9, N 4 (4) 257. (1989).
17. E.A.Pushkar, Oblique magnetohydrodynamic shock waves, Fluid Dynamics (Izvestiya Akademii Nauk SSSR. Mekhanika Zhidkosti i Gaza) N 4, 106 (1978).
18. E.A.Pushkar, The generalized polars for the planepolarized stationary flows in the magnetohydrodynamics, Fluid Dynamics (Izvestiya Akademii Nauk SSSR. Mekhanika Zhidkosti i Gaza) N 3, 111 (1979).
19. P.D.Hudson, Planet. Space Sci., 21, 475 (1973).
20. S.A.Grib and B.A.Khrapov, The anisotropic interplanetary plasma parameters change across the MHD shock wave front. Space Res. (Kosmitch. Issled.), 27, N 2, 258 (1989).
21. A.A.Barmin and E.A.Pushkar. Oblique interaction of Alfvenic and contact discontinuities in the magnetohydrodynamics, Fluid Dynamics (Izvestiya Akademii Nauk SSSR. Mekhanika Zhidkosti i Gaza) N 1, 131 (1990).
22. W.S.Kurth, D.A.Gurnett, F.L. Scar, and R.L.Poynter, Detection of radio emission of 3 KHZ in the outer heliosphere. Nature. 312, 27 (1984).
23. R.L.Mc.Nutt, Jr., Remote sensing of the termination of the solar wind via in situ plasma measurements. Adv. Space Res. 9, 235 (1989).

NUMERICAL STUDY OF COLLIDING ASTROSPHERICAL STELLAR WIND FLOWS IN BINARY SYSTEMS

J. Kallrath

Institut für Extraterrestrische Forschung und Astrophysik der Universität Bonn, Auf dem Hügel 71, D–5300 Bonn 1, F.R.G.

ABSTRACT

The dynamical and geometrical aspects of two quasi radial supersonic counterstreaming gas flows are investigated. Analytical results derived from a trajectorial model are compared with pressure equilibrium surfaces derived with the help of the Newtonian approximation and extended models which account for centrifugal effects. These results are used as initial data for the hyperbolic system of equations describing the conservation of mass, momentum and energy. The equations are solved by using a two-dimensional axial-symmetric second order Godunov scheme set up as a finite volume method. In a model WR/O-binary system a subsonic region with an extension of a few solar radii is established. Our procedure, here used to derive the geometry of steady shock waves near an astropause, might also be appropriate to investigate a variety of phenomena near the heliopause.

INTRODUCTION

Under strong radiation pressure the physics of stellar atmosphere and stellar envelopes requires the existence of a radiatively driven wind as observed in O-, B- or WR-stars. At typical parameters characterizing such winds, mass loss rates of 10^{-5}-10^{-6} M_0/year and asymptotic wind velocities of about *1000-3000* km/s are to be expected. The details of the acceleration mechanism of this wind is not discussed in this paper. If there are two stellar winds in hot (WR+O-star) binary systems as for example *V444 Cygni*, γ^2Vel or *HD 152270* one has to model colliding supersonic flows with Mach numbers of about *100* leading to shock- and contact surfaces. Due to the orbital motion of the binary stars those surfaces are subject to non-inertial forces.

In /1/ the dynamical and geometrical properties of the region near the contact surfaces are analyzed in the framework a) of a trajectorial model, b) in terms of pressure equilibrium models, and finally, c) by numerical integration of the hyperbolic system of hydrodynamical physical equations describing the conservation of mass, momentum and energy, which govern the stellar wind fluid dynamics.

In the next sections these three different approaches and their links to eachother are briefly discussed. The numerical integration procedure, a second-order, axisymmetric Godunov scheme (GS), is applied to a set of parameters appropiate for *HD 152270*.

TRAJECTORIAL MODELS (TMs)

TMs are only valid if the pressure gradient may be neglected. If each star produces a spherically-symmetric wind with a quadratically decreasing acceleration (QDA), a situation similar to the classical two-fixed centre problem (TFCP) arises and the trajectories of test-particles in the superimposed potential field of both stars neglecting non-inertial forces are represented by analytical representations based on elliptical functions /2/. As shown in /3/ the contact surface between both wind flows, here defined as the envelope of all trajectories of equal energy emitted from the WR-star, may also be calculated analytically by solving one transcendental equation. If non-inertial forces are included, the discussion is analogous to the restricted three-body problem. The analysis of this case provides the basic argument for an estimation of the influence of the centrifugal- and Coriolis force. These forces are neglected and only axisymmetric geometries are considered from now on.

PRESSURE EQUILIBRIUM MODELS (PEMs)

PEMs are based on the *Newtonian Approximation* /4/ and define the contact surface by equalizing the projections of the stress tensors to the surface normal or the orthogonal momentum fluxes on both sides of the (unkown) surface by assuming that the interface between both wind flows is of zero-thickness. Utilizing axial symmetry the surface can be calculated by an ordinary 1st order differential equation (ODE). A more enhanced model, accounting for mass conservation and both orthogonal and parallel momentum flux leads to a system of three coupled ODEs which we integrated by Bulirsch-Stoer extrapolation /5/. The input data to this procedure are given density- and velocity fields in the region associated with either the WR- or O-star. In particular the integrations are performed for constant velocity fields and for models with isotropically QDA. In both cases, the thermal pressure is neglected.

NUMERICAL GAS DYNAMICS

Although a steady solution for the geometry of the interface between both wind flows is expected, the hydrodynamical equations are set up with full time-dependence. Following that they are integrated by a 2*nd*-order axisymmetric Godunov scheme which advances a given initial solution at $t=t_0$ through a sequence of time-steps controlled by the *Courant-Friedrich-Lewis condition* finally to the steady state at $t=\infty$. The 1*st*-order Godunov scheme /6/ can be derived in a natural way from the basic conservation laws of mass, momentum and energy written in divergence form enhanced by an equation of state which, in our case, is that of an ideal gas. The discretization leads to a mesh of cells which in cylindrical coordinates are defined by lines of constant r and constant z parallel and orthogonal to the symmetry axis (SA). The construction of the numerical flux function and its evaluation is based on one-dimensional sequences of *Riemann Problems* each describing the flux across the boundary between adjacent cells. A Riemann problem /7/ for a system of conservation laws is an ini-

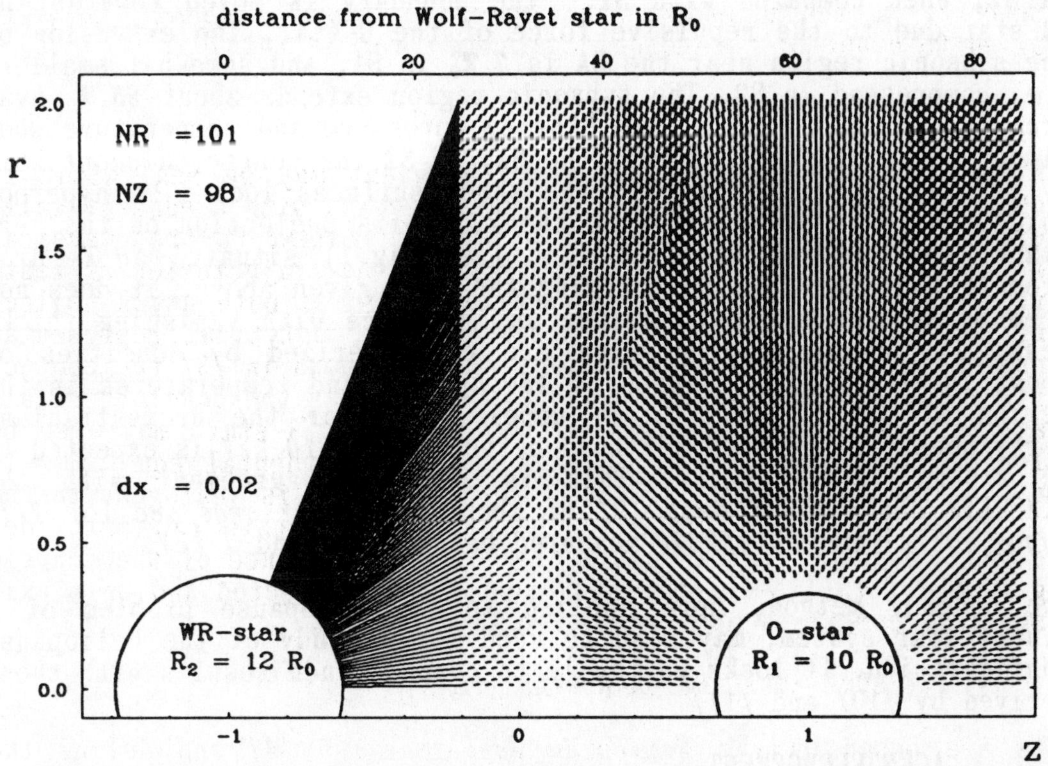

Fig. 1. The flow pattern resulting from S2 is indicated by small lines proportional to the absolute value of velocity. From the shadow effect in this picture the shock surface and the contact surface can be imagined. *NZ* and *NR* give the number of cells in the *z*- and *r*-direction while *dx* is the linear size of each cell.

tial value problem with scale-invariant initial date. In one dimension, these initial data consist of distinct constant states on the left and right of a jump discontinuity. The solution is constructed by elementary scale-invariant waves,i.e. rarefaction waves, shock waves and contact discontinuities. The GS, used as a finite-volume method taking into account the difference between cell centre-points and nodal-points, proves to be a very robust shock-capturing method. By choice of a flux source term we have modeled a free flow pattern (S1), and the flow in the field of the TFCP (S2).

RESULTS AND DISCUSSION

Assuming some typical parameters describing the WR- and O-star wind in *HD 152270* (the stars are *61.6* R_0 separated from each other), the trajectorial analysis leads to the result that the interface is shifted by an angle of about *10* degrees when non-inertial forces are included. Therefore, the effect of these forces is treated as known from now on and we concentrate on the axially-symmetric case only .

The PEMs within the framework of the enhanced model and using the QDA ansatz leads to a shape of the contact surface whichs is nearly identical with that derived from the trajectorial model. This result, and the flow field around the stars are used as initial data for the Godunov procedure running on our VAX/780 and CONVEX C1.

In S2, when compared with S1, the boundary is moved towards the WR-star due to the repulsive force of the O-star. The extension of the subsonic region near the SA is *7* $\mathcal{R}_0$ in S1, and somewhat smaller, i.e. compressed in S2. The subsonic region extends about *25* $\mathcal{R}_0$ away from the SA. While in S1 the density, pressure and temperature show almost constant profiles near the SA, in S2 the contact discontinuity is well developed. In S1, the shock surfaces look like hyperboloids, while in S2 the WR-side shock, or its projection on the *r-z*-plane, is nearly a straight line (see Fig.1) situated *30* $\mathcal{R}_0$ away from the WR-star /8/. Summarizing the data given above, it does not seem to be appropiate to assume an interface with vanishing thickness. Its physical properties are characterized by densities of about 10^{-13} g/cm^3, pressures of 150 dyn/cm^2 and temperatures in the region of 10^7-10^8 *Kelvin.* An upper limit for the Bremsstrahlung x-ray power in the hard keV band between *1* and *10 keV* is expected to be around 10^{35} *ergs/sec.* This is in rough agreement with the EINSTEIN OBSERVATORY data /9/ of about 10^{32}-10^{34} *ergs/sec* for *V444 Cygni*, γ^2 *Vel* and *HD 152270* measured between *0.8* and *4 keV*.

The Godunov method, here applied to an astropause problem of a binary star system, may also be used for a study of the heliopause configuration. It looks worthwile to compare such results with those derived by /10/ and /11/.

ACKNOWLEDGEMENT

The author greatly acknowledges support through a stipend of the "Cusanuswerk (Bonn,FRG)".

REFERENCES

1. J.Kallrath,Dynamik kollidierender Doppelsternwinde,PhD thesis, Bonn,1989

2. W.Neutsch and H.Schmidt,Expanding envelopes of binary stars, Part 2.:Motion in extremely hot envelopes,Astrophys.Sp.Sci. 110, 257(1985)

3. J.Kallrath,Dynamics of colliding binary stellar winds: I. Trajectorial Models,M.N.R.A.S,in preparation

4. H.J.Fahr,S.Grezdzielski and R.Ratkiewicz,MHD modeling of the 3-dimensional heliopause by using the Newtonian approximation,Annales Geophysicae 6,#4,337(1986)

5. R.Bulirsch and J.Stoer,Numerical Treatment of Ordinary Differential Equations by Extrapolation Methods,Num.Mathematik 8, 1(1965)

6. S.K.Godunov,A finite difference method for the numerical computation of discontinuous solutions of the equations of fluid dynamics,Mat.Sbornik 47,357(1959)

7. R.Menikoff,The Riemann problem for fluid flow of real materials,Reviews of Modern Physics 61, 75 (1989)

8. J.Kallrath,Dynamics of colliding binary stellar winds: III. Numerical simulations of colliding astrospherical flows,M.N.R.A.S,in preparation

9. R.L.White and K.S.Long,X-ray emission from Wolf-Rayet stars,ApJ 310,832(1986)

10. V.B.Baranov,M.G.Lebedev and M.S.Ruderman,Structure of the Region of Solar-Wind Interstellar Medium Interaction and its Influence on H Atoms penetrating the Solar Wind,Astrophys.Sp.Sci. 66,441(1979)

11. T.Matsuda,Y.Fujimoto,E.Shima,K.Sawada and T.Inaguchi,Numerical Simulation of Interaction between Stellar Wind and Interstellar Medium,Kyoto University,(1989)

FILTRATION OF THE INTERSTELLAR NEUTRALS AT THE HELIOSPHERIC INTERFACE AND THEIR COUPLING TO THE SOLAR WIND

Hans J. Fahr
Institut für Astrophysik und Extraterrestrische Forschung der Universität Bonn, Auf dem Hügel 71, D-5300 Bonn 1, F.R.G.

ABSTRACT

Neutrals and ions are strongly coupled in interstellar/interplanetary space by means of charge exchange reactions. Over distance scales of the order of charge exchange mean free paths the thermodynamical states of neutral and ionized media thereby is communicated. In the following paper we study the problem of the penetration of neutral interstellar gas species through the perturbed plasma region ahead of the solar system into the inner heliosphere where their existence has been proven by their resonance glows. First we give an introduction to the main outlines of the theory connected with this problem and then discuss in some detail the question of the neutral hydrogen and oxygen presence in the heliosphere. It turns out that for interstellar ion densities of $1o^{-2}cm^{-3}$ or greater neutral LISM hydrogen or oxygen will be depleted at least by a factor of 0.5 or 0.1, respectively. This raises the question whether or not the anomalous cosmic ray components showing up as 1o MeV/nucleon He^+, O^+, Ne^+, N^+-ions in the inner heliosphere can be ascribed to LISM neutrals entering the heliosphere and then becoming ionized and reenergized there. We analyze the theories on the origin of the anomalous cosmic rays and investigate their quantitative results respecting the prediction for the LISM penetration into the heliosphere. In our calculations we come to the conclusion that the ratios of the production rates for He^+, O^+, Ne^+, N^+-ions by no means can be brought into accordance with those values needed to explain the anomalous cosmic ray data.

I) PENETRATION OF LISM NEUTRALS INTO THE HELIOSPHERE

The problem of the penetration of neutral atoms from the local interstellar medium into the heliosphere has been treated in a series of papers in the recent past (Ripken and Fahr, 1983, Wallis, 1984; Fahr and Ripken, 1984; Błeszinski, 1986; Fahr, 1986). The main result in all of these papers was that the neutral LISM components when passing over the perturbed plasma regions ahead of the solar system will experience specific changes of their velocity distribution functions by means of charge exchange processes and electron impact ionizations. As a consequence it is expected by all above mentioned authors that within the heliosphere the moments of the distribution functions of these LISM neutrals, like the particle density, the bulk velocity and the temperature, will deviate specifically from their unperturbed LISM values, however different in magnitude these deviations may have been judged by the authors. Here we only would like to present the main theoretical lines along which these results can be gained. We follow the kinetic approach by means of the Boltzmann equation used by Fahr and Ripken(1983) and Fahr(1986, 1989).

In the plasma interface region ahead of the inner "supersonic" heliosphere plasma media of solar and interstellar origin are present with organized MHD-determined plasma flows. At a place $\vec{r}$ the distribution function $f(\vec{r},\vec{v})$ of any neutral species in the stationary case is given by the following form of Boltzmann's integro-differential equation:

(1) $$\partial f(\vec{r},\vec{v})/\partial s = (1/v)\left[P(\vec{r},\vec{v}) - L(\vec{r},\vec{v})\right]$$

where s is the line element on the trajectory of particles reaching $\vec{r}$ with a velocity $\vec{v}$. The functions P and L in equation (1) describe the rates by which particles at $\vec{r}$ with velocity $\vec{v}$ are produced and removed, respectively. Whereas the function P usually contains a velocity-space-integral over distribution function $f(\vec{r},\vec{v})$ itself, the function L in nearly all cases allows a representation with f only occurring as a factor in the form:

(2) $$L(\vec{r},\vec{v}) = f(\vec{r},\vec{v}) \cdot \nu^{-}(\vec{r},\vec{v})$$

with ν^{-} being the total destruction frequency for particles of the considered species. Consideration of the physical properties in the interface plasma (electron density $n_e \cong 10^{-2}\,cm^{-3}$, temperature $T_e \cong T_p \cong 10^4 K$) reveals that the productive or destructive processes counting there are due to resonant charge exchanges with the plasma ions, mainly hydrogen ions (i.e. $\nu^{-} = \nu^{-}_{ex}$). Counting the particles appearing at $\vec{r}$ according to collision hierarchies i, and introducing f_i as the distribution function for particles that have resonantly exchanged their charge i-times before appearing at $\vec{r}$ (with the evident definition of $f = \sum f_i$), one can convert the integro-differential equation (1) into a hierarchical system of differential equations of the form:

(3) $$\partial f_i/\partial s = (1/v)\left[P_{i-1} - f_i\, \nu^{-}_{ex}\right]$$

where the function P_{i-1} now is defined by:

(4) $$P_{i-1}(\vec{r},\vec{v}) = f_p(\vec{r},\vec{v}) \int^3 \sigma_{ex}(v_{rel})\, v_{rel}(\vec{r},\vec{v},\vec{v}')\, f_{i-1}(\vec{r},\vec{v}')\, d^3v'$$

and thus is independent on f_i. The solution of the inhomogeneous differential equation (3) therefore can be given in the form:

(5) $$f_i(\vec{r},\vec{v}) = f_{i,\infty}(\vec{v}_\infty)\exp\left(-\int_\infty^s \nu'_{ex} ds'/v'\right) + \int_\infty^s P'_{i-1}\exp\left(-\int_{s'}^s \nu''_{ex} ds''/v''\right) ds'/v'$$

Here σ_{ex} and v_{rel} denote the charge exchange cross section and the relative velocity, respectively, and $f_{i,\infty}$ is the unperturbed LISM distribution only existing for i = o (i.e. $f_{i,\infty} = 0$ for i>0).

Using now as basis for the description of the interface plasma Parker's subsonic plasma-plasma model (Parker, 1963, pp. 115 - 12o) one is advised to represent the problem in cylindrical coordinates z, ρ, ϕ with the z-axis oriented parallel to the LISM wind vector $\vec{V}_\infty$ (for illustration purposes see Fig. 1). The proton distribution function $f_p(\vec{r},\vec{v})$ can then be represented in a reasonably well approximation as a Maxwellian shifted by the local bulk flow $\vec{V} = \{V_z, V_\rho, 0\}$ of the protons. Fig. 1 serves for illustrative purposes and clarification of denotations. The flow vector $\vec{V}$ is then given by Parker (1963) in the following form:

(6) $$V_z = V_\infty(1+\bar{z}/\bar{r}^3),\ V_\rho = V_\infty(\bar{\rho}/\bar{r}^3).$$

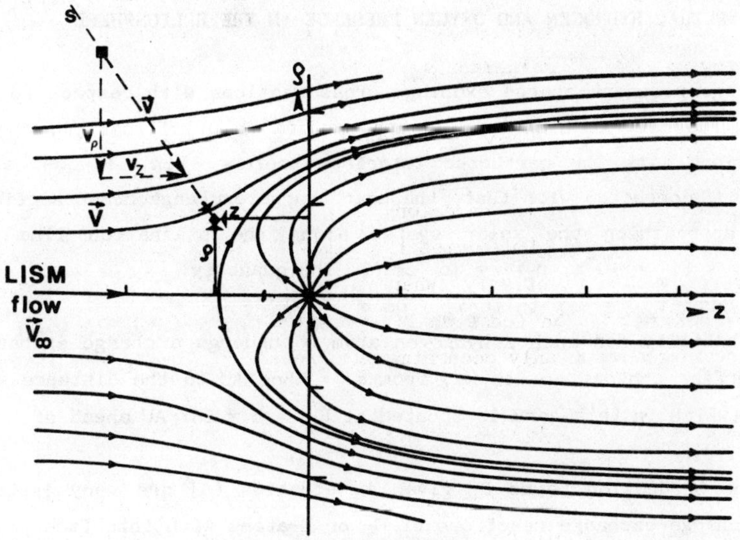

Fig. 1: Illustration of the plasma flow and denotation of variables

Here the bars on top of the variables indicate a normalization with the heliopause stand-off distance L (e.g. see Fahr, 1986).

Thus with equation (6) the function $f_p(\vec{r},\vec{v})$ would attain the form:

(7) $$f_p(\vec{r},\vec{v}) = (m/2\pi KT_p)^{3/2}\exp(-m/2KT_p\ [(v_z-V_z)^2+(v_\rho-V_\rho)^2+v_\phi^2])$$

Here $T_p = T_p(r)$ is the local proton temperature which in Parker's interface is given by (Fahr and Neutsch, 1983):

(8) $$T_p(\vec{r}) = m/K\ [\ 1/2\ V_\infty^2\ +KT_\infty/m\ -\ 1/2\ (\ V_z^2+V_\rho^2)\ ].$$

We now consequently approximate the production term P_{i-1} in the following form:

(9) $$P_{i-1}(\vec{r},\vec{v}) \cong \sigma_{ex}(\bar{v}_{rel})\ \bar{v}_{rel} n_p(\vec{r})\ n_{i-1}(\vec{r})\ f_p(\vec{r},\vec{v})$$

where n_p and n_{i-1} are the local densities of protons and of neutral particles of the order i-1, respectively, and where $\bar{v}_{rel}(\vec{r},\vec{v})$ is the average relative velocity of all neutral particles with respect to protons of the velocity $\vec{v}$. In a satisfactory approximation the latter velocity is given by:

(10) $$\bar{v}_{rel}(\vec{r},\vec{v}) = |\ (\vec{v}_p-\vec{v})\ | \cong [\ (v_z-V_z)^2+(v_\rho-V_\rho)^2+v_\phi^2\ ]^{1/2}$$

With the equations (5) through (10) one is prepared to calculate the distribution function $f(\vec{r},\vec{v})$ by consecutively proceeding from lower to higher order indices i, in each higher order i taking advantage of the predetermined distribution function f_{i-1}.

II) NEUTRAL HYDROGEN AND OXYGEN PRESENCE IN THE HELIOSPHERE

Due to their large resonant charge exchange cross sections with respect to protons , H- and O-atoms amongst the neutral LISM species are subject to specifically pronounced charge exchange interactions with the perturbed interface protons. Fig. 2 shows the mean relative velocity v_{rel} and, connected with that, the mean charge exchange cross section $\sigma_{ex}(v_{rel})$ of a hydrogen atom approaching the solar system along the stagnation line with a velocity $v_{H,\infty} = V_\infty = 23$km/s ($T_\infty = 10^4$K, $n_{p,\infty} = 1o^{-2}$cm^{-3}). The quantity

$$Ex = \exp(-\tau_{ex}) = \exp\left(-(n_\infty/V_\infty)\int ds\,\sigma_{ex} v_{rel}\right)$$

denotes the probability for such a hydrogen atom to undergo a charge exchange reaction with one of the interface protons on its way from s_∞ = 65oo AU to the distance s (in 1o^{3}AU) from the heliopause (which in this case is located at R_{HP} = L = 287 AU ahead of the sun.).

For hydrogen the production terms P_i given in equation (4) are very important since as a consequence of charge exchange reactions of O- or H-atoms with interface protons always neutralized protons, i.e. H-atoms, reappear as reaction products. For oxygen atoms the situation concerning the relevance of the production term P_i is completely different. Oxygen atoms are only produced from charge exchange reactions with O^+-ions as collision partners which in view of their cosmic abundance relative to hydrogen ($\alpha_{O/H} = 6.7\cdot10^{-4}$) are very rare. Furthermore due to their large masses ($m_O = 16m_H$) their thermal velocities compared to the local ion bulk flow $\vec{V}$ in the interface plasma are fairly small. Therefore the velocity vectors $\vec{v}$ needed to bring O-atoms into the inner part of the heliosphere only at a very limited region near the stagnation line are contained in the O^+-velocity distribution function. Thus one may reach a reasonably good number for the oxygen transmissivity of the interface plasma not taking into account any O-production terms there. On the basis of this assumption in Fig. 3 we have shown as function of the LISM proton density $n_{p\infty}$ comparative results for the hydrogen and oxygen transmissivities Tr_{HP} in the interface plasma ahead of the heliopause (R_{HP} = 287 AU). For

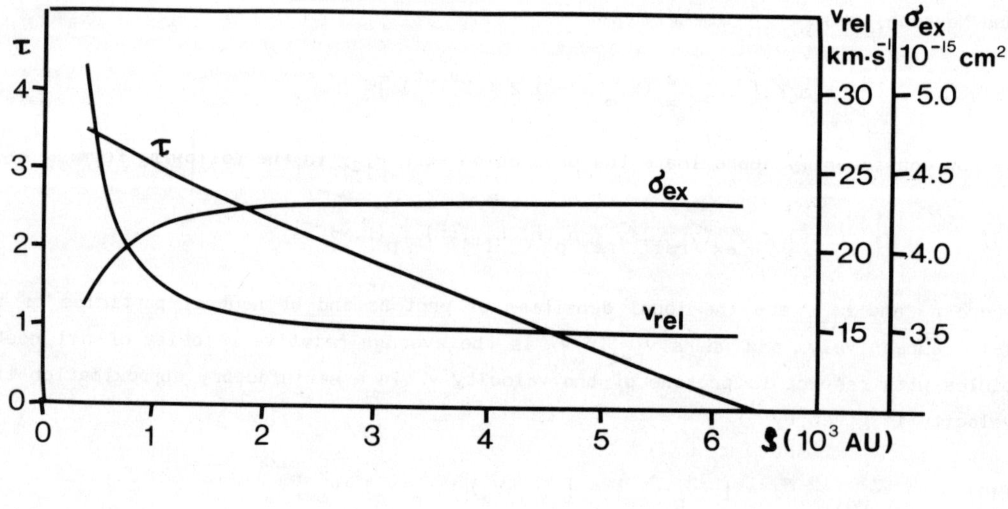

Fig. 2: Shown are mean relative velocity v_{rel}, the charge exchange cross section σ_{ex}, and the optical thickness τ_{ex} for primary charge exchange for bulk H- atoms as function of the distance ρ on the stagnation line in a Parker-type interface flow (L = 287 AU).

the most probable values of about $n_{p\infty} \cong 10^{-2} cm^{-3}$ (Cox and Reynolds, 1987, Reynolds, 1989) from this Fig. one draws the conclusion that about 50 percent of the LISM hydrogen flow and only 10 percent of the LISM oxygen flow after penetration through the interface plasma reaches the heliopause.

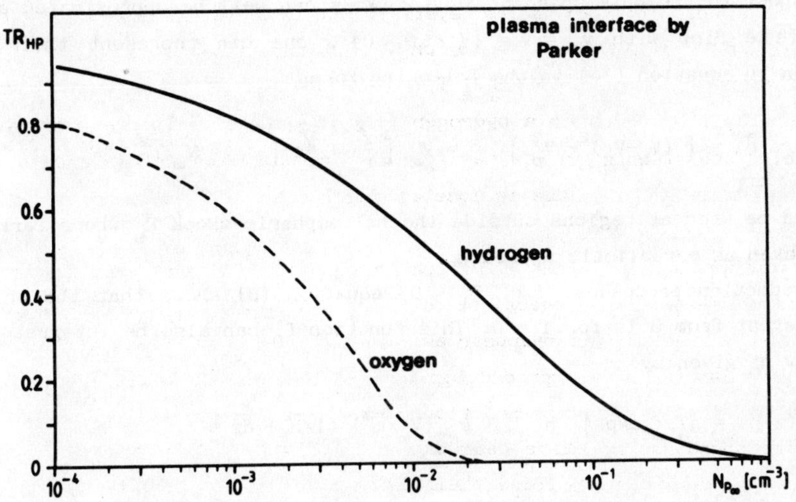

Fig. 3: Shown as function of the LISM proton density $N_{p,\infty}$ is the transmission probability of LISM H- and O- atoms on their way from the "local infinity" to the heliopause ($L = L(N_{p,\infty})$ according to Parker(1963)).

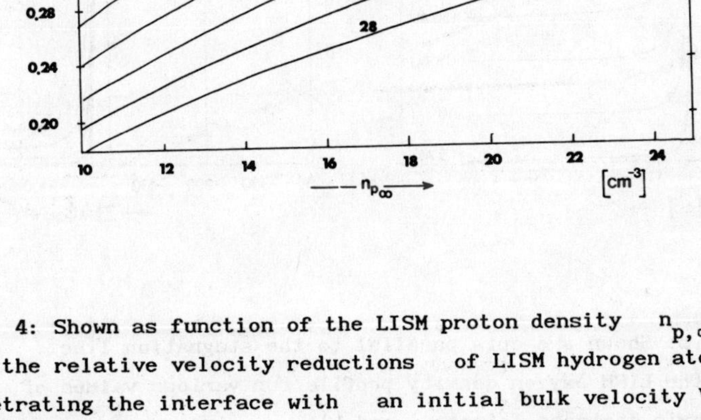

Fig. 4: Shown as function of the LISM proton density $n_{p,\infty}$ are the relative velocity reductions of LISM hydrogen atoms penetrating the interface with an initial bulk velocity V_∞ up to the heliopause.

Also as a function of $n_{p\infty}$ we have given in Fig. 4 the relative velocity reduction $\Delta V_\infty/V_\infty$ in the bulk flow of hydrogen atoms arriving at the heliopause which has to be understood as an imprint of the stagnating LISM plasma flow on the neutrals via charge exchange.

A very interesting, since analytic solution for the density distribution of oxygen atoms in the heliosphere can be obtained due to some logical simplifications that are self suggestive in this case (Fahr, 1989). Realizing that oxygen atoms with a mean thermal velocity of 3.5 km/s with respect to the bulk speed of V_∞ = 23 km/s can well be approximated as a mono energetic particle e flow with $\vec{v}_\infty = \vec{v} = \{V_\infty, 0, 0\}$, one can represent their mean relative velocity given in equation (1o) in the following form:

$$(11) \qquad v_{rel}(\vec{r}) = \left[(V_\infty - V_z)^2 + V_\rho^2 \right]^{1/2} = V_\infty \left[\frac{\bar{z}^2}{\bar{r}^6} + \frac{\bar{y}^2}{\bar{r}^6} \right]^{1/2} = V_\infty / \bar{r}^2$$

which only can be used at regions outside the heliospheric shock R_s where Parker's interface flow can be taken as a realistic approach.

Neglecting production processes, i.e. $P_i = 0$, equation (5) shows that the only function f_i which is different from 0 is for i = 0. This function f_0 can also be integrated to yield the oxygen density n_O given by:

$$(12) \qquad n_O(z, \rho) = n_{O,\infty} \exp \left[-n_{p\infty} L \int_{-\infty}^{z} \sigma_{ex} (V_\infty / \bar{r}^2)\, (1/\bar{r}^2)\, d\bar{z} \right]$$

Taking an average value $\bar{\sigma}_{ex}$ which is characteristic for the interface (see Fig. 2) one can then explicitly carry out the integration in equation (12) and finally come to the result:

$$(13) \qquad n_O(z, \rho) = n_{O,\infty} \exp \left[- \frac{n_{p\infty} L \bar{\sigma}_{ex}}{\rho} \left(\pi/2 + \operatorname{arctg} (z/\rho) \right) \right].$$

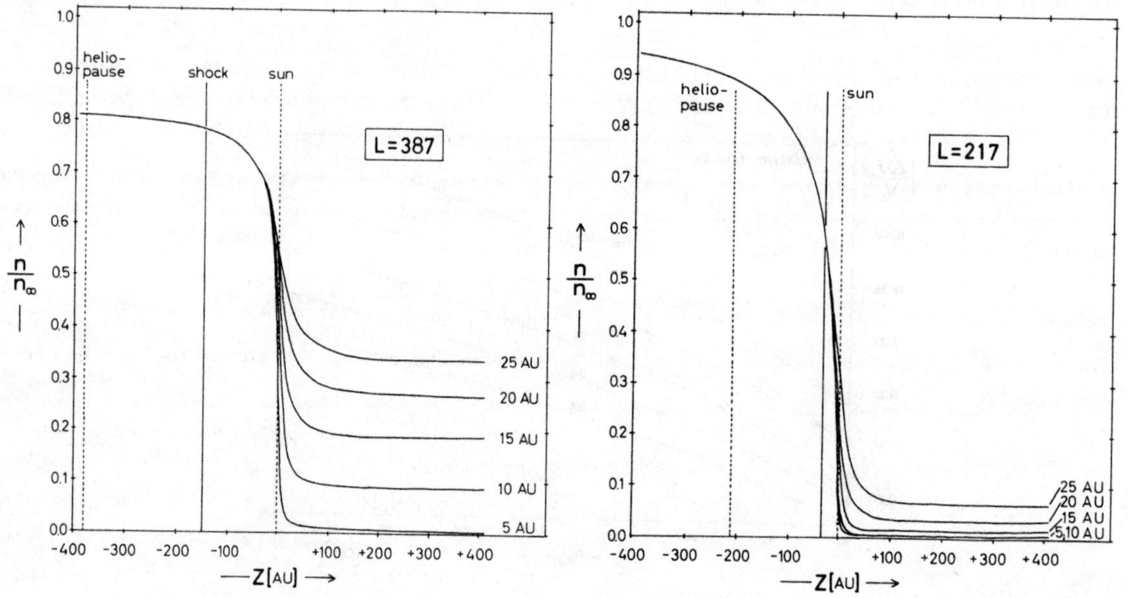

Fig. 5 a,b: Shown are cuts parallel to the stagnation line through the LISM oxygen density profile for various values of the off-axis distance ρ. (case a and b)

In Figs. 5 a,b we have presented as function of z cuts through the oxygen density profile for different distances ρ to the $\vec{V}_\infty$- axis calculated with the above expression (13) and for different stand-off distances L.

a) The stand-off distance L_1 = 387 AU used in Fig. 5a results from the definition of L given in Fahr(1986) if the following LISM properties are adopted:

$$n_{p\infty} = 1.35\cdot 10^{-2}cm^{-3},\ T_\infty = 1\cdot 10^4 K,\ V_\infty = 23\ km/s,\ B_\infty = 0,\ n_s = 8\cdot 10^{-4}cm^{-3},\ V_s = 1\cdot 10^7 Km/s$$

where the quantities with suffix "s" are proton density and velocity of the postshock solar wind at R_{s1} = 173 AU.

b) The stand-off distance L_2 = 217 AU used in Fig. 5b results in connection with the same LISM data as above except that $B_\infty^2/8\pi$ has been adopted with a value such that it dominates the total LISM pressure by a factor of 1o (i.e. $B_\infty = 5\cdot 10^{-6}$Gauss). Under these conditions R_{s2} turns out to be 54 AU.

III) INTERSTELLAR NEUTRALS IN THE HELIOSPHERE AND THE ORIGIN OF THE ANOMALOUS COSMIC RAY COMPONENT:

It is now generally taken for granted that the so-called "anomalous" cosmic ray component which essentially represents He, O, Ne, and N particles at medium energies of 1o - 5o MeV/nucleon is due to interstellar neutral atoms moving into the inner heliosphere from the upwind side of the solar system. There they become ionized via photo-ionization, charge exchange, or electron/proton impact ionization, are convected outwards with the solar wind magnetic fields and thereby as suprathermal ions are subject to acceleration processes most probably due to transit-time damping of magnetosonic interplanetary field fluctuations (Fisk,1974; Fisk,1976a/b/c). Thus these particles on their way to the outer fringes of the heliosphere are systematically shuffled to higher and higher energies and, as shown in Fisk(1976b) and Klecker(1977), at heliocentric distances of about 5o AU maybe described by distribution functions peaking at energies of about 1o MeV/nucleon. As originating from ionizations of LISM neutrals in the heliosphere these particles must be expected to be singly ionized (i.e. Z = 1) and thus to have large rigidities $\rho = B\, r_g$ = pc/Ze compared to usual cosmic ray particles (Z » 1) of the same momentum p (or energy E). Therefore these energetic particles with energies E ≅ 1o MeV are considered to be able to diffuse backwards into the inner solar system and then to show up there as anomalous cosmic ray particles (for a review see Fisk,1986, or Jokipii,1986).
The different processes relevant within this scenario, i.e. energization by Fermi acceleration and/or transit-time damping mechanisms, convection, adiabatic cooling, and diffusion in configuration space, if simultaneously taken into account, lead to a very complicated partial differential equation of the Fokker-Planck-type for the particle distribution function $f_j(r,p,t)$ of species j. However, the theoretical representation of these processes seem to allow for the following split-up into temporal sequences (Fisk,1976b, Klecker,1977) :

a) After pick-up the original interstellar ions are at energies of ≃ 1KeV/nucleon. At these conditions diffusion processes essentially can be neglected. Thus on their way out of the inner heliosphere these interstellar ions only need to be treated as being subject to convection, adiabatic cooling and energization . The corresponding change in the distribution function $f_i(r,p)$ with heliocentric distance, as calculated from a simplified Fokker Planck equation (Fisk, 1976a), was obtained by Fisk(1976b), Klecker(1977) and also by Isenberg(1987) with some assumptions deviating from the first mentioned authors.

b) At the outer ranges of the heliosphere (i.e. at ≃ 5o AU) the energized particles now develop non-negligible diffusive flows that require a different differential equation for adequate description derived from the full Fokker Planck equation now by neglect of the acceleration term. In principle this implies a description of the modulation of energized interstellar ions by solar-wind-convected MHD fluctuations like in the case of the galactic cosmic ray penetration.

Following the method of Gleeson and Urch(1971), Klecker(1977) has numerically solved this modulation equation starting at R_∞ = 5o AU with the distribution function f_i given by Fisk(1976a). By doing so, he was able to achieve a satisfactory fit to anomalous cosmic ray data obtained at 1 AU by Garcia-Munoz et al.(1973,1975) and Klecker et al.(1977) if he started with relative abundances of 15.o , o.15 , and o.o5 for He^+ , N^+, and Ne^+ ions relative to O^+ ions at 5o AU and assumed that only a fraction of $\varepsilon = 2\cdot 1o^{-3}$ of each interstellar ion species present in the outer heliosphere is energized up to 1o MeV energies.

1) Flux balance for the anomalous cosmic ray particles.

The question now to be put here is whether or not this picture for the anomalous cosmic ray particles (ACR) can essentially be supported in view of the responsible source and sink terms. In the framework of a stationary situation the rate by which fresh ACR particles are gained in the heliosphere should just balance the corresponding loss rate, mainly represented by diffusive and convective losses of ACR particles out of the outer heliosphere.

A) The ACR gain rates

Here we may first consider the gain rate for the ACR particles. As this rate is taken to be due to the penetration of LISM neutrals into the heliosphere connected with successive ionization one may calculate it on the basis of typical solar ionization target cross sections F_{ij} for the different elements j. These cross sections are easily found by checking out that specific solar impact parameter P_{ij} of a Keplerian trajectory an LISM atom "j" can attain such that it marginally suffers ionization at its traversal through the heliosphere. This critical impact parameter P_i can be obtained from the interplanetary extinction function $Ex(\vec{r},\vec{v})$ as given e.g. in Fahr et al.(1987) which for time-independent loss rates $\nu(r)$ with a fall-off according to $1/r^2$ attains the following form:

$$(14) \qquad Ex(\vec{r},\vec{v}) = \exp\left[- \frac{\nu_E r_E^2\, \theta}{P_\infty\, v_\infty} \right]$$

and describes the probability of a neutral LISM atom entering the heliosphere with a velocity v_∞ and an impact parameter P_∞ to reach the space point $\vec{r}$ with a velocity $\vec{v}$ without being ionized. Here ν_E is the total loss rate at r_E = 1 AU and θ is the total angle swept over by the particle (i.e. between $\vec{r}$ and and $-\vec{v}_\infty$). The critical impact parameter P_i now is determined such that with it a particle passing over the whole heliosphere may just become ionized with a probability of exp(-1). Thus from equation (14) one can derive the following condition:

(15) $1 = 2\,\nu_E\, r_E^2\, \theta_{per} / P_i v_\infty$

where θ_{per} is the angle swept over by the particle up to the perihelion of its trajectory. As it turns out from elementary Keplerian relations this angle is an implicit function of P_i given by:

(16) $\theta_{per} = \mathrm{arctg}(-v_\infty^2 P_i / GM_o) = \pi - \mathrm{arctg}(v_\infty^2 P_i / GM_o)$

where G and M_o are the gravitational constant and the effective solar mass.
With equations (15) and (16) we thus obtain the following algebraic equation for the critical impact parameter P_i:

(17) $\bar{P}_i = 2\,\nu_E\, r_E\, (\pi - \mathrm{arctg}\, (v_\infty^2 r_E \bar{P}_i / GM_o)) / v_\infty$

where $\bar{P}_i$ is the critical impact parameter in r_E-units. From this equation now the parameters P_{ij} for the different relevant LISM atoms j can be determined if the corresponding loss rates ν_{Ej} are known.
What concerns appropriate values for the relevant loss rates we would like to follow those given in the papers by Siscoe and Mukherjee(1972) and Axford(1972). Since these values are based on the assumption of solar wind proton fluxes of $\Phi_E = (n_E\, v_E) = (6\cdot 5oo)\ cm^{-2}s^{-1} = 3\cdot 1o^8 cm^{-2}s^{-1}$ which are fairly on the high side, and on solar EUV/UV-fluxes taken from Hinteregger(1965) typical for solar minimum conditions, we have decided to consider their loss rates as rather characteristic for solar activity minimum phases.
As shown by Torr et al.(1979) and Torr and Torr(1985) the solar EUV-fluxes may vary by factors 2.5 through 3.o between solar minimum and solar maximum. Thus we have decided to multiply the photoionization rates from Siscoe and Mukherjee(1972) by an average factor of 2.5 in order to arrive at typical solar maximum rates.
On the other hand, the solar wind fluxes are shown to be highest during solar minimum, whereas during solar maximum they appear to be lower by an average factor of (4/5) = 0.8 (Lazarus and Belcher, 1988). Therefore the loss rates for charge exchange and proton/electron impact ionization are likely to be lower during solar maximum. Since in first order they all are proportional to the solar wind flux, we have decided to apply a factor 0.8 to all of these rates given by Siscoe and Mukherjee(1972) in order to arrive at corresponding solar maximum rates. By proceeding in this way we may start out from the following loss rates at $r = r_E$ (in units of $1o^{-7}s$):

TABLE 1: SOLAR MINIMUM CONDITIONS:

Element	Photo-ion.	Particle-ion.	Total ion.
He	0.39	0.08	0.47
Ne	0.95	0.15	1.1
O	2.49	2.58	5.33
N	1.89	1.41	3.5
C	4o.o	0.0	4o.o

TABLE 2: SOLAR MAXIMUM CONDITIONS:

Element	Photo-ion.	Particle-ion.	Total ion.
He	0.98	0.06	1.o4
Ne	2.38	0.12	2.5
O	7.00	2.o0	9.44
N	4.73	1.13	5.86
C	100.0	0.0	100.0

With these rates and formula (17) one obtains the following critical impact parameters P_{ij} (given in units of AU):

TABLE 3:

Element	Solar maximum	Solar minimum
He	3.11	1.67
O	21.27	13.1o
Ne	6.88	3.55
N	14.1o	8.88
C	115.52*	55.oo*

(Here the values with an asterisk have to be considered as idealistic in case the supersonic solar wind region terminates at 5o AU. It may be noticed that in Axford(1972) smaller values are given for the extinction distances at solar minimum conditions. This is explained by the fact that these latter values only determine the upwind distances which LISM atoms can reach unionized when approaching the sun along the LISM wind axis (i.e. P = 0!).

To determine the total production per unit of time of suprathermal ions j available for the stationary conversion into anomalous cosmic ray particles of this species one may therefore simply calculate the number of particles collected per unit of time by the corresponding solar ionization target $F_{ij} = \pi P_{ij}^2$. These numbers are given by:

(18) $$\Pi_j = n_\infty V_\infty \alpha_j (1-\xi_j) \tau_j F_{ij}$$

where V_∞ is the LISM wind velocity relative to the sun and n_∞ is the total LISM hydrogen density (neutral and ionized component). The quantity α_j is the relative cosmic abundance of the element j with respect to hydrogen, ξ_j is the ionization degree of this element in the LISM, and τ_j denotes the probability by which neutrals of the species j are transmitted through the heliospheric plasma interface (Ripken and Fahr, 1983; Fahr and Ripken, 1984; Wallis, 1984; Błeszynski, 1987).
What concerns the values for α_j we would like to follow the work by Cameron(1973) who gives solar system abundances which we take to be representative also for the LISM composition. Thus we find:

TABLE 4:

	Cosmic abund. (α_j)	Atomic mass (A_j)	Ionization potential (eV)	Fractional ioniz. ($1-\xi_j$)
H :	$\alpha_H = 1.o$	1	13.6	0.54
He:	$\alpha_{He} = o.o7$	4	24.6	1.oo
N :	$\alpha_N = 1.18 \cdot 1o^{-4}$	14	21.6	o.75
O :	$\alpha_O = 6.76 \cdot 1o^{-4}$	16	13.6	o.68
Ne:	$\alpha_{Ne} = 1.o8 \cdot 1o^{-4}$	2o	14.5	1.oo
C :	$\alpha_C = 3.71 \cdot 1o^{-4}$	12	11.3	0.o1

According to recent work by Frisch et al.(1987, 1989) our local interstellar environment may be characterized by a very low degree of ionization connected with an average electron density $n_e \simeq 1o^{-3}cm^{-3}$ in the 5 pc-sphere around the sun. However, this result obtained from MgI-absorption line interpretations suffers from the very poor knowledge of the relevant recombination rates in the LISM as was raised by Reynolds(1989). He and also Cox and Reynolds(1987) on the basis of adopted local photoionization equilibria come to the conclusion that the local hydrogen photo-ionization rate at present should be at least $\Gamma_H = 2.5 \cdot 1o^{-16}s^{-1}$ leading to a lower ionization fraction ξ_H for hydrogen of 0.1 and an electron density of $n_e \geq 1o^{-2}cm^{-3}$. This, however, would also point to a local recombination time period of $\tau_r = 1 \cdot 1o^6 \ (o.1/n_e)$ years = $1o^7$years meaning that whenever a supernova event influenced our local LISM environment during the last 1o million years the actual value of ξ_H , and also of n_e , may be much higher in fact compared to the stationary values given by Reynolds(1989).

Thus, in view of all these uncertainties, in the above table 4 we have taken as a leading idea for the values of $(1-\xi_j)$ the Saha-Eggert equilibrium ionization fractions ξ_j calculated for an interstellar environment characterized by $n_e = 1o^{-2}cm^{-3}$ and $T_e = T_o = 1o^4K$.

B) The ACR loss rates

To calculate the ACR loss rates from the heliosphere is not trivial. A broad knowledge of the ACR modulation, especially in the outer heliospheric regions, would be needed for this purpose. Despite the number of uncertainties in the theoretical description of ACR particles we may dare to derive an as reliable number for the ACR loss rates as possible on the basis of nowadays knowledge in this field.

We start out from the ACR particle streaming S(R,p) describing the flux of ACR particles in a fixed reference system like that of the sun. As explained in many papers (e.g. see Jokipii,1971) this flux is given by:

$$S(R,p) = U(R,p)\ V\ C(R,p) - \kappa(R,p)\ \partial U/\partial R \tag{19}$$

where U(R,p) is the pitch angle averaged differential density at a solar distance R and a momentum p. The solar wind velocity is denoted by V. The function C(R,p) is the Compton Getting function explained in detail e.g. by Gleeson and Axford(1968a,b), and $\kappa(R,p)$ denotes the ACR space-diffusion coefficient.

Since we are interested here in the total flux of ACR particles out of the heliosphere we have to integrate equation (19) over the momentum space. For this integrated flux a conservation law can be expected to be valid at the heliospheric shock, unless the structure of the shock itself serves as an ACR source. Thus we obtain the following equivalence:

$$\int p^2 dp\ \Big[\ U(R,p)V\ C(R,p) - \kappa(R,p)\partial U/\partial R\ \Big]_1 = \tag{20}$$

$$\int p^2 dp\ \Big[\ U(R,p)V\ C(R,p) - \kappa(R,p)\partial U/\partial R\ \Big] = \Phi_{out}$$

where the suffixes 1 and 2 at the brackets mean preshock and postshock conditions and where Φ_{out} means the total ACR flux out from the heliospheric shock front.

It is now interesting to note that the quantity given in equation (2o) under radially symmetric heliospheric conditions should be related with the total ACR production rates Π_{ij} calculated earlier in this paper by:

(21) $$\Phi_{out,j} = \int p^2 dp \left[U_j V C_j - \kappa_j \partial U_j / \partial R \right]_2 = \frac{\Pi_{ij}}{4\pi R_o^2}$$

where the heliospheric shock has been located at R_o.
Following Gleeson and Axford(1968) the Compton Getting function can be given by:

(22) $$C(R,p) = \left(1 - \frac{1}{3}\frac{1}{U} \partial/\partial T(\beta T U) \right)$$

where T is the kinetic energy of the ACR particles. The function β is written as:

(23) $$\beta(T) = (T + 2\, m_o c^2)/(T + m_o c^2)$$

and for weakly relativistic particles simplifies to $\beta \cong 2$. In the subrelativistic limit ($T \cong p^2/2m_o$) from equation (22) we can easily obtain:

(24) $$C(R,p) = \frac{1}{3} \left(1 - \frac{p}{U} \partial/\partial p\, (U) \right)_R$$

In order to calculate Φ_{out} one would need the distribution U(R,p). Here we would like to make use of the function first derived by Fisk(1976b) and also used by Klecker (1977) as a basis for his calculations of ACR energy spectra at 1 AU. This function was derived from the ACR Fokker Planck equation when only acceleration (momentum space diffusion) and convection is taken into account. At some place R_c in the outer heliosphere this leads to an energization of the original pick-up ions such that for their distribution function $U_c(R_c,p)$ build up there space diffusion starts to dominate over acceleration processes, leading to an outward diffusion at $R \geq R_c$ and to the usual inward diffusion at $R \leq R_c$ such that the following relation may roughly be valid:

(25) $$\left| -\kappa(R,p) \partial U/\partial R \right|_{R_c + \Delta R} = \left| + \kappa(R_c,p)\, \partial U_c/\partial R \right|_{R_c}$$

With the distribution $U_c(R,p)$ given by Fisk(1976a,b) and Klecker(1977) in the form:

(26) $$U_j(R,p) = C_j\, n_j\, R^4 \exp(-V\, B/3R\, (p^2/D_{pp}))$$

and the relation (25) we can now calculate the flux Φ_{out} in equation (22).
In order to achieve the form (26) for the distribution function U_j a fluctuation power index of $\gamma = 1.5$ for magnetosonic turbulence distributions in the solar wind was used by the above authors. The quantities C_j and B are constants, n_j is the local ACR density of species j, and D_{pp} is the momentum space diffusion coefficient which is considered to attain the following dependence on p:

(27) $$D_{pp}/p = D_o\, p^{\alpha-2} = A_j D_{oo}\, p^{\alpha-2}$$

with $\alpha = \gamma - 1 = 0.5$ and D_{oo} as a constant reference value. Adopting as a typical value for the diffusion coefficient of particles with 1 MeV/nucleon

(28) $$(D_{pp}/p)_r = 3 \cdot 10^{-8}\ s^{-1}$$

one retains equation (27) in the following numerical form:

$$(29) \qquad (D_{pp}/p) = 3\cdot 1o^{-8}\ A_j\ (p_r/p)^{1.5}\ s^{-1}$$

where p_r is the momentum for a 1 MeV particle with $A_j = 1$ and mass m_o.
Introducing the expression (29) into equation (26) for U_c we can then calculate the derivatives $\partial U/\partial R$ and $\partial U/\partial p$ and obtain the following results:

$$(3o) \qquad \partial U_c/\partial R = U_c \left[4/R + \frac{V\ \tau_r\ B}{3R^2 A_j} (p/p_r)^{1.5} \right]$$

and:

$$(31) \qquad (p/U_c)\ \partial U_c/\partial p = -\ 1.5\ \frac{V\ \tau_r\ B}{3\ R\ A_j} (p/p_r)^{1.5}$$

where the period $\tau_r = 1/3\cdot 1o^8$s. As done by Klecker(1977) we may assume here that the ACR acceleration terminates at R_c = 5o AU. Then we find for a constant solar wind velocity $V = 4\cdot 1o^7$ cm/s a numerical value of 1.58 for $Z = V\tau_r B/3R$. With these results we then obtain the ACR flux $S_{cj}(R_c,p)$ in the following form:

$$(32) \qquad S_{cj}(R_c,p) = 1/3 \left[1 + 1.5\ Z/A_j\ (p/p_r)^{1.5} \right] V\ U_c\ +$$
$$\kappa(R_c,p)\ U_c \left[4/R_c + Z/R_c A_j\ (p/p_r)^{1.5} \right]$$

The space diffusion coefficient κ is usually given in the following form (see e.g. Jokipii, 1971):

$$(33) \qquad \kappa(R,p) = \kappa_o\ g(R)\ \beta\ \overset{*}{R} = (5\cdot 1o^{21})\ (p/p_r)^2 \frac{1}{A_j}\ cm^2/s$$

where again κ_o is a reference value for 1 MeV/nucleon particles, $\beta = v/c$, and $\overset{*}{R}$ is the particle rigidity. The function g(R) is to take into account the observational facts on the interplanetary CR-intensity gradient and according to Jokipii(1971) at $R = R_c$ attains the following value:

$$g(R) = \exp(-(5o-1)/3o) = \exp(-49/3o) = o.2\ .$$

Now we can integrate $S_{cj}(R_c,p)$ over momentum space and keeping in mind that the constant C contains the atomic mass number A_j in the form: $C_j = C/A_j^2$ we then obtain:

$$(34) \qquad \Phi_{out,j} = \int S_{cj}(R_c,p)\ p^2 dp = \Phi_1 + \Phi_2 + \Phi_3 + \Phi_4$$

with the following definitions:

$$(35) \qquad \Phi_1 = (1o^7/1.5A_j)\ 4\ p_r^3 \int_o^\infty \chi^4\ U_{cj}(\chi)\ d\chi$$

$$(36) \qquad \Phi_2 = (1o^7/1.5A_j)\ \frac{Z}{A_j} p_r^3 \int_o^\infty \chi^{3.5}\ U_{cj}(\chi)\ d\chi$$

$$(37) \qquad \Phi_3 = (4\cdot 1o^7/3)\ p_r^3 \int_o^\infty \chi^2\ U_{cj}(\chi)\ d\chi$$

$$(38) \qquad \Phi_4 = (4\cdot 1o^7/3)\ \frac{1.5\ Z}{A_j} p_r^3 \int_o^\infty \chi^{1.5}\ U_{cj}(\chi)\ d\chi$$

where the distribution function U_{cj} of species j is given by:

(39) $$U_{cj}(\chi) = C\, n_j \frac{1}{A_j^2} \exp\left(-\frac{Z}{A_j}\chi^{1.5}\right)$$

Introduction of $\xi = (Z/A_j)\,\chi^{1.5}$ and the Gamma function $\Gamma(x) = \int_0^\infty e^{-\xi}\xi^{x-1}d\xi$ then finally leads to the following expression for $\Phi_{out,j}$:

(4o) $$\Phi_{out,j} = C\, n_j\, (p_r^3/A_j^2)\cdot 1o^7 \cdot \Big[(8/4.5)\, A_j^4/Z^5\, \Gamma(1o/3) + (2/1.5)\, A_j^{4.5}/Z^{5.5}\, \Gamma(13/3) + (8/9)\, A_j^3/Z^3\, \Gamma(2) + (8/6)\, A_j^{3.5}/Z^{3.5}\, \Gamma(3) \Big]$$

which after numerical evaluation of the Gamma functions yields:

(41) $$\Phi_{out,j} = C\, n_j\, (p_r^3/A_j^2)\cdot 1o^7 \left[o.5\, A_j^4 + 1.o\, A_j^{4.5} + 0.23\, A_j^3 + 0.54\, A_j^{3.5} \right]$$

With equations (41) and (21) we can find the ACR particle density of species j by the expression:

(42) $$n_j = \frac{\Pi_{ij}}{4\pi R_c^2} \frac{(o.5\, A_j^4 + 1.o\, A_j^{4.5} + 0.23\, A_j^3 + 0.54\, A_j^{3.5})^{-1}}{C\, p_r^3\, 10^7} A_j^2$$

The ratio of ACR particle densities n_j and n_k can thus be given by:

(43) $$(n_j/n_k) = (\Pi_{ij}/\Pi_{ik})\, \overset{*}{A}_{jk}$$

where the expression $\overset{*}{A}_{jk}$ has the form:

(44) $$\overset{*}{A}_{jk} = (A_j/A_k)^2 \frac{(o.5\, A_k^4 + 1.o\, A_k^{4.5} + 0.23\, A_k^3 + 0.54\, A_k^{3.5})}{(0.5\, A_j^4 + 1.0\, A_j^{4.5} + 0.23\, A_j^3 + 0.54\, A_j^{3.5})}$$

Making use of the values given in Tables 3 and 4, and adopting no filtering effect in the heliospheric interface plasma (i.e. $Tr_j \equiv 1$), with equations (43) and (44) we can calculate the following ACR density ratios:

TABLE 5: ACR density ratios at 5o AU

Ratio	Solar maximum	Solar minimum	Klecker(1977)
He/O	85.51 (58.15)	65.oo (44.19)	15.o
Ne/O	0.01 (0.0068)	0.007 (0.0047)	0.05
N/O	0.11 (0.10)	0.12 (0.1o2)	0.15

Here the values given in brackets are valid for an unionized interstellar medium (i.e. $(1-\zeta_j) \equiv 1$), whereas the ones without brackets are calculated on the basis of the values given for $(1-\zeta_j)$ in Table 3.

one retains equation (27) in the following numerical form:

(29) $$(D_{pp}/p) = 3\cdot 10^{-8} A_j (p_r/p)^{1.5} s^{-1}$$

where p_r is the momentum for a 1 MeV particle with $A_j = 1$ and mass m_o.
Introducing the expression (29) into equation (26) for U_c we can then calculate the derivatives $\partial U/\partial R$ and $\partial U/\partial p$ and obtain the following results:

(30) $$\partial U_c/\partial R = U_c \left[4/R + \frac{V \tau_r B}{3R^2 A_j} (p/p_r)^{1.5} \right]$$

and:

(31) $$(p/U_c)\, \partial U_c/\partial p = - 1.5 \frac{V \tau_r B}{3 R A_j} (p/p_r)^{1.5}$$

where the period $\tau_r = 1/3\cdot 10^8$s. As done by Klecker(1977) we may assume here that the ACR acceleration terminates at $R_c = 50$ AU. Then we find for a constant solar wind velocity $V = 4\cdot 10^7$cm/s a numerical value of 1.58 for $Z = V\tau_r B/3R$. With these results we then obtain the ACR flux $S_{cj}(R_c,p)$ in the following form:

(32) $$S_{cj}(R_c,p) = 1/3 \left[1 + 1.5\, Z/A_j\, (p/p_r)^{1.5} \right] V U_c + \kappa(R_c,p)\, U_c \left[4/R_c + Z/R_c A_j\, (p/p_r)^{1.5} \right]$$

The space diffusion coefficient κ is usually given in the following form (see e.g. Jokipii,1971):

(33) $$\kappa(R,p) = \kappa_o\, g(R)\, \beta\, \overset{*}{R} = (5\cdot 10^{21})\, (p/p_r)^2 \frac{1}{A_j}\ cm^2/s$$

where again κ_o is a reference value for 1 MeV/nucleon particles, $\beta = v/c$, and $\overset{*}{R}$ is the particle rigidity. The function g(R) is to take into account the observational facts on the interplanetary CR-intensity gradient and according to Jokipii(1971) at $R = R_c$ attains the following value:

$$g(R) = \exp(-(50-1)/30) = \exp(-49/30) = 0.2\ .$$

Now we can integrate $S_{cj}(R_c,p)$ over momentum space and keeping in mind that the constant C contains the atomic mass number A_j in the form: $C_j = C/A_j^2$ we then obtain:

(34) $$\Phi_{out,j} = \int S_{cj}(R_c,p)\, p^2 dp = \Phi_1 + \Phi_2 + \Phi_3 + \Phi_4$$

with the following definitions:

(35) $$\Phi_1 = (10^7/1.5A_j)\ 4\ p_r^3 \int_0^\infty \chi^4\, U_{cj}(\chi)\, d\chi$$

(36) $$\Phi_2 = (10^7/1.5A_j)\ \frac{Z}{A_j} p_r^3 \int_0^\infty \chi^{3.5}\, U_{cj}(\chi)\, d\chi$$

(37) $$\Phi_3 = (4\cdot 10^7/3)\ p_r^3 \int_0^\infty \chi^2\, U_{cj}(\chi)\, d\chi$$

(38) $$\Phi_4 = (4\cdot 10^7/3) \frac{1.5\, Z}{A_j} p_r^3 \int_0^\infty \chi^{1.5}\, U_{cj}(\chi)\, d\chi$$

where the distribution function U_{cj} of species j is given by:

(39) $$U_{cj}(\chi) = C\, n_j \frac{1}{A_j^2} \exp\left(-\frac{Z}{A_j}\chi^{1.5}\right)$$

Introduction of $\zeta = (Z/A_j)\,\chi^{1.5}$ and the Gamma function $\Gamma(x) = \int_0^\infty e^{-\xi}\xi^{x-1}d\xi$ then finally leads to the following expression for $\Phi_{out,j}$:

(4o) $$\Phi_{out,j} = C\, n_j\, (p_r^3/A_j^2)\cdot 1o^7 \cdot \Big[\, (8/4.5)\, A_j^4/Z^5\, \Gamma(1o/3) + (2/1.5)\, A_j^{4.5}/Z^{5.5}\, \Gamma(13/3) + (8/9)\, A_j^3/Z^3\, \Gamma(2) + (8/6)\, A_j^{3.5}/Z^{3.5}\, \Gamma(3)\,\Big]$$

which after numerical evaluation of the Gamma functions yields:

(41) $$\Phi_{out,j} = C\, n_j\, (p_r^3/A_j^2)\cdot 1o^7 \big[\, o.5\, A_j^4 + 1.o\, A_j^{4.5} + 0.23\, A_j^3 + 0.54\, A_j^{3.5}\big]$$

With equations (41) and (21) we can find the ACR particle density of species j by the expression:

(42) $$n_j = \frac{\Pi_{ij}}{4\pi R_c^2}\, \frac{(\, o.5\, A_j^4 + 1.o\, A_j^{4.5} + 0.23\, A_j^3 + 0.54\, A_j^{3.5})^{-1}}{C\, p_r^3\, 10^7}\, A_j^2$$

The ratio of ACR particle densities n_j and n_k can thus be given by:

(43) $$(n_j/n_k) = (\Pi_{ij}/\Pi_{ik})\, \overset{*}{A}_{jk}$$

where the expression $\overset{*}{A}_{jk}$ has the form:

(44) $$\overset{*}{A}_{jk} = (A_j/A_k)^2\, \frac{(\, o.5\, A_k^4 + 1.o\, A_k^{4.5} + 0.23\, A_k^3 + 0.54\, A_k^{3.5})}{(\, 0.5\, A_j^4 + 1.0\, A_j^{4.5} + 0.23\, A_j^3 + 0.54\, A_j^{3.5})}$$

Making use of the values given in Tables 3 and 4, and adopting no filtering effect in the heliospheric interface plasma (i.e. $Tr_j \equiv 1$), with equations (43) and (44) we can calculate the following ACR density ratios:

TABLE 5: ACR density ratios at 5o AU

Ratio	Solar maximum	Solar minimum	Klecker(1977)
He/O	85.51 (58.15)	65.oo (44.19)	15.o
Ne/O	0.01 (0.0068)	0.007 (0.0047)	0.05
N/O	0.11 (0.10)	0.12 (0.1o2)	0.15

Here the values given in brackets are valid for an unionized interstellar medium (i.e. $(1-\zeta_j) \equiv 1$), whereas the ones without brackets are calculated on the basis of the values given for $(1-\zeta_j)$ in Table 3.

As a first result one may extract from Table 5 that especially for the (Ne/O)- and the (N/O)-ratio the solar cycle variation is indicated to be fairly small. For the (He/O)-ratio it seems to be more pronounced, however, one should keep in mind that the above values are calculated for solar maximum and minimum with identical values for the ACR modulation indices α and γ. Due to the solar cycle variation in these indices possibly a slight change in the coefficients A^*_{jk} given in equation (44) may be caused, however, not of great relevance. Much more relevant these changes may be for the absolute values of the ACR densities n_j, but these we do not want to analyze here.

Comparing our values in Table 5, irrespective of the fact whether they are for solar maximum or minimum, with those given by Klecker(1977) one may recognize substantial differences. These differences are smallest for the (N/O)-ratios where we obtain 0.11 instead of 0.15 , however, they seem to be of relevance what concerns the (He/O)- and the (Ne/O)-ratios. On the other hand, the values taken by Klecker(1977) for him serve as the basis of ACR energy spectra calculated with CR-modulation theory. Since these spectra are shown to nicely fit ACR-data obtained at 1 AU, the input values for the ACR density ratios used by Klecker have to be taken as some serious constraints.

The easiest way to reconcile our ratios with those of Klecker is to take into account a differential transmission Tr_j through the heliospheric interface. Due to their relatively large charge exchange cross-sections with respect to heliospheric protons mainly LISM oxygen and nitrogen would be influenced by this effect, whereas helium and neon would be nearly unaffected. Trying to ascribe the factor 1.5 between the (N/O)-ratio given by Klecker and by us to the differential filtering effect, and extracting a factor 1.83 between the mean charge exchange rates of oxygen and nitrogen from the values given by Siscoe and Mukherjee (1972), one is lead to:

$$(45) \qquad Tr_N/Tr_O = \exp(-\int N_p \sigma_{HN} v_{rel} \frac{ds}{v_\infty}) \;/\; \exp(-\int N_p \sigma_{HO} v_{rel} \frac{ds}{v_\infty}) = 1.5$$

where N_p is the proton density in the heliospheric interface, v_{rel} their mean relative velocity with respect to LISM O, N-atoms, v_∞ the LISM wind velocity, and σ_{HN}, σ_{HO} the corresponding charge exchange cross sections. From equation (45) with the above made assumption $\sigma_{HO} \cong 1.83\ \sigma_{HN}$ one then arrives at the result:

$$(46) \qquad Tr_O \cong (1.5)^{-1/0.45} = 0.41$$

Though the above number for Tr_O does seem reasonable in view of an interface plasma density N_p of the order of $1o^{-2}cm^{-3}$(Reynolds, 1989), one may run into more troubles with it when looking at the other ACR ratios of Table 5. Since helium is not likely to suffer extinction in the interface (i.e. Tr_{He} = 1.o) the ratio in Table 5 by applying Tr_O = 0.41 may even be raised to (n_{He}/n_O) = 85.51/o.41 = 2o8.56 which is far off Klecker's value. For neon the discrepancy would nicely be diminished, since adopting also for neon no interface extinction (i.e. Tr_{Ne} = 1.o) and applying Tr_O = o.41 would reconcile our ratio with that given by Klecker within a factor of 2.

The discrepancy really hard to overcome is in the (He/O)-ratios. As it seems to us the only way out is to assume a principal difference in the acceleration of He^+-pick-up ions to high energies compared to the other LISM pick-up ions. Klecker(1977) has adopted an efficiency of $\varepsilon = 2\cdot 1o^{-3}$ for all ion species to become ACR particles. There may be, however, a good reason why helium behaves different from the other ions. This is connected with the fact that LISM helium is ionized only fairly close to the sun. On their much longer way out to R_c these ions may partly have a chance to assimilate to the solar wind ion population and escape from acceleration to ACR energies. Also these ions are convected outwards preferably in the down-wind region as shown by observations (Möbius et al.,1985,1986,1988) suggesting the possibility that in this region the acceleration efficiency is lower, perhaps because of a larger distance to the heliospheric shock (i.e. R_c) as indicated by Baranov (1989),Baranov and Lebedev (1989), Shima et al.(1986), and Matsuda et al.(1989).

REFERENCES:

Axford,W.I.(1972):In "SOLAR WIND II",NASA-Special Report SP-3o8, pp 6o9-653

Axford,W.I.(1977):"The three dimensional structure of the interplanetary medium",in "Study of Travelling interplanetary processes, pp. 326

Baranov,V.B.(1989) COSPAR-symposium on"Physics of the outer heliosphere",Warsaw(Poland),Sept. 1989,Pergamon Press,(this volume) to appear

Baranov,V.B.,and Lebedev,M.G.(1989) Space Science Reviews, in press

Błeszinski,S.(1987) Astron.Astrophys.,**18o**,2o1

Cameron,A.G.W.(1973) Space Sci.Rev.,**14**,392

Cox,D.P. and Reynolds,R.J.(1987):In" Ann.Rev.Astron,Astrophys.",**25**,3o3

Fahr,H.J.(1986) Advances Space Res.,**6(2)**,13

Fahr,H,J.(1989):"The LISM oxygen presence in the heliosphere in analytic representation", paper submitted to Astron.Astrophys.

Fahr,H.J. and Neutsch,W.(1983) Astron.Astrophys.,**118**,57

Fahr,H.J.,and Ripken,H.W. (1984) Astron.Astrophys.,**139**,551

Fisk,L.A.(1976a) J.Geophys.Res.,**81**,4633

Fisk,L.A.(1976b) J.Geophys.Res.,**81**,4641

Fisk,L.A.(1976c) Astrophys,J.,**2o6**,333

Fisk,L.A.(1986) In"The sun and the heliosphere in three dimensions",pp.4o1-413, Ed.by R.G.Marsden, D.Reidel Publ.Comp.,Doordrecht Holland

Fisk,L.A.,Kozlovsky,B.,and Ramaty,R.(1974): Astrophys.J.,**19o**,L35

Frisch,P.C.,York,D.G.,Fowler,J.R.(1987) Astrophys.J.,**32o**,842

Frisch,P.C.,Welty,D.E.,York,D.G.,and Fowler,J.R.(1989) Astrophys.J.(submitted)

Garcia-Munoz,M.,Mason,G.M.,and Simpson,J.A.(1973) Astrophys.J.,**182**,L81

Garcia-Munoz,M.,Mason,G.M. and Simpson,J.A.(1975),Astrophys.J.,**2o2**,265

Gleeson,L.J. and Axford,W.I.(1968a) Astrophys.J.,**154**,1o11

Gleeson,L.J. and Axford,W.I.(1968b) Astrophys.Space Sci.,**2**,431

Gleeson,L.J. and Urch,I.H.(1971) Astrophys.Space Sci.,**11**,288

Isenberg,P.A.(1987) J.Geophys.Res.,**92**,1o67

Jokipii,J.R.(1971) Rev.Geophys.Space Res.,**9**,27

Jokipii,J.R.(1986) In:"The sun and the heliosphere in three dimensions",pp.375-389, Ed.by.R.G.Marsden,D.Reidel Publ.Comp.,Doordrecht Holland

Klecker,B.(1977) J.Geophys.Res.,**82**,5287

Klecker,B.,Hovestadt,D.,Gloeckler,G. and Fan,C.Y.(1977),Astrophys.J.,**212**,29o

Lazarus,A.and Belcher,J.(1988) SOLAR WIND VI ,pp. 533 ,(Boulder,Colorado(USA)),Ed.by T.E.Holzer

Matsuda,T.,Fujimoto,Y.,Shima,E.,Sawada,K. and Inaguchi,T. (1988), Preprint,Kyoto University, Dept.of Aeronautical Engineering,

Möbius,E.D.(1986) Advances Space Res.,**6**,199

Möbius,E.D.,Hovestadt,D.,Klecker,B.,Scholer,M.,Gloeckler,G.,Ipavich,F.M., and Lühr,H.(1986) Nature,**318**,426

Möbius,E.D.,Klecker,B.,Hovestadt,D., and Scholer,M.(1988):Astrophys.Space Sci.,**144**,487

Parker,E.N.(1963):In" Interplanetary dynamical processes",Interscience Publishers,New York

Reynolds,R.J.(1989):"Lower limit on the ionization fraction of the very local interstellar medium",Preprint, Symposium on COSMIC GAS DYNAMICS,Moscow(USSR),Oct.1989

Ripken,H.W.and Fahr,H.J.(1983) Astron.Astrophys.,**122**,181

Siscoe,G.L. and Mukherjee,N.R.(1972) J.Geophys.Res.,**77**,6o42

Shima,E.,Takuya,M. and Inaguchi,T.(1986) Monthly Not.Roy.Astron.Soc.,**221**,687

Torr,D.G. and Torr,M.R.(1979) J.Atmosph.Terr.Phys.,**41**,797

Torr,M.R. and Torr,D.G.(1985) J.Geophys.Res.,**9o**,6675

Wallis,M.K.(1984) Astron.Astrophys.,**13o**, 2oo

THE INTERACTION OF INTERSTELLAR PICK-UP IONS WITH THE SOLAR WIND — PROBING THE INTERSTELLAR MEDIUM BY *IN-SITU* MEASUREMENTS

E. Möbius

Max-Planck-Institut für Physik und Astrophysik, Institut für extraterrestrische Physik, 8046 Garching, F.R.G.

ABSTRACT

The local interstellar medium (LISM) and its interaction with the heliosphere has been investigated by using the resonant scattering of solar UV at the interstellar gas. The results seem to suggest two different temperatures for H and He and an abundance ratio different from the cosmic abundances, thus indicating a selective filtering at the heliospheric boundary. Measurements of the pick-up ion distribution in the heliosphere provide a new technique to determine the LISM characteristics and the interaction with the solar wind on the basis of totally independent information: the ionization rate can be directly obtained from the velocity distribution of pick-up ions as an image of the radial neutral distribution, and the ion flux and its variation along the Earth's orbit reveal the dynamics, temperature and density of the interstellar gas. After a comparison of the two methods recent results from the He pick-up ion measurements as obtained with AMPTE/IRM in the year 1985 are reviewed. The He temperature and density seem to be in accordance with the H temperature from UV backscattering results and with the cosmic abundance ratio. With future space missions like SOHO, CRAF and CASSINI the capabilities of the pick-up ion method with respect to the study of the elemental and isotopic composition of the LISM and the long term variations will be greatly enhanced.

INTRODUCTION

The interaction between the interstellar neutral gas and the interplanetary medium has been a topic of wide interest for many years in theory as well as experimental investigation. In the early 1970s the first observational evidence of interstellar hydrogen in the solar system was found by means of spaceborn UV scattering measurements /1, 2/, followed by measurements of helium /3/. These observations initiated a rapid development in the study of the interaction processes. For an overview on this topic the reader may be referred to the reviews by Axford /4/, Fahr /5/, Holzer /6/ and Thomas /7/. A very recent review on the local interstellar medium (LISM) within about 100 pc around the sun has been presented by Cox and Reynolds /8/. The solar system is moving with respect to the LISM with a relative velocity of about 20 km/s. Thereby the interstellar gas is streaming through the heliosphere like an interstellar wind, which - when approaching the sun - is subject to the forces of solar gravitation and radiation pressure. On the other hand the interstellar neutrals are ionized by solar UV radiation, by charge exchange with the solar wind ions and by electron collisions. These newly created ions are then picked up by the solar wind via interaction with the interplanetary magnetic field and finally swept out of the heliosphere. Thus the spatial distribution of the interstellar neutral gas and the fluxes of pick-up ions in the interplanetary space contain the information on the temperature, the relative velocity and the abundance of each species of the interstellar gas in the LISM just outside the solar system.

The spatial distribution of interstellar neutral gas in the heliosphere has been extensively probed in the past using UV scattering methods. These techniques have suggested the puzzling result that the temperatures of H and He in the LISM seemed to be different by about a factor of 2. Also the measured H and He densities seemed to describe an abundance ratio different from cosmic abundances, although with large uncertainties. Models of the interaction between the LISM and the heliosphere had difficulties in explaining the different gas temperatures, and the recent results on elemental abundances in the anomalous component of cosmic rays, which is interpreted in terms of interstellar gas penetrated into the solar system being ionized and then accelerated, seem to contradict a significant selective filter effect at the heliospheric boundary. The direct measurement of interstellar pick-up ions, developed very recently /9/, constitutes a very capable method to derive the parameters of the interstellar medium independent from the informations gained with UV scattering techniques. In this report we will discuss the outstanding problems, compare the various techniques, review the first results of the pick-up ion technique and finally give an outline of its future perspectives taking advantage of improved instrumentation and long duration deep space missions.

OUTSTANDING PROBLEMS OF THE LISM-HELIOSPHERE INTERACTION

To understand the physical processes involved as well as the outstanding problems of the state of the LISM and its interaction with the heliosphere it is necessary to note that the interstellar medium is a (still unknown) mixture of a neutral and ionized component. Figure 1 indicates the situation for a partially ionized gas flowing with respect to the solar system. The solar wind with its intrinsic magnetic field creates a bubble which cannot be penetrated by a plasma but is transparent to neutral gas. Therefore, only the neutral component of the interstellar gas can be studied within the solar system and deviations from the cosmic abundances in this sample could, in principle, be interpreted as an image of the degree of ionization of the various species in the LISM. However, it is even not clear yet, whether all the neutral gas reaches the solar system unimpeded with its original composition and velocity distribution. In fact, the only information on the LISM available to us in the past as derived from UV scattering measurements seemed to suggest a higher temperature for interstellar He as compared with H and an abundance ratio which differs from the cosmic abundance of these species (c.f. /10, 11/). Several attempts were made to explain these deviations from the expected undisturbed LISM by selective filtering and heating in the heliospheric boundary layer (e.g. /12, 13, 14/). While the neutral gas follows a flow which is determined by the relative velocity of the LISM and the sun with a deviation only close to the sun due to gravitational forces, the plasma flow is diverted around the system already outside the heliopause. Because of the interplanetary magnetic field the interaction with the interstellar plasma shows similarities to that of the solar wind with the magnetosphere. Depending on whether the flow of the interstellar gas is super- or subsonic, this zone of diverted flow is restricted to a region which is limited by a bow shock or which gradually approaches the undisturbed LISM.

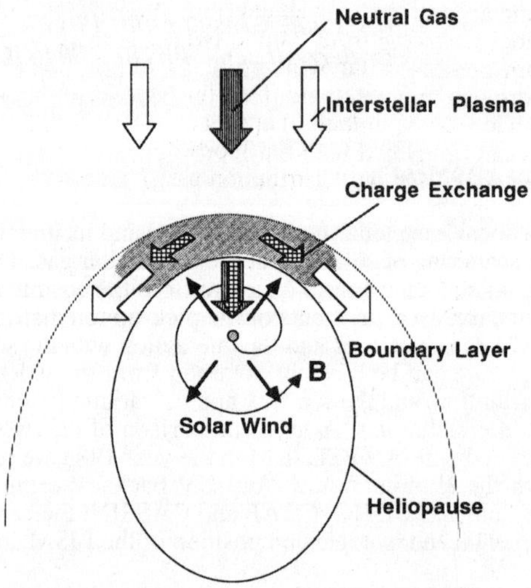

Fig. 1: Interaction of the heliosphere with a partially ionized interstellar gas. Interstellar plasma is diverted around the heliosphere (open arrows), while the heliopause is transparent for neutral gas (hatched arrows). Significant charge exchange with H^+ ions may push part of the neutral gas into the plasma flow (cross-hatched arrows), possibly applicable for H and O.

Via charge exchange collisions the neutral interstellar gas can also be affected by the flow pattern of the interstellar plasma at the heliospheric boundary /12, 14/. Depending on their charge exchange cross-sections with hydrogen, the main component of the interstellar gas, the abundance of various constituents may differ significantly from the original abundance in the LISM. Resonant charge exchange (with cross-sections as high as $\approx 2\cdot10^{-15}$ cm^2) occurs for hydrogen and also for oxygen, since its ionization potential is almost identical to that of hydrogen. For all the other species including gases like He, Ne and N the cross-sections are lower by at least one order of magnitude. Therefore, significant abundance variations of neutral hydrogen and oxygen may occur across the heliospheric boundary, if the effect of charge-exchange in the boundary layer is sufficiently strong. In addition, the effect of charge-exchange can be disentangled from abundance reductions due to a high degree of ionization of the species already in the LISM, since noble gases like He and Ne with a high ionization potential as well as gases with an ionization potential close to that of H like N will experience no effect from the boundary layer.

In order to give a crude estimate of the expected effect, the mean free path for charge exchange is compared with the potential dimensions of the heliopshere and its boundary layer. To remove about 50% of the neutral gas from the original flow the condition $d/\lambda \geq -\ln 0.5$ is required, where d is the thickness of the bounday layer and λ is the mean free path length for charge exchange. With a charge exchange cross-section $\sigma \approx 2\cdot10^{-15}$ cm^2 and a H^+ density of 0.05 cm^{-3} (assuming a high degree of ionization) $d \geq 400$ AU results. This value is of the order of the current estimate for the dimensions of the heliosphere and could as well apply to the thickness of the boundary layer. However, if the thickness of the layer is thinner by more than a factor of 2 and/or the degree of ionization is significantly lower, the effect of the boundary layer on the composition of the interstellar gas in the solar system drops below 20% and thus will be almost impossible to derive from measurements with the current accuracy.

In summary, detailed measurements of the spatial distribution of interstellar gas and pick-up ions in the interplanetary space - beyond the temperature, relative velocity and densities of the LISM - will also contain information on the degree of ionization and the importance of the heliospheric boundary layer. Therefore, a comprehensive study to achieve the necessary pieces of information on our immediate neighbourhood in space should include all the methods which are available to us. This paper will in particular elaborate on the relative importance and the potential future development of the measurement of interstellar pick-up ions.

METHODS TO STUDY THE LOCAL INTERSTELLAR MEDIUM

Interstellar Absorption Lines

A well-known method to study the undisturbed LISM within the distance of up to 100 parsec from the solar system is the measurement of interstellar absorption lines in the light of nearby stars. This method allows the determination of abundances, temperatures and ionization degree integrated over the line-of-sight to the individual stars. As a result a map of the LISM within $\approx$ 100 AU emerged with the sun being close to one of the dense cold interstellar clouds /8, 15/. The ionization degree for H is estimated to $\leq$10%. However, because of the integration over the line-of-sight no conclusion about the situation in the immediate neighbourhood of the solar system can be drawn.

Scattering of Solar UV Radiation at the Interstellar Gas

The method which has been developed to the greatest detail and which has been used for a long period, is based on the resonant scattering of solar UV line radiation at interstellar atoms in the heliosphere /8, 11, 16, 17/. With a photometer either sensitive to Lα (for hydrogen atoms) or He 584 Å (for helium atoms) radiation, the full sky is scanned by using the spacecraft spin and a slow variation of the spin axis, as indicated in Fig. 2. As a result the integral scattered light intensity is obtained in each direction, from which the spatial distribution of the scattering centers (interstellar atoms) is derived by a fit of a model distribution to the measured UV intensities.

Pick-up Ions in the Solar Wind

In contrast to the UV scattering which is a remote sensing technique with an integral result, the detection of pick-up ions allows a local in-situ measurement of interstellar particles. When freshly ionized due to solar UV, charge exchange or solar wind electrons, the ions of interstellar orgin are injected into the solar wind frame of reference with the solar wind velocity, i.e., in this rest frame they start gyrating around the interplanetary magnetic field at an oblique angle according to the actual field direction. For the observer these ions initially perform a cycloidal motion with the velocity varying from 0 up to twice the solar wind velocity. Due to this behaviour they can be clearly differentiated from genuine solar wind (Fig. 3). While most of the constituents of the interstellar gas, including the bulk of hydrogen, are already ionized far beyond the orbit of the Earth, noble gases like helium and neon (because of their high ionization potential) approach the sun to much less than 1 AU /18/. Thus only a limited sample of the interstellar gas can be studied with Earth orbiting spacecraft. The first positive identification of interstellar He^+ pick-up ions and their actual velocity distribution in the solar wind has been reported by Möbius et al. /9/.

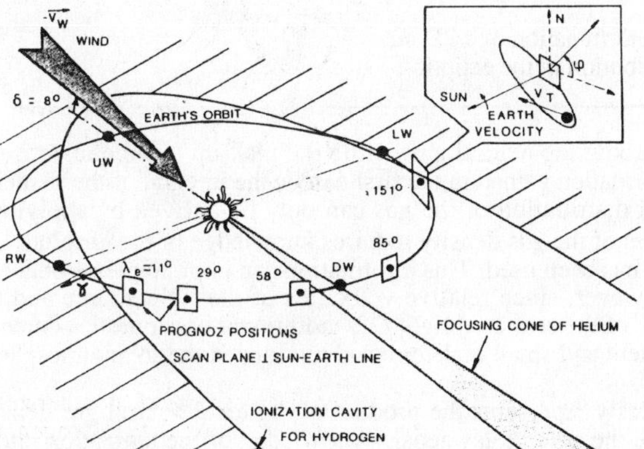

Fig. 2: Schematic for the interstellar gas distribution measurements based on resonant scattering of solar EUV lines (from /17/).

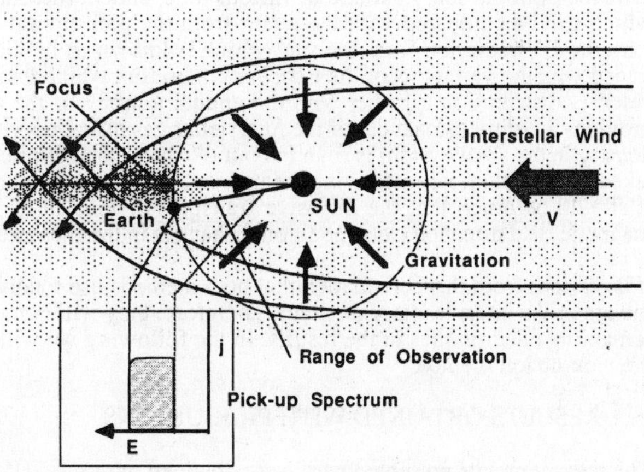

Fig. 3: Schematic for the pick-up ion measurements using the annual journey of the Earth around the sun.

Anomalous Component of Cosmic Rays

On reaching the termination shock of the solar wind the interstellar pick-up ions can be further accelerated to cosmic ray energies due to their interaction with the shock /19/. Therefore, these pick-up ions have been discussed as the seed population for the well-known anomalous component of cosmic rays /19, 20/. Based on this assumption also the measurement of the flux ratios of the anomalous cosmic rays can be used to derive the composition of the interstellar gas in the heliosphere, provided that an adequate model on the ion acceleration is applied. By reproducing the full energy spectra of the anomalous cosmic rays /21/ it has been demonstrated that the abundance of species like N, O and Ne would be consistent with their cosmic abundance ratios within the accuracy of this method.

Comparison of the Different Techniques

Since no local values on the LISM applicable to the location of the solar system can be derived from the interstellar absorption lines, and the anomalous cosmic rays only lead to global abundance ratios after inferring various model assumptions, the direct comparison of the measurement techniques will be restricted to the UV scattering and the pick-up ion measurements. With these methods two almost totally independent methods to determine the distribution of interstellar gas in the solar system are now available. Although the pick-up ions can only be detected after the interstellar neutral particles have been ionized and then accelerated by the combined forces of the interplanetary magnetic and convection electric fields, this constitutes the first direct particle measurement of interstellar matter except for high energy cosmic rays which penetrate into the solar system. Measuring the distribution of particles is a totally different approach as compared with the collection of scattered UV light which has several advantages but also introduces some new complications. The major differences of both methods have been compiled in a direct comparison in Table 1.

TABLE 1: Comparison of the Methods to Determine the LISM Parameters

UV Scattering	Pick-Up Ions
Neutrals measured	Ions measured
$\int N_o I \, ds$ measured	$N_o \cdot \nu_{ion}$ measured
Light intensity and line profile needed	Ionization rate needed (from the velocity distribution)
Information from scattered light	Direct particle measurement
Integral along the line of sight	Differential measurement along the S/C - sun line and the Earth's orbit
-> inference of a 3D distribution after deconvolution of the integral	-> direct determination of a 2D cut of the distribution in the ecliptic

The UV scattering measurements provide information about the neutral gas, while the pick-up technique concerns ions, a secondary product of the interstellar gas. The information gained in the first case is the integral of the scattered light along the actual line-of-sight (Fig. 2). The spatial distribution of the gas can only be derived by applying a thorough deconvolution of the data, and the determination of the gas density requires knowledge of the absolute UV flux and the shape of the specific resonance line which is has been used. This information can in principle be obtained from independent measurements of the solar lines. However, since relative velocities of the light source and the probed gas atoms play an important role for the intensity of the scattered light /22/, additional uncertainties come in from the variable line profiles and line shifts due to turbulent and small scale directed motions in the solar atmosphere.

In the case of pick-up ions, the measured ion flux basically represents the product of the local neutral gas density times the ionization rate ($N_o \cdot \nu_{ion}$), i.e., in order to derive the neutral gas density knowledge of the ionization rate is required. As has been demonstrated recently /23/, the ionization rate can be determined from the velocity distribution of the pick-up ions, which is an image of the neutral gas distribution upstream of the observer as a function of the distance from the sun. In addition, various independent measurements of the ionization rate of e.g. He are available, which can be compared with the result from the pick-up data (e.g. /24/). Short time variations of the intensity of extreme UV radiation from the sun, which is known to be the major source of ionization for He (e.g. /6/), can be traced with the variations of the F10.7 cm radio flux which varies almost proportional to the EUV intensity /24/. The velocity distribution together with the annual journey of the observer (an Earthbound spacecraft) around the sun, provides a 2-dimensional cut of the interstellar gas distribution in the ecliptic plane (Fig. 3). The maximum of the He density in the solar system is found about 6° above the ecliptic plan - this orientation of the observer plane has to be taken into account, when deriving the interstellar gas temperature and relative velocity from the measured distribution. It should be noted that the diffusive transport of ions in the solar wind and the variations of the interplanetary magnetic field direction cause additional changes in the measured distribution.

It is obvious from this compilation that both methods use totally independent pieces of information to derive the parameters of the interplanetary medium. Therefore, they will complement each other in an ideal way in order to remove remaining uncertainties in the results. In the following we will concentrate on the results and future perspectives of the pick-up ion method.

RESULTS OBTAINED WITH PICK-UP IONS

The measurements presented here were obtained with the SULEICA (Suprathermal Energy Ionic Charge Analyzer) instrument of the Max-Planck-Institut and the University of Maryland onboard the AMPTE/IRM spacecraft. The IRM was launched on August 16 1984 into a highly elliptical orbit with an apogee of $\approx 18.7 \, R_E$ and was operational till August 12, 1986. Between July and December the satellite spent a large fraction of each orbit in the solar wind upstream of the Earth's bow shock. The SULEICA instrument combines the selection of incoming ions according to their energy per charge by a spherical section electrostatic analyzer with a subsequent time-of-flight analysis and the final measurement of the residual ion energy in a silicon surface barrier detector. With these techniques the ions are differentiated according to their mass, ionic charge, and energy. The energy range of the instrument from 5 to 270 keV/e is covered by stepping the analyzer voltage in up to 24 logarithmically spaced voltage steps synchronously with the spacecraft spin. The fan-like aperture of the electrostatic analyzer covers a solid angle of 10° in azimuth and 40° in polar angle symmetric to the plane perpendicular to the spin axis. The directional information in azimuth is provided by a sectoring scheme with 16 sectors for H^+ and He^{2+} and eight sectors for all other ions, including He^+ which is of interest here. A more detailed description of the instrument may be found elsewhere /25/.

<u>Informations from the Velocity Distribution</u>

New-born ions in the interplanetary space are immediately subjected to the combined forces of the interplanetary magnetic field and the convective (V_{SW} x B) electric field. Thus the ions form a conical velocity distribution in the rest frame of the solar wind with the pitch-angle α, which corresponds to the angle between the interplanetary magnetic field and the solar wind velocity. By means of a very efficient pitch-angle scattering at Alfvén waves in the solar wind, a spherical shell distribution in velocity space is formed with the radius V_{sw} within less than 0.1 AU convective transport by the solar wind. Finally, the shell shrinks according to the adiabatic cooling in the expanding solar wind, while being convected away from the sun. Thus the pick-up ions fill a sphere in velocity space with radius V_{SW} in the rest frame of the solar wind according to the spatial distribution of the ion production along the sun-spacecraft line. A detailed model calculation for the pick-up ion distribution and its comparison with the measured energy spectra (example in Fig. 4) showed that this idea leads to a good quantitative description and can be used to derive the spatial distribution of the interstellar gas upstream of the observer /23/. The crucial results of this analysis are the following:

1) The shape of the measured energy spectrum limits indeed the mean free path length for pitch-angle scattering to ≤ 0.1 AU, and an efficient energy diffusion on time scales similar to that of the adiabatic cooling can be ruled out.
2) The measured value of the differential energy flux density of pick-up ions is insensitive to variations of the solar wind velocity.
3) The variation of the neutral gas density with distance from the sun which is determined by the actual ionization rate is mapped to the energy spectrum of the pick-up ions. Therefore, the ionization rate can be directly derived from the spectral slope.

It should be emphasized here that all the experimental results on pick-up ions are presented in terms of differential energy flux density, i.e. the energy flux density of pick-up ions as seen per energy and angular interval. It is indeed a differential flux value and does <u>not</u> represent the total pick-up He flux integrated from the sun to the observer, as is being discussed in /26/. The reason that the measured differential energy flux density is of the same order of magnitude as the total particle flux is due to the fact that the energy spectra are flat and that the energy range for the integration is identical with the cut-off energy, i.e. the typical energy with which to multiply the differential ion flux to obtain the energy flux density. A quantitative comparison with the source strength of the pick-up ions ($N_o \cdot v_{ion}$) and thus the neutral gas density, requires a detailed modelling of the ion distribution and exact integration over the instrument response /23/, since the above relationship is not an exact one to one relation, but contains explicitly the transformation due to the adiabatic cooling and the variation of the solar wind velocity with the distance from the sun.

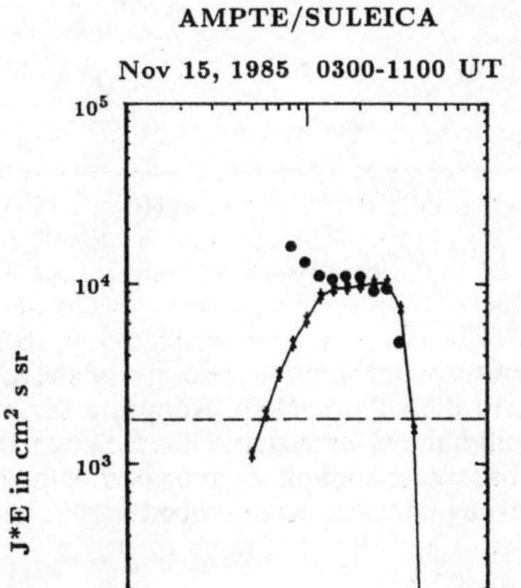

Fig. 4: Measured pick-up ion energy spectrum together with a model calculation based on the combination of pitch-angle scattering and adiabatic deceleration (adapted from /23/).

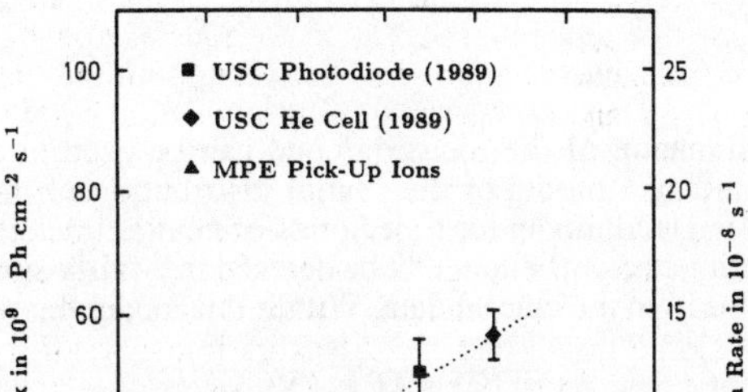

Fig. 5: Comparison of ionization rates for He at 1 AU as derived from pick-up ion spectra /23/ and from direct measurements of the solar EUV radiation /24, 27/.

From the analysis of the pick-up ion energy spectra an ionization rate ν_{ion} for He at 1 AU ranging from 6.5 - 9.1·10^{-8} s^{-1} has been derived for November 15, 1985 /23/. The actual F10.7 cm radio flux was 76 $Wm^{-2}s^{-1}$. This result can be compared with other independent determinations of the ionization rate of He by measuring the UV flux in the appropriate wave length range /24, 27/. Fig. 5 shows these results together with our result from the pick-up ions as a function of the F10.7 cm flux. The pick-up ion result seems to particularly extend the range of the EUV measurements to lower solar activities. This emphasizes the fact that the pick-up ion data were taken during the last solar minimum in 1984 and 1985. It should be noted here that from the pick-up ions the total ionization rate is derived. There is no discrimination between different sources for the ionization. Therefore, pure UV ionization may fall at the lower limit of the range given here. According to /28/, the contribution from charge exchange and electron collisional ionization are of the order of 20% of the total ionization rate. Since the total ionization rate rules the pick-up ion distribution, the further use of this value is justified. In contrast to UV ionization and charge exchange, which vary as $1/r^2$ with the distance from the sun, electron collisional ionization is more efficient closer to the sun /29/. However, the change of the neutral gas distribution in the vicinity of the sun due to this effect is so small that it cannot be resolved with the current accuracy of the pick-up ion velocity distribution.

Variation of the Ion Flux along the Orbit of the Earth

A spacecraft, which is located near the Earth within the solar wind, in principle can determine the full two-dimensional cut of the neutral gas distribution in the ecliptic inside 1 AU. In reality the coverage is restricted to those time periods when the satellite orbit leads beyond the bow shock, except for a spacecraft like ISEE-3 in a halo orbit around the Lagrangian point L1. The orbit of the AMPTE/IRM spacecraft reached into the solar wind during the 2 years of operation (1984 through 1986) between July and January. The variation of the pick-up ion energy flux density with time for 1985 is shown in Fig. 6 as a function of ecliptic longitude. For each data point the mean value of the flat part of the energy spectra has been obtained. A distinct maximum in the pick-up ion flux is seen in early December, which corresponds to about 70° ecliptic longitude, i.e., when the Earth crosses the gravitational focusing cone of the interstellar gas on the downwind side of the solar system. This measurement results in a local cut through the focusing cone within the Earth´s orbit.

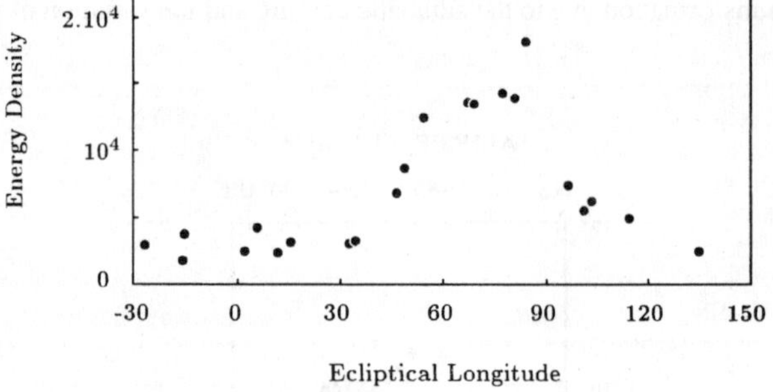

Fig. 6: Variation of the differential energy flux of interstellar He^+ pick-up ions with ecliptical longitude, as obtained during September to December 1985, after correction for the daily variation of the EUV flux and for the initial transport perpendicular to the interplanetary magnetic field.

The signature of the focusing cone can be used to derive the temperature and relative velocity of the LISM by introducing a model of the spatial distribution of the interstellar gas in the solar system. Adopting the so-called modified cold model for trajectories of neutral interstellar particles through the solar system /30/, the key parameters of the interstellar helium can be derived in a fairly simple way, but still accurate enough as compared with the errors induced from the current data. Within this model the width Θ of the focusing cone can be described by

$$\tan(\Theta/2) \propto (T_{ISM}/V_{ISM})^{1/2} \quad (1)$$

where T_{ISM} and V_{ISM} are temperature and relative velocity of the undisturbed interstellar medium. Thus from the width only the combination of the temperature and relative velocity can be determined. Therefore, it is advantageous to use the ratio of the maximum to the minimum density (in the cone and base level of the density) which reads

$$N_{Max}/N_{Min} \propto (2\pi G M_o m/RkT_{ISM})^{1/2} \quad (2)$$

where M_o and m are the mass of the sun and the atom and R is the distance from the sun. It is described by the ratio of the potential and the kinetic energy of the interstellar atoms. With the knowledge of the ionization rate, which describes the depletion of the cone close to the sun (already derived from the energy spectrum of the pick-up ions), the temperature of interstellar helium follows directly from the density ratio, which corresponds to the measured pick-up ion flux ratio. In addition, the undisturbed helium density in the LISM can be derived for a known ionization rate.

However, the quantitative determination of the spatial distribution of interstellar neutrals depends on a careful correction of the measured pick-up ion fluxes to variations of the interplanetary magnetic field and the solar UV flux. Only under the idealized assumption of a purely radial transport of the pick-up ions with the solar wind and an ionization rate which does not vary with time, would the measured cone structure be a one-to-one image of the density distribution. In reality the cone structure is significantly broadened and reduced by the following effects:

1) The solar extreme UV radiation varies significantly with time according to the solar activity, which leads to a corresponding variation of the ionization rate and thus the pick-up ion flux with time. Therefore, the fluxes were corrected using the published daily variation of the solar F10.7 cm radio flux, which is considered to vary in accordance with the EUV radiation. This is confirmed by the comparison of the various measurements of the ionization rate for helium (Fig. 5).

2) Initially the pick-up ions are transported perpendicular to the interplanetary magnetic field, until the pick-up distribution has been isotropized by pitch-angle scattering. This leads to an apparent shift of the individual data points in longitude, which has been corrected according to the actual magnetic field direction assuming a scattering mean free path length of 0.1 AU.

The measured variation of the flux of pick-up ions with ecliptic longitude after these corrections are shown in Fig. 6.

3) The pitch-angle scattering also leads to a spatial diffusion which broadens the focusing cone. The broadening has been calculated under the same assumptions as above.

4) Finally, it is known from UV scattering measurements that exact cone position is located about 6° south of the ecliptic. The reduction of the maximum ion flux has been calculated under the assumption of a Gaussian distribution with the measured width.

Corrections 3) and 4) have been applied to the flux ratio derived from Fig. 6 before calculating the LISM parameters from this result. The corrected value of N_{Max}/N_o is ≈ 7. According to relation (2) the measured the range of the He temperatures is T_{He} = 7000 - 12500 K and the undisturbed He density is $N_{He} = 5 - 7 \cdot 10^{-3}$ cm^{-3}. The error bar results from the uncertainty of the ionization rate, which currently is the major uncertainty factor of the pick-up ion method. The determination of the LISM parameters from the pick-up measurements are described in more detail elsewhere /31/. The temperature is consistent with the values derived from these pick-up ion results using a numerical model of the neutral He distribution in the solar system /28/.

Comparison with Results from Other Methods

In comparison with the determination of the helium temperature using UV scattering techniques /32, 33, 34/ it is apparent that the value derived from the pick-up ion distribution is significantly lower than most of the values given earlier, which vary around 16000 K. However, it is quite consistent with the temperatures (8000 K) found for interstellar Hydrogen from the Doppler width of the scattered UV light /17/. Only very recently it was outlined that a consistent parameter set for interstellar helium with a temperature of ≈ 10000 K may be found, assuming a red shift of the solar UV lines /22/. These results seem to indicate that finally there may be no temperature difference between H and He in the LISM, as has been extensively debated in the literature.

The interstellar He density derived from the pick-up ion flux falls at the lower limit of the density range reported from UV measurements ($3.5 \cdot 10^{-3}$ - $3.5 \cdot 10^{-2}$ cm^{-3}) /35, 32, 36, 37, 33, 34/, and somewhat below the value of 0.01 ± 0.0045 cm^{-3}, quoted most recently /22/ in agreement with the earlier results found in /32, 36, 37/. On the other hand the densities of H cover the range 0.02 - 0.7 cm-3 /2, 32, 37, 22/ with a most recent value of 0.065±0.01 cm-3. The most recent values for both species, although still with a large uncertainty, would indicate a H/He ratio of ≈6.5, i.e., somewhat short of the cosmic abundance ratio. Taking the He density derived from the pick-up ion distribution, would lead to abundance ratios of ≈ 8 - 13, which would be compatible with the cosmic abundance ratio. However, given the current error bars this result cannot close the debate on the interaction of the LISM with the heliosphere, although the arguments calling for a filter effect at the heliospheric boundary seem to be somewhat weaker than in the past. The latter result is also in accordance with the abundance ratios found in the anomalous cosmic rays /21/. In order to arrive at conclusive results strong efforts should be made to reduce the error bars and to extend the capabilities of the pick-up ion method to other species of the interstellar gas. It can then be more sensitive to the effects of the degree of ionization in the LISM and to a possible filtering at the heliospheric boundary.

FUTURE DEVELOPMENTS

To achieve the goals mentioned above and to go significantly beyond the limitations of the existing data sets, the instrumentation has to be improved with respect to sensitivity, energy range and resolution in velocity space and the measurements have to be extended in their temporal and spatial coverage.

Velocity Distribution of the Pick-up Ions

The coarse angular and energy resolution of the current data set limits the study of the pick-up velocity distributions in their ability to discriminate between various physical processes involved, e.g. adiabatic focusing and energy diffusion. The CIS experiment /38/, which is under development for CLUSTER to measure 3-dimensional distribution functions of ions in the energy range from a few eV to 40 keV/charge throughout the magnetosphere, will be able to produce detailed pick-up ion distributions during the periods in the solar wind. Adjacent energy samples will be 15% apart and the angular resollution will be 22.5^o in one direction and 11.25^o in the other direction. This should be sufficient to delineate the individual physical processes in the development of the distribution function in addition to pitch angle scattering and adiabatic deceleration.

Variability of the Ionization and the Gas Distribution

In order to delineate the temporal variation of the ionization and the related interstellar gas distribution in the interplanetary space, a continuous monitoring of both the pick-up ion distribution and the solar EUV flux, preferably over one complete solar cycle, has to be carried out. The tools for this task are basically set up by the WIND and SOHO spacecraft with the appropriate particle instrumentation /39/ located in the Lagrangian point L1. Although SOHO also carries EUV instrumentation for detailed studies of the solar atmosphere, the EUV continuum essential for the ionization of He is not particularly covered by the payload. Therefore, an augmentation of the SOHO payload by a simple total EUV flux measurement, e.g., by using a photodiode /27/, or at least the continuous monitoring of the total solar EUV flux from other platforms, possibly in Earth orbit, should be seriously considered. Such an augmentation would serve several needs in one:

1) The simultaneous measurement of pick-up ions and solar EUV allows a precise determination of the helium distribution and its absolute density in the heliosphere and therefore the paramenters of the LISM.
2) The fact that the ionization rate can also be determined directly from the pick-up distribution allows a cross-calibration with the long term measurement of the EUV radiation. Since the SOHO spectrometer instruments like SUMER /40/ have the capability to occasionally scan the total EUV range including the ionization continuum of

He, this recalibration would be particularly useful. It is well-known that long term drifts in the sensitivity due to degradation of the optical surfaces may occur under the strong solar irradiation.

Composition of the Interstellar Gas

Up to now only He^+ pick-up ions of interstellar origin have been detected. This is due to the fact that helium is the element with the highest ionization potential and therefore comes closest to the sun still as neutral. Flying time-of-flight spectrometers on deep space missions like CRAF and CASSINI will give access to the elemental composition of the interstellar gas including species like H, He, Ne, O and N, and even molecules like H_2. The deep space missions will not only allow the direct determination of the radial distribution of these gases: because of the unique constellation of Saturn in the solar system during the time of the CASSINI mission the spacecraft will encounter the gravitational focusing cone of O and N during the approach and the orbital tour in the Saturnian system (Fig. 7). Although the cone width shrinks with the square root of the particle mass, the CASSINI spacecraft will comfortably cut the cone due to the low inclination of Saturn´s orbit. The penetration distances and the expected cone widths, assuming equal interstellar temperatures for all species, are shown for a sample of elements in Table 2.

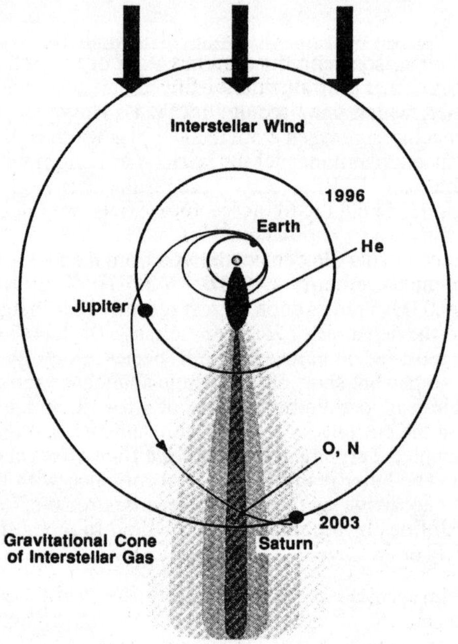

Fig. 7: Schematic view of the gravitational focusing cone structure of interstellar species like He, O and N together with the trajectory of CASSINI assuming a 1996 launch.

TABLE 3: Characteristic Distance from the Sun and Cone Width for Various Species of the Interstellar Medium

Element	Ionization Potential in eV	Λ in AU	Atomic Mass	Cone Width
H	13.6	7.5	1	-
He	24.58	0.35	4	12°
Ne	21.56	0.85	20	5.4°
N	14.54	2.65	14	6.4°
O	13.16	4	16	6°

The abundance of interstellar O, Ne and N pick-up ions in the heliosphere is particularly interesting, since they constitute key elements to decide on the potential effects of an interface layer between the heliosphere and the interstellar medium. If an interface layer with significant charge exchange between interstellar plasma and neutral gas does exist, it should lead to a strong depletion of interstellar oxygen in the heliosphere, since charge exchange between H and O is resonant. Nitrogen and neon should not be affected at all. Therefore, the comparison of the abundances of these three species, so close together in mass, should allow a clear decision on whether there is a partially opaque interface layer or whether the interstellar gas flow into the solar system is unimpeded. A clear decision here will then allow an accurate determination of the elemental abundances and will set constraints on the degree of ionization in the LISM.

Isotopic Abundances

Finally, it is tempting to determine various key isotope ratios in the interstellar gas. Comparing isotope ratios like D/H, $^3He/^4He$, and $^{22}Ne/^{20}Ne$ in the interstellar medium with those in the solar wind means a comparison of current ratios with such of a sample of interstellar matter from ≈ 4.6 billion years ago, thus giving an insight into the chemical evolution in the galaxy. In addition, the isotope ratio of He is important for modeling the nucleosynthesis in the Big Bang /41/.

In order to obtain accurate isotope ratios for the elements mentioned above the instrument background has to be fairly low because of the low abundance of the minor isotopes. With time-of-flight spectrometers this is easily achieved because of their coincidence requirement of consecutive start and stop signals. In the case of helium the minor isotope (^3He) has the shorter time-of-flight, i.e. the signal is measured where the instrumental no background is created from the main constituent (^4He) (upper panel of Fig. 8). In addition, any remaining background can be determined at the upper end of the pick-up energy spectrum, since the cut-off energy of ^3He is lower than that of ^4He. The latter argument can even be used in a stronger way for the cases in which the heavier isotope constitutes the minor constituent (H and Ne): here the pick-up distribution of the minor constituent extends significantly beyond the cut-off energy of the main isotope and thus can be measured without interference (lower panel of Fig. 8). A particular problem may arise for the determination of the D abundance due to the fact that a significant amount of H_2 is expected in the interstellar medium. This problem may be solved by the application of an isochronous time-of-flight spectrometer which can make use of the fact that molecular ions split up during the passage of the carbon foil at the entrance and then measure the mass of the fragments.

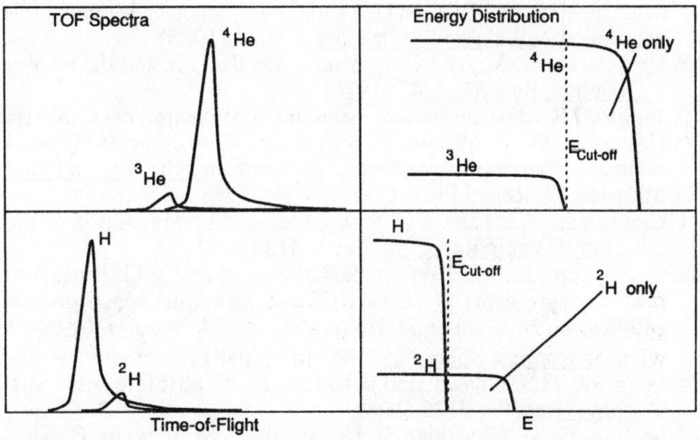

Fig. 8: Separation of isotopes using the combination of the time-of-flight information of a time-of-flight mass spectrometer and the variation of the cut-off energy of pick-up ions with mass.
Upper panel: the lighter isotope is less abundant (e.g. ^3He and ^4He)
Lower panel: the heavier isotope is less abundant (e.g. ^2H and ^1H)

CONCLUSIONS

The measurement of pick-up ions with time-of-flight spectrometers in the heliosphere has developed into a quantitative diagnostic tool to determine the parameters of the local interstellar medium with an in-situ technique. It ideally complements the methods based on remote sensing like measuring the backscattering of solar UV radiation at the interstellar gas and the absorption of light from nearby stars. The study of the pick-up ion spectra allows an independent determination of the ionization rate of He in the interplanetary space, which is crucial for derivation of the helium density and temperature in the interstellar medium. The measured ionzation rate is compatible with the values given from solar EUV measurements. The interstellar temperature and density of helium derived under these circumstances are consistent with the temperature given for hydrogen and the cosmic abundance ratio of H and He. However, the error bars still allow a variation of the values of $\approx\pm 30\%$ and are thus still not sufficient to decide between the different models of the interaction between the heliosphere and the interstellar medium.

Future simultaneous long time measurements of the pick-up ions and the solar EUV radiation will reduce the error bars significantly. In addition, the prospect of deep space missions like CRAF and CASSINI will add more elements to the interstellar sample and thus allow the determination of the composition of the interstellar medium as well as a decision on the interaction processes at the heliospheric boundary. In addition, the determination of isotope ratios of the insterstellar medium will be made possible.

ACKNOWLEDGEMENTS

The author wishes to thank H. Fahr, D. Hovestadt, B. Klecker, M.A. Lee and M. Scholer for valuable discussions. He is grateful to D. Judge and H. Ogawa for making the solar EUV flux data available prior to publication.

REFERENCES

1. Bertaux, J. I., and J. E. Blamont, Evidence for a source of an extraterrestrial hydrogen Lyman Alpha emission, Astron. Astrophys. 11, 200 (1971)
2. Thomas, G. E., and R. F. Krassa, OGO-5 measurements of the Lyman Alpha sky background, Astron. Astrophys. 11, 218 (1971)
3. Weller, C. S., and R. R. Meier, Observation of helium in the interplanetary/interstellar wind: the solar wake-effect, Astrophys. J. 193, 471 (1974)
4. Axford, W. I., The interaction of the solar wind with the interstellar medium, In: The Solar Wind, ed. C. P. Sonnett, P. J. Coleman, Jr., J. M. Wilcox, NASA-SP 308, 609 (1972)
5. Fahr, H. J, The extraterrestrial UV-background and the nearby interstellar medium, Space Sci. Rev. 15, 483 (1974)
6. Holzer, T. E., Neutral hydrogen in interplanetary space, Rev. Geophys. Space Phys. 15, 467 (1977)

7. Thomas, G. E., The interstellar wind and its influence on the interplanetary environment, Ann. Rev. Earth Planet. Sci. 6, 173 (1978)
8. Cox, D.P., and R.J. Reynolds, The local interstellar medium, Ann. Rev. Astron. Astrophys. 25, 303 (1987)
9. Möbius, E., D. Hovestadt, B. Klecker, M. Scholer, G. Gloeckler and F. M. Ipavich, Direct observation of He^+ pick-up ions of interstellar origin in the solar wind, Nature 318, 426 (1985)
10. Baranov, V.B., Interaction of the solar wind with external plasmas, this issue
11. Chassefiere, E., J.L. Bertaux, R. Lallement, and V.G. Kurt, Atomic hydrogen and helium densities of the interstellar medium measured in the vicinity of the sun, Astron. Astrophys. 160, 229 (1986)
12. Ripken, H.W., H.J. Fahr, Modification of the local interstellar gas properties in the heliosperic interface, Astron. Astrophys. 122, 181 (1983)
13. Wallis, M.K., Local interstellar medium, Nature 254, 202 (1975)
14. Fahr, H.J, and H.W. Ripken, The physics of the heliospheric interface and its implications for LISM diagnostics, Astron. Astrophys. 139, 551 (1984)
15. Frisch, P.C., Spectroscopic data on the LISM, this issue
16. Lallement, R., Scattering of solar UV on local neutral gases, this issue
17. Bertaux, J.L., R. Lallement, V.G. Kurt, and E.N. Mironova, Characteristics of the Local Interstellar Hydrogen determined from PROGNOZ 5 and 6 interplanetary Lyman a line profile measurements with a hydrogen absorption cell, Astron. Astrophys. 150, 1 (1985)
18. Vasyliunas, V. M., and G. L. Siscoe, On the flux and the energy spectrum of interstellar ions in the solar system, J. Geophys. Res. 81, 1247 (1976)
19. Jokipii, J.R., The anomalous component of cosmic rays, this issue
20. Hovestadt, D., E. Möbius, B. Klecker, G. Gloeckler, F.M. Ipavich, and M. Scholer, Pick-up ions in the solar wind as a source of suprathermal particles, in: The Sun and the Heliosphere in Three Dimensions, ed. R.G. Marsden, D. Reidel Publ. Comp., 413 (1986)
21. Cummings, A.C., and E.C. Stone, Elemental composition of the anomalous cosmic-ray component, Proc. of the 20th Intl. Cosmic Ray Conf. SH 3, 413 (1987)
22. Chassefiere, E., J.L. Bertaux, R. Lallement, and V.G. Kurt, A new estimate of interstellar helium parameters from measurements of VOYAGER 1/2 ultraviolet spectrometers, Astron. Astrophys. 199, 304 (1988)
23. Möbius, E., B. Klecker, D. Hovestadt, and M. Scholer, Interaction of interstellar pick-up ions with the solar wind, Astrophys. Space Sci. 144, 487 (1988)
24. Feng, W., H.S. Ogawa, and D.L. Judge, The absolute solar soft X ray flux in the 20- to 100-Å region, J. Geophys. Res. 94, 9125 (1989)
25. Möbius, E., G. Gloeckler, D. Hovestadt, F. M. Ipavich, B. Klecker, M. Scholer, H. Arbinger, H. Höfner, E. Künneth, P. Laeverenz, A. Luhn, E. O. Tums, and H. Waldleben, The time-of-flight spectrometer SULEICA for ions of the energy range 5-270 keV/charge on the AMPTE/IRM, IEEE Trans. on Geoscience and Remote Sensing GE-23, 274 (1985)
26. Fahr, H.J., and D. Rucinsky, Modelling of interplanetary pick-up ion fluxes and relevance for LISM parameters, Planet. Space Sci. 37, 555 (1989)
27. Ogawa, H.S., L.R. Canfield, D. McMullin, and D.L. Judge, Sounding rocket measurement of the absolute solar EUV flux utilizing a silicon photodiode, J. Geophys. Res., in press (1990)
28. Ratkiewicz, R., D. Rucinski, and W.-H. Ip, Estimates of the fluxes of new-born He ions due to ionization of the interstellar helium gas, Astron. Astrophys., in press (1989)
29. Rucinski, D., and H.J. Fahr, The influence of electron impact ionization on the distribution of interstellar helium in the inner heliosphere; possible consequences for determination of interstellar helium parameters, Astron. Astrophys. 224, 290 (1989)
30. Feldman, W. C., J. R. Asbridge and S. J. Bame, Upper limits for the solar wind He^+ content at 1 AU, J. Geophys. Res. 79, 1808 (1974)
31. Möbius, E., B. Klecker, and D. Hovestadt, The helium characteristics in the local interstellar medium as obtained from observations of the pick-up ion distribution inside the Earth´s orbit, Astron. Astrophys., subm. (1990)
32. Ajello, J.M., An interpretation of MARINER 10 helium (584Å) and hydrogen (1216Å) interplanetary emission observations, Astrophys. J. 222, 1068 (1978)
33. Weller, C.S., and R.R. Meier, Characteristics of the helium component of the local interstellar medium, Astrophys. J. 246, 386 (1981)
34. Dalaudier, F., J.L. Bertaux, V.G. Kurt, E.N. Mironova, Characteristics of interstellar helium observed with PROGNOZ 6 58.4-nm photometers, Astron. Astrophys. 134, 171 (1984)
35. Paresce, F., S. Bowyer, and S. Kumar, Further evidence for an interstellar source of nighttime He I 584Å radiation, Astrophys. J. 188, L71 (1974)
36. Fahr, H.J., G. Lay, and C. Wulf-Mathies, Derivation of interstellar helium gas parameters from an EUV-rocket observation, Space Res. XVIII, 393 (1977)
37. Wu, F.M., and D.L. Judge, PIONEER 10 UV photometer observations of the interplanetary glow at heliocentric distances from 2 to 17 AU, EOS 60, 365 (1979)
38. Réme, H., et al., The ion spectrometry experiment to measure the properties of the ionic plasma composition and distribution functions at energies between ≈ 0 and 40 keV/e, in: The CLUSTER Mission, ed. R. Schmidt, ESA-SP1103, 65 (1988)
39. Hovestadt, D., et al., CELIAS - Charge, Element and Iostope Analysis System for SOHO, in: The SOHO Mission, ed. V. Domingo, ESA-SP1104, 69 (1989)
40. Wilhelm, K., et al., SUMER - Solar Ultraviolet Measurement of Emitted Radiation, in: The SOHO Mission, ed. V. Domingo, ESA-SP1104, 31 (1989)
41. Geiss, J., and H. Reeves, Cosmic and solar system abundances of deuterium, Astron. Astrophys. 18, 126 (1972)

NEUTRAL SOLAR WIND EXPERIMENT

M.A. Gruntman,* S. Grzedzielski** and V.B. Leonas*

*Institute for Problems in Mechanics, Academy of Sciences, prosp. Vernadskogo 101, Moscow 117526, U.S.S.R.
**Space Research Centre, Polish Academy of Sciences, Ordona 21, Warsaw 01–237, Poland

INTRODUCTION

Interaction between ionized and neutral components of interstellar matter is a common process in the astrophysics. And our heliosphere in this case is no exception. The properties of the interacting neutrals are not in all respects well studied though they can serve as an indicator of important physical phenomena. One of such examples is the neutralized component of Solar Wind – i.e. the Neutral Solar Wind (NSW) - which originates by charge exchange of Solar Wind ions with neutral atoms in interplanetary space.

It was speculated that neutral atoms can be present in the Solar Wind as transients due to ejection of solar matter in violent events /1/. But only realization of the fact, that due to the substantial relative velocity of the Sun with respect to the surrounding Local Interstellar Medium (LISM) the interplanetary space is filled with neutral atoms of interstellar origin, led to the development of the concept of a permanently existing NSW. Solar Wind ions can capture electrons in collisions from neutrals thus giving rise to the NSW. Hence the velocity of NSW atoms equals that of Solar Wind. At first, it was believed that charge exchange only occurred between Solar Wind protons and interplanetary hydrogen /2/, and one could safely assume that at large distances from the Sun, the NSW consists of hydrogen atoms only. Later, however, it was recognized that charge exchange of protons with interplanetary helium atoms can provide comparable sources of NSW in the region within several a.u. from the Sun /3/. Finally, when it was realized in this context /4/, that the cross section for the simultaneous transfer of both electrons from interplanetary helium atoms to Solar Wind alpha particles is fairly high, it became clear, that helium atoms are also a non-negligible component of NSW close to the Sun.

NSW, which builds up gradually as Solar Wind ions travel further and further from the Sun, may play important role in shaping the global features of the interaction between the Sun and LISM. The most prominent of such features is the heliosphere itself. Some effects due to the NSW influence were already considered. Among them are i) transition from supersonic interstellar plasma flow to subsonic flow ahead of the heliospheric bowshock and thus pushing out the "last frontier" of the Solar system a few hundred a.u. further into interstellar space /5/, and ii) creation of anomalous hot component in the interstellar plasma flow with corresponding effects on interface region /6,7/.

THE NSW EXPERIMENT

By now there is no experimental information at all on NSW properties. That is why one may expect that the experiment to study NSW properties will give opportunity to test existing concepts about global interactions between neutrals and ions in Solar Wind plasma. The most feasible place for such an experiment is the Earth orbit, and further discussion thus will be restricted to this region. Serious experimental difficulties are responsible for the total lack of experimental measurements. The number density of NSW component is rather small ($< 10^{-4}$ cm^{-3} at Earth orbit) proving the conventional technique based on ionization and subsequent analysis of ions as unacceptable. However, these relatively high energy atoms can be detected by devices based on the effects of neutral particle interaction with surfaces, e.g. by secondary electron emission. The major obstacle to implement this technique is the highly superior background radiation in EUV (mainly in Lyman-alpha). EUV photons are counted also rather efficiently resulting in noise count rates several orders of magnitude higher than those expected from the NSW. Our attempt to study NSW uses the development of the experimental technique based on the coincidences realised by application of ultrathin foils and measurements of time-of-flight (TOF) of atoms in the instrument / 8,9,10 /. Such an approach is widely used in nuclear physics for identification of particles with much higher (at least by several orders of magnitude) energies.

The NSW experiment (included in the scientific package of GAS experiment to study fluxes of energetic neutral atoms in interplanetary space) is planned to be launched in the framework of the RELICT-2 mission scheduled for 1993. The spacecraft, after a three months cruise phase, will reach the collinear libration point L2 (Sun-Earth system) which is situated at the distance of $1.5\cdot10^6$ km from the Earth in antisunward direction. The spacecraft will perform then two revolutions around this point shifting at most $1.5\cdot10^5$ km from L2 in sun/antisunward direction, $8\cdot10^5$ km in perpendicular direction in ecliptic plane and $1.5\cdot10^5$ km out of ecliptic. Period of each revolution is 6 months. The spacecraft will be spinning around an axis pointed at the Sun. The NSW instrument will be looking at the off-set angle 6° from the spin axis - i.e. scan a ring around the Sun. The instrument field of view is 4°x4°. Spacecraft trajectory gives opportunity to measure NSW itself and also, in certain parts of the trajectory, neutrals originating in charge exchange of Solar Wind ions with the terrestrial exosphere, as well as energetic neutral atoms from terrestrial magnetosphere.

For the observer moving together with the Earth along its orbit around the Sun the average direction of NSW will be shifted 4-5° from the direction to the Sun due to the aberration. Detailed computer simulations /10/ show that the expected NSW flux is in the range 10^3 - 10^5 cm^{-2} s^{-1} sr^{-1} depending on the parameters of interstellar gas as well as on the position of the observer at Earth orbit.

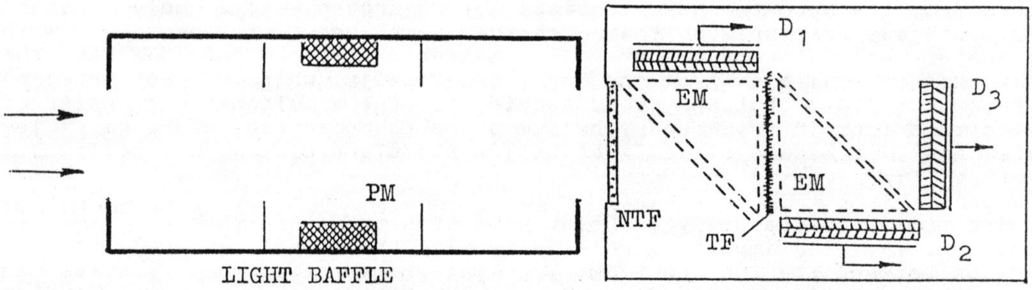

Fig.1. Neutral solar wind detector-energy analyser. D_1,D_2,D_3 - microchannel plate detectors; TF - thin foil; EM - electrostatic mirrors; NTF - nuclear track filter (optional); PM - permanent magnet.

INSTRUMENT FOR NSW DETECTION

The instrument to study NSW (sensor GAS-2 within GAS instrument package) is shown in Fig.1. A solar baffle prevents illumination of the sensor by photons from the Sun directly. Permanent magnets deflect charged particles with energies up to 30-50 keV. When a background UV photon comes to the instrument, it may be either absorbed by the thin carbon foil (with possible emission of electron followed by its acceleration by the nearest grid of electrostatic mirror and by the triggering of detector D_1 or D_2), or it may penetrate through the foil and impinge on (with possible triggering) detector D_3. The foil is 80 Å thick and its diameter is 27 mm. Detectors are built as chevron stacks of microchannel plates. Arrival of the neutral particle may, on the other hand, result in both above mentioned "simultaneous" events, i.e. emission of electrons from the foil and triggering detector D_3. Hence the particles can produce coincidences (time gate 250 ns) which give opportunity to extract the very weak signal due to NSW from the superior background count rates / 8,9,10 /. Also time intervals between pulses from D_1 (D_2) and D_3 are measured, and accumulated TOF spectrum contains information on velocity distribution of the incoming atoms. Typical TOF spectra obtained by the instrument, when registering (in laboratory) monoenergetic hydrogen atom fluxes, are shown in Fig.2. The energy resolution is $E / \Delta E = 2$. Detection efficiency (i.e. probability of coincidences) of the instrument goes down from 20 to 1-3 % with the decrease of registered atom energy from 3000 eV to 600 eV. Detailed estimates based on the expected fluxes of both NSW and UV photons show that NSW can be successfully studied / 10 /.

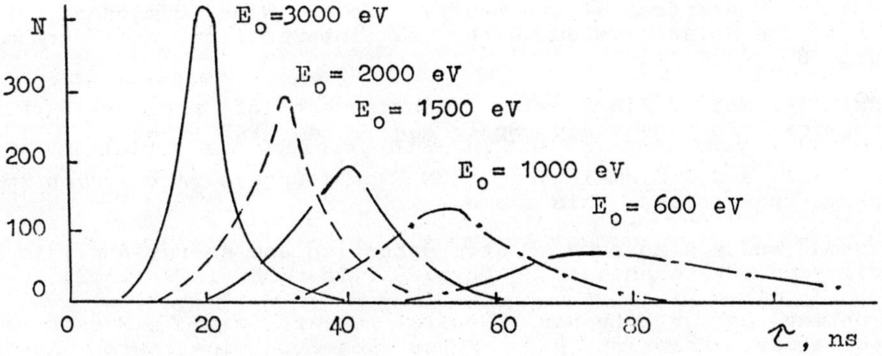

Fig.2. Typical time-of-flight spectra for the detection of monoenergetic (E_0) hydrogen atom fluxes.

It should be mentioned that a nuclear track filter (NTF), which is optional, though reducing 10-20 times absolute value of NSW flux in instrument, increases the ratio of neutrals to UV photons from 3 to 6 orders of magnitude at the sensor entrance. The use of NTF to filter neutrals from photons was proposed in / 9,10 / and it is still under development. The NTF will consist of 1 μm thick film with parallel straight channels (1000 Å diameter) and geometrical transparency 4-10 %.

ENERGETIC NEUTRALS FROM THE HELIOSPHERIC INTERFACE

Solar Wind plasma after passage through the inner shock is in a rather hot state. Due to the charge exchange of its protons on inflowing interstellar gas, this region is an extended source of energetic (0.2-0.8 keV) neutral hydrogen atoms. According to present estimates / 7 /, the expected flux number density is 10^2 cm^{-2} s^{-1} sr^{-1} at Earth orbit. There is another sensor - GAS-3 - in the GAS package of Relict-2 mission. It is identical to GAS-2 except for the baffle which is much simpler and smaller and defines a field of view of 15° x 15° . GAS-3 is pointed in antisunward direction and its main purpose is to try to detect the energetic neutrals from the interface region. Giant planets, Jupiter and Saturn, when in the instrument field of view, may give also detectable count rates due to fluxes of energetic neutral atoms from their magnetospheres.

EXPECTED SCIENTIFIC RETURN

The Neutral Solar Wind experiment will be the first attempt specially dedicated to the study of fluxes of energetic neutral atoms in heliosphere. It will give opportunity to test by a qualitatively new way our basic ideas about interaction (and its consequences) between neutrals and plasmas of Solar system and LISM. Measurements of NSW may lead to determination of interstellar gas parameters, concentration of the interplanetary dust around the Sun and Solar Wind characteristics at distances <0.4 a.u. to the Sun. Detection of the energetic neutral atom flux from the interface region may be the first direct evidence of the existence of the widely discussed termination shock and heliospheric structure in general. The experiment may also provide new information on the magnetospheres of the planets like Earth, Jupiter and Saturn. The latter may lead to a rather promising approach to the remote study of magnetospheres of giant planets.

REFERENCES

1. S.-I.Akasofu, A source of the energy for geomagnetic storms and auroras, Planet. Space Science, 12, #9, 801-833 (1964).

2. P.W.Blum and H.J.Fahr, Interaction between interstellar hydrogen and the Solar Wind, Astron. Astrophys. 4, 280-290 (1970).

3. M.A.Gruntman, Neutral component of Solar Wind at the Earth's orbit, Kosm. Issled. 18, 649-651 (1980) [Cosmic Res. 18, #4 (1981)].

4. V.B.Leonas, unpublished (1987).

5. M.A.Gruntman, The effect of the neutral Solar Wind component upon the interaction of the Solar system with the interstellar gas stream, Sov. Astron. Lett. 8, 24-26 (1982).

6. S.Grzedzielski and J.Ziemkiewicz, Confinement of the heliosphere and relevant physical processes, Adv. Space Res. 9, #4, 247 (1989).

7. S.Grzedzielski and J.Ziemkiewicz, LISM - heliosphere interaction mediated by suprathermal particles , this issue.

8. M.A.Gruntman and V.A.Morozov, H atom detection and energy analysis by use of thin foils and TOF technique, J. Phys. E v.15, 1356-1359 (1982).

9. M.A.Gruntman and V.B.Leonas, Neutral Solar Wind. Possibility of experimental study, Preprint 825, Space Research Institute, Academy of Sciences, Moscow (1983).

10. M.A.Gruntman and V.B.Leonas, Possibility of experimental study of energetic neutral atoms in interplanetary space, Proceed. of Inter. Workshop on Physics of Neutral Particles in the Solar system, Zakopane, Poland, 1985, 99-120 (also preprint 1109, Space research Institute, Academy of Sciences, Moscow 1986).

11. S.Błeszynski,S.Grzędzielski, J.Jakimiec, and D.Ruciński, Expected fluxes of ~ 1 keV Neutral H Atoms in Interplanetary Space: Comparison with the UV Background and Possibility of Detection, (in preparation).

AN INVERSE METHOD OF DETERMINATION OF THE INTERSTELLAR NEUTRAL GAS DISTRIBUTION FUNCTION

M. Banaszkiewicz,* H. Rosenbauer** and M. Witte**

**Space Research Centre, Bartycka 18, 00–716 Warszawa, Poland*
***Max-Planck-Institut für Aeronomie, D-3411 Katlenburg-Lindau, F.R.G.*

ABSTRACT

The Tarantola's method has been applied to solve an inverse problem of determining parameters of the shifted Maxwellian distribution function of interstellar helium from informations on specific cuts through velocity space. Count rates that are expected be measured during the Ulysses mission were simulated and used as an input to an inverse problem solver. The obtained solutions for helium density n_∞, bulk velocity v_∞ and temperature T_∞ are satisfactory. At a statistical noise level of about 20% the resulting values of these parameters differ by less than 8% from the values used to simulate the data.

INTRODUCTION

Helium atoms of interstellar origin pass over the solar system together with other neutral constituents of LISM. Contrary to hydrogen, helium particles are much less affected by radiation pressure and their trajectories are gravity determined hyperbolas. LISM atoms outside the heliosphere are assumed to be thermalized, therefore their velocity distribution function should be close to a Maxwellian characterized by the density, bulk velocity vector and temperature. During the passage through the heliopause and later on, in the solar wind region, LISM neutrals can charge–exchange with ions /1, 2, 3 /. This effect is very probable for hydrogen /4/ but perhaps negligible for helium because of small charge–exchange cross section. Therefore we assume that the only factors which can modify the Maxwellian distribution of helium atoms at the inflow from outside of the heliopause to the inner Solar System are gravitational focusing of orbits by the Sun and photo–ionization losses /5, 6/.

Parameters of interstellar helium were determined from observations of resonant scattering of solar He I 584 Å photons /7, 8, 9/. Density was estimated to be in the range 0.001 - 0.008 cm^{-3}. Velocity vector coincides well with the H–LISM–apex direction. The value v_∞ and the temperature T_∞, as deduced from observations of the cone of enhanced helium density in the downwind direction, are coupled by the relation /10, 11/: $T_\infty^{-1/2}/v_\infty$=const. One thus can fit the measurements with $v_\infty = 27\ m\,s^{-1}$ and $T_\infty = 16000°\ K$ as well as with $v_\infty = 21.5\ m\,s^{-1}$ and $T_\infty = 7000°\ K$ /12/. According to recent analysis by Chasefière *et al.* /13/ there are arguments in favor of the later set.

The GAS experiment on board of the Ulysses spacecraft has been developed to measure fluxes of neutral atoms. One of its objectives is to determine more accurately parameters of interstellar helium, especially the temperature. The detector measures count rates of neutrals coming from different directions. The count rate is represented by a complicated integral over the distribution function at the point of observation convoluted with the sensitivity function of the detector. In this paper an inverse method of determining interstellar helium density, bulk velocity and temperature from count rates measured at different spacecraft positions and for different looking directions is proposed.

SET-UP OF THE EXPERIMENT

Technical description of the GAS device has been given elsewhere /14, 15/. In short, the instrument is installed on the platform attached to the spinning spacecraft. There is a possibility of varying the elevation angle of the detector. Together with the continuous changes of the azimuthal angle due to rotation of Ulysses it enables a scan over almost the whole celestial sphere. An additional degree of freedom in observing the flux of neutrals is attained due to the spacecraft's orbital motion. The spin axis of Ulysses (z-axis) is always directed towards the Earth. The elevation angle is determined with respect to this axis. The x-axis lies in the plane determined by the Sun, the spacecraft and the Earth and points inward as seen from the orbit. The azimuthal angle is measured anti-clockwise from the y-axis. The maximum field of view of the two detectors is 7.4°, resp. 4.9° with 4.7°, resp. 2.0° full width at half maximum (FWHM). Particles that enter the detector from the same direction, but with different velocities, have different probabilities to be registered. This probability is described by an energy–dependent sensitivity function and ranges from 0.001 for low energy, 40 eV atoms to 0.02 for 1 keV neutrals.

FORWARD PROBLEM

The distribution function of helium atoms $f(v, T, n; r)$ at the point r can be calculated from the distribution function at infinity $f(v_\infty, T_\infty, n_\infty)$ by applying the Liouville theorem and accounting for loss processes. The count rate integral includes, besides the distribution function, the sensitivity function $h(v_{sp})$ which depends on the velocity of atoms v_{sp} in the spacecraft system of coordinates:

$$Z = \int_{field-of-view} d\Omega_{sp} \int_{v_{min}}^{v_{max}} h(v_{sp})\, v_{sp}\, f(v, T, n; r)\, L(v, r)\, dv_{sp}. \tag{1}$$

Here v_{sp} is the velocity of atoms v transformed from the ecliptic to the spacecraft coordinate system, $d\Omega_{sp}$ is the solid angle element in the direction of view and L describes ionization losses along the trajectory.

An atom moving along a hyperbolic trajectory can reach the detector before (direct orbit) or after (indirect orbit) its perihelion. When the temperature is low, and the thermal velocity of atoms is much smaller than the bulk velocity, helium gas flows through the point r in two narrow streams directed along the bulk velocity vectors of direct and indirect orbits. In Figure 1 azimuth and elevation angles of these directions are shown for the ecliptic part of Ulysses trajectory. In Figure 2 count rates (1) calculated for the cold model with T_∞=0 are presented as a function of time. It occurs that the most intense fluxes can be observed in the interval between 40 and 80 days after launch in 5 Oct 1990. During this time, the direction of the major flux changes from indirect to direct orbits. In the following part of the paper we will consider the results of simulations only for this interval connected with the highest count rates.

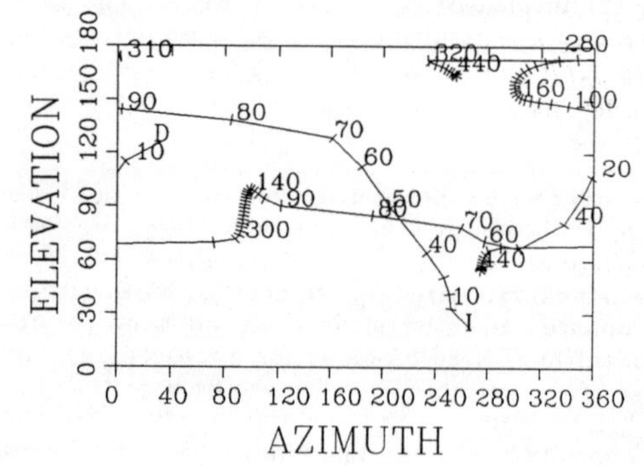

Fig. 1. Azimuth and elevation angles of He fluxes in the cold model ($T_\infty = 0°\ K$) as seen from the moving spacecraft during the ecliptic part of its trajectory. Direct (D) and indirect (I) orbits are indicated. Numbers near the tick marks describe the day after launch date (5 Oct 1990).

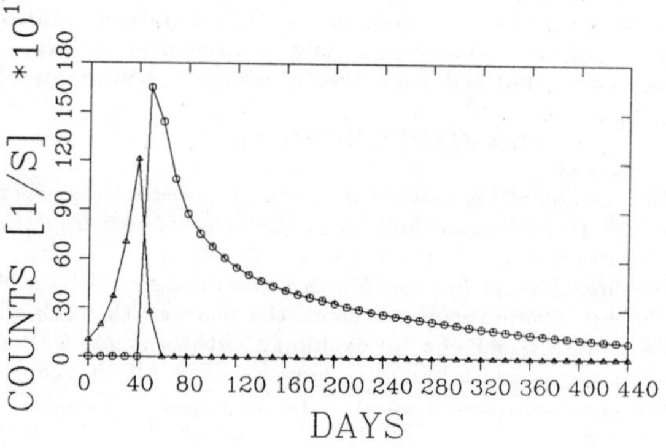

Fig. 2. Count rates Z [$cm^{-2}\, s^{-1}$] for direct (⊙) and indirect (Δ) fluxes in the cold model as functions of time.

Our calculations of count rates for the hot model with realistic helium temperatures are slightly idealized. We assume that the detector is able to measure count rates for any chosen direction, i.e. for any azimuth and elevation angles. Subsequently we assume that the field of view of the detector $\Delta\,\Omega$ is small enough to perform the integration over $d\Omega_{sp}$ by simply multiplying the integrand by $\Delta\,\Omega$. The integrand is calculated for Ω corresponding to the symmetry axis of the field of view. Consequently, the count rate can be expressed by a single integral over v_{sp}. In Figure 3 count rates are shown for the ±30° range in azimuth and elevation about the bulk velocity for direct and indirect orbits. Time of observations corresponds to the 50th day after launch.

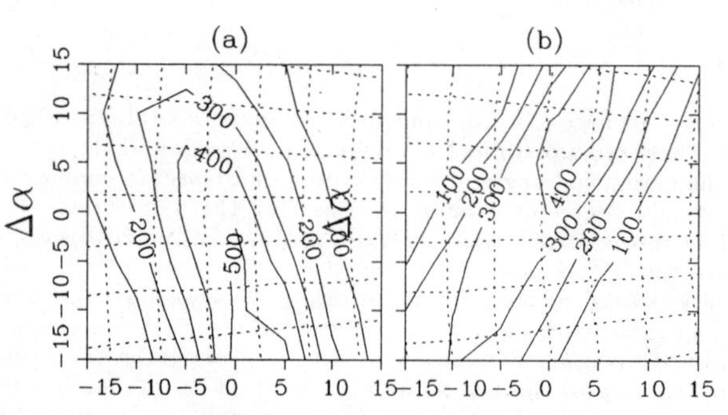

Fig. 3. Simulated count rates Z [$cm^{-2}\, s^{-1}\, sr^{-1}$] for direct (a) and indirect (b) fluxes in the hot model of Maxwellian distribution function with parameters: $n_\infty = 1 \times 10^{-2}\ cm^{-3}$, $v_{\infty x} = 5.85\ km/s$, $v_{\infty y} = 21.08\ km/s$, $v_{\infty z} = -2.30\ km/s$, $T_\infty = 10000°\ K$. The centers of plots ($\Delta\alpha = \Delta\epsilon = 0°$) correspond to azimuth and elevation angles ($\alpha = 303°, \epsilon = 67°$) for direct orbits 50 days after launch (a) and ($\alpha = 233°, \epsilon = 64°$) for indirect orbits 40 days after launch. Dashed lines indicate the azimuth– elevation grid with a step of 5°.

INVERSE PROBLEM

The problem of finding the distribution function from measured count rates according to equation (1) can be considered either for a general form of $f(v_\infty, T_\infty, n_\infty)$ or for a particular distribution characterized by a number of parameters, e.g. the Maxwellian function. In this paper we limit our considerations to the second case, firstly because we expect that the form of distribution function in fact is close to the Maxwellian and secondly because generalization of the presented method though not difficult yet is much more expensive in its numerical application.

As was mentioned in an earlier paper /14/ the problem of reconstructing the velocity distribution function from count rates resembles the computer tomography problem in the way of determining properties of a medium by looking at it from different view-points. Geophysics and seismology, in particular, are other fields of application of inverse methods. There are, at least, two approaches to solve this kind of problems: The Backus-Gilbert /16/ formalism and Tarantola's method /17/. We have chosen the later approach. Tarantola introduces the data space D which contains possible results of measurements d and the model space M where functions, or parameters to be determined, belong to. In our problem the data vector d_{ijk} ; $i = 1, \ldots, I$; $j = 1, \ldots, J$; $k = 1, \ldots, K$ consists of count rates (1) measured for different azimuth (i) and elevation (j) as well as for different positions of the spacecraft (k). The model space for the Maxwellian distribution function is five dimensional with components being simple functions of $n_\infty, v_\infty, T_\infty$. The relation (1) can be expressed in terms of the data and model vectors as $d = g(m)$, where $g_{ijk}(m)$ is the count rate integral. The main idea of Tarantola's approach is to represent the actual state of knowledge about the data, the model and the relation between them as probability densities. Consequently, we assume that the data values are uncertain due to statistical noise and *a priori* values of model parameters are known only with some probability. The relation $d = g(m)$ between model and data spaces is usually an approximation which takes into account only the most important physical effects neglecting the other ones. In our case, for example, the distribution function of helium atoms at infinity needs not to exactly attain a Maxwellian form, the sensitivity function of the detector could be slightly different in space than in laboratory, etc. According to Tarantola /17/ *a posteriori* probability density of the model parameters $\sigma_M(m)$ can be described in terms of three functions: $\rho_D(d)$ which is *a priori* probability density in the data space, $\rho_M(m)$ - the *a priori* probability density in the model space and $\Theta(d|m)$ which is the conditional probabilty density in d for given m. The fundamental formula reads:

$$\sigma_M(m) = \rho_M(m) \int \frac{\rho_D(d)\,\Theta(d|m)\,dd}{\mu_D(d)}, \tag{2}$$

where $\mu_D(d)$ represents the state of null information. Now we can find the most probable model parameters m according to probability density $\sigma_M(m)$. They correspond to the maximum of $\sigma_M(m)$. The task of maximization is easy to perform when the densities $\rho_D(d)$, $\rho_M(m)$ and $\Theta(d|m)$ are normal:

$$\frac{\rho_D(d)}{\mu_D(d)} = \left((2\,\pi)^{ND}\ det\,C_d\right)^{-1/2}\ exp\left(-\frac{1}{2}(d - d_{obs})^t\, C_d^{-1}(d - d_{obs})\right), \tag{3}$$

$$\Theta(d|m) = \left((2\,\pi)^{ND}\ det\,C_T\right)^{-1/2}\ exp\left(-\frac{1}{2}(d - g(m))^t\, C_T^{-1}(d - g(m))\right), \tag{4}$$

$$\rho_M(m) = \left((2\,\pi)^{NM}\ det\,C_M\right)^{-1/2}\ exp\left(-\frac{1}{2}(m - m_{prior})^t\, C_M^{-1}(m - m_{prior})\right), \tag{5}$$

where C_d, C_T and C_M are covariance matrices, ND and NM their dimensions, d_{obs} is the output data vector from the measuring device with the instrumental errors included, as opposed to the "true", input data vector d and m_{prior} is the *a priori* model. The assumption about Gaussian probability densities seems to be very restrictive but in the vicinity of maximum almost any function can be approximated more or less accurately by the Gaussian. To avoid complications with parameters which, according to their physical meaning, are always positive – such as count rate, density and temperature – we will instead use logarithms of them so that the log-normal distributions of these new variables are determined on the whole real axis. Therefore we will have $d_{ijk} = log\, Z_{ijk}$ and $m = (log\,(n_\infty), v_{\infty x}, v_{\infty y}, v_{\infty z}, log\,(1/kT_\infty))$. Covariance matrices are taken to be diagonal which means that components of the data vector as well as components the model parameter vector are uncorrelated. Substituting formulae (4), (5) and (6) into (3) we obtain after some transformations that $\sigma_M(m) = const \times exp(S(m))$, where

$$S(m) = \frac{1}{2}(g(m) - d_{obs})^t\, C_D^{-1}(g(m) - d_{obs}) + (m - m_{prior})^t\, C_M^{-1}(m - m_{prior}), \tag{6}$$

and $C_D = C_d + C_T$.
Instead of maximizing $\sigma_M(m)$ we can find equivalently the minimum of $S(m)$ either by using simple least-square method or by applying more general optimization techniques.

The algorithm of solving the inverse problem is as follows. First we simulate the noiseless data (1) making use of the forward problem solutions. Statistical noise with the dispersion equal to $Z_{ijk}^{1/2}$ is then generated and added to the data. The logarithm of data is calculated and together with some initial values of the model parameters they are provided as an input to an optimization procedure. In each step of optimization the forward problem is solved and the function $S(m)$ is calculated. The solution is obtained when the minimum of $S(m)$ is found within a given tolerance. The above process is visualized in Figure 4.

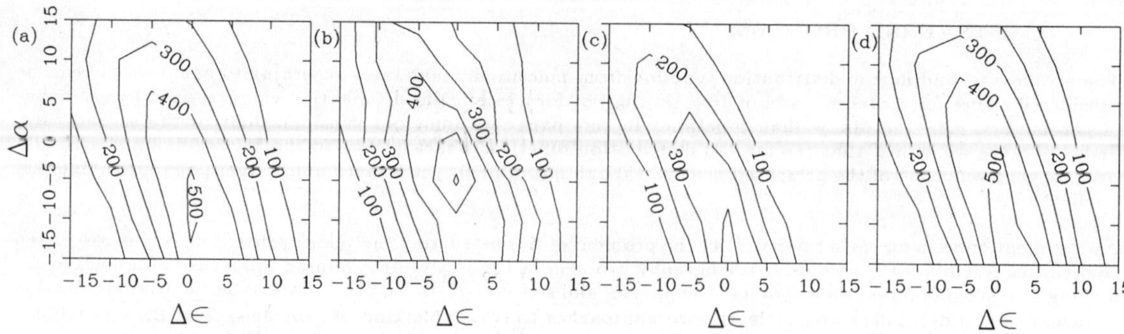

Fig. 4. Four stages of the inverse problem. Simulated count rates Z without (a) and with (b) statistical noise. Parameters of the distribution function are: $n_\infty = 1 \times 10^{-2}\ cm^{-3}, v_{\infty x} = 5.85\ km/s, v_{\infty y} = 21.08\ km/s, v_{\infty z} = -2.30\ km/s, T_\infty = 10000°\ K$. Noise level is 15% at the maximum of Z and increases for smaller Z. Count rates corresponding to the *a priori* model parameters: $n_\infty = 3.7 \times 10^{-3}\ cm^{-3}, v_{\infty x} = 7.0\ km/s, v_{\infty y} = 23.0\ km/s, v_{\infty z} = -0.8\ km/s, T_\infty = 5010°\ K$ are presented in (c) while (d) shows count rates for the best set of parameters found in optimization process: $n_\infty = 9.7 \times 10^{-3}\ cm^{-3}, v_{\infty x} = 5.86\ km/s, v_{\infty y} = 21.54\ km/s, v_{\infty z} = -2.69\ km/s, T_\infty = 9340°\ K$.

RESULTS

The data set for each orbital position and one of the two bulk directions of He fluxes consisted of 7 × 7 grid in azimuth and elevation with a step of 5°. One, two or three data sets were simulated in each case. Values of model parameters used to simulate the data were: $n_\infty = 1 \times 10^{-2}\ cm^{-3}, v_{\infty x} = 5.85\ km/s, v_{\infty y} = 21.08\ km/s, v_{\infty z} = -2.30\ km/s, T_\infty = 10000°\ K$. We applied three optimization procedures: the Powell method (P), the Polak-Ribere (PR) method and the Davidon-Fletcher-Powell approach (DFP) /18/. In the second and the third method the gradient of $g(m)$ is needed in addition to the function itself. The gradient was calculated numerically using symmetric differences. All methods converged towards the minimum but the gradient procedures usually give better values of m. The results are summarized in Table 1. The best values of parameters were obtained for the case 8 in which three data sets corresponding to different positions of the spacecraft were used.

TABLE 1 Inverse Problem Solutions

Case	Noise[a]	Data sets[b]	Method	Solution: n_∞ $[cm^{-3}]$, v_∞ $[km/s]$, T_∞ $[K]$
1	5 %	50 D	P	1.10×10^{-2}, (4.97, 20.38, −1.27), 8903°
2	5 %	50 D	PR	0.97×10^{-2}, (5.95, 21.40, −2.70), 9938°
3	5 %	50 D	DFP	0.98×10^{-2}, (5.84, 21.32, −2.56), 9845°
4	10 %	50 D	DFP	0.97×10^{-2}, (5.77, 21.50, −2.70), 9552°
5	15 %	50 D	DFP	0.97×10^{-2}, (5.85, 21.54, −2.69), 9340°
6	15 %	40 I, 50 D	DFP	1.01×10^{-2}, (5.68, 21.25, −2.43), 9499°
7	15 %	50 D, 90 D	DFP	1.05×10^{-2}, (5.43, 20.83, −1.96), 9618°
8	15 %	40 I, 50 D, 90 D	DFP	1.01×10^{-2}, (5.92, 21.18, −2.34), 9865°

[a]noise to signal ratio for the maximum count rate in the data set
[b]number denotes a day after launch, D stands for direct and I for indirect flux

REFERENCES

1. H.J. Fahr, *Astron. Astrophys.* **66,** 103–117 (1978).
2. J.A. Kunc, F.M. Wu and D.L. Judge, *Pl. Space Sci.* **31,** 1157–1164 (1983).
3. E. Chassefière and J.L. Bertaux, *Astron. Astrophys.* **176,** 121–130 (1987).
4. S. Błeszynski, *Astron. Astrophys.* **180,** 201–206 (1987).
5. T.E. Holzer, *Rev. Geophys. Sp. Phys.* **15,** 467–490 (1977).
6. F.M. Wu and D.L. Judge, *Astrophys. J.* **231,** 594–605 (1979).
7. C.S. Weller and R.R. Meier, *Astrophys. J.* **193,** 471–476 (1974).
8. F. Dalaudier, J.L. Bertaux, V.G. Kurt and E.N. Mironova, *Astron. Astrophys.* **134,** 171–184 (1984).
9. R. Lallement, J.L. Bertaux, V.G. Kurt and E.N. Mironova, *Astron. Astrophys.* **140,** 243–250 (1984).
10. W.C. Feldman, J.J. Lange and F. Scherb, *Solar Wind, NASA SP-308,* 684–695 (1972).
11. H.J. Fahr, *ESLAB Symposium,*, Reidel Publ. Company, ed. by R.G. Marsden, 421–430 (1985).
12. E. Chassefière, J.L. Bertaux, R. Lallement, B.R. Sandel and L. Broadfoot, *Astron. Astrophys.* **199,** 304–311 (1988).
13. E. Chassefière, F. Dalaudier and J.L. Bertaux, *Astron. Astrophys.* **201,** 113–122 (1988).
14. H. Rosenbauer, M.Witte, J.F. Pietsch, H.J. Fahr and G. Lay, in: *Proceedings of the International Workshop on Problems of Physics of Neutral Particles in the Solar System,* 139–149 (1987).
15. H. Rosenbauer, H.J. Fahr, E. Keppler, M. Witte, P. Hemmerich, H. Lauche, A. Loidl and R. Zwick, *ESA Special Publications SP-1050,* 125–129, (1983).
16. G. Backus and F. Gilbert, *Geophys. J. R. astron. Soc.* **13,** 247–276 (1967).
17. A. Tarantola, *Inverse Problem Theory,* Amsterdam: Elsevier, 1987.
18. E. Polak, *Computational Methods in Optimization,* New York: Academic Press, 1971.

LISM–HELIOSPHERE INTERACTION MEDIATED BY SUPRATHERMAL PARTICLES

S. Grzedzielski and J. Ziemkiewicz

Space Research Centre, ul. Ordona 21, 01–237 Warsaw, Poland

ABSTRACT

Problems of the deceleration and termination of the solar wind are discussed taking into account the interactions with neutral interstellar H and He, galactic (GCR) and anomalous (ACR) cosmic ray populations, interstellar magnetic field and effects due to pickup ions in LISM. It is assumed that the shocked solar wind is dominated by pressure of the reflected ions and that the magnetic field energy density in the interstellar medium is equal to the GCR energy density ($B^2/4\pi$=1.5 eV cm^{-3}). The resulting distance to the termination shock is 55-65 AU depending on the energy of ACR produced at the shock.

INTRODUCTION

Various aspects of the solar wind physics in the presence of cosmic rays and pickup ions have been intensively studied during the last few years. There still, however, remains a number of unresolved problems connected with this subject. Among these are :exact modeling of production of the anomalous cosmic ray component (ACR),behavior of the pickup ions and in particular of the so-called reflected (suprathermal) ions at the shock and in the post-shock region, influence of the presence of cosmic rays and pickup ions on the distance to the termination shock.

Here we address the last of these questions. In full generality the problem is exceedingly complex : it requires a calculation of interacting flows of LISM and solar wind, both treated as multicomponent fluids. Recently Ko et al /1/ considered the galactic-cosmic-ray (GCR) mediated solar wind model in a perturbation approach. The fully nonlinear equations describing the expansion of the solar wind in the presence of cosmic rays have been solved by Lee and Axford /2/. As our main task is to estimate the location of the shock ard of the heliopause, we shall just concentrate on those effects that seem to be most important from that point of view.

THE SOLAR WIND

The solar wind is treated as a spherically symmetric, three component medium consisting of cold background plasma, GCR and ACR. The model, first introduced by Axford et al /3/ assumes energy and momentum to be directly exchanged between cosmic ray gas and background fluid. Thus denoting the background plasma density ρ and velocity V we have in the steady state :

$$\frac{\partial}{\partial r} (r^2 \rho V) = Q \tag{1}$$

$$\rho V \frac{\partial V}{\partial r} = - \frac{\partial}{\partial r} (P_G + P_A) + S \tag{2}$$

$$\frac{1}{r^2} \frac{\partial}{\partial r} (r^2 F_i) = V \frac{\partial P_i}{\partial r} \tag{3}$$

where i = G,A (G-GCR , A-ACR), P_i is the cosmic ray pressure and the cosmic ray energy flux F_i is defined by :

$$F_i = 1/(\gamma_i -1) \left[\gamma_i V P_i - K_i \partial P_i / \partial r \right] ,$$

and γ_G= 4/3 , γ_A= 4/3-5/3.

The source terms Q and S, describing the mass and momentum addition/removal due to pickup ions are determined by summing over all relevant charge exchange and photoionisation processes for H and He (cf. reactions 1-7 from Table 1 in /4/).

We shall assume that the presence of ACR is important only in the vicinity of the shock, i.e. $\partial P_A/\partial r$ is negligible in (2) for r not too close to R_{SH}. The average interstellar energy density of GCR E_G = 1–1.5 eV cm^{-3} and the relative gradient (assumed to be constant) is 1.0–1.5 % / AU /5/.
With the above assumptions equations (1) and (2) may be solved to give :

$$V(r) = (V_o + 2Y(\lambda/r_o)^3)e^{-r/\lambda} - Y(\lambda/r_o)^3((r/\lambda)^2 - 2(r/\lambda) + 2) \quad (4)$$

where $Y = (1.5\%\ E_g)/(3\rho_o V_o)$

$r_o = 1$ AU, $\rho_o = \rho(r_o) = m_p n_o + m_\alpha n_\alpha = m_p\ 5\ cm^{-3} + m_\alpha\ 0.2\ cm^{-3}$

$V_o = V(r_o) = 400$ km/s,

$\lambda = 4.4\cdot 10^2$ AU, is the characteristic distance for interaction of solar wind ions with neutral H and He in interplanetary space /4/.

For $r \ll \lambda$ we finally have (accuracy up to $(r/\lambda)^3$)

$$V(r) = V_o\ x(r) \quad ; \quad (5)$$

$$x(r) = 1 - r/\lambda + 1/2(r/\lambda)^2 - 1/6(r/\lambda)^3 - 2.6\ 10^{-7}(r/r_o)^3$$

This means that the combined action of pickup processes and GCR slows down the solar wind (without changing its mass flux) by ~20 % for 50AU $< R_{SH} <$ 60 AU.

The ACR are assumed to be produced at the shock ($r=R_{SH}$). As we are mainly interested in the situation at the shock rather than in the details of the solar wind flow, we simply assume that some part η of the solar wind momentum flux is transformed at the shock into the cosmic ray pressure. A small fraction of the pickup ions is accelerated to energies of 10–30 MeV/nucleon to become ACR while the large majority of them is transmitted through the shock. We may thus assume that the process of diffusive shock acceleration conserves the mass flux of the background fluid and hence write the upstream velocity at $r=R_{SH}$ as

$$V_1(R_{SH}) = V_o\ x(R_{SH})\ (1-\eta) \quad (6)$$

Lee and Axford /2/ suggested that all the downstream pressure is provided by the so called reflected ions (with $\gamma_2=2$), i.e. those ions in the upstream pickup distribution which have the smallest velocity in the shock frame and thus may be reflected back upstream by the shock potential. In this approach the Rankine-Hugoniot equations become :

$$\rho_1 V_1 = \rho_2 V_2 = r_o^2/R_{SH}^2\ \rho_o V_o = A \quad (7)$$

$$AV_1 = AV_2 + P_2 \quad (8)$$

$$\frac{1}{2} AV_1^2 = \frac{1}{2} AV_2^2 + \frac{\gamma_2}{\gamma_2 - 1} V_2 P_2 \quad (9)$$

where subscripts 1,2 denote the upstream and downstream values, respectively. The momentum and energy fluxes of GCR and ACR have been excluded from Eq.(8)–(9).The galactic cosmic rays interact with the background plasma on length scales much larger than the shock thickness so that :

$$P_{G1} = P_{G2} \qquad F_{G1} = F_{G2} \quad (10)$$

From \6\ we have for ACR :

$$P_{A1} = P_{A2} = \eta\ x(R_{SH})\ A\ V_o \quad (11)$$

$$F_{A1} + Q_A = F_{A2} \quad , \quad Q_A = 1/2\ \eta(2-\eta)\ x^2(R_{SH})\ A\ V_o^2\ . \quad (12)$$

Finally the shock compression ratio r_c :

$$r_c = \frac{\rho_1}{\rho_2} = \frac{V_2}{V_1} = (\gamma_2+1)/(\gamma_2-1) = 3 \quad (13)$$

Numerical studies of modulation of the ACR show that, with the assumed shock compression ratio r=4, the average radial gradient is too small by a factor of 2 in comparison with observations. Potgieter and Moraal /7/ note that the proper value is obtainable only when the compression ratio is decreased by a factor of ~1.6 i.e. when r_c = (4 / 1.6) = 2.5 .

We now proceed to the shocked solar wind region. In order to estimate the influence of the cosmic rays on the flow in this region we treat the shocked solar wind as an incompressible medium characterized by :

$$\rho = \rho_2 = \text{const} \quad , \quad V = V_2\ (R_{SH}/r)^2 \quad (14)$$

In the test particle approximation the cosmic ray equation :

$$\frac{\partial}{\partial r}(\ R_{SH}^2 V_2 P_i - r^2 K_i\ \partial P_i/\partial r\ ) = 0 \quad , \quad i = G, A \quad (15)$$

may be solved with the boundary conditions (10)-(12) to give :

$$P_i(r) = C_{i1} \exp(-V_2 R_{SH}^2 / K_i r) + C_{i2} \qquad (16)$$

where the modulation parameter $(V_2 R_{SH})/K_1$ should be calculated separately for both species with the diffusion coefficient in the form /7/ :

$$K_i = 0.4 \cdot 3 \cdot 10^{22} \quad v_{th} / c$$

One can easily check that with the heliopause distance R_H : $R_{SH} \leq R_H \leq 2 R_{SH}$ \8\ the decrease of both cosmic ray pressures at $r = R_H$ is of the order of few percents. Thus the gradient forces $-\nabla P_1$ may be neglected in the post-shock region. If this is the case the pressure P_2 behind the shock and the pressure P_{st} at the stagnation point are related by /9/ :

$$P_{st} = (1 + (\gamma-1)^2/4\gamma)^{\gamma/(\gamma-1)} \quad P_2 = 1.2656 \, P_2 \qquad (17)$$

for the compressible fluid (with $\gamma=2$). This corresponds to $\rho_{st} = 1.125 \, \rho_2$, which is not far from incompressibility.

THE LOCAL INTERSTELLAR MEDIUM

The typical observed values for LISM neutral species ($n_H = 0.07$ cm^{-3}, $n_{He} = 0.01$ cm^{-3} , $T_L = 10^4$ K /10/) suggest that $n_L = n_p = n_e = 0.03$ cm^{-3} if the standard cosmic ratio [H]/[He]=10 is to be preserved. This implies that the LISM ram and thermal pressures amount to :

$$\rho_L V_L^2 = 0.16 \text{ eV cm}^{-3} \quad , \quad P_L = 2 \, n_L k \, T_L = 0.05 \text{ eV cm}^{-3} \qquad (18)$$

The number density of the neutral solar wind based on calculation in /4/ is

$$N_n(R_{SH}) = 1.08 \; 10^{-2} \; (r_o/R_{SH}) \text{ cm}^{-3} \; . \qquad (19)$$

These neutrals become ionised in LISM by charge exchange with the LISM protons. The effective cross section $\sigma = 1.8 \; 10^{-15}$ cm^2 , which yields the mean free path $\lambda_L = 1/(\sigma \, n_L) = 1200$ AU. In each act of ionisation a pickup ion is created in LISM. These pickup ions are removed by charge exchange with LISM neutrals. The mean free path for this reaction is $\lambda'_L = 1/(\sigma \, n_H) = 500$ AU.
As $\lambda_L \ll \lambda'_L$, in the limiting case of very slowly moving LISM the equilibrium between production and destruction rates yields :

$$n_{pu}(R_H) = N_n(R_H) \; n_L/n_H = 4.6 \; 10^{-3} (r_o/R_{SH}) \; (R_{SH}/R_H)^2 \text{ cm}^{-3} \qquad (20)$$

where the value of (R_{SH}/R_H) following Baranov /8/ is 1/1.3.
The behavior of the pickup distribution in the solar wind has been intensively studied during last few years /11,12/. The effect of the plasma waves is to produce rapid pitch angle scattering (to the point of isotropy) but negligible change in energy. This should also be true in the case of the LISM pickup ions : each pickup proton has now the energy $\simeq$ 1 keV . Thus the pressure of the above distribution is :

$$P_{pu} = (2/3) \; n_{pu}(R_H) \; 1 \text{ keV} \cong (\text{at } R_{SH} \cong 60 \text{ AU}) \; P_L = 0.05 \text{ eV cm}^{-3} \qquad (21)$$

Recently several arguments were forwarded /5,13/ that the size of the heliosphere may not be much larger than 50-60 AU. Simple calculation shows that this value of R_{SH} is obtainable only with $\Pi_L \cong 2\text{–}3$ eV cm^{-3}. Two possibilities arise. One is that this excess pressure is provided by viscous interaction between LISM and heliosphere. This, however, requires very high and therefore unrealistic values of the viscosity coefficient /14/.
The second possibility is to look for the additional pressure in the magnetic field pressure. One could argue that over very large scale a kind of energy equipartition between the magnetic field and the GCR energy at infinity prevails. Therefore $P_B = B^2/8\pi = 1/2 \cdot 1.5$ eV cm^{-3} , what corresponds to B = 4.3 μG. Because of the draping of the magnetic field lines the magnetic field energy in the vicinity of the heliopause should be increased by factor α ; $1 \leq \alpha \leq 2.25$ /15/ depending on the magnetic field direction.
With the LISM momentum flux in the form

$$\Pi_L = \rho_L \; V_L^2 + P_L + P_{pu} + P_B \qquad (22)$$

we are now able to find the value of R_{SH} from the equation :

$$P_{st} + P_A = 1.26 \; P_2 + P_A = \Pi_L \qquad (23)$$

With P_2 calculated from Eq (10), the above equation may now be written in the form :

$$\rho_o V_o^2 \left[\frac{r_o}{R_{SH}}\right]^2 \left[1.26 \; x(R_{SH})(1-\eta)(1-1/r_c) + \eta\right] - \Pi_L = 0 \qquad (24)$$

LHS of equation (24) is the polynomial of the 3rd order in R_{SH}/r_o, which may be solved for a given η. The values of R_{SH} as function of η are shown in Fig.1 for $B^2/8\pi$ = 0.75 eV cm^{-3} and, for comparison, for $B^2/8\pi$ = 0.05 eV cm^{-3} (what corresponds to LISM plasma parameter β=1).

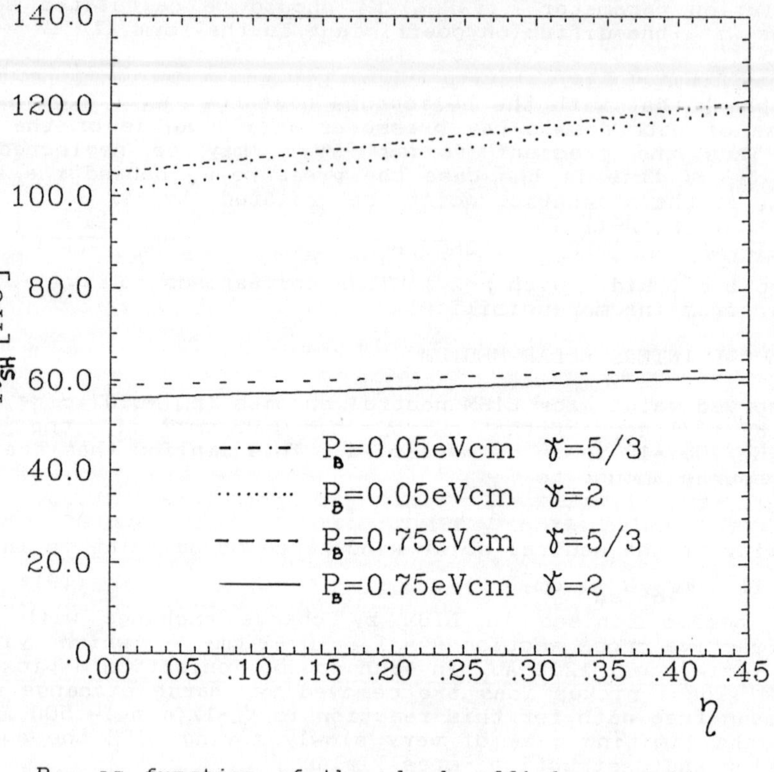

Fig. 1. R_{SH} as function of the shock efficiency ratio η .

SUMMARY

The (upwind) radius of the termination shock has been recalculated with the following assumptions :
1. the shocked solar wind plasma is a medium with γ = 2 i.e. all the downstream pressure is provided by the reflected ions.
2. the LISM magnetic field energy is equal to the GCR energy at infinity i.e. B = 4.3 μG.
In the above calculation the following effects have been taken into account : /a/ loss of momentum by the mass loading of the solar wind, /b/ influence of GCR, /c/ influence of ACR, /d/ pressure of pickup ions in LISM.
While the effects /a/,/b/,/d/ diminish the size of the heliosphere, **the production of the anomalous component cosmic rays leads to the increase of the shock radius.** With reasonable value of the shock efficiency ratio η 0.2 and relatively small magnetic field (B ≅ 4 μG) we obtain R_{SH} = 55–60 AU .

REFERENCES

1. C.M.Ko and G.M.Webb, Ap.J. 323,657 (1987)
C.M.Ko,J.R.Jokipii and G.M.Webb, Ap.J. 326, 761 (1988)
2. M.A.Lee and W.I.Axford, Astron.Astrophys. 194, 297 (1988)
3. W.I.Axford,E.Leer,G.Scadron, Proc.15th Int.Cosmic Ray Conf.2,223 (1977)
4. S.Grzedzielski, D.Rucinski, this issue.
5. W.R.Webber, Astron.Astrophys. 179,277 (1987)
6. L.J.Gleeson and W.I.Axford, Ap.J. 149,L115 (1967)
7. M.S.Potgieter and H.Moraal, Proc.20th Int.Cosmic Ray Conf. 3, 464 (1987)
8. V.B.Baranov,M.G.Lebedev, M.S.Ruderman, Astrophys.Sp.Sci. 66, 441 (1979)
9. E.Parker, *Interpl. Dynamical processes* (New York:Interscience), (1963)
10. E.Chassefiere, J.L.Bertaux, R.Lallement and V.G.Kurth, Astron.Astrophys. 160,229 (1986)
11. M.A.Lee and W.H.Ip, JGR 92,1104 (1987)
12. H.J.Fahr and J.Ziemkiewicz, Adv.Sp.Res. 9, (4)251 (1989)
13. W.S.Kurth, D.A.Gurnett and F.L.Scarf, Nature 312, 27 (1984)
14. S.Grzedzielski and J.Ziemkiewicz, Adv.Sp.Res. 9, (4)247 (1989)
15. W.I.Axford, Solar Phys. 100, 575 (1985)

EXPECTED BEAMS OF ENERGETIC NEUTRAL ATOMS IN THE OUTER HELIOSPHERE

Stanislaw Grzedzielski and Daniel Ruciński

Space Research Centre of the Polish Academy of Sciences, ul. Ordona 21, 01–237 Warsaw, Poland

ABSTRACT

Estimates of the fluxes of energetic neutral atoms (ENA) in the distant heliosphere are calculated taking into account all important interactions between solar wind plasmas and neutral gases in the heliosphere. Effects of charge-exchange, double charge-exchange (for helium component), electron impact ionization, photoionization, recombination in thermal and non-thermal plasmas are calculated for various conditions over the solar cycle. The resulting neutral beams of energetic atoms emerging from the heliosphere may carry into the local interstellar medium (LISM) up to ≈ 0.1–0.3 of the total solar wind energy, depending mainly on the location of the termination shock ("small" or "large" heliosphere), spatial direction and rather weakly on adopted plasma and gas parameters (densities, temperatures). The free energy thus transferred to LISM may significantly affect the state of the interstellar plasma in the immediate vicinity of the heliosphere.

THEORETICAL APPROACH

The physical conditions in the Very Local Interstellar Medium (VLISM) flowing past the heliosphere should be influenced by the mass/momentum/energy flux lost by the heliosphere in form of electromagnetic radiation, solar wind plasma, energetic neutral atoms (ENA), cosmic rays (anomalous component) etc. The aim of the present paper is to provide estimates of mass and energy fluxes emitted by the heliosphere in form of ENA. The heliospheric production of ENA is mainly due to solar wind plasma interactions with neutral species of galactic origin (interstellar neutral wind). Relevant reactions with cross sections corresponding to energies of 1 keV/nucleon are given in Table 1.

Assuming that the solar wind consists initially of protons and alpha particles with corresponding densities at Earth: n_p = 7 cm^{-3} and n_α = 0.3 cm^{-3}; one obtains the abundance of any species by integrating the equation:

$$\frac{1}{r^2}\frac{d}{dr}\left[r^2 n_k v\right] = \sum_l \xi^{(1)}_{kl} - \sum_l \xi^{(2)}_{kl} \qquad (1)$$

where n_k is the density of the k-th species, $\xi^{(1)}_{kl}$, $\xi^{(2)}_{kl}$ describe the source and loss terms for the k-th species in the l-th reaction and v is here the single fluid bulk velocity of the solar wind. Note, that the assumed spherical symmetry may not apply to solar wind conditions prevailing at higher heliolatitudes.

The neutral background involved in reactions listed in Table 1 leading to the direct (or indirect) production of energetic neutral atoms consists of interstellar hydrogen and helium gas penetrating the heliosphere. To study quantitatively the expected fluxes of energetic H and He atoms in the outer heliosphere, as well as the transfer of energy into VLISM by neutral atoms, we have assumed that the density distribution of both (hydrogen and helium) original neutral components is determined according to the well-known "hot" kinetic model /1–3/ in which all important ionization processes (as charge-exchange, double charge-exchange, photoionization, electron impact) and gravitational interaction of the Sun have been taken into account. In our case, this distribution has been calculated for the "average" solar activity conditions, which correspond to the following ratios μ of the radiation pressure to solar gravitational force and total ionization rates β_0 at 1 AU:

$$\mu^{H} = 0.75; \quad \beta_{o}^{H} = 5.5\cdot 10^{-7}\,s^{-1} \tag{2}$$

$$\mu^{He} = 0; \quad \beta_{o}^{He} = 9.0\cdot 10^{-8}\,s^{-1} \tag{3}$$

The interstellar wind parameters have been adopted for both components as follows:

$$n_{\infty}^{H} = 0.07\ cm^{-3}; \quad V_{\infty}^{H} = 20\ km/s; \quad T_{\infty}^{H} = 10000\ K \tag{4}$$

$$n_{\infty}^{He} = 0.01\ cm^{-3}; \quad V_{\infty}^{He} = 23\ km/s; \quad T_{\infty}^{He} = 13000K \tag{5}$$

TABLE 1 Reactions in Solar Wind

cx = charge-exchange, sw = solar wind, ci = collisional ionization, is = interstellar, pup = pick-up, 2cx = double charge exchange, f = fast = 400 km/s, el = electron impact ionization, r = recombination, phi = photoionization

		Reaction	Reference No.	σ (cm^2)
(1)	cx	$p(sw)+H^{o}(is) \longrightarrow H^{o}(f)+p(pup)$	/4/	$1.75\cdot 10^{-15}$
(1a)	ci	$p(sw)+H^{o}(is) \longrightarrow p(sw)+p(pup)+e$	/5/	$2.30\cdot 10^{-18}$
(2)	cx	$p(sw)+He^{o}(is) \longrightarrow H^{o}(f)+He^{+}(pup)$	/6/	$5\cdot 10^{-18}$
(2a)	ci	$p(sw)+He^{o}(is) \longrightarrow p(sw)+He^{+}(pup)+e$	/7/	threshold at 7 keV
(2b)	2cx	$p(sw)+He^{o}(is) \longrightarrow H^{-}(sw)+\alpha(pup)$	/8/	$\sim 2\cdot 10^{-20}$ (extrapol.)
(3)	cx	$\alpha(sw)+H^{o}(is) \longrightarrow He^{+}(sw)+p(pup)$	/9/	$2.3\cdot 10^{-16}$
(4)	cx	$\alpha(sw)+He^{o}(is) \longrightarrow He^{+}(sw)+He^{+}(pup)$	/10/	$1.8\cdot 10^{-17}$
(5)	2cx	$\alpha(sw)+He^{o}(is) \longrightarrow He^{o}(f)+\alpha(pup)$	/10/	$\sim 3\cdot 10^{-16}$
(6)	phi	$H^{o}(is)+h\nu \longrightarrow p(pup)+e$	/11/	$\left(\beta = 8.8\cdot 10^{-8}\left(\frac{1AU}{r}\right)^{2} s^{-1}\right)$
(7)	phi	$He^{o}(is)+h\nu \longrightarrow He^{+}(pup)+e$	/11/	$\left(\beta = 8.0\cdot 10^{-8}\left(\frac{1AU}{r}\right)^{2} s^{-1}\right)$
(8)	phi	$He^{+}(sw/pup)+h\nu \longrightarrow \alpha(sw/pup)+e$	/12/	$\left(\beta = 9.0\cdot 10^{-9}\left(\frac{1AU}{r}\right)^{2} s^{-1}\right)$
(9)	r	$p(sw)+e \longrightarrow H^{o}(f)$		
(10)	r	$\alpha(sw)+e \longrightarrow He^{+}(sw)$		
(11)	r	$He^{+}(sw)+e \longrightarrow He^{o}(f)$		
(12)	el	$H^{o}(is)+e \longrightarrow p(pup)+2e$	/13,14/	$6.5\cdot 10^{-17}$ for 50 eV
(13)	el	$He^{o}(is)+e \longrightarrow He^{+}(pup)+2e$	/13,14/	$2.5\cdot 10^{-17}$ for 50 eV
(14)	cx	$He^{+}(sw/pup)+H^{o}(is) \longrightarrow He^{o}(f)+p(pup)$	/15/	$4.6\cdot 10^{-17}$ for 7 keV
(15)	cx	$He^{+}(sw/pup)+He^{o}(is) \longrightarrow He^{o}(f)+He^{+}(pup)$	/15/	$6.5\cdot 10^{-17}$ for 7 keV

RESULTS AND DISCUSSION

The important consequence of the interactions between neutral interstellar gas and solar plasma is that fluxes of energetic neutral atoms will be generated. These fluxes depend noticeably on heliocentric distance R, as well as (at least in the plane of ecliptic) on the ecliptic longitude. The dominant part of the flux (>99%) as well as of the energy (97%) is carried by energetic hydrogen component. At R = 50 AU the expected fluxes of energetic H atoms vary from $7.7\cdot 10^{3}\,cm^{-2}s^{-1}$ (upwind axis) to $4.4\cdot 10^{3}\,cm^{-2}s^{-1}$ (downwind axis). This noticeable upwind/downwind asymmetry, which smears out only at large heliocentric distances (R > 150 AU), results from the asymmetric distribution of interstellar hydrogen in the circumsolar region and reflects the remnants of the hydrogen "wake" mostly pronounced in the vicinity of the Sun. The corresponding fluxes of energetic helium atoms (at 50 AU) are much lower and range from $\approx 12\ cm^{-2}s^{-1}$ (upwind axis) up to $28\ cm^{-2}s^{-1}$ (downwind axis). This opposite behaviour of the helium fluxes with respect to ecliptic longitude is caused by solar gravitational focusing resulting in a region of enhanced helium density ("helium cone"). The detailed dependencies of the fluxes of both constituents on heliocentric distance are shown in Fig.1 for upwind and downwind directions. The calculations include the effect of slowing down the solar wind by the pick-up process.

If the heliosphere is small (R_B = 50 AU) the energy transferred by neutrals into VLISM amounts to $8.4 \cdot 10^{-6}$ ergs cm^{-2} s^{-1} i.e. ≈ 6% of the total energy carried by the solar wind. If the heliospheric boundary is more distant this fraction increases approximately linearly with R_B and attains 10%, 20%, 28% for R_B = 80 AU, 150 AU, 200 AU (upwind), respectively (see Fig.2).

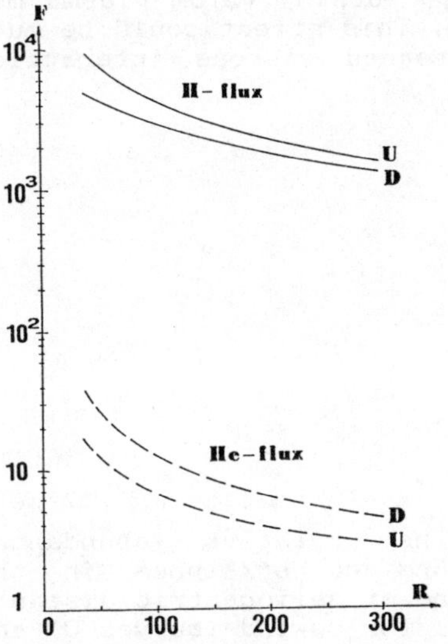

Fig.1. The fluxes of energetic H and He atoms ($cm^{-2}s^{-1}$) as a function of heliocentric distance R (in AU). The curves U correspond to the upwind direction, whereas the curves D - to the downwind direction.

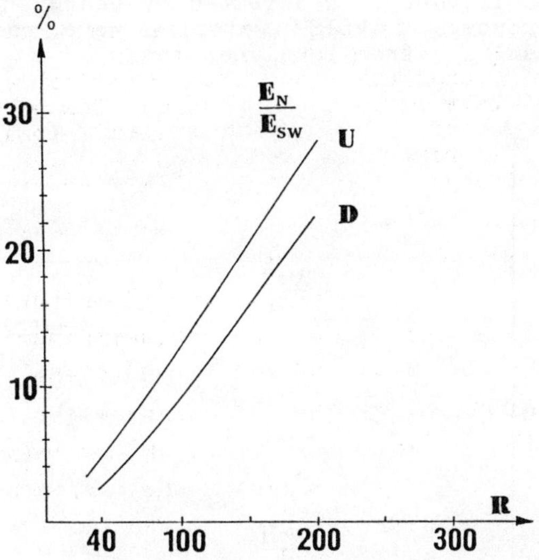

Fig.2. The ratio of energy carried by energetic neutrals to the total solar wind energy as a function of heliocentric distance R (in AU) for upwind (curve U) and downwind (curve D) directions.

Since hydrogen is the major component of the neutral energetic population, the solar cycle variability of the ENA-related energy flux transferred into VLISM is determined mainly by the variation of the solar wind proton flux. According to the Helios data this long-term variability is of the order of ~ 30% in the ecliptic plane /16/. The variability of the expected ENA fluxes at large heliocentric distances (R > 50 AU) due to the solar cycle variations of the radiation pressure exerted on original interstellar H atoms are of importance only in the downwind region and does not exceed 15%.

The uncertainties of the determination of interstellar neutral wind parameters: T and V_∞ (especially for helium) do not significantly affect the determination of ENA fluxes at large heliocentric distances; the expected error is well below 5 % .

The possible ecliptic/pole asymmetries were only poorly studied and thus are difficult to estimate. The charge-exchange cross section per solar wind proton decreases by ~ 18 % for ecliptic/pole solar wind velocity ratio 400 km/s to 700 km/s. Also, polar solar wind densities and corresponding densities in the polar neutral hydrogen cavity may differ from the ecliptic values. For the values estimated by Witt et al., /17/ from the Mariner-10 Lyman-α data the expected polar energetic hydrogen flux should be smaller than the ecliptic value by about 30 %.

Besides the creation of the energetic neutral population the other effect caused by the reactions in Table 1 is strong enrichment of the solar wind in He^+ ions, particularly at large heliocentric distances R (see Fig.3). In the downwind region at distances 50-200 AU from the Sun the abundance of He^+ may reach 10-30 % of the abundance of solar wind alpha particles. The energy carried by energetic neutral species becomes deposited far away (therefore also ahead) of the heliopause: the spatial range of ~ 1 keV hydrogen atoms in VLISM will mainly be determined by the probability of reionization of the created fast neutrals. This probability is given by reactions 1, 1a supplemented by ionization resulting from collisions of two neutral species

(relevant cross section for H-H and H-He collisions are given by Gilbody /9/ and Tawara /8/). For instance, for the assumed VLISM proton density n^p_∞ = 0.03 cm^{-3} one obtains the ionization time of $4.7 \cdot 10^8$s which yields the mean-free path against reionization equal to 1250 AU. The fraction of solar wind energy carried by the fast neutrals is then distributed over a huge volume of ~ 10^3 AU in radius. If all energy deposited by the fast H atoms in a moving element of VLISM plasma were locally conserved each VLISM proton would acquire ~ 4 eV before reaching the stagnation point at ~ 50 AU. This indicates that the thermodynamical state of the approaching VLISM plasma may significantly be affected by energy carried by ENA. This effect could be much stronger if VLISM velocity were reduced by increased viscous interaction resulting from local heating.

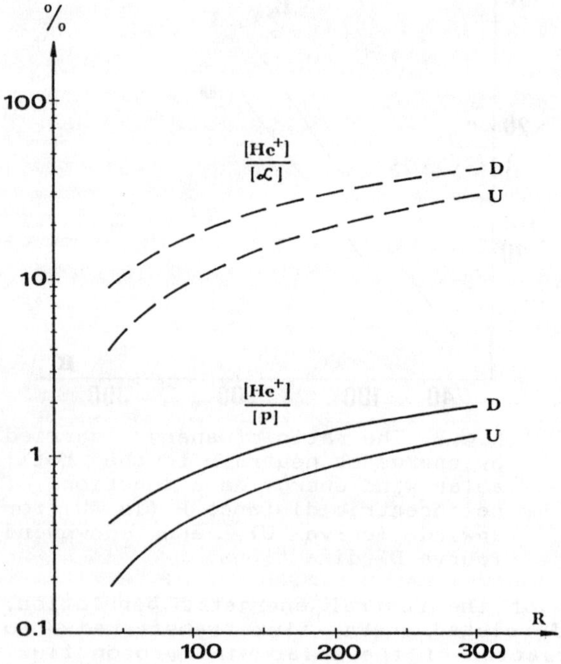

Fig.3. The relative abundances He^+/protons and He^+/alphas in the solar wind at heliocentric distance R (in AU) for upwind (curves U) and downwind (curves D) directions.

REFERENCES

1. J.M.A.Danby and G.L.Camm, M.N.R.A.S. 117, 50 (1957)
2. H.J.Fahr, Astron.Astrophys. 14, 263 (1971)
3. G.E.Thomas, Ann.Rev.Earth Planet.Sci. 6, 173 (1977)
4. L.J.Maher and B.A.Tinsley, J.Geophys.Res. 82, 689 (1977)
5. N.F.Mott and H.S.W.Massey, The theory of atomic collisions, Oxford, 1965
6. M.Kimura and G.D.Lin, Phys.Rev.A. 34, 176 (1986)
7. W.Keever and E.Everhart, Phys.Rev.A. 1, 1083 (1970)
8. H.Tawara, Atomic and Nuclear Data Tables 22, 491 (1978)
9. H.B.Gilbody, Physica Scripta 24, 712 (1981)
10. K.H.Barkner,R.V.Pyle,J.W.Stearns and J.C.Warren, Phys.Rev. 166, 44 (1968)
11. T.E.Holzer, Revs.Geophys.Space Phys. 15, 467 (1977)
12. P.M.Banks and G.Kockarts Aeronomy Part A, Academic Press, 1973
13. L.J.Kieffer and G.H.Dunn, Rev.Mod.Phys. 38, 1 (1966)
14. W.Lotz, Zeitschrift f. Physik 206, 205 (1967)
15. S.Allison, Rev.Mod.Phys. 30, 1137 (1958)
16. R.Schwenn ,The "Average" Solar Wind in the Inner Heliosphere: Structures and Slow Variations , in Solar Wind Five, ed.M.Neugebauer, Jet Propulsion Laboratory,Pasadena,California, 1983, NASA CP-2280, p.489
17. N.Witt,J.M.Ajello and P.W.Blum, Astron.Astrophys. 95, 80 (1981)

Special Session

Deep Space Missions

PIONEERS 10 AND 11 DEEP SPACE MISSIONS

P. Dyal

Space Research Directorate, NASA Ames Research Center, Moffett Field, CA 94035, U.S.A.

ABSTRACT

Pioneers 10 and 11 were launched from Earth March 2, 1972, and April 5, 1973, respectively. They were the first spacecraft to explore the asteroid belt and the first to encounter the giant planets, Jupiter and Saturn. The Pioneer 10 spacecraft is now the most distant man-made object in our solar system and is farther from the Sun than all nine planets. It is 47 AU from the Sun and is moving in a direction opposite to that of the Sun's motion through the galaxy. Pioneer 11 is 28 AU from the Sun and is traveling in the direction opposite of Pioneer 10, in the same direction as the Sun moves in our galaxy. These two Pioneer spacecraft have provided the first large-scale, in situ measurements of the gas and dust surrounding a star, our Sun. Since launch, the Pioneers have measured large-scale properties of the heliosphere during more than one complete 11-year solar sunspot cycle, and have measured the properties of the expanding solar atmosphere, the transport of cosmic rays into our heliosphere, and the high-energy trapped radiation belts and magnetic fields associated with the planets Jupiter and Saturn. Accurate Doppler tracking of these spin-stabilized spacecraft has been used to search for differential gravitational forces from a possible trans-Neptunian planet and to search for gravitational radiation. Future objectives of the Pioneer 10 and 11 missions are to continue measuring the large-scale properties of our heliosphere and to search for its boundary with interstellar space.

GENERAL MISSION DESCRIPTION

The Pioneer 10 spacecraft launched March 2, 1972, is the most distant man-made object in our solar system. Mission studies initiated in the early 1960s came to fruition in 1969 when Ames Research Center was assigned the exploratory Pioneer Jupiter Project. Three spacecraft were built to achieve the mission objective; the first, for testing, which is now in the National Air and Space Museum, the second, named Pioneer 10, and the third launched April 5, 1973, named Pioneer 11. These two spacecraft were the first to be launched with enough energy to escape the solar system. After exploring the regions of space beyond the orbit of Mars, both spacecraft encountered Jupiter and then Pioneer 11 was targeted for Saturn. Pioneer 11 was launched one year after Pioneer 10 so its trajectory could be modified depending on the condition of Pioneer 10 after it passed through the asteroid belt and Jupiter's radiation belts.

The missions are expected to last three decades and the following objectives have evolved:

(1) Explore the interplanetary medium beyond the orbit of Mars.

(2) Investigate the nature of the asteroid belt and assess its hazard to future missions.

(3) Explore the environment of Jupiter.

(4) Explore the environment of Saturn.

(5) Explore the outer heliosphere and potentially pass into interstellar space.

Many individuals and institutions were required to implement these two missions. The Pioneer 10 and 11 Project was managed by Ames Research Center and NASA Headquarters, the Atlas Centaur booster by Lewis Research Center, launch operations by Kennedy Space Center, the Deep Space Network tracking by Jet Propulsion Laboratory (JPL), the radioisotope thermoelectric power generators by the Atomic Energy Commission, and the spacecraft was built by TRW Systems Group. Six universities and four NASA centers provided the scientific experiments for the project; they are listed in Table 1.

TABLE 1 Experiments and Principal Investigators, Pioneer 10 and 11

Helium Vector Magnetometer	Dr. Edward J. Smith Jet Propulsion Laboratory
Plasma Analyzer	Dr. Aaron Barnes Ames Research Center
Charged Particle Instrument	Dr. John A. Simpson University of Chicago
Cosmic Ray Telescope	Dr. Frank B. McDonald Goddard Space Flight Center
Geiger Tube Telescope	Dr. James A. Van Allen University of Iowa
Trapped Radiation Detector	Dr. R. Walker Filluis University of California, San Diego
Ultraviolet Photometer	Dr. Darrell L. Judge University of Southern California
Celestial Mechanics	Dr. John D. Anderson Jet Propulsion Laboratory
Imaging Photopolarimetry	Dr. Tom Gehrels University of Arizona, Tucson

The two spacecraft and most of the scientific instruments are in excellent condition. Data are being acquired on a daily basis at 16 bits per second with the 70-m diameter antennas of the JPL Deep Space Network. This telemetry command and receiving capability should permit the spacecraft and experiments to be operated to the end of the century. In August 1989, Pioneer 10 was 47.16 AU from the Sun and the round-trip light time was 13 hr 6 min. Pioneer 11 was 28.99 AU from the Sun and the corresponding command and receive time was 7 hr 58 min.

Figure 1 shows the prominent physical features of the spacecraft. It is a highly reliable spacecraft of relatively simple design in which many of the components and subsystems have successfully performed on earlier missions. It has a thermally controlled equipment compartment with two sections, one is hexagonally shaped and contains electronic units and the propellant tank, and the other is a bay which contains most of the scientific sensors and their associated electronics (the magnetometer sensor and two meteoroid detectors are external). Forward of the equipment compartment is a 2.7-m diameter parabolic reflector for the high-gain antenna. Mounted on a tripod structure forward of the reflector are the medium-gain antenna and the feed for the high-gain antenna. Three appendages are stowed within a 2.7-m cylindrical envelope at launch; they are shown in their deployed positions attained within an hour after launch. Two pairs of radioisotope thermoelectric generators (RTGs) are extended approximately 1.8 m at 120° spacing. The RTGs are retained in a stowed position for launch next to the equipment compartment and under the antenna reflector. The magnetometer sensor is located on the end of a long folding boom which, in the deployed condition, extends 5.2 m radially from the instrument side of the equipment compartment.

Six 4-N hydrazine thrusters are located in three clusters near the perimeter of the 2.7-m reflector. Two pairs of thrusters are aligned parallel to the spin axis for precession and velocity correction maneuvers; two thrust tangentially for spin control. Other external features include a mast-mounted, omnidirectional antenna directed aft, and a Sun sensor mounted near one of the thruster assemblies which determines the spacecraft's position in the spin cycle. Two large light shields are associated with the stellar-reference assembly, and with an optical asteroid/meteroid detector.

The spacecraft electronic assemblies are located in the central hexagonal portion of the compartment, surrounding a 42-cm diameter hydrazine tank. Most of the scientific instruments' electronic units and internally mounted sensors are in an instrument bay mounted on one side of the central hexagon. This equipment compartment is contained within a structure of aluminum honeycomb which provides support and meteroid protection. It is covered with insulation which, together with louvers under the mounting platform, provide thermal control. Adequate temperature is provided by dissipation of the 90 watts of electrical power within the compartment and by use of radioisotope and electrical heaters.

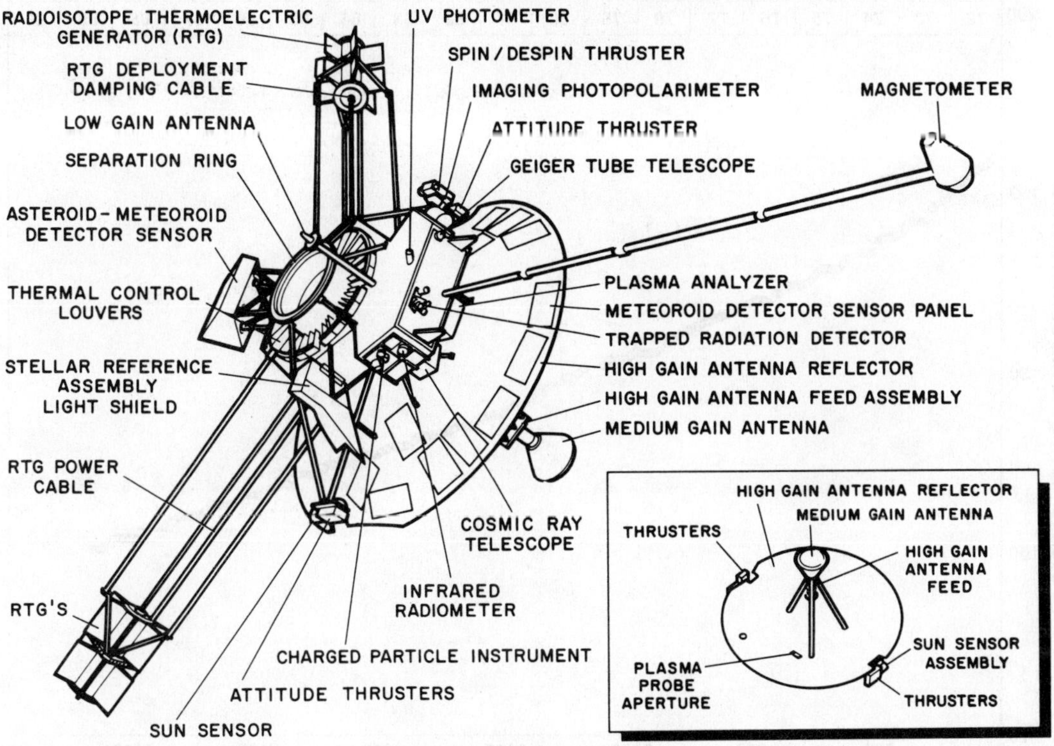

Fig. 1. Pioneer/Jupiter spacecraft.

The spacecraft is attitude-stabilized by spinning about an axis which is parallel to the axis of the 2.7-m reflector; the nominal spin rate for Pioneer 10 is 4.8 rpm. Pioneer 11 spins at approximately 7.8 rpm because of an inoperative thruster. During the mission, an Earth-pointing attitude is required to illuminate the Earth with the narrow-beam, high-gain antenna. Periodic attitude adjustments are required throughout the mission to compensate for the variation in the heliocentric longitude of the Earth-spacecraft line. In addition, correction of launch vehicle injection errors were required to provide the desired Jupiter encounter trajectory and Saturn (Pioneer 11 only) encounter trajectory. These velocity vector adjustments involve reorienting the spacecraft to direct the thrust in the desired direction.

Thrusters are used to control the spin of the spacecraft, its attitude, and its velocity, as shown in Figure 1. The precession maneuvers can be open-loop, for orientations toward or away from Earth-pointing, or closed-loop, for homing on the uplink radio frequency transmission from Earth.

For attitude control, a star sensor (Canopus) and two sunlight sensors provide reference, and three pairs of thrusters located near the rim of the antenna dish provide for orientation and roll maneuver. The star sensor on Pioneer 10 has been inoperative since the Jupiter encounter. About 44.7 lb of hydrazine propellant remain on Pioneer 10 and about 11.3 lb remain on Pioneer 11. The magnetometer boom incorporates a hinged viscous-dampening mechanism at its attachment point, for passive nutation control. The remaining hydrazine propellant on Pioneer 10 and 11 is adequate to support the attitude control maneuvers well into the next century and will not limit the mission lifetime.

A unique characteristic of this spacecraft is the use of RTGs as the primary source of electrical power, making it independent of the Sun. Each of the four space-proven SNAP-19 RTGs converts 5 to 6 percent of the heat released from plutonium dioxide radioactive decay to electrical power. The RTG power is greatest at 4.2 volts and an inverter boosts this to 28 volts for distribution. The RTG lifetime is degraded at lower currents, therefore, voltage is regulated by shunt dissipation of excess power. At launch, the four RTGs supplied about 160 watts on each spacecraft as shown in Figures 2 and 3. Examination of all subsystems on both spacecraft to date indicates that RTG power is the most critical in determining mission lifetime. Approximately 80 watts minimum are required to operate the spacecraft and a small subset of the scientific instruments. As seen in Figures 2 and 3, the RTGs decay at different rates. It is projected that Pioneer 11 will last until 1994 and Pioneer 10 until 1997 before the power drops to 80 watts and the shunt current decreases to zero. Presently the scientific instruments on Pioneer 11 are sequentially turned on and off to time-share the remaining power. It is expected that Pioneer 10 will be required to start a power-sharing program during September 1989.

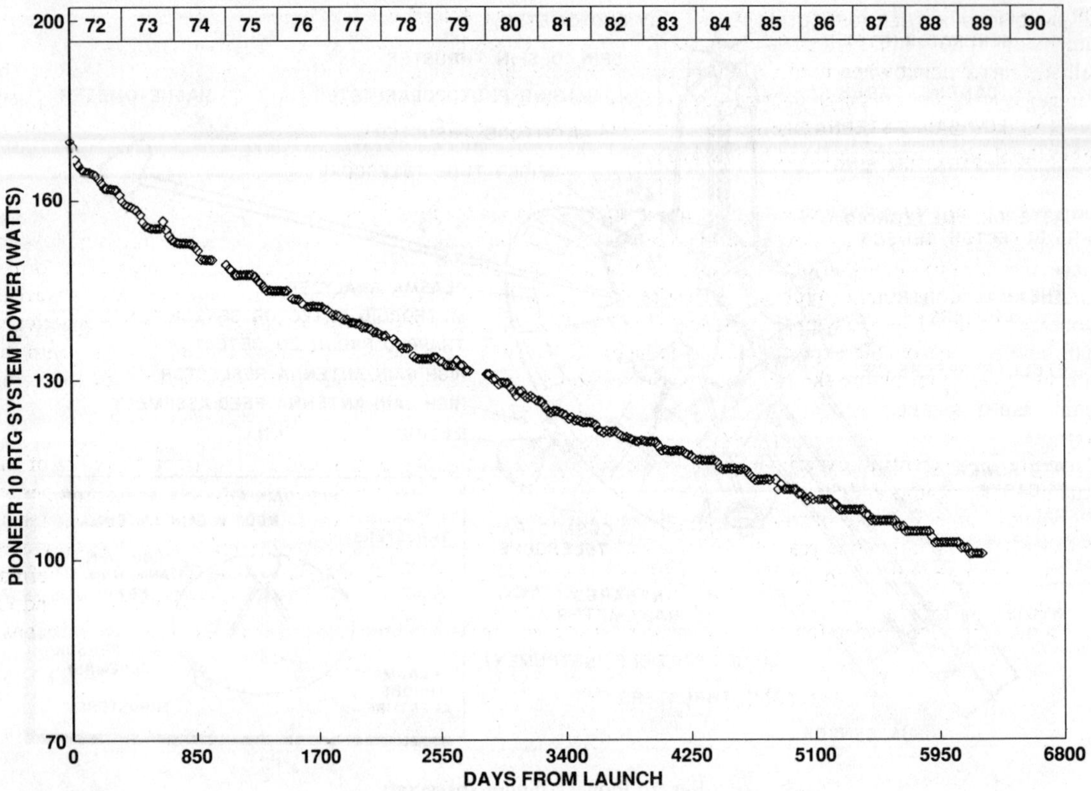

Fig. 2. Pioneer 10 RTG system power history.

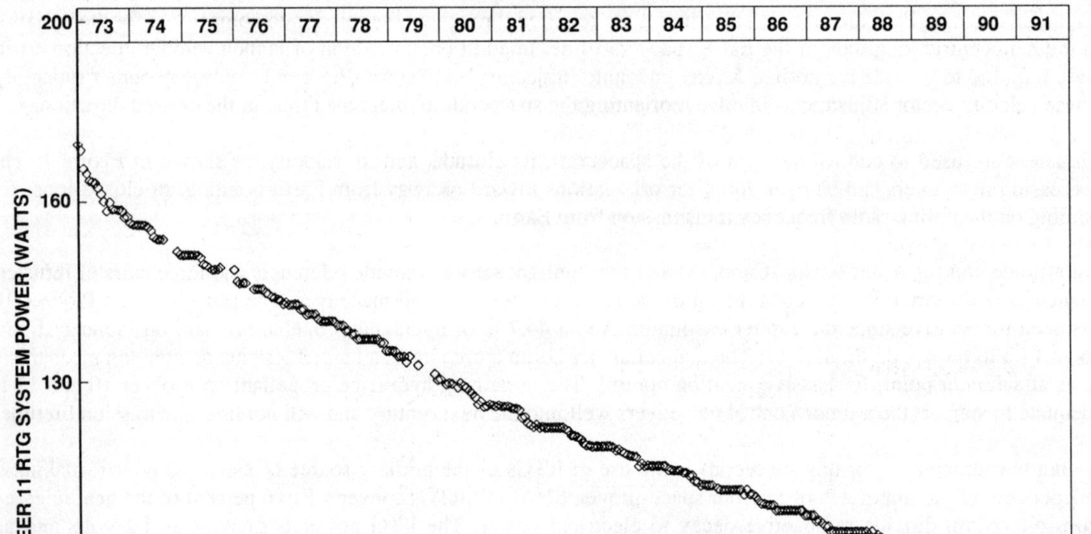

Fig. 3. Pioneer 11 RTG system power history.

The launch vehicle for Pioneer 10 and 11 was the three-stage Atlas/Centaur/TE-364-4, shown in Figure 4 with dimensions in feet and inches. The long-tank Atlas is a space booster version of the Series D intercontinental ballistic missile, and when used within the second-stage Centaur vehicle is designated the "Atlas SLV-3C." The Centaur is an operational high-energy space vehicle which uses liquid hydrogen for fuel. The Atlas/Centaur System required the TE-364-4 third stage for Jupiter injection velocities. This launch vehicle boosted each spacecraft to begin the flight to Jupiter at about 51,680 km/hr.

Pioneers 10 and 11 are proceeding in approximately opposite directions, the latter toward the nose of the heliosphere, and the former down the hypothetical tail of the heliosphere, as shown in Figure 5. Both Voyager spacecraft are proceeding in the same quadrant of the solar system as Pioneer 11. These four spacecraft form a unique set of measurement probes to measure the three-dimensional, large-scale structure of the Sun's corona. Pioneers 10 and 11 have asymptotic escape velocities from the solar system of 2.4 and 2.2 AU/year, respectively, and there is a reasonable expectation that Pioneer 10 will penetrate the heliosphere boundary and pass into the interstellar medium before the radioisotope power supply outputs decrease below the level to sustain spacecraft requirements.

Currently nine scientific experiments are being conducted on the two nearly identical spacecraft. A list of the principal investigators and their experiments is shown in Table 2 and an abbreviated list of the instrument properties is given in Table 3. These instruments are all operating normally except for the helium vector magnetometer on Pioneer 10 which failed one year after the Jupiter encounter and the imaging photopolarimeter which is only used for aspect determination on Pioneer 10. Two meteoroid detection experiments and a Jovian Infrared Thermal Structural Experiment were also turned off after their primary objectives were completed. References /1/ and /2/ give a more detailed description of the experiments and their operation during the early phases of the two missions.

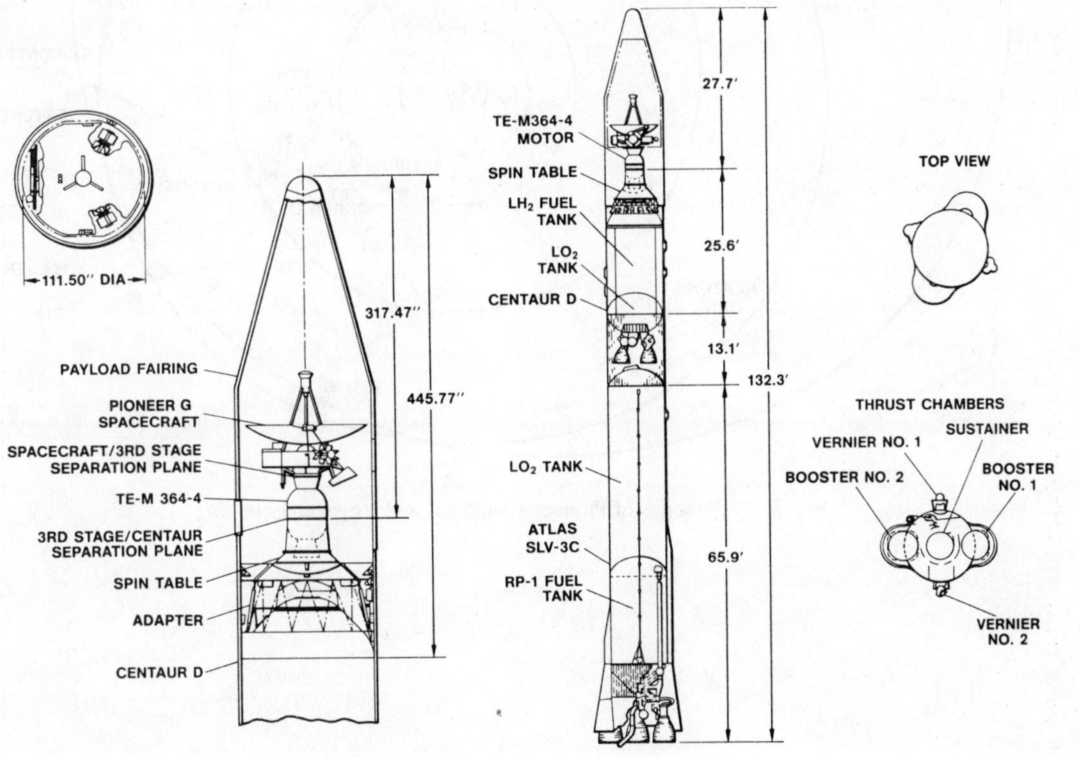

Fig. 4. Pioneer 10/11 launch vehicle.

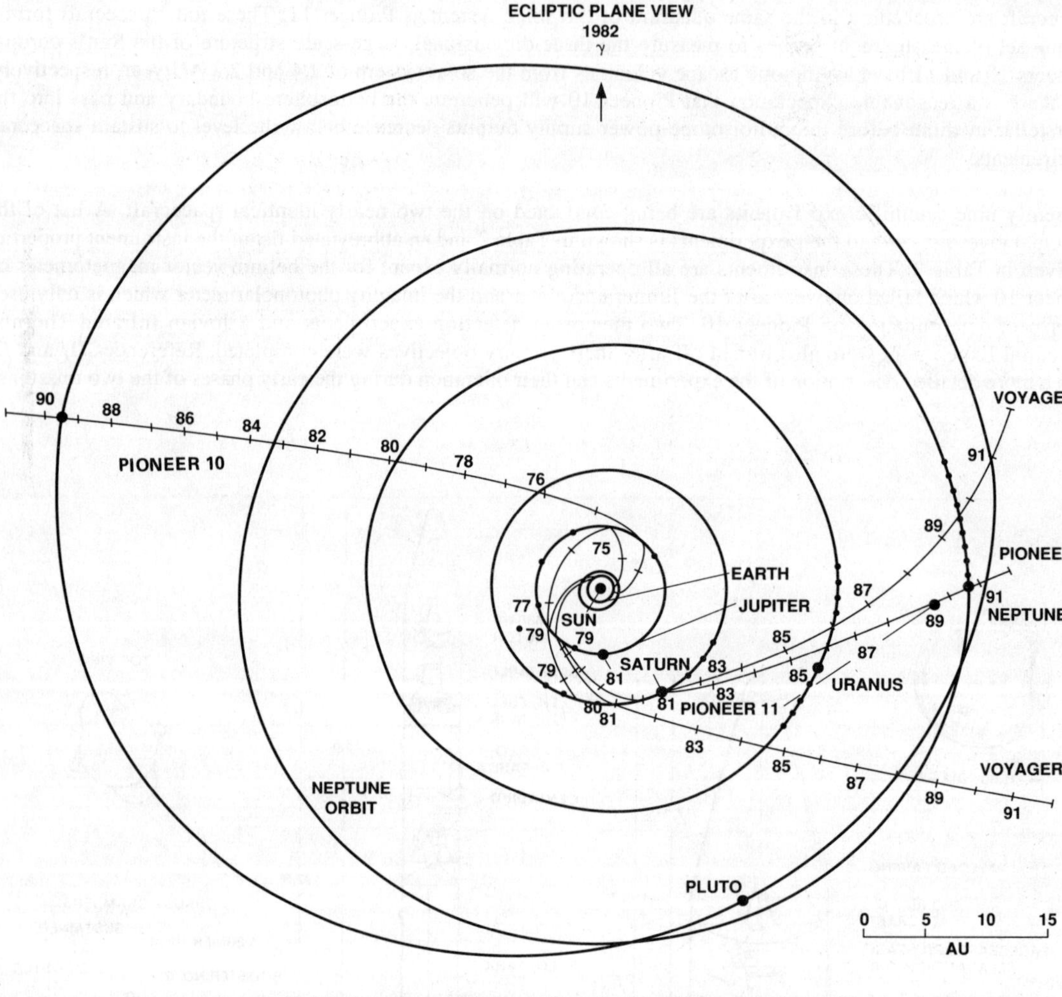

Fig. 5. Trajectories of Pioneer 10 and 11, and Voyager 1 and 2.

TABLE 2 Current Scientific Experiment Properties

Experiment	Detection Principle	Sensitivity
1. Helium Vector Magnetometer	Helium vapor; Zeeman effect	0.01 gamma; range ±4.0 gamma to ±1.41 gauss
2. Plasma Analyzer	Electrostatic analyzer; channeltrons	0.1 to 18 KeV protons, 0.001 to 0.5 KeV electrons; 10^2 to $10^9 P^+ CM^{-2} SEC^{-1}$
3. Charged Particle Instrument	Two Si detector multi-element telescopes electron detector and fission cell	Discriminates protons through oxygen; 0.5 to 500 MeV/nucleon; electrons 3 to 30 MeV; integral fluxes for P^+ through Fe^+ > 500 MeV/nucleon
4. Cosmic Ray Telescope	Three Si solid state telescopes	I. P^+ to Ne; 3 to 22 MeV/nucleon II. electrons 0.5 to 1.0 MeV; P^+ 0.05 to 20 MeV III. P^+ and He; 22 to 800 MeV/nucleon
5. Geiger Tube Telescope	Geiger-Mueller tubes, 7 each Si solid-state detector	P^+ > 70 MeV; 0.2 to 10^6 counts/sec; e^- > 0.06 MeV; P^+ 0.6 to 3.4 MeV
6. Trapped Radiation Detector	Cerenkov detector: alcohol	e^- 0.5 to 12 MeV; P^+ > 80 MeV; Z^+ > 500 MeV/nucleon
7. Ultraviolet Photometer	A1 filter with channeltron for hydrogen Lyman alpha and LiF target cathode for 584Å helium	200 to 1400Å; field of view 1.15° × 9.3° H and He lines at 1216Å and 584Å
8. Celestial Mechanics	Doppler tracking of spacecraft at 2110 MHZ uplink and 2292 MHZ downlink	Acceleration > 1 mm/sec^2
9. Imaging Photopolarimeter	2.54 cm Maksutov telescope Wollaston prism	Field of view 0.028° square, red and blue filters. 64 shades of gray. Used spacecraft spin for scanning.

SYNOPSIS OF SCIENTIFIC RESULTS

Pioneer 10 entered the center of the asteroid belt in September 1972 and the meteoroid detectors indicated much less material at small particle sizes than had been anticipated. This spacecraft then made its closest approach to Jupiter on December 3, 1973, at 130,354 km above the cloud tops and then proceeded out to deep space in a direction opposite to that of the Sun's motion in the galaxy. Pioneer 11 followed about one year later and passed through the asteroid belt by March 1974. Its meteoroid experiments showed a decline in small-sized particles (0.001 mm) in the belt, an even distribution of larger particles (0.01 to 0.1 mm), and an increase by threefold of particles 0.1 to 1.0 mm in the center of the belt compared to near-Earth regions. Then Pioneer 11 made its closest encounter with Jupiter on December 2, 1974, at 43,000 km above the cloud tops and proceeded on a trajectory toward Saturn. A brief summary of the experimental results obtained by both spacecraft from the encounter with Jupiter is listed in Table 3. On September 2, 1979, the Pioneer 11 spacecraft made its closest approach to Saturn at 20,930 km above the cloud tops /3,4/. A synopsis of these results is given in Table 4. After its Saturn encounter, the Pioneer 11 spacecraft is proceeding out to deep space in the same direction as the Sun's motion in our galaxy.

The next phase of the Pioneer 10 and 11 missions involves the longest period of their lifetime and is devoted to studying the heliosphere /5/. This region is dominated by the plasma envelope of the Sun and is thought to be on the order of 100 AU in distance before this solar plasma interacts with the interstellar plasma. More than one complete solar cycle has elapsed during the 17 years that the two Pioneers have been measuring the solar wind ions and magnetic field and the cosmic ray energetic particles in this extremely large region of space. Interpretation of these measurements depends upon baseline measurements in the region near the Sun obtained by three other spacecraft:

TABLE 3 Jupiter Encounter

1. Measurements of the magnetic field indicated 4.28 gauss at the equatorial 1 bar level, and the dipole tilted 11° with respect to the spin axis.

2. Plasma and magnetic field measurements show that a strong ring current distorts the magnetosphere. A bow shock was detected at $100R_J$.

3. The electron flux trapped in the Jovian magnetic field was 10^4 times greater than the Earth's and peaked at $3R_J$. The trapped protons were 10^3 times more intense than the Earth's and peaked in two regions; $3.4R_J$ and $1.9R_J$.

4. The energetic particle experiments discovered that Jupiter is the dominant source of low-energy relativistic electrons in the heliosphere and the intensity and spectrum varies with the 10-hr rotation period of Jupiter.

5. Infrared radiometer measurements showed that Jupiter emits 1.7 times more heat than it receives from the Sun. The helium to hydrogen ratio was measured to be 0.14 ± 0.08.

6. Close-up observations of Jupiter's atmosphere showed that the cloud tops at the poles are lower than those at the equator and that atmospheric convection cells are smallest in the polar regions.

7. Energetic particle absorption measurements indicated the existence of a ring of particulate matter around Jupiter.

8. Spacecraft Doppler tracking significantly improved our knowledge of the gravity field of Jupiter and determined the masses of the Galilean satellites with a 20% improvement in accuracy.

TABLE 4 Saturn Encounter

1. The magnetic field was measured to be 0.22 gauss at the equatorial 1 bar level with the dipole axis tilted less than 1° to the spin axis.

2. Saturn's magnetosphere contained a plasma that was rigidly corotating with the planet in regions less than $10R_S$. Beyond this region the plasma was corotating at speeds less than that of the planet.

3. Energetic particle experiments discovered that the trapped radiation belts had comparable intensity to those of Earth, but extended over much larger regions. The maximum intensity of electrons and protons were found to be between $4R_S$ and the outer edge of the rings at $2.292R_S$. The proton flux peaked at $2.67R_S$.

4. The infrared radiometer measurements showed that Saturn emits 2.8 ± 0.9 times more heat than it receives from the Sun.

5. A new ring designated F was discovered by imaging and energetic particle absorption measurements.

6. A new inner Saturnian satellite named Epimethens was detected by imaging and energetic particle absorption.

7. The close-up imaging of Saturn showed that the clouds are thicker than those of Jupiter and have much less contrast. The cloud belts are also narrower and more numerous than those of Jupiter.

8. Accurate Doppler tracking of the spacecraft permitted the masses of Saturn, Iapetus, Rhea, and Titan to be determined more precisely.

the Interplanetary Monitoring Platform (IMP-8), the International Cometary Explorer (ICE), and the Pioneer Venus Orbiter (PVO). The main scientific objectives of these spacecraft are to examine the following: (1) solar wind structure and dynamics, (2) interaction of the solar wind with the interstellar medium, and (3) acceleration, transport, and modulation of energetic particles. A brief list of the results obtained during the last 17 years in the heliosphere is given in Table 5. These results have significantly changed the models of the heliosphere developed in the early 1960s and have produced a much better understanding of the origin and transport of cosmic radiation.

Future objectives of the Pioneer 10 and 11 missions are listed in Table 6. Accomplishment of these objectives is strongly dependent on the RTG spacecraft power supplies and the continued allocation of the Deep Space Network of 70-m antenna stations required to command and track the two spacecraft. The continued operation of the Pioneer 10 and 11 spacecraft in conjunction with Voyager 1 and 2, and later, Galileo and Ulysses, provides a unique capability to study the heliosphere during this era.

TABLE 5 Heliospheric Measurements

1. Measurements of the speed, number density, and temperature of the solar wind out to 47 AU show that the mean velocity is constant ~430 km/sec, the mean density falls off as R^{-2}, and that the temperature gradient is complicated but always subadiabatic.

2. The average structure of the heliosphere substantiates the Parker model with a magnetic field drawn out to form Archimedean spirals in the ecliptic plane and dipole-like asymmetry in the polar directions. However, the azimuthal magnetic component shows a 20% departure from the model at 20 AU.

3. The magnetic field strength observed in the outer heliospheres is a minimum at solar minimum and the inclination of the current sheet is near equatorial at solar minimum.

4. The solar wind stream morphology evolves from corotating interaction regions at 5 to 10 AU into lower velocity stream with broad high-density regions beyond 20 AU.

5. Strong repeated flaring regions on the Sun produce very strong shocks at large distances ($R > 27$ AU).

6. Energetic particle measurements of the intensity and radial gradient of galactic cosmic rays indicate a “modulation boundary” at 40 to 80 AU.

7. An “anomalous component” of the cosmic ray spectrum between 5 to 50 MeV/nucleon was discovered for He, C, N, O, and Ne. The source is thought to be interstellar neutrals which penetrate the inner heliospheres, become ionized, and are transported to the outer heliosphere where the ionized particles are accelerated by a factor of 10^4.

8. The energetic particle radial gradient changes only slightly over the solar cycle while the intensity varies by a significant factor. This indicates that the distance of the modulation boundary from the Sun depends upon the phase of the solar cycle.

9. Energetic particle modulation in the outer heliosphere takes place in a series of step decreases propagating outward at ~50 km/sec and are produced by merged interaction regions beyond 8 AU.

10. UV photometer measurements on Pioneer 10 of the hydrogen and helium density values in the distant heliosphere show an abrupt drop in the flux of Lyman alpha photons. This has been interpreted as being caused by enhanced charge exchange between solar wind protons and the interstellar hydrogen.

TABLE 6 Future Plans

1. Utilize spacecraft power sharing and enhanced capabilities of the deep space network to keep the Pioneer experiments working until the year 2000.

2. Measure the radial evolution of the solar wind.

3. Determine the latitudinal variation in the plasma and magnetic field.

4. Measure the interaction of the solar wind with the incoming interstellar neutral particles.

5. Determine the existence and location of a terminal shock and the heliopause.

6. Determine the location and acceleration mechanism for the anomalous component of cosmic rays.

7. Measure the temporal, radial, and latitudinal variations in the strength of cosmic ray modulation.

8. Use accurate Doppler tracking to search for gravitational radiation and differential gravitational forces from a possible trans-Neptunian planet.

REFERENCES

1. G. D. Mead, Pioneer 10 Mission: Jupiter Encounter Special Issue, J. Geophys. Res. 79 #A25, 3487-3886 (1974).

2. T. G. Northrop, A.G. Opp, and J. H. Wolfe, Pioneer 11 Saturn Encounter Special Issue, J. Geophys. Res. 85 #A11, 5651-5958 (1980).

3. R. D. Fimmel, J. Van Allen, and E. Burgess, Pioneer First to Jupiter, Saturn, and Beyond, NASA SP-446, 1980.

4. D. Morrison, Voyages to Saturn, NASA SP-451, 1982.

5. P. Dyal and R. O. Fimmel, Pioneer 10 and 11 Heliospheric Mission, Ames Research Center, Moffett Field, CA 94035, (1982).

ULYSSES: A STATUS REPORT

D.E. Page,* R.G. Marsden,** E.J. Smith*** and
K.-P. Wenzel

**ESA Ulysses Office, Jet Propulsion Laboratory, Pasadena, CA 91109, U.S.A.*
***Space Science Department of ESA, ESTEC, 2200 AG Noordwijk, The Netherlands*
****Jet Propulsion Laboratory, California Institute of Technology, Pasadena, CA 91109, U.S.A.*

ABSTRACT

Ulysses, a joint ESA/NASA mission often delayed and now scheduled for launch in October 1990, will be the first to explore the heliosphere through the full range of heliospheric latitudes. The mission, the spacecraft and the science instrumentation are outlined and references are given indicating where these areas are described in detail. Successful completion of some recent project activities is reported and it is concluded that the spacecraft should be ready as required for delivering to the Kennedy Space Centre in the Spring of 1990.

INTRODUCTION

It has long been realized that the narrow slice of the heliosphere in which the earth moves is unlikely to be representative of the heliosphere at other latitudes /1,2/. A joint exploration out of the ecliptic plane was therefore embarked upon by ESA and NASA in 1977 and, after many programmatic and schedule changes, a launch of the Ulysses spacecraft is now (September 1989) foreseen in October 1990. The mission, first called Out-of-Ecliptic, later became ISPM (International Solar Polar Mission) and later still was renamed Ulysses. The story is told in Dante's Inferno that Ulysses, King of Ithica, becoming restless for further adventure, wanted to explore beyond the known world which then ended at Gibraltar. He exhorted his companions "To venture the uncharted distances...of the uninhabited world beyond the Sun...to follow after knowledge and excellence". "Ulysses" therefore seemed to be an apt name for man's first adventure over the solar poles.

The mission has been outlined on several occasions, for example in /3,4/, and the spacecraft and scientific instruments to be carried are described in detail in /5/. A fresh look at the scientific aims of the mission was taken in 1985, at which time launch was foreseen in 1986. Relevant papers can be found in the proceedings of the 19th ESLAB Symposium on The Sun and the Heliosphere in Three Dimensions /6/. Those who might wish to contact Ulysses investigators and co-investigators can find a listing of names, addresses, etc. of all 124 in the Ulysses Science Team Directory /7/.

SCIENTIFIC GOALS AND INSTRUMENTATION

By, for example, studying interplanetary scintillations or by using galactic cosmic rays as tracers, guesses can be made from the ecliptic plane about the properties of the heliosphere at high solar latitudes. Such remote sensing techniques rarely yield unambiguous answers. With the Ulysses mission it will be possible for the first time to get direct in-situ measurements of the solar wind, its associated magnetic field and solar energetic particles through the complete range of solar latitude. At the same time, the energy spectrum, elemental and isotopic composition and arrival direction of the cosmic radiation will be recorded, making it possible to examine the dependence of cosmic ray modulation on variations in heliospheric structure at high latitudes.

Ulysses will also monitor solar radio bursts and solar X-ray emission. The instrument which looks at solar X-rays will be able also to record the arrival of cosmic gamma rays and together with data from other spacecraft to pinpoint the origin of cosmic gamma bursts. Interstellar dust and neutral helium gas will be measured too. The interstellar neutral helium measurement uses the gravity field of the sun as a mass analyser. Combining measurements made in the ecliptic plane with those in a perpendicular plane also containing the sun should make it possible to determine the velocity vector of neutral helium arriving from the Local Interstellar Medium. Two further investigations will make use of the spacecraft transmitter and receiver. In the first, the corona will be "sounded" by radio waves. The second will look for very small shifts in the Doppler tracking of the spacecraft in order to detect the passage of gravitational waves whose existence is predicted by the general theory of relativity.

On the way to high solar latitudes via Jupiter, many instruments will monitor basic interplanetary parameters in the ecliptic and, of course, study the Jovian magnetosphere.

The complement of scientific instruments to be carried is given in Table 1.

TABLE 1 Scientific investigations on ULYSSES.

Investigation/ Principal Investigator	Measurement	Collaborating Institutes
Magnetic Fields (HED) A. Balogh, Imperial College, London, UK	Magnetic fields: ± 0.01 - ± 44000 aT	JPL
Solar Wind Plasma (BAM) S. J. Bame, Los Alamos Nat Lab, USA	Ions : 257 eV/Q - 35 keV/Q e^- : 1-903 eV	NASA/ARC JPL HAO, Boulder UCLA NASA/MSFC MPAe, Lindau
Solar Wind Ion Composition (GLG) G. Gloeckler, U. Maryland, USA J. Geiss U. Bern. CH	Ion composition: 145 km/s (H) - 1352 km/s (Fe^{+8})	NASA/HQ NASA/GSFC TU Braunschweig U. Natal MPAe, Lindau
Low Energy Ions and Electrons (LAN) L. Lanzerotti Bell Laboratories, USA	Ions: 50 keV -5 MeV e-: 30-300 keV ion composition: 0.2 - 15 MeV/n(Fe)	UCB U. Kansas APL Obs. Paris U. Thrace U. Birmingham
Energetic Ion Composition and Interstellar Gas (KEP) E. Keppler, MPAe, Lindau, FRG	Ion Composition: 80 keV - 15 MeV/n Neutral He	Imperial College Geophys. Obs. Kiruna MPI, Heidelberg Aerospace Corp. U. Bonn
Cosmic Rays and Solar Particles (SIM) J. A. Simpson, U. Chicago, USA	Nuclei : 0.3 - 600 MeV/n e^- : 1-2000 MeV	Imperial College ESA/SSD NRC, Canada U. Kiel CEN, Saclay Danish Sp. Res. Inst. NRC, Milan MPI, Heidelberg U. Maryland MPAe, Lindau U. Bonn

Radio and Plasma Waves (STO) R. G. Stone NASA/GSFC, USA	Plasma Waves: 0-60 kHz Radio: 1-940 kHz Magnetic field: 10-500 Hz	Obs. Paris U. Minnesota CNET/CRPE, St. Maur
Solar X-Rays and Cosmic γ-Ray Bursts (HUS) K. Hurley, Space Sciences Lab, Berkeley, USA and CESR, Toulouse, F. M. Sommer, MPI Garching, FRG	X-, γ-rays: 5-150 keV	Obs. Paris SLR, Ultrecht NASA/GSFC CESR, Toulouse
Cosmic Dust (GRU) E. Grun, MPI, Heidelberg, FRG	Dust Particles 10^{-16} - 10^{-9}g	U. Canterbury NASA/JSC ESA/SSD
Radio Science/ Coronal Sounding (SCE) H. Volland, U. Bonn, FRG	Spacecraft dual-frequency ranging and Doppler data	U. Bochum
Radio Science/ Gravitational Waves (GWE) B. Bertotti U. Pavia, I	Spacecraft Doppler data	CNR, Frascati U. Uppsala JPL
Interdisciplinary/ Directional Discontinuities J. Lemaire, Inst. d'Aeronomie Spatiale de Belgique, Brussels, B		Aerospace Corp.
Interdisciplinary/Mass Loss and Ion Composition G. Noci, Obs. Arcetri, Florence, I		

SPACECRAFT TRAJECTORY

Ulysses will be launched in the ecliptic plane toward Jupiter by a combined two-stage IUS/PAM-S upper stage solid rocket motor, carried on board the NASA space shuttle (Figure 1). The Jovian gravity field will deflect the suitably directed spacecraft so that it travels southward from the ecliptic plane to pass over the south solar pole about 3-1/2 years after launch. At this point Ulysses will be approximately 2.3 AU below the ecliptic. Perihelion at 1.4 AU will be reached about 6 months later still. The trajectory is illustrated in Figure 2. Launch in October 1990 (the window is 18 days long) will mean that most of the time spent at high solar latitudes should occur during a fairly quiet part of the solar cycle. This should mean that heliospheric structure will be relatively stable and well ordered and that it will be less difficult to differentiate between genuine latitude variations and random time variations.

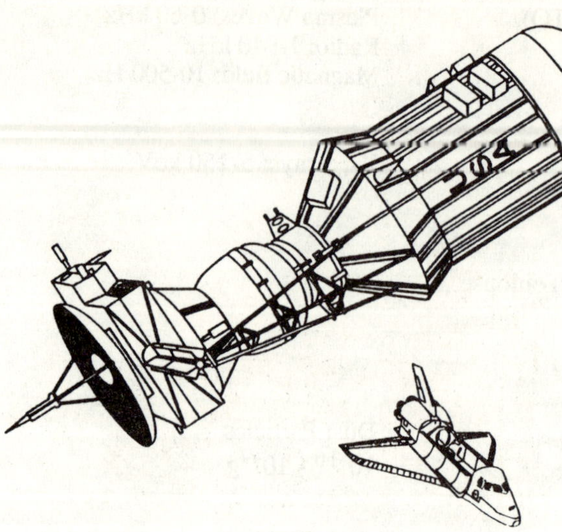

Figure 1 The ULYSSES spacecraft attached to the Inertial Upper Stage (IUS)/PAM-S after deployment from the Shuttle and prior to interplanetary injection.

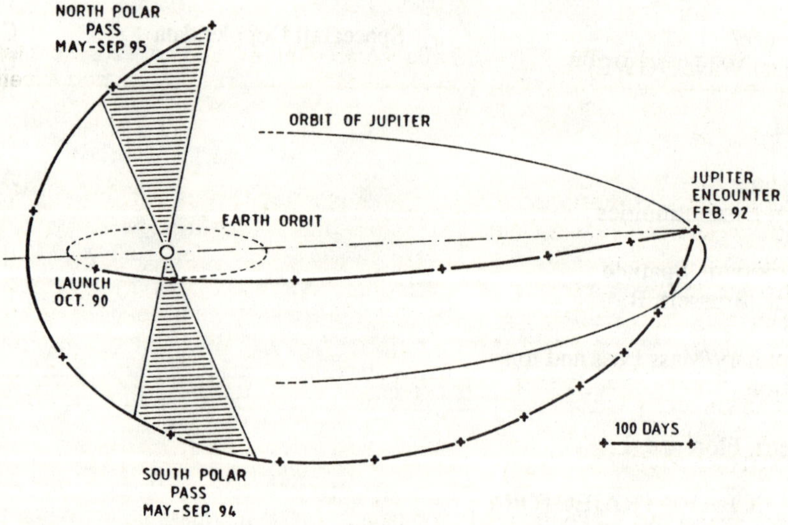

Figure 2 Typical ULYSSES trajectory viewed from 15° above the ecliptic plane. Crosses are shown at 100-day intervals.

OPERATIONS AND DATA RECOVERY

Ulysses will be tracked by NASA's Deep Space Network, operations being directed by a small ESA team resident at the Jet Propulsion Laboratory in Pasadena. Real time data will be transmitted at 1024 bits per second for 8 hours each day. During the remaining 16 hours, data continuity will assured by using on board tape recorders which will record scientific data at 512 bits per second.

Tapes (Experiment Data Records) of cleaned and sorted data will be provided by JPL to each Principal Investigator, who carries the responsibility to pass on the relevant data to his co-investigators. A Common Data Record will also be prepared to assist data correlation between instruments and to help identify when events of special note occurred. This Record contains selected channels of relatively raw data extracted from each of the main experiments carried on the spacecraft.

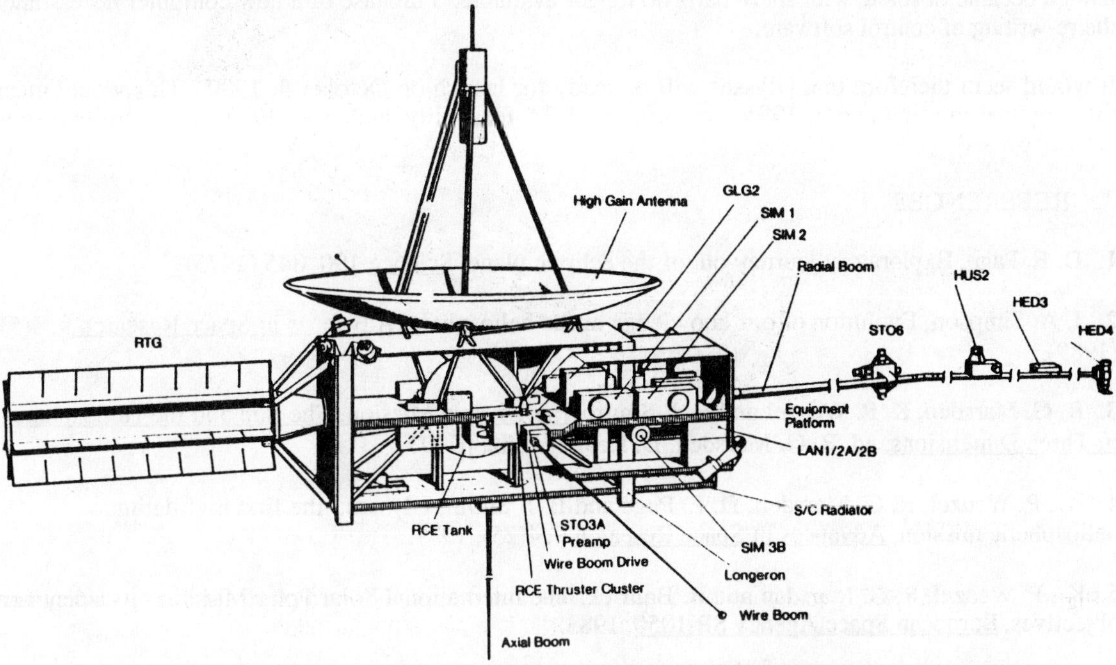

Figure 3 Schematic of the ULYSSES spacecraft. The experiment codes are listed in Table 1: the number following the code identifies the sub-unit of a specific experiment

THE SPACECRAFT

The spacecraft layout is illustrated in Figure 3. It measures about 3 meters across and about 2 meters in height with booms folded. At launch it will weigh approximately 370 kg including the 55 kg of scientific instrumentation. Because Ulysses must operate out to 5.4 AU from the sun, power is obtained from a Radioisotope Thermoelectric Generator (RTG) which will provide close to 290 watts at launch. The spacecraft will be spin stabilized at 5 rpm with information for logic operations coming from a sun sensor. At some parts of the trajectory, it will not be easy to keep the solar aspect angle above the minimum acceptable and at the same time to point the high gain parabolic antenna at the earth. S band telemetry is used for the up-link and both S and X band for the down-link.

LAUNCH READINESS

The Ulysses spacecraft has been ready for launch for some years now. When the programme began formally in 1977 launch was foreseen in 1983. The spacecraft was actually at the launch site heading for a mid-1986 launch when the Challenger accident occurred. Adjustments to the shuttle, changes in upper stages dictated by safety considerations and competition with other programmes have led to a situation when launch is now planned for October 1990, within a window which opens on October 5 and closes on October 23.

Activities since 1986 have been at low level with only skeleton teams remaining in many areas. This has had the consequence that new people have had to be found. Therefore, when major work began again in early 1989, one of the first activities was the integration of flight spare scientific instruments on the qualification model of the spacecraft, largely with the aim of familiarizing new people with the hardware and procedures. The exercise was completed successfully and was followed by integration of the prime scientific instruments on the flight spacecraft. This was completed without significant problems and the results were discussed with the investigators on September 12, 1989. Testing of the fully integrated spacecraft - including a thermal vacuum test - will be carried out at ESTEC before the end of 1989 and Ulysses should be shipped to the Kennedy launch site in May 1990.

At this time also the small ESA operations team is scheduled to arrive at JPL to take its place alongside NASA colleagues involved in space operations. Some adjustments in mission operations planning have been necessary since launch was moved from 1986. For example, the computer purchased for the 1983 launch became obsolete with spare parts no longer available. Purchase of a new computer necessitated the re-writing of control software.

It would seem therefore that Ulysses will be ready for launch on October 5, 1990. This would mean that nominal mission end in 1995 would follow ESA feasibility studies of 1973 by a mere two solar activity cycles!

REFERENCES

1. D. E. Page, Exploratory journey out of the ecliptic plane, Science 190, 845 (1975).

2. J. A. Simpson, Evolution of our knowledge of the heliosphere, Advances in Space Research 9, 4(5), (1989).

3. R. G. Marsden, K. P. Wenzel and E. J. Smith, The Ulysses Mission. The Sun and the Heliosphere in Three Dimensions, ed. R. G. Marsden, D. Reidel, 1986, p. 477.

4. K-. P. Wenzel, R. G. Marsden, D. E. Page and E. J. Smith, Ulysses: the first high-latitude heliospheric mission, Advances in Space Research 9, 4(25), 1989.

5. K-. P. Wenzel, R. G. Marsden and B. Battrick, The International Solar Polar Mission - its scientific objectives, European Space Agency SP-1050, 1983.

6. R. G. Marsden, The Sun and the Heliosphere in Three Dimensions, D. Reidel, 1986.

7. Ulysses Science Team Directory, compiled by R. G. Marsden, Space Science Department of ESA, ESTEC, 2200 AG, Noordwijk, Netherlands. Phone (31) 1719-83583, Telex 39098, FAX (31) 1719-84698, SPAN ESTCS1::RMARSDEN.

Poster Session

MOTION OF THE STRONG DISTURBANCES IN THE INTERPLANETARY MEDIUM

N.L. Borodkova, Yu.I. Yermolaev and G.N. Zastenker

Space Research Institute, Academy of Sciences, U.S.S.R.

ABSTRACT

Features of the strong disturbances, mainly shock waves, moving through the interplanetary medium are reviewed on basis of the solar wind measurements onboard the Soviet highapogee satellites Prognoz-7,8 in 1978-81. Some estimations for the distances from the Sun where the equalization of solar wind proton and α-particle velocities and temperatures take place (from 5-10 to 20-30 AU) are discussed.

INTRODUCTION

The disturbances propagated in interplanetary medium, can be divided into two classes: events (the strongest of them are interplanetary shock waves (ISW)) connected with some solar origins and the events resulting from distortion of thermodynamic equilibrium between various ion components in space [1,2,].

CHARACTER OF DISTURBANCES ON INTERPLANETARY SHOCK WAVES

20 ISW's were selected among all registered on Prognoz-7,8 on which there existed more complete solar, plasma and magnetic field data. The flow of solar wind plasma after the shock front is of a considerably interest, because it is manifestation of the large-scale wave-interplanetary medium interaction. Fig.1 gives the time variations of hourly average values of solar wind proton and α-particle parameters obtained by epoch superposition in the -5hrs to +30hrs interval relative to the shock arrival moment (dashed line). Strong plasma compression downstream goes on ~ 3 hrs, and after ~18 hrs it decreases by a factor of 2-3 and during the large period interplanetary medium is largely rarefied. The strong plasma heating is preserved during ~12 hrs. By contrast to the solar wind velocity downstream decreases much slower and in ~ 30 hrs it is appreciably higher (~10%) than that upstream. The comparison of the values of proton and α-particle temperature shows that α-particle heating is greater on the front than that of protons. Thus if ratio T_α/T_p before the front was equal to 3-4 it grows up to 5-7 beyond the front in the compressed plasma sheet (● -all, x -single and "the first", o -"the second" events).

PROPAGATION AND ENERGETIC CHARACTERISTICS OF THE SHOCKS.

Fig. 2 shows dependence of shock wave local velocity on the mean velocity of their propagation. It is clearly seen that in the majority of cases local velocities are considerably lower than the mean ones, i.e., shock waves strongly decelerated in their motion in the interplanetary space.
The considerable part of energy released from a strong flare can be taken away by the shock waves. The connection of mean shock velocity and the dynamic pressure jump on the shock front with the acquired energetic values is shown in Fig.3. Though the number of measurements is not too high and there can be some uncertainties in attribution of the observed shocks to the solar flares, the given data do not clearly prove the drastic dependence of the mean velocity and the pressure jump on shock energy, obtained in [3] by numerical simulation. The discrepancy with model can be a consequence of the fact, that the shock parameters are considerably determined by the conditions, existing in the interplanetary medium before the shock. Thus the shock velocity is higher, if it moves through the medium, disturbed by the preceding wave, that passed 2- 3 days before, i.e., is unified into compact groups.

DYNAMICS OF DEVIATIONS FROM THE THERMODYNAMIC EQUILIBRIUM.

Experimental dependences of the velocity difference $V_\alpha - V_p$ and temperature ratio T_α/T_p on ratio of time of the solar wind expansion τ_e to the times of momentum and energy exchange due to Coulomb collisions τ_s, τ_c [4,5] show, that at $R \cong 1$AU, on the average $\tau_e/\tau_s \cong 3 \times 10^{-2}$ and $\tau_e/\tau_c \cong 5 \times 10^{-2}$ [5]. In the assumption of adiabatic change ($n \sim R^{-2}$, $T_p \sim R^{-4/3}$), parameters $\tau_e/\tau_s \sim R$ and $\tau_e/\tau_c \sim R$. Then equalization of ion velocities due to Coulomb collisions will occur on the average, at the distance $R_v \cong 7$ AU, and one for temperatures at $R_t \cong 20$ AU [5]. Fig. 4 shows schematically summarized change in the average values V_α/V_p and T_α/T_p in wide range of heliocentric distances.

REFERENCES

1. Zastenker G.N., Borodkova N.L. Adv. Space Res., Vol.4, No.7, 347, (1984)
2. Yermolaev Yu.I. et al. Adv. Space Res., Vol.9, No.4, 123, (1989)
3. Smith Z., M.Dryer, in: Proceedings of the Sixth International Solar Wind Conference ed. by V.J.Pizzo, T.E.Holzer, D.G.Sime, Vol.1, Boulder, Colorado, May 1988, pp.309-314.
4. Neugebauer M. Cosmic Phys., Vol.7, 131, (1981)
5. Yermolaev Yu.I., Stupin V.V., in: Sixth IAGA Scientific Assembly Bulletin N53 (Part B: Abstracts), United Kingdom, 1989, p.268.

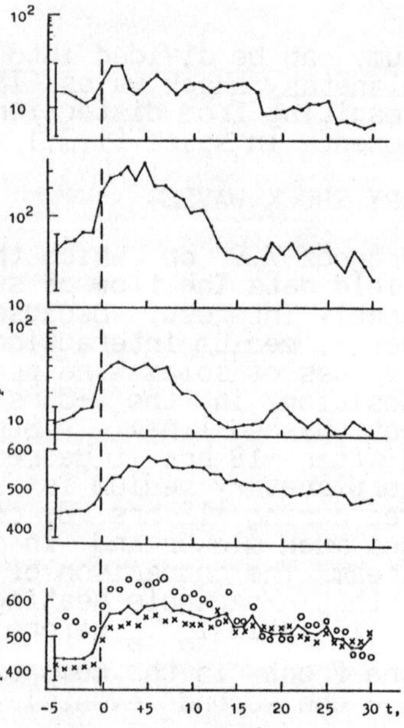

Fig. 1.

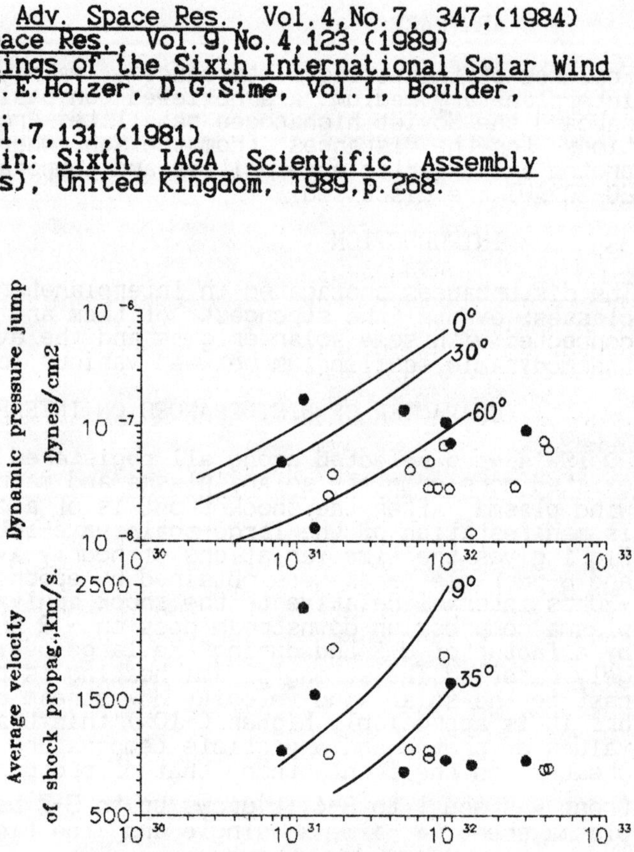

Fig.3. Energy W, ergs.

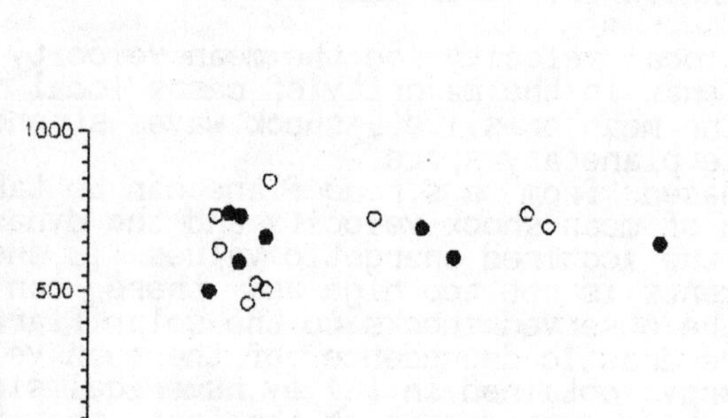

Fig.2. Average velocity of shock propagation V_{av}, km/s.

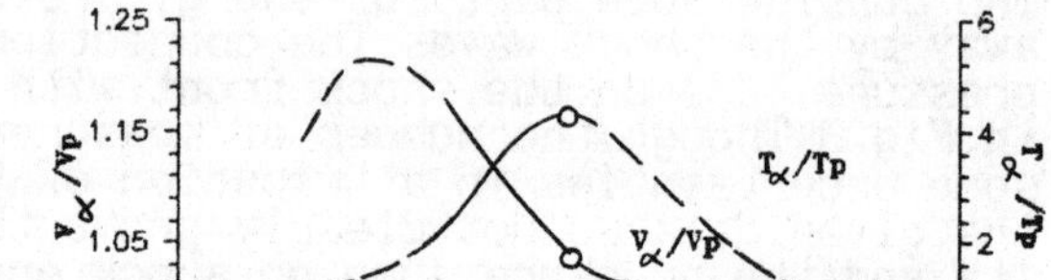

Fig.4. Solid lines —Helios 1,2 data;
o — 1AU data; dash — extrapolations.

STATISTICAL ANALYSIS OF SOLAR FLARE, SOLAR NEUTRINO FLUX AND SUNSPOT DATA

Godhuli Chattopadhyay and Probhas Raychaudhuri

Department of Applied Mathematics, Calcutta University, Calcutta-700009, India

It is known that ^{37}Cl solar neutrino experimental data show that neutrino production ($\gamma_e \geq 0.814$ MeV) is about 1/3 less than that from the standard solar model /1/. Kamiokande experiment on γ_e(solar) + $e^- \rightarrow \gamma_e + e^-$ confirmed the above discrepancy. The ^{37}Cl detector responds to the ^8B and ^7Be neutrino fluxes from the centre of the Sun and production of neutrinos is a sensitive function of the temperature in the solar interior. Raychaudhuri /2,3,4/ suggested that the sunspot cycle is due to the pulsating character of the nuclear energy generation inside the core of the Sun. Raychaudhuri /4,5/ and others showed that the solar neutrino flux data are varying with 11 year solar cycle. Raychaudhuri /4,5,6/, Davis et al /7/ and others have shown that there exists an anticorrelation of solar neutrino flux with the 11 year solar cycle. Raychaudhuri /8/ and others have shown by statistical analysis that both the solar neutrino data and sunspot data exhibit multiple periods apart from 155 days and 11 year periods. Ichimoto et al /9/ have found 5 and 17 months periodicity in the solar diameter measurements. If the 11 year solar cycle is due to the pulsating character of the nuclear energy generation then the variation of solar neutrino flux must be related to some phenomena in the solar photosphere which are causally connected with those taking place in the core of the Sun /2,3/. Despite the fact that the solar neutrino data pulsate with the solar cycle, there still appears to be some doubt on this point. Here we will confirm the same periodic behaviour of the solar neutrino flux, sunspot number, and solar flares by Jurkevich's method /11/.Using the Jurkevich's method we see that solar flare data, sunspot data and solar neutrino data from 1970-1985 exhibit almost the same period of 11 years. Solar neutrino data, sunspot data and solar flare data exhibit 132, 142 and 133 months period respectively.

To study the correlation let us take the sunspot data and solar flare data for the period 1970-1985 /12/ and then adjust the sunspot data and solar flare data so that they correspond to the solar neutrino data which are not evenly spaced. We then find out the sunspot data and solar flare data corresponding to run numbers in solar neutrino flux data. We use Spearman's rank correlation coefficient $r_s=1-6(\Sigma d^2)/(n(n^2-1))$,where d is the difference in the ranks and n is the number of the observation pairs. To test the significance of the sample correlation coefficient r_R/13/ (see the definition of r_R), we note that $Z = (n-1)^{1/2} r_R$ is asymptotically normal with mean zero and standard deviation unity. From the results for run numbers 18-70 ($r_R = -0.371$, $Z = 2.57$ where $n = 49$), run numbers 32-70 ($r_R = -0.51$, $Z = -3.11$), run numbers 46-70 ($r_R = -0.791$, $Z = -3.88$), run numbers 45-77 ($r_R = -0.76$, $Z = -4.08$), we see that there is a significant (more than 99.5%) negative correlation between the sunspot & solar neutrino data. Only for the runs 46-70 ($r_R = -0.55$, $Z = -2.7$) there is a significant (99.5%) negative correlation between the solar neutrino and flare data. The above anticorrelation suggests that the solar activity cycle is due to the instability of the solar core. Raychaudhuri /2,3/ suggested a perturbed solar model (due to density and T dependence of nuclear energy generation and the energy emitted by neutrino emission and photon emission) which generates cyclic variation of energy production inside the core and can explain the above mentioned anticorrelation. The most important neutrino process is the plasma neutrino process ($\Gamma \rightarrow \nu + \bar{\nu}$) in the weak interaction theory /14/ and the rate of energy loss is $E_\nu = (3.59/\rho)\, 10^{13} T_9^5\, x^5\, F(x)$, where $T_9=T/10^9K$, $x = \hbar w_0/KT$, $\hbar w_0 = 3.265\times10^{-11}(1+6.413\times10^{-5}\rho^{2/3})^{-1/4}\rho^{1/2}$

$$F(x)=\int_0^{\infty} \frac{\sinh^2\xi \ \cosh\xi \ d\xi}{\exp(x \cosh\xi)-1}$$

and is important in the temperature range of 10^7K to 10^9K. From the equation $E_N-E_\nu-E_{ph} = 0$ we can find the temperature, density and composition of the Sun where pp-cycle is in operation. Imbalance between the nuclear energy generation E_N and the energy radiated by neutrinos and photons E_ν and E_{ph} plays a role in the excitation of the oscillations. Following Milne /15/ and Ledoux /16/, let t=0 be the time when the imbalance between E_N, E_ν and E_{ph} was started and assume that with any subsequent time t it is possible to associate a later time t + T(t) such that, apart from a constant energy, all the energy generated from t = 0 to t (i.e. $E_N t$) has been radiated away by the time t + T(t). Let f(t) denotes the total energy lost at the time t. Then we have

$$f(t+T(t))=\int_{t=0}^{t+T(t)}\{E(t)+E_{ph}(t)\}dt=E_N(t+c)$$

where $T(0)=T_o=c$. Whatever be the origin of time, in steady state we always have $f(t)=E_N(t)$. Total energy I(t) at time t is given by

$$I(t)=E_N t-f(t)+I_o=E_N t-[E_N(t-T+c)]+I_o=E_N(T-c)+I_o$$

When t=0, $I(0)=E_N.(t-c)+I$. Thus $f(T_o)=E_N\cdot c=E_N T_o$. If T(t) may be treated as a single valued function of I(t) and $f(t)-E_N\cdot t$ is small compared to T_o, we can write $T=T_o+(k/E_N)\{f(t)-E_N.(t)]=T_o+\phi(t)$ where $k=-E_N(\partial T/\partial I)_{t-o}$ and $f(t) = E_N\{t+(1/k)\phi(t)\}$. From the above after some calculation we have $k\phi(t) = -\phi(t+T_o)$. Since $\phi(t)$ is small we neglect $\phi^2(t)$, $\phi^3(t)$ etc. and the equation is linear. In the nonlinear case the variations are anharmonic. Again $E_{ph}(t)+E(t)=E_N\{1+ \phi'(t)/k\}$.

If $k\neq1$, the solution of the equation (1) is of the form $\phi(t)=K^{t/T_o}\cdot \sin(2\pi t/2To)$ and the motion is damped if k<1. If k>1 the motion is unstable and the amplitude increases continuously. In fact, however, the amplitude will be changed due to mixing between the solar core and the outer solar layers at the stage of expansion of the solar core during the solar activity .cycle. If k=1, equation (1) corresponds to a stable periodic motion with constant amplitude. As far as the interpretation of P (period) $=2\cdot T_o$ is concerned, T_o is not necessarily the time that the energy would take to reach the surface if transmitted by radiation alone. This is much too long to have any significance for the mechanical oscillation of the Sun. But the release of extra energy at the centre will create a pressure perturbation that will reach the surface in a time of the order $R/\bar{c}$, where $\bar{c}$ is an appropriate mean value of the sound velocity in the Sun (e.g. $R/\bar{c}$ = 5 years), and with this interpretation $P=2T_o$ is the right order of magnitude. The onset of instability associated with changing temperature and pressure necessarily produces large scale flow of matter inside the Sun. This type of large scale flow of matter inside the sun has been observed by Howard and La Bonte /17/. From the above it is clear that all the activities of the Sun must be periodic in character (e.g. solar neutrino flux, sunspot numbers, solar flares, solar luminosity, solar diameter etc.) and that is due to the pulsating character of thee nuclear energy generation inside the Sun. We suggest also that the solar core instability is mainly controlling the interplanetary dynamical processes.

1. B. Kuchwicz, Report Prog. Phys. 39, 291 (1976)
2. P. Raychaudhuri, Ap. Sp. Sci. 13, 231(1971)
3. P. Raychaudhuri, Ap. Sp. Sci. 18, 425(1972)
4. P. Raychaudhuri, Solar Physics, 93, 397(1984)
5. P. Raychaudhuri, Solar Physics 104, 415(1986)
6. P. Raychaudhuri, Mod. Phys. Lett A3, 1319(1988)
7. J.K. Rowley, B.T.Cleveland and R.Davis, Jr., Am. Inst. of Physics conference proceedings. 126, 1(1985)
8. P. Raychaudhuri, Solar Physics, 106, 421(1986)
9. K. Ichimoto, J. Kubota, M. Suzuki, I.Toshimura, H.Kurokawa, Nature, 316, 422(1985)
10. P.Delache, F. Laclare and H. Sadsoud, Nature, 317, 426(1985)
11. I.Jurkevich, Ap. Sp. Sci. 13, 154(1971)
12. Solar Geophysical Data, NOAA, Boulder, Colorado, USA(1970-1985)
13. M.G. Kendall, Rank Correlation Methods, C. Griffin & Co, London (1948)
14. P. Raychaudhuri, Physics Essays, 2, 118(1989)
15. E. Milne Q.J.Math. 3, 258(1933)
16. P. Ledoux, Handbuch der Phys. 51, 657, Springer Verlag (1958)
17. R. Howard and B.J. La Bonte Ap. J. 239, L33(1980).

NUMERICAL SIMULATION OF THE COMETARY SHOCKS

A.S. Lipatov

Space Research Institute, U.S.S.R. Academy of Sciences,
Profsouznaya 84/32, St., 117810 Moscow GSP-7, U.S.S.R.

ABSTRACT

Research on structure of cometary shock wave is of interest not only for analyzing the processes accompanying the interaction of solar wind with comets, but also for studing the shock which are excited in the interection of solar wind with interstellar gas, the mechanisms for the acceleration of cosmic rays. Now a series of global and local simulations of cometary plasma processes are known (see for example review /6/). But the present paper is directed to the investigation of a kinetic effect in local structures — quasiperpendicular and quasiparallel shocks. In order to simulate the shock structure we use the ordinary hybrid code with massless electrons. The main dimensionless parameters which controls the solution are: ϵ, M_A, M_{MS}, γ, Re —ions Larmor radius, Alfven and magnetosonic Mach number, effective adiabatic index and effective magnetic Reynolds number. The ratio of the ion and proton densities was $N_iM_i = 0.30$ and cyclotron gyration velocity the cometary ions on the boundary was $V = 1$ and $M_{MS} = 1.8$. At initial moment the parameters profiles are specified in accordance with the Rankine–Hugoniot relations. The velocity distribution of proton is Maxwellian, and the velocity distribution of the ions is ring (a) or sphere (b) distribution.

In case of perpendicular cometary shock (a) the main results are following /2/. Profile of the basic plasma parameters shows the jump of the electron-proton plasma component, in front of which is the "foot" associated with the outflow of part of the cometary ions into the region preceeding the wave front. The outflow of ions into the region ahead of the shock front in our case occurred not because ions are reflected under the influence of the electrostatic field at the wave front and the Lorentz force, but owing to the high cyclotron gyrating velocity of the cometary ions behind the wave front. Moving in the region in front of the jump, these ions are accelerated by the electric field $E = U \times B/c$ and after dropping behind the wave front they have an energy substantially greater than their initial energy. The contours of the velocity distribution function of the accelerated ions form a crescent-shaped structure in the plane $(V_{\perp 1}, V_{\perp 2})$, on the left above the ring of the basic distribution of ions in the foot region.

The "foot" of the shock is not steady, but pulsates with a characteristic time on the order of half the cyclotron gyroperiod of the cometary ions. After its formation, the smooth foot again begins to accumulate in a narrow region and finally merges with the proton jump. The pulsation are associated with the fact that the times for the cometary ions to stream into the region in front of the electron-proton jump and to return back to the jump add up to approximately the same total for most of these ions. As the spread in the cyclotron gyration velocities decreases the local minimum of the plasma velocity and maximum of the density in the region of the "foot" are enhanced significantly and can even give rise to the formation of a second jump there when such a velocity spread is absent.

In case of quasiparallel shock (b) the main results are following /5/. The profile of the basic plasma parameters shows the shock structure with the formation of a small proton-electron jump. Profile of density has a diffusive structure with a small pulsation with wavelength λ=Rci. The characteristic scale of this structure is 4–5 Rci in accordance with quasilinear estimation /3/. The jump of the average magnetic field is 2.5. Profiles of velocity, Alfven and magnetosonic Mach number show a change of parameters in the region of proton jump with characteristic scale 4–5 Rci. A jump of value of velocity is 0.4. The calculated distribution of perpendicular $B_\perp$, and parallel $B_\parallel$ magnetic field component shows that in case of quasiparallel shocks ($\theta_{bn} = 30^o$) intensive Alfven oscillations are excited which has maximum amplitute about 3, 1, 1.5 B_∞ in the ramp, foreshock and downstream. The wavelength of this oscillation is 0.9, 3 Rci in the foreshock and downstream and it is right hand polarized in accordance with dynamics of earth's foreshock and cometary pick up processes. This oscillations are corresponded to the excitation of the fire-hose instability in upstream and Alfven resonant mode in downstream. $B_\parallel$ distribution characterizes the excitation of magnetosonic oscillations. At ramp of the shock the tail of the accelerated and reflected cometary ions is formed with $V_{max} = -2U_\infty$. Analize of temperature ion and proton shows that at shock ramp we have Tp_2/Tp_1=10–13 and Ti_2/Ti_1=1.6. The high jump of proton temperature due to proton reflection at subshock and intensive heating by the electromagnetic oscillations.

In accordance with results of kinetic global simulation /1/ the consentration of cometary ions Ni at shock front is 0.01 Ni_∞/N_{sw}, that approximately corresponds to experimental value received by "Giotto" /9/. The calculated jumps near quasiperpendicular shocks are /2/: Ni_2/Ni_1=1.33; Np_2/Np_1=1.5-2; $(B_2/B_1)^2 = 4-6$; Ti_2/Ti_1=2.2; Pi_2/Pi_1=2.9. But experimental data give the following value /9/: Ni_2/Ni_1=1.7; Np_2/Np_1=1.5; $(B_2/B_1)^2$=3.3; Pi_2/Pi_1=2. The smaller value of cometary ion concentration is due to absence of cometary ion injection in calculation box (compare with /7/). But the greater value Pi_2/Pi_1 in simulation is due to greater heating and acceleration at shock front which has a stronger intensity compare with experimental value. The shock scale near Halley comet is 24000km that is about 2 Larmor raduis of cometary ions.

In case of quasiparallel shock calculated magnetic field profile is in accordanse with experimental data /4,8,3/. Calculated scale of shange velocity $5Rci = 5\cdot10^4$km corresponds to result produced by "VEGA-2"/3/, $6\cdot10^4$km for quasiparallel shock ($\theta_{bn} = 45^o$). The scale of diffusive shock is determined by relaxation length for ion velocity under scattering on Alfven turbulence and equals 5–10 Rci. In case of quasiparallel shock lets the magnetic field jump is about 5.2 under $\theta_{bn} = 28.7^o$, that approximately corresponds to calculated value.

Thus in the quasiperpendicular and quasiparallel cases an electron–proton subshock is established inside the shock front on a scale comparable to the Larmor radius of a cometary ion.

The author wishes to thank R.Z.Sagdeev and A.A.Galeev for useful discussions.

REFERENCES

1. A.A.Galeev, A.S.Lipatov, in: Adv. Space Res. 4, n.9,229(1984)

2. A.A.Galeev, A.S.Lipatov, R.Z.Sagdeev, Sov. Phys. JETP, 62, 866(1985)

3. A.A.Galeev, B.E.Gribov, T.Gombosi, K.I.Gringauz, S.I.Klimov, P.Oberz, A.P.Remizov, W.Riedler, R.Z.Sagdeev, S.P.Savin, A.Yu.Sokolov, V.D.Shapiro, V.I.Shevchenko, K.Szego, M.I.Verigin, Ye.G.Yeroshenko, Geophys.. *Res. Lett.* 13, 841(1986)

4. D.E.Jones, E.J.Smith, J.A.Slavin, B.T.Tsurutani, G.L.Siscoe, D.A.Mendis, Geophys. Res. Lett. 13, n.3,243 (1986)

5. A.S.Lipatov, I.N.Syrovatskii, Kosm. Issled. 25, 925(1987)

6. A.S.Lipatov, Numerical simulation of the interaction of solar wind with comets, in: Interaction of comets with the solar wind, Plasma processes in space, vol.1, ed. R.Z.Sagdeev, VINITI, Moscow 1989, p.100

7. N.Omidi, D.Winske, J. Geophys. Res. 92, 13409(1987)

8. B.T.Tsurutani, E.J.Smith, Geophys. Res. Lett. 13, 259(1986)

9. A.J.Coates, R.P.Lin, B.Wilken, E.Amata, K.A.Anderson, H.Borg, D.A.Bryant, C.W.Carlson, D.W.Curtis, V.Formisano, K.Jockers, A.D.Johnstone, A.Korth, D.A.Mendis, H.Reme, A.K.Richter, H.Rosenbauer, J.A.Sauvand, W.Studemann, M.F.Thomsen, C.d'Uston, J.D.Winningham, Nature, 327, n.6122,489(1987)

INVESTIGATION OF THE SOLAR WIND TRANSONIC REGION AT METER WAVELENGTH

N.A. Lotova

Solar Radio Laboratory, IZMIRAN, Troitsk, Moscow Region, 142092, U.S.S.R.

The solar wind starts as a subsonic flow at the coronal base, but very little is known about the sub- to supersonic transition. It is shown here that this transition occurs in an extended region located at about 10 to 40 $R_\odot$. Several independent occultation studies using both spacecraft and natural sources indicate the existence of an about 10 $R_\odot$ wide region of enhanced electron density fluctuations associated with the transonic transition in the solar wind.

A new series of coronal occultation experiments have been performed in the strong scattering regime at the wavelength of 2.9 m using quasars. The measurable quantity in this case is the apparent source size, also called the scattering angle θ. The transonic region is revealed in these observations by the enhanced values of θ above the asymptotic power law given by $\theta \sim R^{-1.6}$, which fits well the data outside the transonic region. Regular occultation experiments were performed in March, June, August, and October in 1986-1987. An example of the results of data processing is given in Fig.1.

Fig. 1

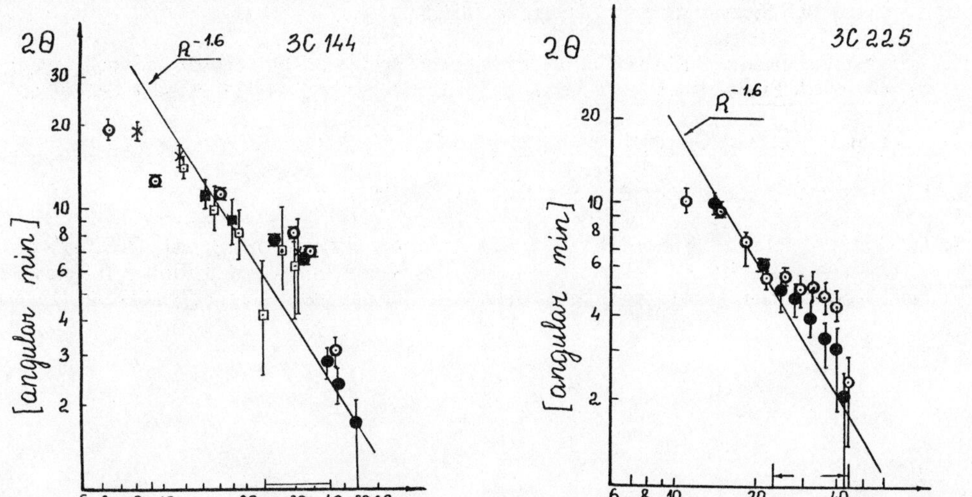

Fig. 1. The radial dependence of the full scattering angle 2θ :
a) 3C144 source in June 1987, ⊙, ● - the cross radio telescope DCR-1000; □, ■, X-the two polarized components, radio telescope URAN-4.
b) 3C225 source in August 1987, observations on the DCR-1000.
⊙, □, X-preconjunction, ●, ■, -post-conjunction phase.

The region of enhanced scattering is apparent in all eight series of observations, making it possible to localize the inner, R_{in}, and the outer, R_{out}, boundaries of the transonic region. One obvious result from this study is that R_{in} and R_{out}, are influenced by changes in the solar activity, more distant positions of the transonic region being typical of solar minima. The characteristic distance and width of the transonic region for the epoch of deep stable solar minimum were determined earlier on - 1985 measurements gave

$R_{in} \simeq 16$, $R_{out} \simeq 30\ R_{\odot}$. Pronounced changes in the geometry of the transonic region occur over the period of 1986-1987. Meeasurements in August and October 1986 gave the results close to those of 1985, but besides that, pronounced short time-scale deviations from the mean location of the transonic region were discovered. So, in March 1986 diminished values $R_{in}=12$. $R_{out}= 25\ R_{\odot}$ were obtained. The most distant location $R=(23-40)R_{\odot}$ was observed in June and August 1987, after the deepest solar minimum had been gone. More distant positions of the transonic region are, as a rule, accompanied by increased turbulence levels. Three of the eight series of experiments were performed with more than one occulted source, permitting the investigation of additional details. Two of the series detected a marked asymmetry of the transonic region, N-S asymmetry was observed in August 1986, and both E-W and N-S asymmetry was seen in October 1987. It is obvious that the asymmetry of the transonic region is a natural consequence of asymmetric streams in the solar corona close to the Sun.

RECONNECTION PATTERN AT THE HELIOPAUSE

W.M. Macek

Space Research Centre, Polish Academy of Sciences, PL-01 237 Warsaw, Ordona 21, Poland
At present: Department of Physics and Astronomy, The University of Iowa, Iowa City, IA 52242, U.S.A.

ABSTRACT

By analogy with the case of planetary magnetospheres it is argued that magnetic field reconnection processes may play an important role for the plasma mixing at the heliopause /1/. Views of the front side heliopause and different areas, where reconnection can take place, are provided for various assumed orientations of the magnetic field of the LISM. Within these reconnection patches the lines of constant angle between the magnetic field of the unperturbed solar wind (standard Parker's spiral field) and the (uniform) LISM field, both projected onto the heliopause surface, are computed. Low frequency (2-3 kHz) interplanetary radio emissions detected by Voyager mission and implications of possible observations of the heliospheric boundary are discussed.

INTRODUCTION

Plasma mixing between the solar wind (SW) and the ambient ionized component of the local interstellar medium (LISM) at the heliopause, the boundary separating both media, is discussed here. The average rate of the plasma mixing resulting from possible reconnection is estimated to be $\sim 10^{-1}$ of the incident mass flow /1/. Moreover, it is suggested that the dependence of the cosmic-ray penetration into the heliosphere on the distribution of these reconnection areas might give some information about the orientation of the LISM magnetic field /1,2/. These questions could be also very important for the interpretation of the measurements of Voyager mission.

SIMPLE MODEL

The total LISM pressure Π^{IS} is taken constant (parallel flow). However, the total solar wind pressure Π^{SW} decreases with the heliocentric distance r as $1/r^2$ (radial flow). Taking the pressure balance one obtains the "nose" of the heliopause to be located at a distance r ($r \gg r_o = 1$ AU):

$$D = r_o \, (\Pi_o^{SW} / \Pi^{IS})^{1/2} \tag{1}$$

Taking only ram pressures on both sides one gets the heliocentric distance $r = R(\delta)$ to the position of the heliopause to be described by a simple analytic formula (see Figure 2):

$$R / D = \delta / \sin \delta \tag{2}$$

where δ is the angle measured from the Sun-apex line. The plasma apex is assumed to be located in the solar equatorial (X,Y) plane (SEP) and is taken to be X-axis ($\Theta = 90^\circ$, $\Phi = 0^\circ$). Here Φ (PHI) and Θ (THETA) are the equatorial longitude, counted from the apex (X-axis) in the direction of solar rotation, and the equatorial co-latitude counted from the Z-axis of solar rotation, respectively. A deflection of the nose from the apex is not considered here. Using the typical values listed in Table 1 one gets D~200 AU.

The simplest draping consists of a projection of the field lines onto the surface given by equation (2) with a unit normal vector $\hat{n}$. The direction of the projected line of force is determined by the vector

$$H = \hat{n} \times (B \times \hat{n}) \quad (3)$$

An angle Ψ between the projected magnetic field H^{SW} of the unperturbed solar wind (standard Parker's spiral field) and the projected uniform LISM field H^{IS} within the reconnection patches (Ψ > 90°) is given by

$$H^{IS} \cdot H^{SW} = |H^{IS}|\ |H^{SW}| \cos \Psi \quad (4)$$

TABLE 1 Typical Average Plasma Parameters

		Local Inter-Stellar Medium	Solar Wind at r_o = 1 AU	and beyond
Bulk velocity	v, km s^{-1}	25	450	≈ const
Ion density	n, cm^{-3}	~ 0.04	8	$\propto r^{-2}$
Temperature	T, 10^4 K	~ 1	10	~ const *
Magnetic field	B, 10^{-10}T	~ 2	50	$\propto r^{-1}$ +

* γ = 1.1 – 1.2

\+ $B_r \propto r^{-2}$ ($B_r \sim B_\varphi$ at 1 AU)

Calculated Pressures (in units of 10^{-12} Pa)

Ram	mnv^2	0.042	2711
Thermal	p=2nkT	0.011	22
Magnetic	$B^2/2\mu_o$	0.016	10
Total	Π	0.069	2743

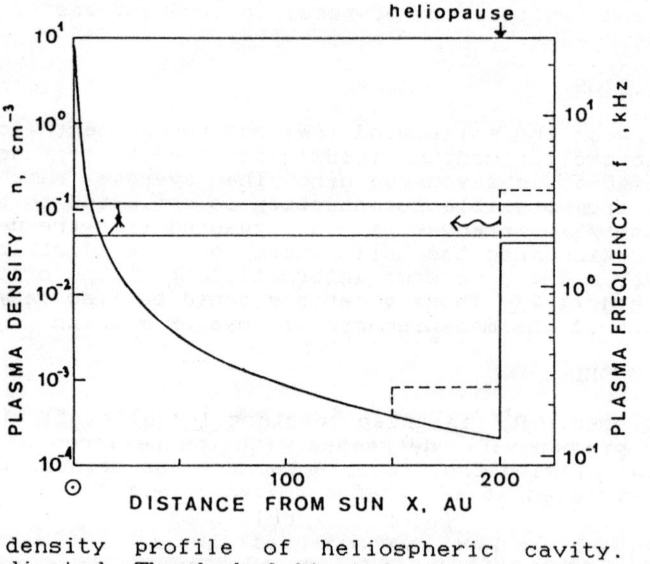

Fig 1. The density profile of heliospheric cavity. The heliopause position is indicated. The dashed lines denote only simplified solutions: steps of possible increase of the plasma density by factor of ≈ 4 due to strongly shocked regions (γ=5/3, cf./3/). The frequency limits 2 to 3 KHz observed by the plasma wave instruments on Voyager 1 and 2 (6-8) are also marked.

RESULTS

The view of the reconnection patches (Ψ > 90°) at the front side heliopause, indicated by dots, depending on the orientation of the LISM magnetic field vector $\mathbf{B}^{IS}$ (in the SEP), is shown in Figure 3 (taken from /2/). They correspond to the outward polarity of the $\mathbf{B}^{SW}$ in the northern solar hemisphere. The observer is thought to be located in the Sun.

The lines corresponding to constant angles Ψ are given in views presented in Figures 4 a, b and c for $\mathbf{B}^{IS}$ lying in the SEP, i.e. for THETA = 90° /11/. The direction of $\mathbf{B}^{IS}$ = (B^{IS}, THETA, PHI) in the frame of reference adopted here is shown in each view. Views for the LISM field deflected from the SEP by 45° are shown in /4/. The procedure presented in /5/ may be more appropriate for THETA ~0° and 180° ("polar caps" of the heliosphere).

The maximum value of the angle Ψ indicates the regions where the reconnection is expected to be most effective. If the direction of the LISM field is changed by 45° and 90° from the direction of the apex in the SEP those regions are shifted from the edge towards the nose of the heliosphere, i.e. the centre of Figure 4 a,b,c.

DISCUSSION

Since 1983 radio emissions in the frequency range of 2 to 3 kHz (marked in Figure 1) were observed by the plasma wave instruments on Voyager 1 and 2 moving in the outer Solar System /6-8/. The detected radio noise was postulated to emanate from the inner shock of the distant solar wind /6/ or from the heliopause itself /2/. Recently, it has been suggested that the interaction of an anomalous high speed stream with the terminal shock may be responsible for the generation of the most intense emission /9, 10/.

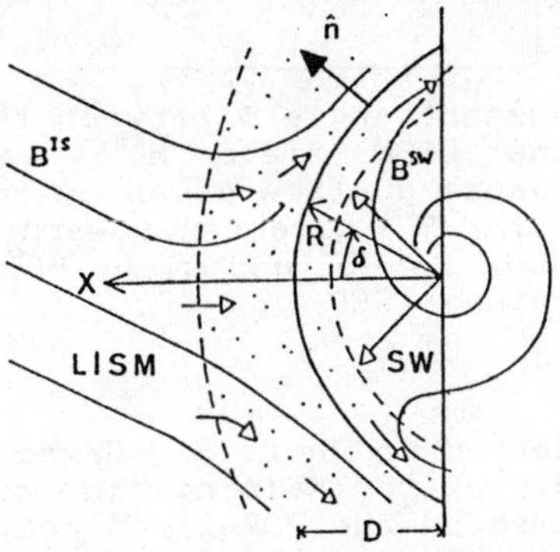

Fig. 2. Sketch of the interfaces between the solar wind (SW) and the supersonic ionized component of the local interstellar medium (LISM): an inner and an outer shocks (dashed) (cf./3/) and heliopause of equation (2). Parker`s spiral at the inner side and an uniform field at the outer side are taken.

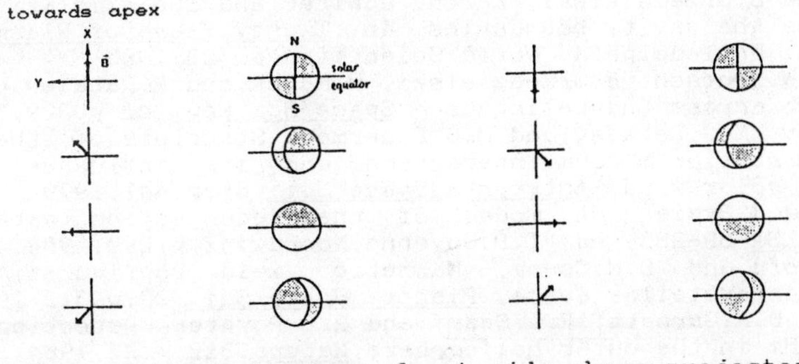

Fig. 3. A view of the heliopause front side shown projected onto Z-Y plane (X-direction, THETA=90°, PHI=0°, is collinear with the LISM wind flow vector) depending on the orientation of the interstellar magnetic field vector $B^{IS}=(B^{IS}$, THETA, PHI) lying in the solar equatorial (X,Y) plane (taken from /2/).

Various estimates of the distance to the terminal shock are throughly discussed in /9/. The estimate of ~135 AU, based on the stream speed inferred from the data /10/ is consistent with the simple picture given in Figure 1, according to Table 1 and equation (1). It is worth to note that the density gradient is expected to be much stronger at the heliopause than at the inner shock front. It indicates that, if the source in question is located at the boundary region of the heliosphere, the *heliopause* rather than the *inner shock front* could be responsible for the generation of these waves shown schematically in Figure 1, as was already suggested in /2/.

It should be noted that the reconnection pattern at the heliopause is expected to be very much sensitive neither to the assumed shape of that boundary surface nor to the model of the draping of the field lines over the heliopause surface /4/. Of course, there is still an open question whether the magnetic fields at the boundary region are sufficiently regular and whether the reconnection between oppositely directed magnetic fields does indeed take place at some regions of the heliopause surface. If so, one should ask: how does reconnection would affect the access of cosmic rays to Solar System? The answer to these questions could be very important for the interpretation of the measurements of Voyager mission.

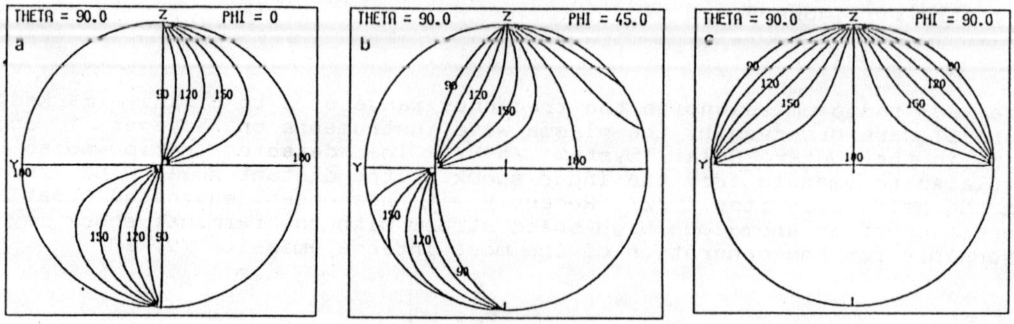

Fig. 4. The lines of constant angle Ψ between the magnetic field H^{sw} of the solar wind and the LISM field H^{is}, both projected (equations (3)-(4)) onto the heliopause surface given by equation (2), within the reconnection pathes: a) for B^{is} directed towards the apex, and for the direction of B^{is} changed b) by 45° and c) by 90° from the apex in the solar equatorial (X,Y) plane (SEP).

ACKNOWLEDGEMENTS

The author is very much indebted to Prof. S. Grzedzielski for illuminating discussions and to Mr. J. Bysiek for helping in numerical calculations of Figure 4, which were performed on the PDP-11/45 computer in the N.Copernicus Astronomical Centre in Warsaw. This research was supported by the Central Program for Fundamental Research under contract 01.20 coordinated by the Space Research Centre of the Polish Academy of Sciences.

REFERENCES

1. W.Macek and S.Grzedzielski, Earth, Jupiter and Sun-similarity of plasma transport across the cavity boundaries, in: Twenty Years of Plasma Physics, ed. B. McNamara, Philadelphia, World Scientific,p.320,1985.

2. H.J.Fahr, W.Neutsch, S.Grzedzielski, W.Macek,and R.Ratkiewicz-Landowska, Plasma transport across the heliopause, Space Sci.Rev. 43,p.329,1986.

3. V.B.Baranov, M.G.Lebedev,and M.S.Ruderman, Structure of the region of solar wind-interstellar medium interaction and its influence on H atoms penetrating the solar wind, Astrophys.Space Sci. 66,p.441,1979.

4. W.Macek,and J.Bysiek, A model of the reconnection pattern at the heliopause, in ESA SP-285, ed. T.D.Guyenne,Noordwijk,p.299,1988.

5. S.-I.Akasofu,and D.N.Covey, Magnetic field configuration of the heliosphere in interstellar space, Planet. Space Sci. 29,p.313,1981.

6. W.S.Kurth, D.A.Gurnett, F.L.Scarf,and R.L.Poynter, Detection of a radio emission at 3 kHz in the outer heliosphere,Nature,312,p.27,1984.

7. W.S.Kurth, D.A.Gurnett,and F.L.Scarf, Recent observation of the very low frequency interplanetary radio emission,Adv. Space Res. 6,p.379,1986.

8. W.S.Kurth, D.A.Gurnett, F.L.Scarf,and R.L.Poynter, Long-period dynamic spectrograms of low-frequency interplanetary radio emission, Geophys. Res. Lett. 14,p.49,1987.

9. R.L.McNutt,Jr., A solar-wind "trigger" for the outer heliosphere radio emissions and the distance to the terminal shock, Geophys.Res.Lett. 15,p.1307, 1988.

10. R.L.McNutt,Jr., Remote sensing of the termination of the solar wind via in situ plasma measurements, Adv.Space Res.9,p.235,1989.

11.W.Macek, Reconnection at the Heliopause,Adv.Space Res. 9,p.257,(1989).

FLARE INSTABILITY AND DRIVING MECHANISM

Probhas Raychaudhuri

Department of Applied Mathematics, Calcutta University, Calcutta–700009, India

Solar flares involve variety of physical processes occurring from the interior to the solar photosphere. Before the flare excess energy becomes stored in the magnetic field of the solar core. The constant interplay between gravitational field, nuclear energy generation and the neutrino process /1/ gives rise to the instability in the core of the Sun. Magnetic energy also keeps accumulating in the system, driving it ever closer to the instability until some event makes the magnetic configuration unstable. Flare results from sudden release of the energy stored in the magnetic field associated with active regions. The energy production in a star is related to the stellar evolution and Raychaudhuri /2/ suggested that the magnetic field of the star can originate from the dynamo action associated with the variations in the nuclear energy generation. Thus the stellar magnetic field is only a byproduct of the star evolution. Similarly sunspots, solar flares, solar wind etc. are the byproducts of the star evolution. Kangas and Raychaudhuri /3/ suggested that the nonrecurrent geomagnetic activity is characterised by the release of solar magnetic energy in the form of bipolar magnetic regions. In this paper we will show that as a result of acceleration of a trapped population of particles in the magnetic field of the active regions, a streaming of the ions relative to the electrons is established in such a way that, for the electrons moving with speed greater than the thermal speed, a Buneman kind of instability leads to excessive heating of the electrons and gives rise to the possibility of the solar flares.

From the analysis of solar neutrino flux data /4,5/ and some of the solar physics data /6/ it appears that the solar activity cycle consists of four different events called 1st / 2nd sunspot minimum / maximum and that the core energy production is a result of superposition of multiperiodic phenomena corresponding to these four events. To explain the higher neutrino flux in the runs 71 and 86 during 1980–1985 and runs 27 and 37 during 1970–1975 in the ^{37}Cl solar neutrino detector Raychaudhuri /5/ suggested that when the Sun is in 1st sunspot maximum, maximum expansion of the core occurs during the solar cycle and at that time mixing occurs (first mixing) between solar core and solar outer layers and, in spite of the decrease of the temperature, the solar neutrino flux is higher due to large amount of ^3He mixed into the core so that the rate of reaction $^3He + {^4He} \rightarrow {^7Be} + \gamma$ is raised. Mixing takes place within 2 to 3 years from the sunspot maximum and hence further expansion of the core is possible. Also there will be mixing between the solar core and solar outer layers and the neutrino production depends on mixing of heavier elements. Thus the higher neutrino fluxes in runs 27, 37, 71 and 86 can be explained.

Suggestion that the great solar flares were responsible for the higher neutrino flux in runs 27, 71 and 86 of the ^{37}Cl solar neutrino detector has been checked by Alexeyev et al. /7/ and by Zerda and O'Brien /8/. Both the groups found no evidence for this mechanism and according to /7/ these high fluxes of neutrinos can probably come from thermonuclear reactions occurring in internal regions of the Sun. This directly supports Raychaudhuri's /5/ suggestion regarding higher neutrino fluxes in runs 27, 71 and 86.

Ichimoto et al. /9/ have found 5 and 17 months periodicities in the solar flare activity. Raychaudhuri /10/ studied both solar neutrino data and sunspot data from 1970–1982 and found 5, 10, 15, 20, 25 and 30 months periodicities in both. The common periodicity in solar flares, sunspot number and solar neutrino data suggests that the 5 months core oscillation may be responsible for the occurrence of hard solar flares. Chattopadhyay and Raychaudhuri /11/ have found anticorrelation of solar neutrino data with sunspot data and solar flare data. This suggests strongly that 5 months period refers to the global oscillation of the Sun. This period may be

related to the time scale for storage and/or escape of the magnetic field in the solar convection zone. Thus the solar neutrino data may also provide important information of solar flare production. It has been found that intense solar flares follow the sunspot cycle and are most frequent during the peak phase. The fast type of PCA events as well as GLE of cosmic rays depend also on the phase of the solar cycle. To explain these we suggest the following two relevant processes: a gradual distortion of the magnetic field and a trigger action by the solar nuclear energy source to release the excess magnetic energy. Then in the maximum phase of the solar cycle, the frequency of the disturbance in the solar atmosphere being very high, the time interval between successive triggering actions is too short for much energy to be stored to cause the most intense eruptions. During the moderate activity triggering is less frequent and the large solar flares of 1972, 1981 have occurred in the Sun due to the possibility of storage of more nuclear energy at the time of 2nd sunspot maximum. Perez Enriquez /12/ proposed that the over-abundance of ^3He was characteristic of the plasma in active regions. This result came as a consequence of assuming that the ratio of the electron to ion temperature could reach values greater than 10 in an active region.
During the sunspot maximum of the 11 year solar cycle, solar magnetic energy is released mainly in the form of URM. After this stage the bipolar magnetic region begins to develop (one or two years before the 1st sunspot maximum in the active region /3/). In this period it is possible that solar active region can be capable of storing particles from the atmospheric plasma continuously for a period dependent on the solar core. We assume that there exists a hydromagnetic acceleration mechanism such that its efficiency is relevant only for the ion component of the plasma. Due to the anisotropy of the ions in the plasma of the active region a drift velocity v_d of the electrons relative to the ions is established so that an electric field E is induced and the electron equation of motion is $m_e dv_d/dt = -eE - m_e v_d \nu(v)$ where $\nu(v)$ is the collision frequency and m_e the electron mass. While $v_d < v_{Te}$ (thermal velocity of electrons), a stationary situation is reached with $v_d = -eE/m_e \nu(v)$ because $v_d = v_{Te}$ and the drag force $m_e v_d \nu(v) \sim v_d/v_{Te}^3$, which increases with v_d. But for the electrons with $v_d > v_{Te}$ the stationary situation is not reached, as $m_e v_d \nu(v) \sim v_d^{-2}$. These electrons, called "runaway electrons", drift anisotropically with respect to the ions with velocity $v_d = -e\,Et/m_e$ and can be capable of producing instabilities in the plasma with excessive heating of electrons. Thus, as the ions in the active region of plasma are accelerated, the electrons are thermalised in such a way that T_e (temperature of the electron) $> T_i$ (temperature of the ions). In a nearly isothermal plasma $T_e \simeq T_i$ and in this case only hydrodynamic and Buneman kind of instability can be present. According to Buneman, this type of instability leads to excessive heating of the electrons in such a way that the plasma can reach the other instability region i.e., $T_e \geq T_i$ (resonant oscillations) and $T_e \gg T_i$ (acoustic oscillations). From the above consideration Perez Enriquez /12/ showed that the active region is capable to evolve towards a flare with temperature of the order of 10^7K or more when the initial temperature of the active region plasma is very high. The solar core is constantly exciting the active regions of the Sun in the expanding phase and the electron density there will be much higher during the sunspot maximum. Thus in the expanding phase excessive heating of the electrons is possible due to the presence of acceleration mechanism and this heating leads to high temperature required for the active region to produce a flare. Thus the inner solar core instability is actually responsible for the activities of the Sun and also affects the physics of the outer heliosphere. We suggested /3/ that the formation and development of UMR in the Sun which magnetic structures can be identified as interplanetary sector structure are directly related to the solar nuclear energy generation.

1. P. Raychaudhuri, Ap. Sp. Sci. 13, 231 (1971)
2. P. Raychaudhuri, Ap. Sp. Sci. 18, 425 (1972)
3. J. Kangas and P. Raychaudhuri, Ap. Sp. Sci. 21, 3 (1973)
4. P. Raychaudhuri, Solar Physics, 93, 397 (1984)
5. P. Raychaudhuri, Solar Physics, 104, 415 (1986)
6. M.N. Gnevyshev, Solar Physics, 51, 175 (1977)
7. E.N. Alexeyev, L.N. Alexeyeva, A.E. Chudakov and L.V. Krivosheina, Proc. 20th ICRC 4, 351 (1987)
8. A. De la Zerda and K. O'Brien, Nature, 330 353 (1987)
9. K. Ichimoto, J. Kubota, M. Suzuki, I. Toshimura, H. Kurikawa, Nature, 316, 422 (1985)
10. P. Raychaudhuri, Solar Physics 106, 421 (1986)
11. G. Chattopadhyay and P. Raychaudhuri, 1st COSPAR Colloquium "Physics of the outer heliosphere", Warsaw, Poland (1989) 19-22 September
12. R. Perez Enriquez, Proc. 20th ICRC, 3, 37 (1987)

AUTHOR INDEX

1st COSPAR Colloquium
PHYSICS OF THE OUTER HELIOSPHERE
List of Participants

Alania, M. V. U.S.S.R.
Axford, W. I. F.R.G.

Badruddin India
Banaszkiewicz, M. Poland
Barabash, S. V. U.S.S.R.
Baranov, V. B. U.S.S.R.
Barnes, A. U.S.A.
Baumgärtel K. D.D.R.
Bird, M. K. F.R.G.
Blecki, J. Poland
Bochorishvilli, T. B. U.S.S.R.
Borodkova, N. L. U.S.S.R.
Bourgin, M. U.S.S.R.
Bowyer, S. U.S.A.
Burger, R. A. South Africa
Bzowski, M. Poland

Chalov, S. V. U.S.S.R.
Cini Castagnoli, G. Italy
Czechowski, A. Poland

Debowski, S. Poland
Dorman, L. I. U.S.S.R.
Dyal, P. U.S.A.

Fahr, H. J. F.R.G.
Ferlet, R. France
Filipiak, W. Poland
Frisch, P. C. U.S.A.

Galazka, R. Poland
Grib, S. A. U.S.S.R.
Grochulska, J. Poland
Gruntman, M. A. U.S.S.R.
Grzedzielski, S. Poland

Hlond, M. Poland
Holynski, R. Poland

Ip, W. -H. F.R.G.

Jokipii, J. R. U.S.A.
Jones, M. V. U.S.A.
Juchniewicz, J. Poland
Judge, D. L. U.S.A.

Kallrath, J. F.R.G.
Kiraly, P. Hungary
Klos, Z. Poland
Kobylinski, Z. Poland
Kömle, N. Austria
Koperski, P. Poland
Kota, J. Hungary
Krainiuk, A. G. U.S.S.R.
Kurth, W. S. U.S.A.
Kyrölä, E. Finland

Lallement, R. France
Lazarus, A. J. U.S.A.
Lee, M. A. U.S.A.
Lehmann, H. F.R.G.
Leopold, M. Poland
Likin, O. V. U.S.S.R.
Lindgren, S. Sweden
Lipatov, A. S. U.S.S.R.
Lotova, N. A. U.S.S.R.

Mann, I. F.R.G.
Mariani, F. Italy
McKenzie, J. F.R.G.
McKibben, R. B. U.S.A.
Möbius, E. F.R.G.
Mostafa, I. Egypt

Ness, N. F. U.S.A.

Oraevskiy, A. A. U.S.S.R.
Ostrowski, M. Poland

Page, D. E. U.S.A.
Parzydlo, W. Poland
Perez-de-Tejada, H. Mexico
Popielawska, B. Poland
Potgieter, M. S. South Africa

Ratkiewicz, R. Poland
Raychaudhuri, P. India
Roelof, E. C. U.S.A.
Rubel, M. Poland
Rucinski, D. Poland
Ruderman, M. S. U.S.S.R.
Rudnicki, K. Poland

Sauer, K. D.D.R.
Scherer, K. F.R.G.
Serio, M. Italy
Smith, E. J. U.S.A.
Somogyi, A. Hungary
Stanislawska, I. Poland

Staniucha, M. S. Poland
Stroganova, L.. I. U.S.S.R.
Summanen, T. Finland
Svirzhevskaya, A. K. U.S.S.R.
Szabelski, J. Poland

Verigin, M. U.S.S.R
Veselovsky, I. S. U.S.S.R.

Wdowczyk, J. Poland
Wenzel, K. -P. The Netherlands
Whang, Y. C. U.S.A.
Wieser, N. F.R.G.
Witte, M. F.R.G.
Wodnicka, B. Poland

Zarnowiecki, T. Poland
Zastenker, G. N. U.S.S.R.
Ziemkiewicz, J. Poland
Zwolakowska, D. Poland

1st COSPAR Colloquium
PHYSICS OF THE OUTER HELIOSPHERE
List of Unpublished Papers

A Lower Limit on the Ionization Fraction of the Very Local Interstellar Medium
R. J. Reynolds

Effect of Solar Activity and Heliospheric Neutral Sheet on the Structure of the Outer Heliosphere as Inferred from the Data on Temporal Cosmic Ray Variations
M. V. Alania, R. G. Aslamazashvili, T. S. Bochorishvili, L. I. Dorman, R. T. Gushchina and K. Iskra

Anisotropic Diffusion of Cosmic Rays Including Solar Wind Deceleration in the Outer Heliosphere
M. V. Alania, R. G. Aslamazashvili, K. Iskra, Z. Kobylinski and L. I. Dorman

Relative Importance of Various Factors Restricting and Forming the Outer Heliosphere
L. I. Dorman and S. Grzedzielski

Solar Particle Events in the Outer Heliosphere
E. C. Roelof

Voyager Studies of the Interplanetary Magenetic Field
N. F. Ness

Hydromagnetic Waves in LISM and the Cosmic Ray Anisotropy at High Energies
J. F. McKenzie

The Plasma Toruses of Phobos and Deimos
S. V. Baarabash, E. M. Dubinin, N. F. Pisarenko, R. Lundin, B. Multqvist, R. Pellinen, I. Liede, W. Reidler, K. Schwingenschuh, K. Moehlmann and K. Sauer

Multi-fluid Model of the Distant Solar Wind
K. Sauer and H. J. Fahr

Voyagers
N. F. Ness

Phobos
M. Verigin

Voyager-2 Neptune Fly-by
N. F. Ness

Fractal Properties of Underground Muon Time Series from the Nusex Experiment
L. Bergamasco, C. Castagnoli, A. Castellina, G. Cini Castagnoli, A. Provenzale, O. Saavedra and M. Serio

